T0192586

# The Method of Moments in Electromagnetics

# The Method of Moments in Electromagnetics

## Third Edition

Walton C. Gibson

CRC Press
Taylor & Francis Group
Boca Raton  London  New York

CRC Press is an imprint of the
Taylor & Francis Group, an **informa** business

A CHAPMAN & HALL BOOK

Third edition published 2022
by CRC Press
6000 Broken Sound Parkway NW, Suite 300, Boca Raton, FL 33487-2742

and by CRC Press
2 Park Square, Milton Park, Abingdon, Oxon, OX14 4RN

© 2022 Taylor & Francis Group, LLC

Second edition published by Chapman & Hall 2014
First edition published by Chapman & Hall 2008

CRC Press is an imprint of Taylor & Francis Group, LLC

Reasonable efforts have been made to publish reliable data and information, but the author and publisher
cannot assume responsibility for the validity of all materials or the consequences of their use. The
authors and publishers have attempted to trace the copyright holders of all material reproduced in this
publication and apologize to copyright holders if permission to publish in this form has not been
obtained. If any copyright material has not been acknowledged please write and let us know so we may
rectify in any future reprint.

Except as permitted under U.S. Copyright Law, no part of this book may be reprinted, reproduced,
transmitted, or utilized in any form by any electronic, mechanical, or other means, now known or
hereafter invented, including photocopying, microfilming, and recording, or in any information storage
or retrieval system, without written permission from the publishers.

For permission to photocopy or use material electronically from this work, access www.copyright.com
or contact the Copyright Clearance Center, Inc. (CCC), 222 Rosewood Drive, Danvers, MA 01923, 978-
750-8400. For works that are not available on CCC please contact mpkbookspermissions@tandf.co.uk

*Trademark notice*: Product or corporate names may be trademarks or registered trademarks and are used
only for identification and explanation without intent to infringe.

*Library of Congress Cataloging-in-Publication Data*

Names: Gibson, Walton C., author.
Title: The method of moments in electromagnetics / Walton C. Gibson.
Description: Third edition. | Boca Raton : C&H/CRC Press, 2021. | Includes
bibliographical references and index.
Identifiers: LCCN 2021007648 (print) | LCCN 2021007649 (ebook) | ISBN
9780367365066 (hardback) | ISBN 9781032042329 (paperback) | ISBN
9780429355509 (ebook)
Subjects: LCSH: Electromagnetism--Data processing. | Electromagnetic
fields--Mathematical models. | Moments method (Statistics) |
Electromagnetic theory--Data processing. | Integral equations--Numerical
solutions.
Classification: LCC QC760.54 .G43 2021 (print) | LCC QC760.54 (ebook) |
DDC 530.14/1015118--dc23
LC record available at https://lccn.loc.gov/2021007648
LC ebook record available at https://lccn.loc.gov/2021007649

ISBN: 9780367365066 (hbk)
ISBN: 9781032042329 (pbk)
ISBN: 9780429355509 (ebk)

DOI: 10.1201/9780429355509

*Publisher's note:* This book has been prepared from camera-ready copy provided by the author.

# Contents

# Preface to the Third Edition

Shortly before the publication of the second edition of this book, I finished work on a new version of the *Serenity* RCS solver code. This new software added support for the Adaptive Cross Approximation (ACA), permitting a direct solution of the matrix system in compressed form. It also took advantage of the incredible computation power afforded by general purpose graphics processing units (GPUs), which can perform matrix operations much faster than traditional CPUs. Unfortunately, given the short timeframe, I was not able to add any material on the ACA to the second edition. Since that time, significant additional progress has been made on *Serenity*, as well as implementation of a Multi-Level Adaptive Cross Approximation (MLACA), which has recently begun to receive significant attention in the literature. Thus, the time now seems right to update the second edition with material on these very exciting, powerful algorithms.

This third edition adds a significant amount of new material. First, I have added a chapter that presents an in-depth treatment of the ACA as applied to the Method of Moments, as well as its software implementation on CPUs and GPUs. Next, I have added a chapter on the Multi-Level Adaptive Cross Approximation (MLACA), an advanced compression technique that uses a recursive, butterfly compression algorithm. This yields a far better level of compression than the regular ACA, at the expense of extra computation time. Finally, I have revisited and updated the existing material on the Fast Multipole Method (FMM), as well as the Multi-Level Fast Multipole Algorithm (MLFMA). This new treatment utilizes elements of the ACA approach to further reduce the the memory requirements of the FMM and MLFMA, and also presents compressed, block versions of the ILUT and SAI preconditioners. I have also done my best to correct known typographical errors from the second edition, and to incorporate feedback I've received from readers.

I began working on *Serenity* in 2000, and the first edition of this book in 2004. After 20 years of very hard work, I feel that it's finally time to take a rest. As a result, this will be the final edition of the book. Please send any errata to me via email at kalla@tripoint.org, and going forward I will maintain an up-to-date errata sheet for those interested.

# Preface to the Second Edition

The first edition of this book was well received, and since its publication I received a lot of feedback and questions about its content. When I began considering a second edition, I knew that a key element had to be the treatment of dielectric materials. After I started working on the new material, it became apparent that to accomplish this goal I would have to re-write a significant amount of the existing material. The result is a book that I believe significantly improves and expands upon the first edition in several key areas:

1. The first edition focused exclusively on integral equations for conducting problems. However, dielectric materials are an important aspect of practical electromagnetic devices, and the ability to include these in computational codes is of increasing importance. In this second edition, we will derive coupled surface integral equations that will treat conducting as well as composite conducting/dielectric objects. This treatment extends to objects having multiple dielectric regions with interfaces and junctions. Though many sections of this book have been changed or updated to reflect new material, we will continue to use the approach in the first edition, where we progress gradually from simple to more complex problems and topics.

2. As my background is in the design and programming of software codes for radar cross section (RCS) prediction, the first edition focused almost exclusively on far-zone radiated fields and RCS. In this second edition, I have added more material involving the calculation of near fields.

3. Technology continues to march ever onward, and I have attempted to keep the material in this edition as up to date as possible. As a result, existing material has been updated, and material deemed to be outdated or no longer relevant has been removed.

4. Known errors from the first edition have been corrected, and concepts that were unclear or poorly explained have been updated for clarity. Many equations have also been reformatted, and figures adjusted or re-generated to fix errors or improve their appearance.

# Preface

The Method of Moments (MoM) as applied to electromagnetic field problems was first described at length in Harrington's classic book[1], and was no doubt in use before that. Since then, computing technology has grown at a staggering pace, and the field of computational electromagnetics (CEM) has followed closely behind. Though many dissertations and journal papers have been dedicated to the MoM in electromagnetics, few textbooks have been written for those who are unfamiliar with it. For graduate students who are just beginning their study of CEM, or seasoned professionals solving real-world problems, this sets the barrier to entry fairly high.

Thus, this author felt that a new book describing the Moment Method as applied to electromagnetics was necessary. The first reason was because of the lack of a good introductory graduate-level text on the Moment Method. Though many universities offer courses in computational electromagnetics, the course material often comprises a disjoint collection of journal papers and copies of the instructor's personal notes. In the few books that do exist on the subject, the material is often presented at a high level with little to no attention paid to implementation details. Additionally, many of these often omit key information or refer to other papers (or worse, private communications), forcing the student to spend their time searching for missing information instead of focusing on course material. The second reason is the hope that a concise, up-to-date reference book will be of significant benefit to researchers and practicing professionals in the field of CEM. This book has several key features that set it apart from others of its kind:

1. *A straightforward, progressive introduction to field problems and the Moment Method.* This book begins by introducing the Method of Moments in the context of simple, electrostatic field problems. We then move to a review of frequency-domain electromagnetic theory, radiation equations, and the Green's functions for two- and three-dimensional objects. The surface equivalence theorem is then used to derive coupled surface integral equations of radiation and scattering for conducting and composite dielectric/conducting objects. Subsequent chapters are then dedicated to solving these integral equations for progressively more dif-

---

[1] R. F. Harrington *Field Computation by Moment Methods*, The Macmillan Company, 1968.

ficult problems involving thin wires, bodies of revolution, and two and three-dimensional bodies. With this material behind them, the student or researcher will be well-equipped for more advanced MoM topics encountered in the literature.

2. *A clear and concise summary of equations.* One of the fundamental problems encountered by this author is that clear expressions for the MoM matrix elements are almost never found in the literature. This book derives or summarizes the matrix elements used in every MoM problem, and all examples are computed using the expressions summarized in the text.

3. *A focus on radiation and scattering problems.* This book is primarily focused on scattering and radiation problems. Therefore, we will consider many practical examples such as antenna impedance calculation and radar cross section prediction, as well as calculation of near fields. Each example is presented in a straightforward manner with a careful explanation of the approach as well as explanation of the results.

4. *An up-to-date reference.* The material contained within is presented in the context of current-day computing technology, and includes up-to-date material on methods such as as the Fast Multipole Method and Adaptive Cross Approximation that are now commonly used in CEM.

5. *"Show your work."* In this work we attempt to avoid a so-called "it can be shown" approach, which can be detrimental to students, and instead focus on step-by-step derivations in key areas that are often glossed over or omitted in other texts.

This book is intended for a one- or two- semester course in computational electromagnetics and a reference for the practicing engineer. It is expected that the reader will be familiar with time-harmonic electromagnetic fields and vector calculus, as well as differential/integral equations and linear algebra. The reader should also have some basic experience with computer programming in a language such as C, C++ or FORTRAN, or a mathematical environment such as MATLAB®. Because some of the expressions in this book require the calculation of special functions, the reader should be aware of what they are and know how to calculate them.

This book comprises nine chapters:

Chapter 1 presents a very brief overview of computational electromagnetics and some of the commonly used numerical techniques in this field. This will show the reader how surface integral equations and the Method of Moments fits into the world of CEM algorithms.

Chapter 2 introduces the Method of Moments in the context of simple electrostatic field problems. We then formalize the MoM, discuss point matching and Galerkin's Method, and present some commonly used two-dimensional basis functions.

Chapter 3 begins with Maxwell's Equations, and then discusses electromagnetic boundary conditions, formulations for radiation, vector potentials, Green's functions, and near and far fields. It then uses the equivalence principle to derive a set of coupled integral equations for scattering by multi-region, composite conducting/dielectric objects. These equations are then discretized in general form using the Method of Moments, and are refined and adapted to specific problems in later chapters.

Chapter 4 discusses the solution of matrix equations, including topics such as Gaussian Elimination, scalar and block-LU Decomposition, matrix condition numbers, and iterative solvers.

Chapter 5 considers radiation and scattering by thin wires. We derive the thin wire kernel and the Hallén and Pocklington thin wire integral equations, and show how these are solved. We then apply the Moment Method to thin wires of arbitrary shape, and then solve practical thin wire problems involving antennas and feedlines.

Chapter 6 applies the Moment Method to two-dimensional problems. The integral equation formulations in Chapter 3 are adapted to general, two-dimensional boundaries, and at TM and TE polarizations. The first half of the chapter considers conducting geometries, and the latter half then moves into dielectric and composite geometries and problems.

Chapter 7 considers three-dimensional objects which can be described as bodies of revolution (BoRs). Examples involving the radar cross section of conducting, dielectric, and composite objects are then presented. We will then discuss junctions between dielectric and conducting regions and their treatment, and consider some example problems having junctions.

Chapter 8 moves to three-dimensional (3D) surfaces of arbitrary shape. We discuss the modeling of surfaces using triangular facets, and devote significant effort to summarizing the expressions used to evaluate singular potential integrals over triangular elements. We will then present a method by which generalized junctions between regions can be treated in a systematic and straightforward manner. Next, we consider radar cross section problems involving conducting and composite objects, as well as some three-dimensional antenna problems.

Chapter 9 introduces the Adaptive Cross Approximation (ACA), which is a fast, kernel-agnostic algorithm that can be used to compress blocks of the MoM system matrix, as well as blocks in the corresponding LU-factored matrix. We will cover the ACA algorithm as well as software implementation details on regular CPUs as well as graphics processing units (GPUs). We will then consider problems of larger electrical size than were considered in Chapter 8.

Chapter 10 introduces the Multi-Level Adaptive Cross Approximation (MLACA), which comprises a hierarchical extension of the ACA presented in Chapter 9, allowing for much greater levels of compression at the expense of more computation. The MLACA algorithm and its implementation details on regular CPUs as well as GPUs is covered in detail. We will then consider problems of much larger electrical size than were considered in Chapter 9.

Chapter 11 discusses the Fast Multipole Method (FMM) and how it is used to accelerate the matrix-vector product in the iterative solution of 3D problems. We will cover the addition theorem, wave translation, and single and multi-level fast multipole algorithms. The treatment is concise and contains all the information required to successfully implement the FMM for conducting or composite geometries in new or existing moment method codes. We will then consider consider problems of even larger electrical size than were considered in Chapter 10.

Chapter 12 discusses some methods of numerical integration including the trapezoidal and Simpson's rule, and then area coordinates and Gaussian quadrature over planar triangular elements.

Throughout this text, an $e^{jwt}$ time convention is assumed and suppressed throughout. We use SI units except in some examples where the test articles had dimensions in inches or feet. Numbers in parentheses ( ) refer to equations, and numbers in brackets [ ] are citations. Scalar quantities are written in italic font ($a$), vectors in bold lowercase font (**a**) with unit vectors denoted using the caret (**â**), and matrices in bold uppercase font (**A**).

# Acknowledgments

This book came about because of my experience as a student in electromagnetics and as a developer of computational electromagnetics codes. My walk down this path undoubtedly began very early with my interest in the power of the computer, as well as the mysteries behind radio waves and antennas. I got my first taste as an undergraduate student at Auburn University, where my professor of electromagnetics, Dr. Thomas Shumpert, encouraged me to pursue the subject at the graduate level. It progressed further during my study at the University of Illinois, where I learned many great things about electromagnetics from Dr. Weng Chew and my graduate advisor Dr. Jianming Jin. During my subsequent work on the code suite known as *lucernhammer*, I encountered many hard problems in mathematics and software algorithms, programming, and optimization. Without the help of friends and colleagues, it is likely I would not have completed this path at all. Therefore, I would like to thank the following people who provided me with useful assistance, advice, and information when it was needed:

Jason Amos
Michael J. Perry
Dr. John L. Griffin
Scott Kelley
Paul Burns
Dr. Stuart DeGraaf
Dr. Andy Harrison
Dr. Bassem R. Mahafza
Jeroen van der Zijp

# *About the Author*

**Walton C. Gibson** was born in Birmingham, Alabama in 1975. He received the Bachelor of Science degree in electrical engineering from Auburn University in 1996, and the Master of Science degree from the University of Illinois Urbana-Champaign in 1998. His professional interests include electromagnetic theory, computational electromagnetics, radar cross section prediction, and computer graphics. He is the author of the industry-standard *lucernhammmer* electromagnetic signature prediction software. He is a licensed amateur radio operator (callsign K4LLA), with interests in antennas and low-frequency HF propagation.

# Chapter 1

## Computational Electromagnetics

[1] In the beginning, the design and analysis of electromagnetic devices and structures was largely experimental. However, once the computer and numerical programming languages were developed, people immediately began using them to solve electromagnetic field problems of ever-increasing complexity. This pursuit of *computational electromagnetics* (CEM) has yielded many innovative, powerful analysis algorithms, and it now drives the development of electromagnetic devices people use every day. As the power of the computer continues to grow, so do the number of available algorithms as well as the size and complexity of the problems that can be solved. While the data gleaned from experimental measurements is invaluable, the entire process can be costly in terms of money and the manpower required to do the machine work and assembly, and to collect data at the measurement range. One of the fundamental drives behind reliable computational electromagnetics algorithms is the ability to simulate the behavior of devices and systems before they are actually built. This allows engineers to engage in levels of optimization that would be painstaking or even impossible if done experimentally. CEM also helps to provide fundamental insights into electromagnetic problems through the power of computation and computer visualization, making it one of the most important areas of engineering today.

## 1.1    CEM Algorithms

The range of electromagnetic problems is extensive, and this has led to the development of different classes of CEM algorithms, each with its own benefits and limitations. In the "early days" of CEM, many problems of practical size could not be solved unless some assumptions were made about the underlying physics and approximations made, usually under the asymptotic or high-frequency limit. These approximate algorithms are now commonly known as "high-frequency" methods. Algorithms that do not make these sorts of approximations are more demanding in terms of CPU and system memory, and historically have been limited to problems of small electrical size. These are usually referred to as "exact" or "low-frequency" algorithms. Both classes of algorithms can be further subdivided into time or frequency-domain algorithms. We will now summarize some of the most commonly used methods to provide some context in how the Moment Method fits in the CEM environment.

## 1.1.1   Low-Frequency Methods

Low-frequency (LF) methods are so-named because they solve Maxwell's Equations with no implicit approximations, and are typically limited to problems of small electrical size due to limitations of computation time and system memory. Though computers continue to grow more powerful and can solve problems of ever-increasing size, this nomenclature will likely remain common in the literature for the foreseeable future.

### 1.1.1.1   Finite Difference Time Domain Method

The Finite Difference Time Domain (FDTD) method [2, 3] uses finite differences to solve Maxwell's Equations in the time domain. The application of FDTD is usually very straightforward: the solution domain is discretized into small rectangular or curvilinear elements, and a "leap frog" in time is used to compute electric and magnetic fields from one another at discrete time steps. FDTD excels at analysis of inhomogeneous and nonlinear media, though its demands for system memory are high due to discretization of the entire solution domain, and it suffers from dispersion issues and the need to artificially truncate of the solution boundary. FDTD is typically applied in EM packaging and waveguide problems, as well as the study of wave propagation in complex (often composite) materials.

### 1.1.1.2   Finite Element Method

The Finite Element Method (FEM) [4, 5] is a method used to solve frequency-domain boundary-valued electromagnetic problems via variational techniques. It can be used with two- and three-dimensional canonical elements of differing shape, allowing for a highly accurate discretization of the solution domain. FEM is often used in the frequency domain for computing the field distribution in complex, closed regions such as cavities and waveguides. As in the FDTD, the solution domain must be discretized and truncated, making the FEM approach often unsuitable for radiation or scattering problems unless combined with a boundary integral equation approach such as the Method of Moments [4].

### 1.1.1.3 Method of Moments

The Method of Moments (MOM) is a technique used to solve electromagnetic surface[1] or volume integral equations in the frequency domain. MOM differs from FDTD and FEM as the electromagnetic sources (surface or volume currents) are the quantities of interest, and so only the surface or volume of the antenna or scatterer must be discretized. As a result, the MOM is widely used in solving radiation and scattering problems. In this book, we focus on the practical solution of surface integral equations of radiation and scattering using MOM.

## 1.1.2 High-Frequency Methods

Electromagnetic problems of large size have existed long before the computers that could solve them. Common examples of larger problems are those involving radar cross section prediction, or the calculation of an antenna's radiation pattern when in close proximity to a large structure. Many approximations have been made to the equations of radiation and scattering to make these problems tractable. Most of these treat the fields in their asymptotic limit, and employ ray-optics and edge diffraction. When the problem is electrically large, many asymptotic methods produce results that are accurate enough on their own, or can be used as a "first pass" before a more accurate though computationally demanding method is applied.

### 1.1.2.1 Geometrical Theory of Diffraction

The Geometrical Theory of Diffraction (GTD) [6, 7] uses ray-optics to determine electromagnetic wave propagation. The spreading and amplitude intensity and decay in a ray bundle are computed using Fermat's principle and the radius of curvature at the bounce points. GTD attempts to account for the fields diffracted by edges, allowing for a calculation of the fields in shadow regions. GTD is fast but the results are often fair to poor for complex geometries.

### 1.1.2.2 Physical Optics

Physical Optics (PO) [8] is a method for approximating the surface currents, allowing a boundary integration to be performed to obtain the fields. As we will see, PO and the MOM are closely related as they use the same equations to integrate the surface currents, however the MOM calculates the surface currents directly instead approximating them. While robust, the PO does not account for the fields diffracted by edges or those from multiple reflections, so supplemental corrections are typically added to it. The PO method is used

---

[1] In this context, it is also referred to as the Boundary Element Method (BEM).

extensively in high-frequency reflector antenna analysis, as well as many radar cross section prediction codes such as *lucernhammer MT*.

### 1.1.2.3  Physical Theory of Diffraction

The Physical Theory of Diffraction (PTD) [9, 10] is a means for supplementing the PO solution by adding to it the fields radiated by nonuniform currents along diffracting edges of an object. PTD is commonly used in high-frequency radar cross section and scattering analysis.

### 1.1.2.4  Shooting and Bouncing Rays

The Shooting and Bouncing Ray (SBR) method [11, 12] was developed to predict the multiple-bounce scattered fields from complex objects. It uses the ray-optics model to determine the path and amplitude of a ray bundle, but uses a physical optics-based scheme that integrates the surface currents induced by the ray at each bounce point. SBR is often used in scattering codes to account for multiple reflections on a surface or those inside a cavity, and as such it supplements the fields computed by PO and the PTD. SBR is also used to predict wave propagation and scattering in complex urban environments to determine the coverage for cellular telephone service.

---

## References

[1] D. E. Amos, "A subroutine package for Bessel functions of a complex argument and nonnegative order," Tech. Rep. SAND85-1018, Sandia National Laboratory, May 1985.

[2] A. Taflove and S. C. Hagness, *Computational Electrodynamics: The Finite-Difference Time-Domain Method*. Artech House, third ed., 2005.

[3] K. Kunz and R. Luebbers, *The Finite Difference Time Domain Method for Electromagnetics*. CRC Press, 1993.

[4] J. Jin, *The Finite Element Method in Electromagnetics*. John Wiley and Sons, 1993.

[5] J. L. Volakis, A. Chatterjee, and L. C. Kempel, *Finite Element Method for Electromagnetics*. IEEE Press, 1998.

[6] J. B. Keller, "Geometrical theory of diffraction," *J. Opt. Soc. Amer.*, vol. 52, pp. 116–130, February 1962.

[7] R. G. Kouyoumjian and P. H. Pathak, "A uniform geometrical theory of diffraction for an edge in a perfectly conducting surface," *Proc. IEEE*, vol. 62, pp. 1448–1461, November 1974.

[8] C. A. Balanis, *Advanced Engineering Electromagnetics*. John Wiley and Sons, 1989.

[9] P. Ufimtsev, "Approximate computation of the diffraction of plane electromagnetic waves at certain metal bodies (i and ii)," *Sov. Phys. Tech.*, vol. 27, pp. 1708–1718, August 1957.

[10] A. Michaeli, "Equivalent edge currents for arbitrary aspects of observation," *IEEE Trans. Antennas Propagat.*, vol. 23, pp. 252–258, March 1984.

[11] H. Ling, S. W. Lee, and R. Chou, "Shooting and bouncing rays: calculating the RCS of an arbitrarily shaped cavity," *IEEE Trans. Antennas Propagat.*, vol. 37, pp. 194–205, February 1989.

[12] H. Ling, S. W. Lee, and R. Chou, "High-frequency RCS of open cavities with rectangular and circular cross sections," *IEEE Trans. Antennas Propagat.*, vol. 37, pp. 648–652, May 1989.

# Chapter 2

## The Method of Moments

In this book, we are concerned with solving surface integral equations. Because no analytic solutions exist for most of these problems, in practice we must apply numerical methods. In this chapter we will introduce the Method of Moments (MoM), a technique that is used to convert integral equations into a linear system that can then be solved numerically via computer. To begin, we will analyze a few simple electrostatic problems to provide a bit of context, and then formally define the MoM. We will next discuss the expansion of an unknown function by a sum of weighted basis functions, and compare and contrast point matching and the Method of Galerkin. In the chapters that follow, we will then discuss electrodynamic problems and numerical methods for solving a general system of linear equations. With this material behind us, we will then be ready to apply the MoM to the more challenging integral equation problems in later chapters.

### 2.1 Electrostatic Problems

Because electrostatic problems are relatively simple compared to the electrodynamic case, they provide a good context for introducing algorithms used to solve integral equations. Recall that the electric potential $\phi_e$ at a point $\mathbf{r}$ due to an electric charge density $q_e$ is given by the integral

$$\phi_e(\mathbf{r}) = \int_V \frac{q_e(\mathbf{r}')}{4\pi\epsilon|\mathbf{r} - \mathbf{r}'|} \, d\mathbf{r}'. \tag{2.1}$$

If we know $q_e(\mathbf{r}')$, we can obtain the electric potential everywhere. If we instead know the electric potential $\phi_e(\mathbf{r})$ but not the charge density, (2.1) becomes an integral equation for $q_e(\mathbf{r}')$. We will now solve this problem numerically for a pair of practical examples, a charged wire and plate.

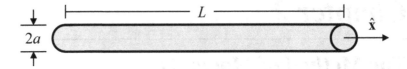

**FIGURE 2.1:** Thin Wire Dimensions

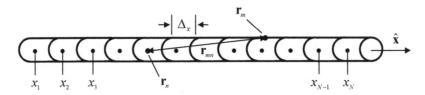

**FIGURE 2.2:** Thin Wire Segmentation

### 2.1.1 Charged Wire

Consider a thin, conducting wire of length $L$ and radius $a$ oriented along the $x$ axis, as shown in Figure 2.1. If the radius of the wire is very small compared to the length ($a \ll L$), the electric potential on the wire can be expressed via the integral

$$\phi_e(\mathbf{r}) = \int_0^L \frac{q_e(x')}{4\pi\epsilon|\mathbf{r} - \mathbf{r}'|} \, dx' \tag{2.2}$$

where

$$|\mathbf{r} - \mathbf{r}'| = \sqrt{(x - x')^2 + (y - y')^2} \, . \tag{2.3}$$

We intend to convert (2.2) into a linear system of equations, so we first subdivide the wire into $N$ subsegments, each having length $\Delta_x$, as shown in Figure 2.2. Within each subsegment, we assume that the charge density has a constant value such that $q_e(x')$ is piecewise constant over the length of the wire. Mathematically, this can be written as

$$q_e(x') = \sum_{n=1}^{N} a_n f_n(x') \tag{2.4}$$

where $a_n$ are unknown coefficients, and $f_n(x')$ is a pulse function having a constant value on segment $n$ but zero on all others, i.e.

$$f_n(x') = \begin{cases} 0 & x' < (n-1)\Delta_x \\ 1 & (n-1)\Delta_x \le x' \le n\Delta_x \\ 0 & x' > n\Delta_x \end{cases} \tag{2.5}$$

We now set the potential of the wire to $\phi_e = 1\text{V}$. Substituting (2.4) into (2.2) then yields

$$1 = \int_0^L \sum_{n=1}^N a_n f_n(x') \frac{1}{4\pi\epsilon |\mathbf{r} - \mathbf{r}'|} \, dx'. \tag{2.6}$$

Using the previous definition of the pulse function, we can rewrite this as

$$1 = \frac{1}{4\pi\epsilon} \sum_{n=1}^N a_n \int_{(n-1)\Delta_x}^{n\Delta_x} \frac{1}{|\mathbf{r} - \mathbf{r}'|} \, dx' \tag{2.7}$$

where we now have a sum of integrals, each over the domain of a single pulse function. Let us now fix the source points so that they are on the wire axis, and the field point to be on the wire's surface. This ensures that we do not encounter a singularity in the integrand. The denominator of the integrand now becomes

$$|\mathbf{r} - \mathbf{r}'| = \sqrt{(x - x') + a^2} \tag{2.8}$$

and (2.7) can be written as

$$4\pi\epsilon = a_1 \int_0^{\Delta_x} \frac{1}{\sqrt{(x - x') + a^2}} \, dx' + a_2 \int_{\Delta_x}^{2\Delta_x} \frac{1}{\sqrt{(x - x') + a^2}} \, dx' + \cdots$$
$$+ a_{N-1} \int_{(N-2)\Delta_x}^{(N-1)\Delta_x} \frac{1}{\sqrt{(x - x') + a^2}} \, dx' + a_N \int_{(N-1)\Delta_x}^{N\Delta_x} \frac{1}{\sqrt{(x - x') + a^2}} \, dx' \tag{2.9}$$

which comprises one equation in $N$ unknowns. If we can now somehow convert this equation to $N$ equations in $N$ unknowns, we can solve it by common matrix algebra routines. To do so, let us choose $N$ independent field points $x_m$ on the surface of the wire, each at the center of a wire segment. Doing so yields

$$4\pi\epsilon = a_1 \int_0^{\Delta_x} \frac{1}{\sqrt{(x_1 - x') + a^2}} \, dx' + \cdots + a_N \int_{(N-1)\Delta_x}^{N\Delta_x} \frac{1}{\sqrt{(x_1 - x') + a^2}} \, dx'$$
$$\vdots$$
$$4\pi\epsilon = a_1 \int_0^{\Delta_x} \frac{1}{\sqrt{(x_N - x') + a^2}} \, dx' + \cdots + a_N \int_{(N-1)\Delta_x}^{N\Delta_x} \frac{1}{\sqrt{(x_N - x') + a^2}} \, dx' \tag{2.10}$$

which comprises a matrix system of the form

$$\begin{bmatrix} Z_{11} & Z_{12} & Z_{13} & \cdots & Z_{1N} \\ Z_{21} & Z_{22} & Z_{23} & \cdots & Z_{2N} \\ Z_{31} & Z_{32} & Z_{33} & \cdots & Z_{3N} \\ \vdots & \vdots & \vdots & \ddots & \vdots \\ Z_{N1} & Z_{N2} & Z_{N3} & \cdots & Z_{NN} \end{bmatrix} \begin{bmatrix} a_1 \\ a_2 \\ a_3 \\ \vdots \\ a_N \end{bmatrix} = \begin{bmatrix} b_1 \\ b_2 \\ b_3 \\ \vdots \\ b_N \end{bmatrix} \tag{2.11}$$

where the matrix elements $Z_{mn}$ are

$$Z_{mn} = \int_{(n-1)\Delta_x}^{n\Delta_x} \frac{1}{\sqrt{(x_m - x') + a^2}} \, dx' \qquad (2.12)$$

and the right-hand side (RHS) vector elements $b_m$ are

$$b_m = 4\pi\epsilon . \qquad (2.13)$$

#### 2.1.1.1 Matrix Element Evaluation

The integral in (2.12) can be evaluated in closed form. Performing this integration yields [1] (200.01)

$$Z_{mn} = \log \left[ \frac{(x_b - x_m) + \sqrt{(x_b - x_m)^2 - a^2}}{(x_a - x_m) + \sqrt{(x_a - x_m)^2 - a^2}} \right] \qquad (2.14)$$

where $x_b = n\Delta_x$ and $x_a = (n-1)\Delta_x$. Note that the linear geometry of this problem yields a matrix that has a symmetric Toeplitz structure of the form

$$\mathbf{Z} = \begin{bmatrix} Z_1 & Z_2 & Z_3 & \cdots & Z_N \\ Z_2 & Z_1 & Z_2 & \cdots & Z_{N-1} \\ Z_3 & Z_2 & Z_1 & \cdots & Z_{N-2} \\ \vdots & \vdots & \vdots & \ddots & \vdots \\ Z_N & Z_{N-1} & Z_{N-2} & \cdots & Z_1 \end{bmatrix} \qquad (2.15)$$

where the elements from just the first row can be used to populate the entire matrix.

#### 2.1.1.2 Solution

In Figures 2.3a and 2.3b is plotted the computed charge density on the wire using 15 and 100 segments, respectively. The density at the lower discretization level is somewhat crude, as expected. The increase to 100 unknowns improves greatly the fidelity of the result. Using this charge density, we then compute the potential at 100 points along the wire using (2.2). The potential using 15 charge segments is shown in Figure 2.4a. While the voltage is near the expected value of 1V, it is not of constant value, particularly near the ends of the wire. Figure 2.4b shows the potential obtained using 100 segments. The voltage is now nearly constant across the entire wire, except at the endpoints. As a uniform segment size was used, the charge density tends to be oversampled in the middle of the wire and undersampled near the ends. As a result, the variation of the charge near the ends of the wire is not represented as accurately as in

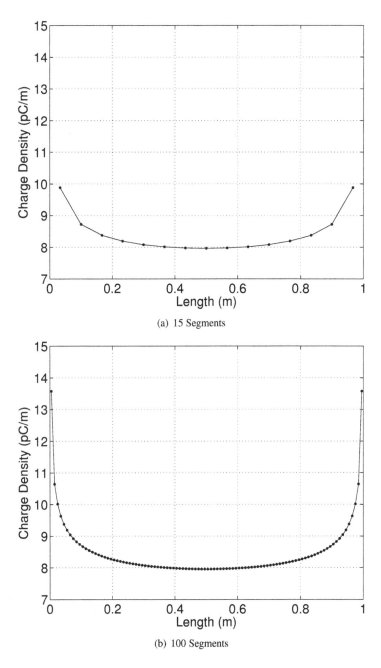

(a) 15 Segments

(b) 100 Segments

**FIGURE 2.3:** Straight Wire Charge Distribution

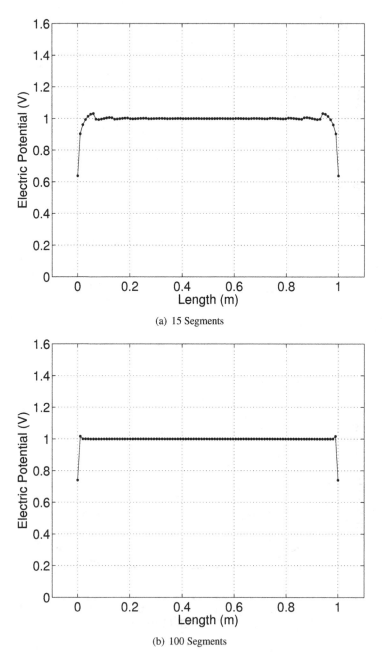

(a) 15 Segments

(b) 100 Segments

**FIGURE 2.4:** Straight Wire Potential

the center, and the computed voltage diverges from the true value. Realistic objects often have irregular surface features such as cracks, gaps, and corners that give rise to a more rapid variation in the solution at those points. To increase accuracy, it is advantageous to employ a denser level of discretization in areas where the most variation is expected.

## 2.1.2 Charged Plate

We next consider a similar problem involving a thin, charged conducting square plate of side length $L$, as shown in Figure 2.5. The potential on the plate can be written as

$$\phi_e(\mathbf{r}) = \int_{-\frac{L}{2}}^{\frac{L}{2}} \int_{-\frac{L}{2}}^{\frac{L}{2}} \frac{q_e(x', y')}{4\pi\epsilon|\mathbf{r} - \mathbf{r}'|} \, dx' \, dy' \tag{2.16}$$

and fixing the plate to a potential of 1V as before, (2.16) becomes

$$4\pi\epsilon = \int_{-\frac{L}{2}}^{\frac{L}{2}} \int_{-\frac{L}{2}}^{\frac{L}{2}} \frac{q_e(x', y')}{\sqrt{(x - x')^2 + (y - y')^2}} \, dx' \, dy' \,. \tag{2.17}$$

We now subdivide the plate into $N$ square patches having an edge length of $2a$ and area $\Delta_s = 4a^2$, and we assume that the charge is constant within each patch. We choose $N$ independent field points, each located at the center $(x_m, y_m)$ of a patch. Doing so yields a matrix equation with elements $Z_{mn}$ given by

$$Z_{mn} = \int_{S_n} \frac{1}{\sqrt{(x_m - x')^2 + (y_m - y')^2}} \, dx' \, dy' \tag{2.18}$$

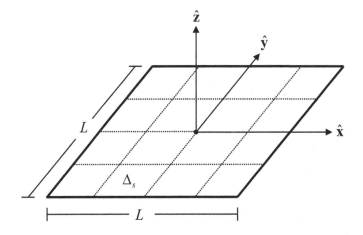

**FIGURE 2.5:** Thin Charged Plate Dimensions

where $S_n$ comprises the extents of patch $n$. Right-hand side vector elements remain the same as in (2.13).

### 2.1.2.1   Matrix Element Evaluation

When the source and field patches are the same ($m = n$), the integrand has a singularity and the integral must be evaluated analytically. These matrix elements are called *self terms*, and represent the most dominant interactions between elements. We will devote significant attention to the evaluation of self-term matrix elements in this book. The self-term integral for the charged plate is

$$Z_{mm} = \int_{-a}^{a} \int_{-a}^{a} \frac{1}{\sqrt{(x')^2 + (y')^2}} \, dx' \, dy' \,. \tag{2.19}$$

Performing the innermost integration yields [1] (200.01)

$$Z_{mm} = \int_{-a}^{a} \log \left[ \frac{\sqrt{a^2 + (y')^2} + a}{\sqrt{a^2 + (y')^2} - a} \right] dy' \tag{2.20}$$

and performing the outermost integration then yields [2]

$$Z_{mm} = 2a \log \left[ y + \sqrt{a^2 + y^2} \right] + y \log \left[ \frac{y^2 + 2a(a + \sqrt{a^2 + y^2})}{y^2} \right] \Bigg|_{-a}^{a} \tag{2.21}$$

which reduces to

$$Z_{mm} = \frac{2a}{\pi \epsilon} \log(1 + \sqrt{2}) \,. \tag{2.22}$$

For patches that do not overlap ($m \neq n$) we will use a simple centroidal approximation to the integral, resulting in

$$Z_{mn} = \frac{\Delta_s}{\sqrt{(x_m - x_n)^2 + (y_m - y_n)^2}} \tag{2.23}$$

where $x_n$ and $y_n$ are at the center of the source patch. This approximation is not very accurate for elements that are near to one other, though it serves to illustrate solving the problem. In such cases, an analytic or adaptive numerical integration should be used instead. Matrix elements involving source and field points that do not overlap but are still sufficiently close are called *near terms*, and are also considered in greater detail in this book.

### 2.1.2.2   Solution

In Figures 2.6a and 2.6b is shown the computed surface charge density obtained using 15 and 35 patches in the $x$ and $y$ directions, which comprises 225 and 1225 unknowns, respectively. In Figures 2.7a and 2.7b is shown the surface charge density on the patches along the diagonal of the plate. As with the

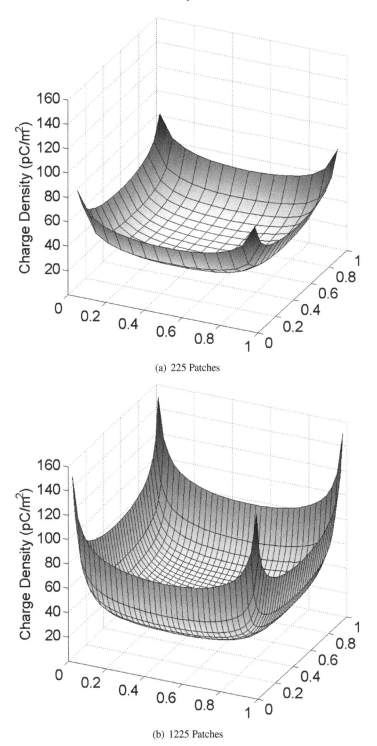

(a) 225 Patches

(b) 1225 Patches

**FIGURE 2.6:** Charge Distribution on Square Plate

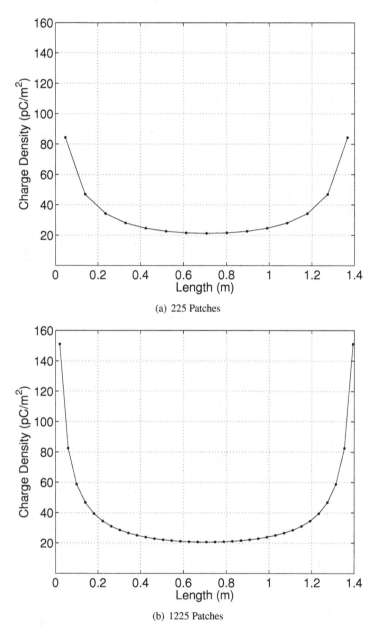

(a) 225 Patches

(b) 1225 Patches

**FIGURE 2.7:** Potential on Square Plate Diagonal

thin wire, the charge accumulates near the corners and the edges of the plate, and our solution would likely benefit from additional discretization density in those areas.

---

## 2.2 The Method of Moments

In the previous section, we considered the problem of computing an unknown charge distribution on a thin wire or plate at a known potential. Our basic approach was to expand the unknown quantity using a set of known functions with unknown coefficients. We then converted the resulting equation into a linear system of equations by enforcing the boundary conditions, in this case the electric potential, at a number of field points on the object. The resulting linear system was then solved numerically for the unknown coefficients. Let us formalize this process by introducing a method of weighted residuals known as the Method of Moments. Consider the generalized problem

$$L(f) = g \qquad (2.24)$$

where $L$ is a linear operator, $g$ is a known forcing function, and $f$ is unknown [3]. In electromagnetic problems, $L$ is typically an integro-differential operator such as $\mathcal{L}$ (3.77) and $\mathcal{K}$ (3.78), $f$ is the unknown function (current) and $g$ is a known driving function (incident field). Let us now expand $f$ into a sum of $N$ weighted basis functions

$$f = \sum_{n=1}^{N} a_n f_n \qquad (2.25)$$

where $a_n$ are unknown coefficients. Because $L$ is linear, substitution of the above into (2.24) yields

$$\sum_{n=1}^{N} a_n L(f_n) \approx g \qquad (2.26)$$

where the residual is

$$R = g - \sum_{n=1}^{N} a_n L(f_n) . \qquad (2.27)$$

The basis functions $f_n$ are chosen so they model the expected behavior of the unknown function throughout its domain, and they may be scalars or vectors depending on the problem. If the basis functions have local support in the domain, they are called *local* or *subsectional* basis functions. If their support spans the entire problem domain, they are called *global* or *entire-domain* basis functions. In this book we focus almost exclusively on local basis functions.

Let us now generalize the method by which the boundary conditions were previously enforced. We define an inner product or *moment* between a basis function $f_n(\mathbf{r}')$ and a *testing* or *weighting* function $f_m(\mathbf{r})$

$$\langle f_m, f_n \rangle = \int_{f_m} f_m(\mathbf{r}) \cdot \int_{f_n} f_n(\mathbf{r}') \, d\mathbf{r}' d\mathbf{r} \qquad (2.28)$$

where the integrals can be line, surface, or volume integrals depending on the support of the basis and testing functions. Requiring the inner product of each testing function with the residual function to be zero yields

$$\sum_{n=1}^{N} a_n \langle f_m, L(f_n) \rangle = \langle f_m, g \rangle \qquad (2.29)$$

which results in the $N \times N$ matrix equation $\mathbf{Z}\mathbf{a} = \mathbf{b}$ with matrix elements

$$Z_{mn} = \langle f_m, L(f_n) \rangle \qquad (2.30)$$

and right-hand side vector elements

$$b_m = \langle f_m, g \rangle . \qquad (2.31)$$

In electromagnetic problems, basis functions interact with all others via the Green's function and the resulting system matrix is full. This can be compared to other algorithms such as the Finite Element Method, where the matrix is typically sparse, symmetric, and banded, and many elements of each matrix row are zero [4].

## 2.2.1   Point Matching

In Sections 2.1.1 and 2.1.2, we enforced the boundary conditions by testing the integral equation at a set of discrete points on the object. This is equivalent to using a delta function as the testing function in (2.28), resulting in

$$f_m(\mathbf{r}) = \delta(\mathbf{r}) . \qquad (2.32)$$

This method, referred to as *point matching* or *point collocation*, offers advantages as well as disadvantages. One benefit is that in evaluating the matrix elements, no integral is required over the range of the testing function, only that of the source function, which may make evaluation of the matrix elements easier. The primary disadvantage is that the boundary conditions are matched only at discrete locations throughout the solution domain. Regardless, the results are often still reasonable, and we will use this method for a few problems in this book for comparative purposes.

## 2.2.2 Galerkin's Method

For testing, we are free to use whatever functions we wish. However, for most problems the choice of testing function is crucial to obtaining a good solution. One of the most commonly used is the *Method of Galerkin*, where the basis functions are used as the testing functions. This has the advantage of enforcing the boundary conditions throughout the solution domain, instead of at discrete points as with point matching. However, this comes at the expense of increased complexity and computation time required to evaluate the matrix elements. Most of the MoM problems encountered in the literature use Galerkin testing as the standard method (with its use often assumed or implied), and we will do so as well for most of the problems in this book.

## 2.3 Common One-Dimensional Basis Functions

The most important characteristic of a basis function is that it can reasonably represent the behavior of the unknown function throughout its domain. If the solution has a high level of variation in a particular region, pulse basis functions may not be as good a choice as a linear or higher-order function. The choice of basis function also determines the level of difficulty in evaluating the MoM matrix elements, which in some cases may be quite high. We will now briefly consider some one-dimensional local basis functions commonly used in Moment Method problems, as well as entire-domain functions.

### 2.3.1 Pulse Functions

A set of pulse basis functions is depicted in Figure 2.8, where the domain has been divided into $N$ points with $N - 1$ subsegments/pulses. In our figure

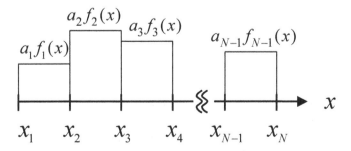

**FIGURE 2.8:** Pulse Functions

the segments all have equal lengths, However, this is not required. The pulse function is defined as

$$f_n(x) = 1 \qquad x_n \le x \le x_{n+1} \tag{2.33}$$

and zero otherwise. Pulse functions comprise a simple and crude approximation to the solution over each segment, but can greatly simplify the evaluation of MoM matrix elements. Note that since the derivative of pulse functions is impulsive, they cannot be used when the integral operator contains a derivative with respect to $x$ [5].

## 2.3.2   Piecewise Triangular Functions

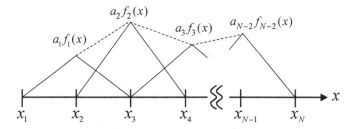

**FIGURE 2.9:** Triangle Functions (End Condition 1)

Where pulse functions are constant on a single segment, a triangle function spans two segments and varies from zero at the outer points to unity at the center. A set of triangle functions is shown in Figure 2.9. The domain has been divided into $N$ points and $N-1$ subsegments, resulting in $N-2$ basis functions. Though we have again shown the segments as having equal length, this is not required. Since adjacent functions overlap by one segment, triangles provide a piecewise linear variation of the solution between segments. The triangle function is defined as

$$f_n(x) = \frac{x - x_{n-1}}{x_n - x_{n-1}} \qquad x_{n-1} \le x \le x_n \tag{2.34}$$

and

$$f_n(x) = \frac{x_{n+1} - x}{x_{n+1} - x_n} \qquad x_n \le x \le x_{n+1} . \tag{2.35}$$

Triangle functions can be used when the integral operator contains a derivative with respect to $x$, which is useful in redistributing vector derivatives.

Note that the configuration in Figure 2.9 forces the solution to zero at $x_1$ and $x_N$. This may be desirable when the value of the solution at the ends of the domain is known to be zero *a priori*, however it should not be used when

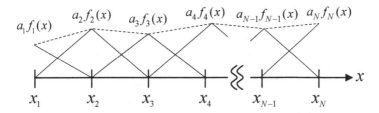

**FIGURE 2.10:** Triangle Functions (End Condition 2)

the solution might be nonzero. In this case, we can assign a *half triangle* to the first and last segments, which allows the solution to take on any value at the ends. This is illustrated in Figure 2.10, where we now have a total of $N$ basis functions.

### 2.3.3 Piecewise Sinusoidal Functions

Piecewise sinusoid functions are similar to triangle functions, as illustrated in Figure 2.11. They are sometimes used in the analysis of wire antennas because of their ability to represent sinusoidal current distributions. These functions are defined as

$$f_n(x) = \frac{\sin k(x - x_{n-1})}{\sin k(x_n - x_{n-1})} \qquad x_{n-1} \leq x \leq x_n \qquad (2.36)$$

and

$$f_n(x) = \frac{\sin k(x_{n+1} - x)}{\sin k(x_{n+1} - x_n)} \qquad x_n \leq x \leq x_{n+1} \qquad (2.37)$$

where $k$ is the wavenumber, and the length of the segments is generally much less than the period of the sinusoid.

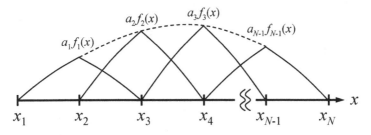

**FIGURE 2.11:** Piecewise Sinusoidal Functions

### 2.3.4 Entire-Domain Functions

Unlike local basis functions, entire-domain functions exist everywhere throughout the problem domain. One might have reason to use entire-domain functions if there exists some *a priori* information about the solution. For example, the solution may be known to comprise a sum of weighted polynomials or sines and cosines. As an example, consider the current $I(x)$ on a thin dipole antenna of length $L$. We might represent this current by the sum [6]

$$I(x) = \sum_{n=1}^{N} a_n (1 - |x/L|)^m \qquad (2.38)$$

where the testing procedure is still performed as before. After solving the resulting matrix equation, only the first few coefficients $a_n$ may be needed to accurately represent the current in the above sum. A disadvantage of entire-domain functions is that it may not be feasible to apply them to a geometry of arbitrary shape. As a result, entire-domain functions are not often encountered in the literature, and we will not consider them further in this book.

### 2.3.5 Number of Basis Functions

For a given problem, the number of basis functions (and thus, unknowns) must be chosen so that local variations in the solution are accurately modeled. As in this book we are concerned with time-harmonic problems, we must model amplitude as well as phase behavior. For linear basis functions such as the triangle in Section 2.3.2, a common "rule of thumb" is that at least ten unknowns per wavelength are needed to represent a sinusoid. This number should be increased in areas where the solution may vary significantly, such as gaps, cracks, and edges on a surface. Also note that in problems having multiple dielectric regions, the wavelength in each material will differ and so must the density of the basis functions on the different material interfaces. Higher-order basis functions also exist that can reduce the unknown count at the expense of increased complexity in describing the surface geometry, as well as the MoM formulation and evaluation of matrix elements. Functions such as these are covered in detail in advanced references on basis functions in electromagnetics.

Though the number of unknowns $N$ always increases with the size of the problem, the rate of the increase depends on whether the problem domain is a line, surface, or volume. In the MoM, $N$ increases linearly for one-dimensional problems and exponentially for the surface problems considered in Chapter 8. The resulting number may be as small as a few hundred, or may grow into the thousands or even millions for very large problems. This has significant consequences in terms of system memory, as it requires the storage of an $N \times N$

matrix, as well as the computation time to solve it. We will consider compressive methods for very large MoM problems in later chapters.

## References

[1] H. B. Dwight, *Tables of Integrals and Other Mathematical Data*. The Macmillan Company, 1961.

[2] *The Integrator*. Wolfram Research, integrals.wolfram.com.

[3] R. F. Harrington, *Field Computation by Moment Methods*. The Macmillan Company, 1968.

[4] J. Jin, *The Finite Element Method in Electromagnetics*. John Wiley and Sons, 1993.

[5] N. Morita, N. Kumagai, and J. R. Mautz, *Integral Equation Methods for Electromagnetics*. Artech House, 1990.

[6] B. D. Popovich, "Polynomial approximation of current along thin symmetrical cylindrical dipoles," *Proc. IEE*, vol. 117, pp. 873–878, May 1970.

# Chapter 3

## Radiation and Scattering

In this chapter we will review Maxwell's equations and the electromagnetic boundary conditions, and derive the integral equations for radiation and scattering and the Green's functions needed to solve those equations. We will next derive the electric and magnetic vector potential and expressions for the radiated electric and magnetic near and far fields. We will then consider the solution of radiation problems by way of the equivalence principle, and derive coupled surface integral equations used to treat multi-region conducting, dielectric, and hybrid conducting/dielectric geometries. Finally, we will discretize these equations in a general sense using the Method of Moments, which will be applied to more specific two- and three-dimensional problems of increasing complexity in later chapters.

## 3.1 Maxwell's Equations

In a homogeneous dielectric region having constituent parameters $\epsilon$ and $\mu$, the electric and magnetic fields must satisfy Maxwell's equations, which in the frequency domain are

$$\nabla \times \mathbf{E} = -\mathbf{M} - j\omega\mu\mathbf{H} , \qquad (3.1)$$

$$\nabla \times \mathbf{H} = \mathbf{J} + j\omega\epsilon\mathbf{E} , \qquad (3.2)$$

$$\nabla \cdot \mathbf{D} = q_e , \qquad (3.3)$$

$$\nabla \cdot \mathbf{B} = q_m , \qquad (3.4)$$

where $\mathbf{D} = \epsilon\mathbf{E}$, $\mathbf{B} = \mu\mathbf{H}$, and the time dependence $e^{jwt}$ is assumed and suppressed throughout this text. Though the magnetic current $\mathbf{M}$ and charge $q_m$ are not physically realizable quantities, they are useful mathematical tools in solving radiation and scattering problems, as we will show in Section 3.6.1.

## 3.2 Electromagnetic Boundary Conditions

At the interface between two regions $R_1$ and $R_2$, each having different dielectric parameters, the generalized electromagnetic boundary conditions can be written as

$$\hat{\mathbf{n}}_1 \times (\mathbf{E}_1 - \mathbf{E}_2) = -\mathbf{M} \,, \tag{3.5}$$

$$\hat{\mathbf{n}}_1 \times (\mathbf{H}_1 - \mathbf{H}_2) = \mathbf{J} \,, \tag{3.6}$$

$$\hat{\mathbf{n}}_1 \cdot (\mathbf{D}_1 - \mathbf{D}_2) = q_e \,, \tag{3.7}$$

$$\hat{\mathbf{n}}_1 \cdot (\mathbf{B}_1 - \mathbf{B}_2) = q_m \,, \tag{3.8}$$

where any of $\mathbf{M}, \mathbf{J}, q_e$, or $q_m$ may be present, and $\hat{\mathbf{n}}_1$ is the normal vector on the interface between $R_1$ and $R_2$ that points into $R_1$. At the interface between a dielectric ($R_1$) and a perfect electric conductor (PEC, $R_2$), the boundary conditions are

$$-\hat{\mathbf{n}}_1 \times \mathbf{E}_1 = 0 \,, \tag{3.9}$$

$$\hat{\mathbf{n}}_1 \times \mathbf{H}_1 = \mathbf{J} \,, \tag{3.10}$$

$$\hat{\mathbf{n}}_1 \cdot \mathbf{D}_1 = q_e \,, \tag{3.11}$$

$$\hat{\mathbf{n}}_1 \cdot \mathbf{B}_1 = 0 \,, \tag{3.12}$$

and between a dielectric ($R_1$) and perfect magnetic conductor (PMC, $R_2$) they are

$$-\hat{\mathbf{n}}_1 \times \mathbf{E}_1 = \mathbf{M} \,, \tag{3.13}$$

$$\hat{\mathbf{n}}_1 \times \mathbf{H}_1 = 0 \,, \tag{3.14}$$

$$\hat{\mathbf{n}}_1 \cdot \mathbf{D}_1 = 0 \,, \tag{3.15}$$

$$\hat{\mathbf{n}}_1 \cdot \mathbf{B}_1 = q_m \,. \tag{3.16}$$

In this book we will consider dielectric and PEC regions only.

## 3.3 Formulations for Radiation

The electromagnetic radiation problem involves the calculation of fields everywhere in space due to a set of electric and magnetic currents. Scattering problems can be treated as radiation problems where these currents are generated by other currents or impressed fields. Let us start from first principles and derive the expression for the electric field in the presence of electric and magnetic currents $\mathbf{J}$ and $\mathbf{M}$. To begin, we take the curl of (3.1) and combine with (3.2) to get

$$\nabla \times \nabla \times \mathbf{E} = -j\omega\mu\nabla \times \mathbf{H} = \omega^2\mu\epsilon\mathbf{E} - j\omega\mu\mathbf{J} - \nabla \times \mathbf{M} \qquad (3.17)$$

or

$$\nabla \times \nabla \times \mathbf{E} - \omega^2\mu\epsilon\mathbf{E} = -j\omega\mu\mathbf{J} - \nabla \times \mathbf{M} . \qquad (3.18)$$

Using the vector identity

$$\nabla \times \nabla \times \mathbf{E} = \nabla(\nabla \cdot \mathbf{E}) - \nabla^2\mathbf{E} , \qquad (3.19)$$

we can write (3.18) as

$$\nabla(\nabla \cdot \mathbf{E}) - \nabla^2\mathbf{E} - k^2\mathbf{E} = -j\omega\mu\mathbf{J} - \nabla \times \mathbf{M} , \qquad (3.20)$$

where the wavenumber $k = w\sqrt{\mu\epsilon} = 2\pi/\lambda$. Substituting (3.3) into the above we get

$$\nabla^2\mathbf{E} + k^2\mathbf{E} = j\omega\mu\mathbf{J} + \frac{\nabla q_e}{\epsilon} - \nabla \times \mathbf{M} . \qquad (3.21)$$

Employing the equation of continuity

$$\nabla \cdot \mathbf{J} = -j\omega q_e \qquad (3.22)$$

allows us to then obtain

$$\nabla^2\mathbf{E} + k^2\mathbf{E} = j\omega\mu\mathbf{J} - \frac{1}{j\omega\epsilon}\nabla(\nabla \cdot \mathbf{J}) - \nabla \times \mathbf{M} . \qquad (3.23)$$

Since Maxwell's equations are linear, we can consider $\mathbf{J}$ and $\mathbf{M}$ to be a superposition of point sources distributed over some volume. Therefore, if we know the response of a point source, we can solve the original problem by integrating that response over the volume. We now make use of this idea to convert (3.23) into an integral equation. To do so, we introduce the *Green's function* $G(\mathbf{r}, \mathbf{r}')$, which satisfies the scalar Helmholtz equation [1]:

$$\nabla^2 G(\mathbf{r}, \mathbf{r}') + k^2 G(\mathbf{r}, \mathbf{r}') = -\delta(\mathbf{r}, \mathbf{r}') . \qquad (3.24)$$

Assuming that we know $G(\mathbf{r}, \mathbf{r}')$, the radiated electric field is then

$$\mathbf{E}(\mathbf{r}) = -j\omega\mu \int_V G(\mathbf{r}, \mathbf{r}') \left[1 + \frac{1}{k^2}\nabla'\nabla'\cdot\right]\mathbf{J}(\mathbf{r}')\,d\mathbf{r}' - \int_V G(\mathbf{r}, \mathbf{r}')\nabla'\times\mathbf{M}(\mathbf{r}')\,d\mathbf{r}' , \qquad (3.25)$$

and by a similar derivation, the radiated magnetic field is

$$\mathbf{H}(\mathbf{r}) = -j\omega\epsilon \int_V G(\mathbf{r}, \mathbf{r}') \left[1 + \frac{1}{k^2}\nabla'\nabla'\cdot\right]\mathbf{M}(\mathbf{r}')\,d\mathbf{r}' + \int_V G(\mathbf{r}, \mathbf{r}')\nabla'\times\mathbf{J}(\mathbf{r}')\,d\mathbf{r}' . \qquad (3.26)$$

We will next solve (3.24) for $G(\mathbf{r}, \mathbf{r}')$ in two and three dimensions. To do so, we will first consider the homogeneous version, and then match the boundary conditions of the inhomogeneous case to obtain a unique solution.

### 3.3.1 Three-Dimensional Green's Function

As $G(\mathbf{r}, \mathbf{r}')$ is the solution for a point electromagnetic source, it must exhibit spherical symmetry in three dimensions. Therefore, we will retain only the radial term of the Laplacian and write

$$\nabla^2 G = \frac{1}{r^2} \frac{d}{dr} \left( r^2 \frac{dG}{dr} \right) = \frac{d^2 G}{dr^2} + \frac{2}{r} \frac{dG}{dr} . \qquad (3.27)$$

Next, we observe that

$$\frac{1}{r} \frac{d^2(rG)}{dr^2} = \frac{1}{r} \frac{d}{dr} \left[ r \frac{dG}{dr} + G \right] = \frac{d^2 G}{dr^2} + \frac{2}{r} \frac{dG}{dr} \qquad (3.28)$$

and if we substitute in the function $rG$ for $r > 0$, we can write

$$\frac{d^2(rG)}{dr^2} + k^2(rG) = 0 . \qquad (3.29)$$

The solution to the above is

$$G(r) = A \frac{e^{-jkr}}{r} + B \frac{e^{jkr}}{r} , \qquad (3.30)$$

where $r = |\mathbf{r} - \mathbf{r}'|$. Noting that the solution must contain only outgoing waves, we retain only the first term in (3.30), yielding

$$G(r) = A \frac{e^{-jkr}}{r} . \qquad (3.31)$$

It should be pointed out that the phase convention of the exponential in the Green's function is not standard throughout the literature. The phase as defined in (3.31) is standard in most electrical engineering references; however, other references [2, 3] define the Green's function with a positive exponent, assuming the time harmonic term $e^{-j\omega t}$. Similarly, the two-dimensional Green's function in (3.49) would instead use the Hankel function of the first kind.

We must now match the boundary conditions to determine a unique solution. First, the amplitude of the outgoing wave must decay to zero with increasing $r$, requiring that $G(r) \to 0$ as $r \to infty$. By inspection, we see that this is already satisfied. Next, we must match the Green's function at the point source ($r = 0$), which will allow us to determine $A$. To do so, we integrate (3.24) over a spherical volume of radius $a$ around the source. Substituting (3.31) for $G(r)$, we get

$$A \int_V \left[ \nabla \cdot \nabla \left( \frac{e^{-jkr}}{r} \right) + k^2 \frac{e^{-jkr}}{r} \right] dV = -1 , \qquad (3.32)$$

and to evaluate the first term, we use the divergence theorem to write

$$\int_V \nabla \cdot \nabla \left( \frac{e^{-jkr}}{r} \right) dV = \int_S \hat{\mathbf{n}} \cdot \nabla \left( \frac{e^{-jkr}}{r} \right) dS . \qquad (3.33)$$

On the sphere, $\hat{\mathbf{n}} = \hat{\mathbf{r}}$, therefore

$$\int_S \hat{\mathbf{r}} \cdot \nabla(\frac{e^{-jkr}}{r}) \, dS = \int_S \frac{\partial}{\partial r}(\frac{e^{-jkr}}{r}) \, dS \qquad (3.34)$$

which is

$$4\pi a^2 \left[\frac{\partial}{\partial r}\left(\frac{e^{-jkr}}{r}\right)\right]_{r=a} . \qquad (3.35)$$

Taking the limit as $a \to 0$ we get

$$\lim_{a \to 0} 4\pi a^2 \left[\frac{\partial}{\partial r}\left(\frac{e^{-jkr}}{r}\right)\right]_{r=a} = -4\pi . \qquad (3.36)$$

The second term is

$$k^2 \int_0^a \frac{e^{-jkr}}{r} 4\pi r^2 dr = 4\pi k^2 \int_0^a r e^{-jkr} dr . \qquad (3.37)$$

By inspection, this integral tends to zero as $a \to 0$. Therefore,

$$A = \frac{1}{4\pi} \qquad (3.38)$$

and finally,

$$G(r) = \frac{e^{-jkr}}{4\pi r} . \qquad (3.39)$$

which is the electrodynamic Green's function in three dimensions.

### 3.3.2 Two-Dimensional Green's Function

Two-dimensional problems are those where the currents, surfaces, and fields extend to $\pm\infty$ along the $\hat{\mathbf{z}}$ axis. As such, there is no wave propagation in the $\hat{\mathbf{z}}$ direction. For these problems we will work in the cylindrical coordinate system $(\rho, \phi, z)$ where $\boldsymbol{\rho} = \rho\cos\phi\,\hat{\mathbf{x}} + \rho\sin\phi\,\hat{\mathbf{y}}$, and surface integrals are performed in the $(x, y)$ plane $(z = 0)$. In two dimensions, (3.24) can be written as

$$\nabla^2 G(\boldsymbol{\rho}, \boldsymbol{\rho}') + k^2 G(\boldsymbol{\rho}, \boldsymbol{\rho}') = -\delta(\boldsymbol{\rho}, \boldsymbol{\rho}') . \qquad (3.40)$$

The solutions to the above for the homogeneous case are the Hankel functions of the first and second kinds of order zero. Since we know the solution contains only outgoing waves, we can then write

$$G(\rho) = AH_0^{(2)}(k\rho) \qquad (3.41)$$

where $\rho = |\boldsymbol{\rho} - \boldsymbol{\rho}'|$. To obtain $A$, we use the small-argument approximation to the Hankel function [4]

$$H_0^{(2)}(k\rho) \approx 1 - j\frac{2}{\pi} \log\left(\frac{\gamma k\rho}{2}\right) \qquad \rho \to 0 , \tag{3.42}$$

and integrate (3.40) over a very small circle of radius $a$ centered at the origin, yielding

$$A \int_S \left[\nabla^2 + k^2\right]\left[1 - j\frac{2}{\pi} \log\left(\frac{\gamma k\rho}{2}\right)\right] dS = -1 . \tag{3.43}$$

Noting that $\nabla^2 f = \nabla \cdot (\nabla f)$, the first term can be converted to a line integral using the divergence theorem, yielding

$$-j\frac{2}{\pi} \int_0^{2\pi} \nabla\left[\log\left(\frac{\gamma k\rho}{2}\right)\right] \rho \, d\phi = -4j . \tag{3.44}$$

The second term is

$$k^2 \int_0^a \left[1 - j\frac{2}{\pi} \log\left(\frac{\gamma k\rho}{2}\right)\right] 2\pi\rho \, d\rho . \tag{3.45}$$

The first part of (3.45) goes to zero as $a \to 0$. Integrating the second part yields

$$4jk^2 \int_0^a \log\left(\frac{\gamma k\rho}{2}\right) \rho \, d\rho = \left[\frac{\rho^2}{2} \log\left(\frac{\gamma k\rho}{2}\right) - \frac{\rho^2}{4}\right]\Bigg|_0^a . \tag{3.46}$$

The above goes to zero, since

$$\lim_{\rho \to 0} \rho^2 \log \rho = 0 . \tag{3.47}$$

Therefore,

$$A = -\frac{j}{4} \tag{3.48}$$

and finally,

$$G(\rho) = -\frac{j}{4} H_0^{(2)}(k\rho) , \tag{3.49}$$

which is the electrodynamic Green's function in two dimensions.

## 3.4 Vector Potentials

In Section 3.3, we derived expressions that allow us to determine the radiated field everywhere in space from electric and magnetic current distributions. In this section we will derive alternative expressions for the radiated fields in terms of vector potentials, also commonly used in radiation and scattering problems. We will then address the differences between the formulations in the following section.

### 3.4.1 Magnetic Vector Potential

We will first derive the magnetic vector potential for a homogeneous, source-free region. We begin by observing that since the magnetic field **H** is always solenoidal, it can be written as the curl of another vector **A**, which is

$$\mathbf{H} = \frac{1}{\mu}\nabla \times \mathbf{A} . \tag{3.50}$$

Substituting the above into (3.1), we get

$$\nabla \times \mathbf{E} = -j\omega\nabla \times \mathbf{A} , \tag{3.51}$$

which can be written as

$$\nabla \times (\mathbf{E} + j\omega\mathbf{A}) = 0 . \tag{3.52}$$

We next use the identity

$$\nabla \times (-\nabla\Phi_e) = 0 \tag{3.53}$$

to write

$$\mathbf{E} = -j\omega\mathbf{A} - \nabla\Phi_e , \tag{3.54}$$

where $\Phi_e$ is an arbitrary electric *scalar potential*. We next use the identity

$$\nabla \times \nabla \times \mathbf{A} = \nabla\nabla \cdot \mathbf{A} - \nabla^2\mathbf{A} , \tag{3.55}$$

and take the curl of both sides of (3.50) allowing us to write

$$\mu\nabla \times \mathbf{H} = \nabla\nabla \cdot \mathbf{A} - \nabla^2\mathbf{A} , \tag{3.56}$$

and combining this with (3.2) leads to

$$\mu\mathbf{J} + j\omega\mu\epsilon\mathbf{E} = \nabla\nabla \cdot \mathbf{A} - \nabla^2\mathbf{A} . \tag{3.57}$$

Substituting (3.54) into the above leads to

$$\mu\mathbf{J} + j\omega\mu\epsilon(-j\omega\mathbf{A} - \nabla\Phi_e) = \nabla\nabla \cdot \mathbf{A} - \nabla^2\mathbf{A} , \tag{3.58}$$

which is

$$\nabla^2 \mathbf{A} + k^2 \mathbf{A} = -\mu \mathbf{J} + \nabla(\nabla \cdot \mathbf{A} + j\omega\mu\epsilon\Phi_e) \,. \tag{3.59}$$

We have already defined the curl of $\mathbf{A}$ in (3.50), however we have not yet defined its divergence. We are therefore free to make it whatever we wish as long as we remain consistent with that definition elsewhere. Therefore, if we make the following assignment

$$\Phi_e = -\frac{1}{j\omega\mu\epsilon}\nabla \cdot \mathbf{A} \,, \tag{3.60}$$

we can simplify (3.59) to yield

$$\nabla^2 \mathbf{A} + k^2 \mathbf{A} = -\mu \mathbf{J} \,, \tag{3.61}$$

which is an inhomogeneous vector Helmholtz equation for $\mathbf{A}$. Using the Green's function, $\mathbf{A}(\mathbf{r})$ can be written as

$$\mathbf{A}(\mathbf{r}) = \mu \int_V G(\mathbf{r}, \mathbf{r}') \, \mathbf{J}(\mathbf{r}') \, d\mathbf{r}' \,, \tag{3.62}$$

and the corresponding electric field is then

$$\mathbf{E}(\mathbf{r}) = -j\omega \left[1 + \frac{1}{k^2}\nabla\nabla \cdot \right] \mathbf{A}(\mathbf{r}) \,. \tag{3.63}$$

### 3.4.1.1   Three-Dimensional Magnetic Vector Potential

Substitution of (3.39) into (3.62) yields the magnetic vector potential in three dimensions, which is

$$\mathbf{A}(\mathbf{r}) = \mu \int_V \mathbf{J}(\mathbf{r}') \frac{e^{-jk|\mathbf{r}-\mathbf{r}'|}}{4\pi|\mathbf{r}-\mathbf{r}'|} \, d\mathbf{r}' \,. \tag{3.64}$$

### 3.4.1.2   Two-Dimensional Magnetic Vector Potential

The corresponding magnetic vector potential in two dimensions is obtained via (3.49), and is

$$\mathbf{A}(\boldsymbol{\rho}) = -j\frac{\mu}{4}\int_S \mathbf{J}(\boldsymbol{\rho}')H_0^{(2)}(k|\boldsymbol{\rho}-\boldsymbol{\rho}'|) \, d\boldsymbol{\rho}' \,. \tag{3.65}$$

## 3.4.2   Electric Vector Potential

Using the symmetry of Maxwell's equations, we can derive similar expressions for the electric vector potential $\mathbf{F}$. We will not repeat the derivation, and simply summarize these expressions below:

$$\mathbf{E} = -\frac{1}{\epsilon}\nabla \times \mathbf{F} \,, \tag{3.66}$$

$$\Phi_m = -\frac{1}{j\omega\mu\epsilon} \nabla \cdot \mathbf{F} , \tag{3.67}$$

$$\nabla^2 \mathbf{F} + k^2 \mathbf{F} = -\epsilon \mathbf{M} , \tag{3.68}$$

$$\mathbf{H}(\mathbf{r}) = -j\omega \left[ 1 + \frac{1}{k^2} \nabla\nabla \cdot \right] \mathbf{F}(\mathbf{r}) , \tag{3.69}$$

$$\mathbf{F}(\mathbf{r}) = \epsilon \int_V \mathbf{M}(\mathbf{r}') \, G(\mathbf{r}, \mathbf{r}') \, d\mathbf{r}' . \tag{3.70}$$

#### 3.4.2.1 Three-Dimensional Electric Vector Potential

Substitution of (3.39) into (3.70) yields the electric vector potential in three dimensions, which is

$$\mathbf{F}(\mathbf{r}) = \epsilon \int_V \mathbf{M}(\mathbf{r}') \frac{e^{-jk|\mathbf{r}-\mathbf{r}'|}}{4\pi|\mathbf{r}-\mathbf{r}'|} \, d\mathbf{r}' . \tag{3.71}$$

#### 3.4.2.2 Two-Dimensional Electric Vector Potential

Similarly, the two-dimensional electric vector potential is obtained via (3.49), and is

$$\mathbf{F}(\boldsymbol{\rho}) = -j\frac{\epsilon}{4} \int_S \mathbf{M}(\boldsymbol{\rho}') H_0^{(2)}(k|\boldsymbol{\rho}-\boldsymbol{\rho}'|) \, d\boldsymbol{\rho}' . \tag{3.72}$$

### 3.4.3 Total Fields

Using the results from Sections 3.4.1 and 3.4.2, we can write the total electric field $\mathbf{E}(\mathbf{r})$ as

$$\mathbf{E}(\mathbf{r}) = -j\omega \left[ \mathbf{A}(\mathbf{r}) + \frac{1}{k^2} \nabla\nabla \cdot \mathbf{A}(\mathbf{r}) \right] - \frac{1}{\epsilon} \nabla \times \mathbf{F}(\mathbf{r}) , \tag{3.73}$$

and the magnetic field $\mathbf{H}(\mathbf{r})$ as

$$\mathbf{H}(\mathbf{r}) = -j\omega \left[ \mathbf{F}(\mathbf{r}) + \frac{1}{k^2} \nabla\nabla \cdot \mathbf{F}(\mathbf{r}) \right] + \frac{1}{\mu} \nabla \times \mathbf{A}(\mathbf{r}) . \tag{3.74}$$

In *operator notation*, we can rewrite (3.73) and (3.74) as

$$\mathbf{E}(\mathbf{r}) = -j\omega\mu(\mathcal{L}\mathbf{J})(\mathbf{r}) - (\mathcal{K}\mathbf{M})(\mathbf{r}) \tag{3.75}$$

and

$$\mathbf{H}(\mathbf{r}) = -j\omega\epsilon(\mathcal{L}\mathbf{M})(\mathbf{r}) + (\mathcal{K}\mathbf{J})(\mathbf{r}) , \tag{3.76}$$

where the operators $\mathcal{L}$ and $\mathcal{K}$ are

$$(\mathcal{L}\mathbf{X})(\mathbf{r}) = \left[1 + \frac{1}{k^2}\nabla\nabla\cdot\right] \int_V G(\mathbf{r},\mathbf{r}')\mathbf{X}(\mathbf{r}')\,d\mathbf{r}' \qquad (3.77)$$

and

$$(\mathcal{K}\mathbf{X})(\mathbf{r}) = \nabla \times \int_V G(\mathbf{r},\mathbf{r}')\mathbf{X}(\mathbf{r}')\,d\mathbf{r}' \ . \qquad (3.78)$$

### 3.4.4 Comparison of Radiation Formulas

The expressions for the electric and magnetic fields in (3.73) and (3.74) are almost identical to those of (3.25) and (3.26) except that the differential operators are distributed differently. It is important that we show their equivalence, as the reader will likely encounter both forms in the literature, as well as expressions where the operators are distributed differently than shown in this book. We will also use techniques from this section to redistribute differential operators in cases where it simplifies evaluation of the MoM matrix elements. To demonstrate this equivalence, we must show that

$$\mathbf{I}_1(\mathbf{r}) = \nabla\nabla\cdot\int_V G(\mathbf{r},\mathbf{r}')\,\mathbf{X}(\mathbf{r}')\,d\mathbf{r}' = \int_V G(\mathbf{r},\mathbf{r}')\nabla'\nabla'\cdot\mathbf{X}(\mathbf{r}')\,d\mathbf{r}' \quad (3.79)$$

and

$$\mathbf{I}_2(\mathbf{r}) = \nabla\times\int_V G(\mathbf{r},\mathbf{r}')\,\mathbf{X}(\mathbf{r}')\,d\mathbf{r}' = \int_V G(\mathbf{r},\mathbf{r}')\nabla'\times\mathbf{X}(\mathbf{r}')\,d\mathbf{r}' \ . \quad (3.80)$$

Let us begin with $\mathbf{I}_1(\mathbf{r})$. Using the vector identity

$$\nabla\cdot\left[G(\mathbf{r},\mathbf{r}')\,\mathbf{X}(\mathbf{r}')\right] = \left[\nabla G(\mathbf{r},\mathbf{r}')\right]\cdot\mathbf{X}(\mathbf{r}') + G(\mathbf{r},\mathbf{r}')\nabla\cdot\mathbf{X}(\mathbf{r}') \qquad (3.81)$$

and the fact that $\mathbf{X}(\mathbf{r}')$ is not a function of the unprimed coordinates, we can write

$$\mathbf{I}_1(\mathbf{r}) = \nabla\nabla\cdot\int_V G(\mathbf{r},\mathbf{r}')\,\mathbf{X}(\mathbf{r}')\,d\mathbf{r}' = \nabla\int_V\left[\nabla G(\mathbf{r},\mathbf{r}')\right]\cdot\mathbf{X}(\mathbf{r}')\,d\mathbf{r}' \ . \quad (3.82)$$

Using the symmetry of the Green's function,

$$\nabla G(\mathbf{r},\mathbf{r}') = -\nabla' G(\mathbf{r},\mathbf{r}') \ , \qquad (3.83)$$

the above becomes

$$\mathbf{I}_1(\mathbf{r}) = -\nabla\int_V\left[\nabla' G(\mathbf{r},\mathbf{r}')\right]\mathbf{X}(\mathbf{r}')\,d\mathbf{r}' \ . \qquad (3.84)$$

Using the previous vector identity we can write the above as a sum of two integrals as

$$\mathbf{I}_1(\mathbf{r}) = \nabla \int_V G(\mathbf{r}, \mathbf{r}') \nabla' \cdot \mathbf{X}(\mathbf{r}') \, d\mathbf{r}' - \nabla \int_V \nabla' \cdot \left[ G(\mathbf{r}, \mathbf{r}') \, \mathbf{X}(\mathbf{r}') \right] \, d\mathbf{r}' . \quad (3.85)$$

We then make use of the divergence theorem to write the second term as

$$\mathbf{I}_1(\mathbf{r}) = \nabla \int_V G(\mathbf{r}, \mathbf{r}') \nabla' \cdot \mathbf{X}(\mathbf{r}') \, d\mathbf{r}' - \nabla \oint_S \hat{\mathbf{n}} \cdot \left[ G(\mathbf{r}, \mathbf{r}') \, \mathbf{X}(\mathbf{r}') \right] \, d\mathbf{r}' . \quad (3.86)$$

Assuming $\mathbf{X}(\mathbf{r}')$ is of finite extent and completely inside $V$, it will be zero on $S$. Thus, the second integral is zero and we are left with

$$\mathbf{I}_1(\mathbf{r}) = \nabla \int_V G(\mathbf{r}, \mathbf{r}') \nabla' \cdot \mathbf{X}(\mathbf{r}') \, d\mathbf{r}' . \quad (3.87)$$

We must now move the remaining gradient operator under the integral sign, and doing so results in

$$\mathbf{I}_1(\mathbf{r}) = - \int_V \nabla' G(\mathbf{r}, \mathbf{r}') \nabla' \cdot \mathbf{X}(\mathbf{r}') \, d\mathbf{r}' , \quad (3.88)$$

where we have again employed the symmetry of the Green's function. We must now move the gradient so that it operates on the $\nabla' \cdot \mathbf{X}(\mathbf{r}')$ term. To do so, we use the vector identity

$$\nabla' G(\mathbf{r}, \mathbf{r}') \nabla' \cdot \mathbf{X}(\mathbf{r}') = \nabla' \left[ G(\mathbf{r}, \mathbf{r}') \nabla' \cdot \mathbf{X}(\mathbf{r}') \right] - G(\mathbf{r}, \mathbf{r}') \nabla' \nabla' \cdot \mathbf{X}(\mathbf{r}') \quad (3.89)$$

to write

$$\mathbf{I}_1(\mathbf{r}) = \int_V G(\mathbf{r}, \mathbf{r}') \nabla' \nabla' \cdot \mathbf{X}(\mathbf{r}') \, d\mathbf{r}' - \int_V \nabla' \left[ G(\mathbf{r}, \mathbf{r}') \nabla' \cdot \mathbf{X}(\mathbf{r}') \right] \, d\mathbf{r}' . \quad (3.90)$$

To show that the second term is zero, we apply the following result of the divergence theorem

$$\int_V \nabla f(\mathbf{r}) \, dV = \oint_S f(\mathbf{r}) \, d\mathbf{S} \quad (3.91)$$

to write the above as

$$\mathbf{I}_1(\mathbf{r}) = \int_V G(\mathbf{r}, \mathbf{r}') \nabla' \nabla' \cdot \mathbf{X}(\mathbf{r}') \, d\mathbf{r}' - \oint_S G(\mathbf{r}, \mathbf{r}') \nabla' \cdot \mathbf{X}(\mathbf{r}') \, d\mathbf{r}' . \quad (3.92)$$

Again making $V$ large enough such that $\mathbf{X}(\mathbf{r}')$ has zero value on $S$, we are left with

$$\mathbf{I}_1(\mathbf{r}) = \int_V G(\mathbf{r}, \mathbf{r}') \nabla' \nabla' \cdot \mathbf{X}(\mathbf{r}') \, d\mathbf{r}' , \quad (3.93)$$

which is the expected result. We next focus our attention on $\mathbf{I}_2(\mathbf{r})$. Using the vector identity

$$\nabla \times \left[G(\mathbf{r}, \mathbf{r}') \, \mathbf{X}(\mathbf{r}')\right] = G(\mathbf{r}, \mathbf{r}')\nabla \times \mathbf{X}(\mathbf{r}') + \nabla G(\mathbf{r}, \mathbf{r}') \times \mathbf{X}(\mathbf{r}') , \quad (3.94)$$

we can write

$$\nabla \times \int_V G(\mathbf{r}, \mathbf{r}') \, \mathbf{X}(\mathbf{r}') \, d\mathbf{r}' = \int_V G(\mathbf{r}, \mathbf{r}')\nabla \times \mathbf{X}(\mathbf{r}') \, d\mathbf{r}' + \int_V \nabla G(\mathbf{r}, \mathbf{r}') \times \mathbf{X}(\mathbf{r}') \, d\mathbf{r}' .$$
$$(3.95)$$

Noting that $\nabla G(\mathbf{r}, \mathbf{r}') = -\nabla' G(\mathbf{r}, \mathbf{r}')$ and $\nabla \times \mathbf{X}(\mathbf{r}') = 0$, we write the above as

$$\nabla \times \int_V G(\mathbf{r}, \mathbf{r}') \, \mathbf{X} \, d\mathbf{r}' = - \int_V \nabla' G(\mathbf{r}, \mathbf{r}') \times \mathbf{X}(\mathbf{r}') \, d\mathbf{r}' . \quad (3.96)$$

Again using the identity (3.94) but on primed coordinates, we write

$$- \int_V \nabla' G(\mathbf{r}, \mathbf{r}') \times \mathbf{X}(\mathbf{r}') \, d\mathbf{r}' = \int_V G(\mathbf{r}, \mathbf{r}')\nabla' \times \mathbf{X}(\mathbf{r}') \, d\mathbf{r}'$$
$$- \int_V \nabla' \times \left[G(\mathbf{r}, \mathbf{r}') \, \mathbf{X}(\mathbf{r}')\right] d\mathbf{r}' , \quad (3.97)$$

and using the following result of the divergence theorem

$$\int_V \nabla \times \mathbf{X}(\mathbf{r}) \, dV = \oint_S \hat{\mathbf{n}}(\mathbf{r}) \times \mathbf{X}(\mathbf{r}) \, dS , \quad (3.98)$$

we can write (3.4.4) as

$$- \int_V \nabla' G(\mathbf{r}, \mathbf{r}') \times \mathbf{X}(\mathbf{r}') \, d\mathbf{r}' = \int_V G(\mathbf{r}, \mathbf{r}')\nabla' \times \mathbf{X}(\mathbf{r}') \, d\mathbf{r}'$$
$$- \oint_S \hat{\mathbf{n}} \times \left[G(\mathbf{r}, \mathbf{r}') \, \mathbf{X}(\mathbf{r}')\right] d\mathbf{r}' . \quad (3.99)$$

Making $V$ large enough such that $\mathbf{X}(\mathbf{r}')$ has zero value on $S$, the second term on the right hand side of (3.4.4) goes to zero and we are left with

$$\nabla \times \int_V G(\mathbf{r}, \mathbf{r}') \, \mathbf{X} \, d\mathbf{r}' = \int_V G(\mathbf{r}, \mathbf{r}')\nabla' \times \mathbf{X}(\mathbf{r}') \, d\mathbf{r}' , \quad (3.100)$$

which is again the expected result.

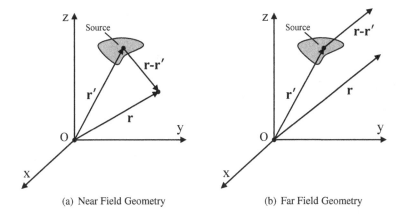

(a) Near Field Geometry          (b) Far Field Geometry

**FIGURE 3.1:** Near and Far Field Geometry

## 3.5 Near and Far Field

The radiation equations presented thus far are general formulas that can be used to compute the electric and magnetic fields anywhere. Usually though, one is interested in the fields in one of two regions: the *near field* region or the *far field* region. The near field region is located close to the source, where the fields comprise a superposition of stored (reactive) as well as radiated power. The near field region is of interest in applications such as antenna design, electromagnetic interference (EMI), and packaging of electronic components. The far field region is that which is far away from the source, where the fields have transitioned into purely radiative plane waves. In this section we will derive expressions for the near and far fields in two and three dimensions.

### 3.5.1 Three-Dimensional Near Field

Let us obtain expressions for the three-dimensional near fields at a point close to a source, as illustrated in Figure 3.1a. Recall that the magnetic field radiated by an electric current can be obtained from (3.50) and (3.62) and is

$$\mathbf{H}_A(\mathbf{r}) = \frac{1}{\mu}\nabla \times \mathbf{A}(\mathbf{r}) = \nabla \times \int_V \mathbf{J}(\mathbf{r}')\, G(\mathbf{r},\mathbf{r}')\, d\mathbf{r}' . \qquad (3.101)$$

Moving the curl operator under the integral sign and using the vector identity (3.94) and the fact that $\nabla \times \mathbf{J}(\mathbf{r}') = 0$, we then write

$$\mathbf{H}_A(\mathbf{r}) = \int_V \nabla G(\mathbf{r},\mathbf{r}') \times \mathbf{J}(\mathbf{r}')\, d\mathbf{r}' . \qquad (3.102)$$

Taking the gradient of the 3D Green's function (3.39) yields

$$\nabla\left(\frac{e^{-jkr}}{4\pi r}\right) = -(\mathbf{r} - \mathbf{r}')\left(\frac{1+jkr}{4\pi r^3}\right)e^{-jkr}\,, \qquad (3.103)$$

and using this result we can then write the magnetic field as

$$\mathbf{H}_A(\mathbf{r}) = -\frac{1}{4\pi}\int_V [(\mathbf{r}-\mathbf{r}') \times \mathbf{J}(\mathbf{r}')]\frac{1+jkr}{r^3}e^{-jkr}\,d\mathbf{r}'\,. \qquad (3.104)$$

Expanding into Cartesian components, the above can be written as

$$H_{A,x}(\mathbf{r}) = \frac{1}{4\pi}\int_V \left[\Delta_z J_y - \Delta_y J_z\right]\frac{1+jkr}{r^3}e^{-jkr}\,d\mathbf{r}'\,,$$

$$H_{A,y}(\mathbf{r}) = \frac{1}{4\pi}\int_V \left[\Delta_x J_z - \Delta_z J_x\right]\frac{1+jkr}{r^3}e^{-jkr}\,d\mathbf{r}'\,, \qquad (3.105)$$

$$H_{A,z}(\mathbf{r}) = \frac{1}{4\pi}\int_V \left[\Delta_y J_x - \Delta_x J_y\right]\frac{1+jkr}{r^3}e^{-jkr}\,d\mathbf{r}'\,,$$

where

$$\Delta_x = x - x'\,, \qquad \Delta_y = y - y'\,, \qquad \Delta_z = z - z'\,, \qquad (3.106)$$

and

$$r = |\mathbf{r} - \mathbf{r}'| = \sqrt{(x-x')^2 + (y-y')^2 + (z-z')^2}\,. \qquad (3.107)$$

Using (3.2), the corresponding electric field is

$$E_{A,x}(\mathbf{r}) = -\frac{j}{4\pi\omega\epsilon}\int_V \left(2C_1 J_x + C_2\left[\Delta_x(\Delta_z J_z + \Delta_y J_y) - J_x(\Delta_y^2 + \Delta_z^2)\right]\right)d\mathbf{r}'\,,$$

$$E_{A,y}(\mathbf{r}) = -\frac{j}{4\pi\omega\epsilon}\int_V \left(2C_1 J_y + C_2\left[\Delta_y(\Delta_x J_x + \Delta_z J_z) - J_y(\Delta_z^2 + \Delta_x^2)\right]\right)d\mathbf{r}'\,,$$

$$E_{A,z}(\mathbf{r}) = -\frac{j}{4\pi\omega\epsilon}\int_V \left(2C_1 J_z + C_2\left[\Delta_z(\Delta_y J_y + \Delta_x J_x) - J_z(\Delta_x^2 + \Delta_y^2)\right]\right)d\mathbf{r}'\,,$$

$$(3.108)$$

where

$$C_1 = \frac{1+1jkr}{r^3}e^{-jkr} \qquad (3.109)$$

and

$$C_2 = \frac{3+3jkr - k^2r^2}{r^5}e^{-jkr}\,. \qquad (3.110)$$

Similarly, the electric field radiated by a magnetic current is

$$\mathbf{E}_F(\mathbf{r}) = -\frac{1}{\epsilon}\nabla \times \mathbf{F}(\mathbf{r}) = -\nabla \times \int_V \mathbf{M}(\mathbf{r}')\,G(\mathbf{r},\mathbf{r}')\,d\mathbf{r}' \qquad (3.111)$$

and its rectangular components are

$$E_{F,x}(\mathbf{r}) = -\frac{1}{4\pi} \int_V \left[ \Delta_z M_y - \Delta_y M_z \right] \frac{1+jkr}{r^3} e^{-jkr} \, d\mathbf{r}' \, ,$$

$$E_{F,y}(\mathbf{r}) = -\frac{1}{4\pi} \int_V \left[ \Delta_x M_z - \Delta_z M_x \right] \frac{1+jkr}{r^3} e^{-jkr} \, d\mathbf{r}' \, , \qquad (3.112)$$

$$E_{F,z}(\mathbf{r}) = -\frac{1}{4\pi} \int_V \left[ \Delta_y M_x - \Delta_x M_y \right] \frac{1+jkr}{r^3} e^{-jkr} \, d\mathbf{r}' \, ,$$

and using (3.1), the corresponding magnetic field is

$$H_{F,x}(\mathbf{r}) = -\frac{j}{4\pi\omega\mu} \int_V \left( 2C_1 M_x + C_2 \left[ \Delta_x (\Delta_z M_z + \Delta_y M_y) - M_x (\Delta_y^2 + \Delta_z^2) \right] \right) d\mathbf{r}' \, ,$$

$$H_{F,y}(\mathbf{r}) = -\frac{j}{4\pi\omega\mu} \int_V \left( 2C_1 M_y + C_2 \left[ \Delta_y (\Delta_x M_x + \Delta_z M_z) - M_y (\Delta_z^2 + \Delta_x^2) \right] \right) d\mathbf{r}' \, ,$$

$$H_{F,z}(\mathbf{r}) = -\frac{j}{4\pi\omega\mu} \int_V \left( 2C_1 M_z + C_2 \left[ \Delta_z (\Delta_y M_y + \Delta_x M_x) - M_z (\Delta_x^2 + \Delta_y^2) \right] \right) d\mathbf{r}' \, .$$

$$(3.113)$$

Note that terms $C_1$ and $C_2$ in these equations become singular for field points located very close to the source, and numerical integration algorithms alone may yield inaccurate results. In such cases, a more advanced singularity treatment may be needed, such as those outlined in [5, 6] for current distributions on planar triangles.

### 3.5.2 Two-Dimensional Near Field

Following (3.101) and using the two-dimensional Green's function (3.49), the magnetic field $\mathbf{H}(\rho)$ can be written as

$$\mathbf{H}_A(\rho) = \frac{1}{\mu} \nabla \times \mathbf{A}(\rho) = -\frac{j}{4} \int_S \nabla H_0^{(2)}(k\rho) \times \mathbf{J}(\rho') \, d\rho' \, . \qquad (3.114)$$

The gradient of the Hankel function in cylindrical coordinates is [4]

$$\nabla H_0^{(2)}(k\rho) = -(\rho - \rho') \frac{k}{\rho} H_1^{(2)}(k\rho) \, , \qquad (3.115)$$

which results in

$$\mathbf{H}_A(\rho) = j\frac{k}{4} \int_S (\rho - \rho') \times \mathbf{J}(\rho') \frac{H_1^{(2)}(k\rho)}{\rho} \, d\rho' \, . \qquad (3.116)$$

Expanding into Cartesian components, the above can be written as

$$H_{A,x}(\boldsymbol{\rho}) = j\frac{k}{4}\int_S \left[\Delta_y J_z\right] \frac{H_1^{(2)}(k\rho)}{\rho}\, d\boldsymbol{\rho}'\, ,$$

$$H_{A,y}(\boldsymbol{\rho}) = j\frac{k}{4}\int_S \left[-\Delta_x J_z\right] \frac{H_1^{(2)}(k\rho)}{\rho}\, d\boldsymbol{\rho}'\, ,$$  (3.117)

$$H_{A,z}(\boldsymbol{\rho}) = j\frac{k}{4}\int_S \left[\Delta_x J_y - \Delta_y J_x\right] \frac{H_1^{(2)}(k\rho)}{\rho}\, d\boldsymbol{\rho}'\, ,$$

where

$$\Delta_x = x - x'\, , \qquad \Delta_y = y - y'\, ,$$  (3.118)

and

$$\rho = |\boldsymbol{\rho} - \boldsymbol{\rho}'| = \sqrt{(x - x')^2 + (y - y')^2}\, .$$  (3.119)

Using (3.2), the corresponding electric field is

$$E_{A,x}(\boldsymbol{\rho}) = \frac{k^2}{8\omega\epsilon}\int_S \left[\left([\Delta_y^2 - \Delta_x^2]J_x - 2\Delta_x\Delta_y J_y\right)\frac{H_2^{(2)}(k\rho)}{\rho^2} - J_x H_0^{(2)}(k\rho)\right] d\boldsymbol{\rho}'\, ,$$

$$E_{A,y}(\boldsymbol{\rho}) = \frac{k^2}{8\omega\epsilon}\int_S \left[\left([\Delta_x^2 - \Delta_y^2]J_y - 2\Delta_x\Delta_y J_x\right)\frac{H_2^{(2)}(k\rho)}{\rho^2} - J_y H_0^{(2)}(k\rho)\right] d\boldsymbol{\rho}'\, ,$$

$$E_{A,z}(\boldsymbol{\rho}) = -\frac{k^2}{4\omega\epsilon}\int_S J_z H_0^{(2)}(k\rho)\, d\boldsymbol{\rho}'\, .$$

(3.120)

Similarly, the electric field radiated by a magnetic current is

$$\mathbf{E}_F(\boldsymbol{\rho}) = -j\frac{k}{4}\int_S (\boldsymbol{\rho} - \boldsymbol{\rho}') \times \mathbf{M}(\boldsymbol{\rho}')\, \frac{H_1^{(2)}(k\rho)}{\rho}\, d\boldsymbol{\rho}'\, ,$$  (3.121)

and its rectangular components are

$$E_{F,x}(\boldsymbol{\rho}) = -j\frac{k}{4}\int_S \left[\Delta_y M_z\right] \frac{H_1^{(2)}(k\rho)}{\rho}\, d\boldsymbol{\rho}'\, ,$$

$$E_{F,y}(\boldsymbol{\rho}) = -j\frac{k}{4}\int_S \left[-\Delta_x M_z\right] \frac{H_1^{(2)}(k\rho)}{\rho}\, d\boldsymbol{\rho}'\, ,$$  (3.122)

$$E_{F,z}(\boldsymbol{\rho}) = -j\frac{k}{4}\int_S \left[\Delta_x M_y - \Delta_y M_x\right] \frac{H_1^{(2)}(k\rho)}{\rho}\, d\boldsymbol{\rho}'\, .$$

The corresponding magnetic field is

$$H_{A,x}(\boldsymbol{\rho}) = \frac{k^2}{8\omega\mu} \int_S \left[ \left( [\Delta_y^2 - \Delta_x^2] M_x - 2\Delta_x\Delta_y M_y \right) \frac{H_2^{(2)}(k\rho)}{\rho^2} - M_x H_0^{(2)}(k\rho) \right] d\rho' ,$$

$$H_{A,y}(\boldsymbol{\rho}) = \frac{k^2}{8\omega\mu} \int_S \left[ \left( [\Delta_x^2 - \Delta_y^2] M_y - 2\Delta_x\Delta_y M_x \right) \frac{H_2^{(2)}(k\rho)}{\rho^2} - M_y H_0^{(2)}(k\rho) \right] d\rho' ,$$

$$H_{A,z}(\boldsymbol{\rho}) = -\frac{k^2}{4\omega\mu} \int_S M_z H_0^{(2)}(k\rho) \, d\rho' .$$

(3.123)

### 3.5.3 Three-Dimensional Far Field

When the field point is located very far away from the source, the vectors **r** and **r** − **r**′ are virtually parallel, as shown in Figure 3.1b. As a rule of thumb, this is typically satisfied when $kr \gg 1$. In this case, the fields are virtually planar, and under this assumption, we make the following approximations for $r$ in the far field region [1]

$$r = \begin{cases} r - \mathbf{r}' \cdot \hat{\mathbf{r}} , & \text{for phase variations,} \\ r , & \text{for amplitude variations.} \end{cases}$$

(3.124)

Consider the total electric field as expressed in (3.73). We first focus on the components of **E**(**r**) due to **A**(**r**), which are

$$\mathbf{E}_A(\mathbf{r}) = -j\omega \left[ \mathbf{A}(\mathbf{r}) + \frac{1}{k^2} \nabla\nabla \cdot \mathbf{A}(\mathbf{r}) \right] .$$

(3.125)

The first term on the right-hand side of (3.125) will result in fields that vary according to $1/r$, and the second term in fields that vary according to $1/r^2, 1/r^3$, etc., due to the differential operators. In the far field, we know that only those fields that vary according to $1/r$ are of significant amplitude, and that these components are planar with no components along the direction of propagation. The far electric field $\mathbf{E}_A(\mathbf{r})$ is therefore

$$\mathbf{E}_A(\mathbf{r}) = -j\omega\mathbf{A}(\mathbf{r})$$

(3.126)

and the far magnetic field $\mathbf{H}_F(\mathbf{r})$ is

$$\mathbf{H}_F(\mathbf{r}) = -j\omega\mathbf{F}(\mathbf{r}) ,$$

(3.127)

where $\mathbf{E}_A(\mathbf{r})$ and $\mathbf{H}_F(\mathbf{r})$ can be written in terms of their $\hat{\theta}$ and $\hat{\phi}$ components as

$$\mathbf{E}_A(\mathbf{r}) = -j\omega \left[ A_\theta(\mathbf{r})\hat{\theta} + A_\phi(\mathbf{r})\hat{\phi} \right]$$

(3.128)

and

$$\mathbf{H}_F(\mathbf{r}) = -j\omega \left[ F_\theta(\mathbf{r})\hat{\theta} + F_\phi(\mathbf{r})\hat{\phi} \right] .$$

(3.129)

To obtain the far electric field $\mathbf{E}_F(\mathbf{r})$ due to $\mathbf{F}(\mathbf{r})$ we can use (3.66), but instead we will use the fact that for plane waves, the electric and magnetic field are related via the relationship

$$\mathbf{H}(\mathbf{r}) = \frac{1}{\eta}\,\hat{\mathbf{r}} \times \mathbf{E}(\mathbf{r}) , \qquad (3.130)$$

where $\hat{\mathbf{r}}$ is the direction of propagation and $\eta = \sqrt{\mu/\epsilon}$ is the impedance of the dielectric medium. Using this relationship we can then write

$$\mathbf{E}_F(\mathbf{r}) = -j\omega\eta\big[F_\phi(\mathbf{r})\hat{\boldsymbol{\theta}} - F_\theta(\mathbf{r})\hat{\boldsymbol{\phi}}\big] \qquad (3.131)$$

and

$$\mathbf{H}_A(\mathbf{r}) = -j\frac{\omega}{\eta}\big[-A_\phi(\mathbf{r})\hat{\boldsymbol{\theta}} + A_\theta(\mathbf{r})\hat{\boldsymbol{\phi}}\big] . \qquad (3.132)$$

The total far fields are then

$$\mathbf{E}(\mathbf{r}) = -j\omega\Big(\big[A_\theta(\mathbf{r}) + \eta F_\phi(\mathbf{r})\big]\hat{\boldsymbol{\theta}} + \big[A_\phi(\mathbf{r}) - \eta F_\theta(\mathbf{r})\big]\hat{\boldsymbol{\phi}}\Big) \qquad (3.133)$$

and

$$\mathbf{H}(\mathbf{r}) = -j\frac{\omega}{\eta}\Big(\big[-A_\phi(\mathbf{r}) + \eta F_\theta(\mathbf{r})\big]\hat{\boldsymbol{\theta}} + \big[A_\theta(\mathbf{r}) + \eta F_\phi(\mathbf{r})\big]\hat{\boldsymbol{\phi}}\Big) , \qquad (3.134)$$

where in the far field, $\mathbf{A}(\mathbf{r})$ and $\mathbf{F}(\mathbf{r})$ can be written as

$$\mathbf{A}(\mathbf{r}) = \mu\frac{e^{-jkr}}{4\pi r}\int_V \mathbf{J}(\mathbf{r}')e^{jk\mathbf{r}'\cdot\hat{\mathbf{r}}}\,d\mathbf{r}' \qquad (3.135)$$

and

$$\mathbf{F}(\mathbf{r}) = \epsilon\frac{e^{-jkr}}{4\pi r}\int_V \mathbf{M}(\mathbf{r}')e^{jk\mathbf{r}'\cdot\hat{\mathbf{r}}}\,d\mathbf{r}' . \qquad (3.136)$$

For scattering problems, where the currents $\mathbf{J}$ and $\mathbf{M}$ induced by the incident fields $\mathbf{E}^i$ and $\mathbf{H}^i$ generate the scattered far fields $\mathbf{E}^s$ and $\mathbf{H}^s$, the three-dimensional radar cross section $\sigma_{3D}$ is defined as [1]

$$\sigma_{3D} = 4\pi r^2 \frac{|\mathbf{E}^s|^2}{|\mathbf{E}^i|^2} . \qquad (3.137)$$

It is common in the literature to see $\hat{\boldsymbol{\theta}}$ polarization referred to as *vertical* polarization and $\hat{\boldsymbol{\phi}}$ polarization as *horizontal* polarization. We will use this nomenclature when analyzing the RCS of three-dimensional objects in later chapters.

### 3.5.4 Two-Dimensional Far Field

To determine the far field expressions for $\mathbf{A}(\boldsymbol{\rho})$ and $\mathbf{F}(\boldsymbol{\rho})$, we use the large-argument approximation of the Hankel function [4]

$$H_n^{(2)}(k\rho) \approx \sqrt{\frac{2j}{\pi k \rho}} \; j^n e^{-jk\rho} \qquad \rho \to \infty \tag{3.138}$$

and the two-dimensional version of (3.124) to write

$$H_0^{(2)}(k\rho) = \sqrt{\frac{2j}{\pi k \rho}} \; e^{-jk\rho} e^{jk\boldsymbol{\rho}' \cdot \hat{\boldsymbol{\rho}}} \; , \tag{3.139}$$

which yields

$$\mathbf{A}(\boldsymbol{\rho}) = -j\mu \sqrt{\frac{j}{8\pi k}} \; \frac{e^{-jk\rho}}{\sqrt{\rho}} \int_S \mathbf{J}(\boldsymbol{\rho}') \, e^{jk\boldsymbol{\rho}' \cdot \hat{\boldsymbol{\rho}}} \, d\boldsymbol{\rho}' \tag{3.140}$$

and

$$\mathbf{F}(\boldsymbol{\rho}) = -j\epsilon \sqrt{\frac{j}{8\pi k}} \; \frac{e^{-jk\rho}}{\sqrt{\rho}} \int_S \mathbf{M}(\boldsymbol{\rho}') \, e^{jk\boldsymbol{\rho}' \cdot \hat{\boldsymbol{\rho}}} \, d\boldsymbol{\rho}' \; . \tag{3.141}$$

In the two-dimensional case, we know that the total fields $\mathbf{E}(\boldsymbol{\rho})$ and $\mathbf{H}(\boldsymbol{\rho})$ comprise $\hat{\mathbf{z}}$ and $\hat{\boldsymbol{\phi}}$ components only. Following similar reasoning as in the three-dimensional case, we can then write the far electric and magnetic fields as

$$\mathbf{E}(\boldsymbol{\rho}) = -j\omega \left( \left[ A_\phi(\boldsymbol{\rho}) + \eta F_z(\boldsymbol{\rho}) \right] \hat{\boldsymbol{\phi}} + \left[ A_z(\boldsymbol{\rho}) - \eta F_\phi(\boldsymbol{\rho}) \right] \hat{\mathbf{z}} \right) \tag{3.142}$$

and

$$\mathbf{H}(\boldsymbol{\rho}) = -j\frac{\omega}{\eta} \left( \left[ -A_z(\boldsymbol{\rho}) + \eta F_\phi(\boldsymbol{\rho}) \right] \hat{\boldsymbol{\phi}} + \left[ A_\phi(\boldsymbol{\rho}) + \eta F_z(\boldsymbol{\rho}) \right] \hat{\mathbf{z}} \right) \; . \tag{3.143}$$

Similarly, the two-dimensional radar cross section $\sigma_{2D}$ is defined as [1]

$$\sigma_{2D} = 2\pi\rho \, \frac{|\mathbf{E}^s|^2}{|\mathbf{E}^i|^2} \; . \tag{3.144}$$

## 3.6 Formulations for Scattering

Scattering problems can be treated as radiation problems where the radiating currents are themselves generated by other currents or fields. For example, to determine the radiation pattern of an antenna, we compute the radiated field due to the current distribution on the antenna. These currents, however, are generated by other sources, such as a voltage applied at the antenna's feedpoint or fields radiated by other antennas nearby. Computation of an object's radar cross section (RCS) is another common example of this problem. In this case, the scattering object (or *target*) is in the far field of an external antenna, and induced currents on the scatterer radiate a secondary or *scattered* field.

Strictly speaking, radiation problems require the integration of known currents $\mathbf{J}$ and $\mathbf{M}$ to obtain the fields by way of either (3.25) and (3.26) or (3.73) and (3.74). Though in some cases the expressions for the currents may be known *a priori*, most practical problems of interest have no analytic solution and the currents must be determined numerically. Thus, these problems usually involve three steps:

1. Formulating an integral equation for the unknown currents $\mathbf{J}$ and $\mathbf{M}$ generated by known fields $\mathbf{E}^i$ or $\mathbf{H}^i$

2. Solving the integral equation

3. Integrating the currents $\mathbf{J}$ and $\mathbf{M}$ to obtain the radiated or scattered fields $\mathbf{E}^s$ and $\mathbf{H}^s$.

The last step is achieved using the methods outlined in Section 3.5 and is relatively straightforward once $\mathbf{J}$ or $\mathbf{M}$ are known. However, the first two steps must be considered carefully, and the remainder of this book is dedicated to this task. We will first review the equivalence principle in Section 3.6.1 and use it to reformulate the scattering problem using equivalent surface currents. Next, we will develop surface integral equations of scattering in Section 3.6.2. We will then apply these integral equations to problems of interest in subsequent chapters, and solve them numerically using the Method of Moments.

### 3.6.1 Surface Equivalent

The expressions for the radiated fields in (3.73) and (3.74) assume that the currents $\mathbf{J}$ and $\mathbf{M}$ radiate in an unbounded, homogeneous dielectric region with parameters $(\mu, \epsilon)$. If we now introduce an obstacle into this region with different material properties, we can no longer use (3.73) and (3.74) to determine the radiated fields. This is a serious problem, as virtually all problems of practical interest involve currents radiating in the presence of an obstacle. Fortunately, it is possible to reformulate these problems in terms of an *equivalent* problem that is more convenient to solve in the region of interest. By introducing certain fictitious electric and magnetic currents, we can create a set of coupled integral equations that eliminate the obstacle. Once the equations are solved for these currents, it is straightforward to determine the field in each region. As we treat conducting, homogeneous dielectric and composite conducting/dielectric objects in this book, we will formulate our equivalent problem in terms of surface currents that exist on the interfaces between dielectric and conducting regions.

The surface equivalence theorem (also called *Huygen's principle*) states that every point on an advancing wavefront is itself a source of radiated waves. Using this theorem, a radiating source can be replaced by a fictitious set of different but *equivalent* sources that lie on an arbitrary closed surface around the original source. By matching the appropriate boundary conditions, these currents will generate the same radiated field outside the closed surface as the original sources. The theorem allows us to determine the far fields radiated by a structure if the near fields are known, or to formulate a surface integral equation that can be solved for the currents induced on an object by an externally incident field.

To introduce the concept, we will develop the integral equations for a small, conceptual scattering problem, and generalize them to general problems later. To begin, consider the geometry of Figure 3.2a, where the electric and magnetic currents $\mathbf{J}$ and $\mathbf{M}$ generate known *incident* fields $\mathbf{E}_1^i$ and $\mathbf{H}_1^i$ in an unbounded region $R_1$ with dielectric parameters $(\mu_1, \epsilon_1)$. A small, bounded region $R_2$ with the same parameters is placed inside $R_1$, where the interface between the two regions comprises the surface $S$ with normal $\hat{\mathbf{n}}_1$ that points into $R_1$. Because $R_1$ and $R_2$ have the same dielectric parameters, $\mathbf{E}_1^i$ and $\mathbf{H}_1^i$ comprise the fields everywhere. If we now change the parameters of $R_2$ to $(\mu_2, \epsilon_2)$ as in Figure 3.2b, it becomes an obstacle inside $R_1$ and the fields are no longer the same. The fields in $R_1$ now comprise $\mathbf{E}_1$ and $\mathbf{H}_1$, given by

$$\mathbf{E}_1 = \mathbf{E}_1^i + \mathbf{E}_1^s \tag{3.145}$$

and

$$\mathbf{H}_1 = \mathbf{H}_1^i + \mathbf{H}_1^s , \tag{3.146}$$

where $\mathbf{E}_1^s$ and $\mathbf{H}_1^s$ are the perturbed or *scattered* fields generated by the obstacle, and the fields $\mathbf{E}_2$ and $\mathbf{H}_2$ inside $R_2$ are referred to as *transmitted* fields. Let us now remove $\mathbf{J}$ and $\mathbf{M}$ and replace them by fictitious but *equivalent* surface currents $\mathbf{J}_1$ and $\mathbf{M}_1$ on the outside of $S$, which we refer to as $S^+$. For the fields in $R_1$ and $R_2$ to remain unchanged, the boundary conditions (Section 3.2) on $S^+$ must satisfy

$$\mathbf{M}_1 = -\hat{\mathbf{n}}_1 \times (\mathbf{E}_1 - \mathbf{E}_2) \tag{3.147}$$

and

$$\mathbf{J}_1 = \hat{\mathbf{n}}_1 \times (\mathbf{H}_1 - \mathbf{H}_2) . \tag{3.148}$$

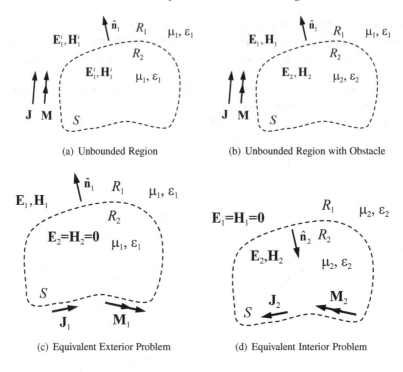

(a) Unbounded Region                    (b) Unbounded Region with Obstacle

(c) Equivalent Exterior Problem          (d) Equivalent Interior Problem

**FIGURE 3.2:** Surface Equivalent

Note that because the obstacle is still present, we cannot use (3.73) and (3.74) in either (3.147) or (3.148). However, the quantities of interest in scattering problems are usually the fields $\mathbf{E}_1^s$ and $\mathbf{H}_1^s$, not the fields inside the obstacle. Therefore, we apply the *extinction theorem* [2] and simply assume that those fields are zero. Doing so, we can now assign to $R_2$ the same dielectric parameters as $R_1$, and the boundary conditions on $S^+$ become

$$\mathbf{M}_1 = -\hat{\mathbf{n}}_1 \times \mathbf{E}_1 = -\hat{\mathbf{n}}_1 \times \left(\mathbf{E}_1^s + \mathbf{E}_1^i\right) \tag{3.149}$$

and

$$\mathbf{J}_1 = \hat{\mathbf{n}}_1 \times \mathbf{H}_1 = \hat{\mathbf{n}}_1 \times \left(\mathbf{H}_1^s + \mathbf{H}_1^i\right). \tag{3.150}$$

We now have an equivalent *exterior* problem, as illustrated in Figure 3.2c, where the fields outside $S$ remain unchanged. We can likewise formulate an equivalent *interior* problem inside $R_2$. To do so, we place the surface currents $\mathbf{J}_2$ and $\mathbf{M}_2$ on the inside of $S$, which we refer to as $S^-$. As before, if we assume that the fields in $R_1$ are zero, the boundary conditions on $S^-$ become

$$\mathbf{M}_2 = -\hat{\mathbf{n}}_2 \times \mathbf{E}_2 \qquad (3.151)$$

and

$$\mathbf{J}_2 = \hat{\mathbf{n}}_2 \times \mathbf{H}_2 \qquad (3.152)$$

where $\hat{\mathbf{n}}_2$ points into $R_2$.

Let us focus again on the fields in $R_1$. Substituting (3.75) and (3.76) into (3.149) and (3.150) yields

$$\mathbf{M}_1(\mathbf{r}) + \hat{\mathbf{n}}_1(\mathbf{r}) \times \left[ -j\omega\mu(\mathcal{L}\mathbf{J}_1)(\mathbf{r}) - (\mathcal{K}\mathbf{M}_1)(\mathbf{r}) \right] = -\hat{\mathbf{n}}_1(\mathbf{r}) \times \mathbf{E}_1^i(\mathbf{r}) \quad (3.153)$$

and

$$\mathbf{J}_1(\mathbf{r}) - \hat{\mathbf{n}}_1(\mathbf{r}) \times \left[ -j\omega\epsilon(\mathcal{L}\mathbf{M}_1)(\mathbf{r}) + (\mathcal{K}\mathbf{J}_1)(\mathbf{r}) \right] = \hat{\mathbf{n}}_1(\mathbf{r}) \times \mathbf{H}_1^i(\mathbf{r}) . \quad (3.154)$$

As $\mathbf{r}$ approaches $\mathbf{r}'$ on $S^+$, the $\mathcal{L}$ operator is well behaved, however the $\mathcal{K}$ operator contains a singularity which must be addressed [2]. To do so, we must determine the value of

$$\lim_{\mathbf{r} \to \mathbf{r}'} \left[ \hat{\mathbf{n}}(\mathbf{r}) \times (\mathcal{K}\mathbf{X})(\mathbf{r}) \right] . \qquad (3.155)$$

Using (3.78), this is

$$\lim_{\mathbf{r} \to \mathbf{r}'} \left[ \hat{\mathbf{n}}(\mathbf{r}) \times \nabla \times \int_S G(\mathbf{r}, \mathbf{r}') \, \mathbf{X}(\mathbf{r}') \, d\mathbf{r}' \right] , \qquad (3.156)$$

and following (3.96), we can move the gradient inside, which yields

$$\lim_{\mathbf{r} \to \mathbf{r}'} \left[ -\hat{\mathbf{n}}(\mathbf{r}) \times \int_S \mathbf{X}(\mathbf{r}') \times \nabla G(\mathbf{r}, \mathbf{r}') \, d\mathbf{r}' \right] . \qquad (3.157)$$

We now break the integral into two parts following [7]

$$-\hat{\mathbf{n}}(\mathbf{r}) \times \left[ \int_{S-\delta S} \mathbf{X}(\mathbf{r}') \times \nabla G(\mathbf{r}, \mathbf{r}') \, d\mathbf{r}' + \int_{\delta S} \mathbf{X}(\mathbf{r}') \times \nabla G(\mathbf{r}, \mathbf{r}') \, d\mathbf{r}' \right] , \quad (3.158)$$

where $\delta S$ is a very small, circular region of radius $a$ in $S$ about $\mathbf{r}$. If we now make the center of $\delta S$ the origin of a local cylindrical coordinate system, we can write

$$|\mathbf{r} - \mathbf{r}'| = \sqrt{(\rho')^2 + z^2} , \qquad (3.159)$$

and for $|\mathbf{r} - \mathbf{r}'| \ll 1$, the Green's function inside $\delta S$ can be written as

$$G(\mathbf{r}, \mathbf{r}') = \frac{e^{-jk|\mathbf{r} - \mathbf{r}'|}}{4\pi|\mathbf{r} - \mathbf{r}'|} \approx \frac{1}{4\pi\sqrt{(\rho')^2 + (z - z')^2}} . \qquad (3.160)$$

The gradient in cylindrical coordinates is

$$\nabla = \frac{\partial}{\partial \rho}\hat{\boldsymbol{\rho}} + \frac{1}{\rho}\frac{\partial}{\partial \phi}\hat{\boldsymbol{\phi}} + \frac{\partial}{\partial z}\hat{\mathbf{z}} , \qquad (3.161)$$

and because $\hat{\mathbf{n}}(\mathbf{r}) = \hat{\mathbf{z}}$ on $\delta S$ and $\mathbf{X}(\mathbf{r}')$ is everywhere tangent to $\delta S$, we can write

$$\hat{\mathbf{z}} \times \mathbf{X}(\mathbf{r}') \times \nabla G(\mathbf{r}, \mathbf{r}') = \mathbf{X}(\mathbf{r}')\left[\frac{\partial}{\partial z}G(\mathbf{r}, \mathbf{r}')\right] , \qquad (3.162)$$

which is

$$\mathbf{X}(\mathbf{r}')\left[\frac{\partial}{\partial z}G(\mathbf{r}, \mathbf{r}')\right] = \mathbf{X}(\mathbf{r}')\frac{-z}{4\pi\left[(\rho')^2 + z^2\right]^{3/2}} . \qquad (3.163)$$

Because $\delta S$ is very small, we will assume that $\mathbf{X}(\mathbf{r}')$ is approximately equal to $\mathbf{X}(\mathbf{r})$. We now write the integral over $\delta S$ as

$$-\hat{\mathbf{n}}(\mathbf{r})\times\int_{\delta S} \mathbf{X}(\mathbf{r}')\times\nabla G(\mathbf{r}, \mathbf{r}')\, d\mathbf{r}' = \frac{\mathbf{X}(\mathbf{r})}{2}\int_0^a \frac{z\rho'}{\left[(\rho')^2 + z^2\right]^{3/2}}\, d\rho' . \qquad (3.164)$$

Integrating this expression we get

$$\frac{\mathbf{X}(\mathbf{r})}{2}\left[\frac{z}{|z|} - \frac{z}{\sqrt{a^2 + z^2}}\right] , \qquad (3.165)$$

and taking the limit as $z$ approaches 0 from above yields

$$\lim_{z\to 0^+}\frac{\mathbf{X}(\mathbf{r})}{2}\left[\frac{z}{|z|} - \frac{z}{\sqrt{a^2 + z^2}}\right] = \frac{\mathbf{X}(\mathbf{r})}{2} . \qquad (3.166)$$

Thus,

$$\lim_{\mathbf{r}\to\mathbf{r}'}\left[\hat{\mathbf{n}}_1(\mathbf{r}) \times (\mathcal{K}\mathbf{X})(\mathbf{r})\right] = \frac{\mathbf{X}(\mathbf{r})}{2} . \qquad (3.167)$$

Note that in cases where the small area $\delta S$ is not locally planar, such as the apex of a cone or an edge shared by two planar facets, a modification of (3.167) is required. It is shown in [7] that the result of (3.167) should be adjusted to take into account the exterior solid angle $\Omega_0$, which results in

$$\lim_{\mathbf{r}\to\mathbf{r}'}\left[\hat{\mathbf{n}}_1(\mathbf{r}) \times (\mathcal{K}\mathbf{X})(\mathbf{r})\right] = \frac{\Omega_0}{4\pi}\mathbf{X}(\mathbf{r}) . \qquad (3.168)$$

For smooth geometries, the exterior solid angle is $\Omega_0 = 2\pi$ at every observation point, and (3.168) reduces to (3.167). When an object has sharp corners or edges, the solid angle can be adjusted as in [8] for three-dimensional objects (see Chapter 8). However, in this book we will use (3.167) with no modifications, which is a common approach in the literature. Thus, (3.153) and (3.154) become

$$\frac{1}{2}\mathbf{M}_1(\mathbf{r}) + \hat{\mathbf{n}}_1(\mathbf{r}) \times \left[ -j\omega\mu(\mathcal{L}\mathbf{J}_1)(\mathbf{r}) - (\mathcal{K}\mathbf{M}_1)(\mathbf{r}) \right] = -\hat{\mathbf{n}}_1(\mathbf{r}) \times \mathbf{E}_1^i(\mathbf{r}) \quad (3.169)$$

and

$$\frac{1}{2}\mathbf{J}_1(\mathbf{r}) - \hat{\mathbf{n}}_1(\mathbf{r}) \times \left[ -j\omega\epsilon(\mathcal{L}\mathbf{M}_1)(\mathbf{r}) + (\mathcal{K}\mathbf{J}_1)(\mathbf{r}) \right] = \hat{\mathbf{n}}_1(\mathbf{r}) \times \mathbf{H}_1^i(\mathbf{r}) \quad (3.170)$$

for $\mathbf{r}$ on $S^+$, where the $\frac{1}{2}\mathbf{M}_1(\mathbf{r})$ and $\frac{1}{2}\mathbf{J}_1(\mathbf{r})$ terms are used when $\mathbf{r} \to \mathbf{r}'$, and the $(\mathcal{K}\mathbf{X})(\mathbf{r})$ operators are used otherwise. Similarly, the fields in $R_2$ can be written as

$$\frac{1}{2}\mathbf{M}_2(\mathbf{r}) + \hat{\mathbf{n}}_2(\mathbf{r}) \times \left[ -j\omega\mu(\mathcal{L}\mathbf{J}_2)(\mathbf{r}) - (\mathcal{K}\mathbf{M}_2)(\mathbf{r}) \right] = 0 \quad (3.171)$$

and

$$\frac{1}{2}\mathbf{J}_2(\mathbf{r}) - \hat{\mathbf{n}}_2(\mathbf{r}) \times \left[ -j\omega\epsilon(\mathcal{L}\mathbf{M}_2)(\mathbf{r}) + (\mathcal{K}\mathbf{J}_2)(\mathbf{r}) \right] = 0 \quad (3.172)$$

for $\mathbf{r}$ on $S^-$. Note that since a limiting argument was used to determine the value of $(\mathcal{K}\mathbf{X})(\mathbf{r})$ as $\mathbf{r} \to \mathbf{r}'$ on $S^\pm$, it is only valid for surfaces that are closed and cannot be used on open surfaces or those that are very thin.

We can now state the surface integral equations for the concept problem outlined in Section 3.6.1. Using the fact that on $S$,

$$\hat{\mathbf{n}} \times \hat{\mathbf{n}} \times \mathbf{A} = (\hat{\mathbf{n}} \cdot \mathbf{A})\hat{\mathbf{n}} - \mathbf{A} = -\mathbf{A}_{tan} , \quad (3.173)$$

we can write (3.169) and (3.170) as

$$\left[ j\omega\mu(\mathcal{L}\mathbf{J}_1)(\mathbf{r}) + (\mathcal{K}\mathbf{M}_1)(\mathbf{r}) \right]_{tan} + \frac{1}{2}\hat{\mathbf{n}}_1(\mathbf{r}) \times \mathbf{M}_1(\mathbf{r}) = \left[ \mathbf{E}_1^i(\mathbf{r}) \right]_{tan} \quad (3.174)$$

and

$$\left[ j\omega\epsilon(\mathcal{L}\mathbf{M}_1)(\mathbf{r}) - (\mathcal{K}\mathbf{J}_1)(\mathbf{r}) \right]_{tan} - \frac{1}{2}\hat{\mathbf{n}}_1(\mathbf{r}) \times \mathbf{J}_1(\mathbf{r}) = \left[ \mathbf{H}_1^i(\mathbf{r}) \right]_{tan} , \quad (3.175)$$

and (3.171) and (3.172) as

$$\left[ j\omega\mu(\mathcal{L}\mathbf{J}_2)(\mathbf{r}) + (\mathcal{K}\mathbf{M}_2)(\mathbf{r}) \right]_{tan} + \frac{1}{2}\hat{\mathbf{n}}_2(\mathbf{r}) \times \mathbf{M}_2(\mathbf{r}) = 0 \quad (3.176)$$

and

$$\left[ j\omega\epsilon(\mathcal{L}\mathbf{M}_2)(\mathbf{r}) - (\mathcal{K}\mathbf{J}_2)(\mathbf{r}) \right]_{tan} - \frac{1}{2}\hat{\mathbf{n}}_2(\mathbf{r}) \times \mathbf{J}_2(\mathbf{r}) = 0 . \quad (3.177)$$

Equations (3.170–3.174) comprise a set of integral equations for the unknown currents $\mathbf{J}_1, \mathbf{M}_1, \mathbf{J}_2$ and $\mathbf{M}_2$, in terms of the known incident fields $\mathbf{E}_1^i$ and $\mathbf{H}_1^i$ in $R_1$. Once the currents are found, the fields everywhere in $R_1$ and $R_2$ can be found using (3.75) and (3.76). As written, these equations do not appear coupled, as they have no quantities in common. However, the currents on each side of $S$ are related via the boundary conditions. We will consider this in detail in the next section.

### 3.6.2   Surface Integral Equations

Let us now replace the simple scatterer from Section 3.6.1 with an object of a more general configuration, as shown in Figure 3.3. This object comprises multiple, piecewise homogeneous dielectric (penetrable) or conducting (PEC) regions $R_1, \ldots, R_N$. The material parameters of dielectric region $R_l$ is $(\mu_l, \epsilon_l)$, where either $\mu$ or $\epsilon$ may be complex (denoting a lossy material) and $R_0$ is the unbounded external region to which we assign the free space parameters $(\mu_0, \epsilon_0)$. The boundary of $R_l$ comprises the surface $S_l$, and the interface between regions $R_l$ and $R_m$ is the surface $S_{lm}$ (or $S_{ml}$), which may comprise

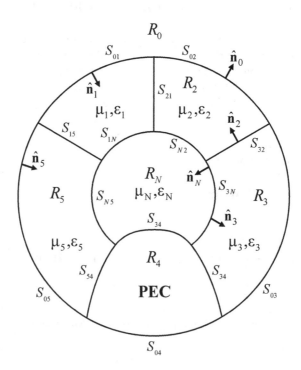

**FIGURE 3.3:** Composite Multi-Region Geometry

a portion or all of $S_l$. The normal vector $\hat{\mathbf{n}}_l$ points into $R_l$ on each of its interfaces. Conductors may be closed regions with finite volume ($R_4$ in Figure 3.3), or open, infinitely thin sheets. For each non-PEC region $R_l$, the surface integral equations on $S_l$ are

$$\left[j\omega\mu(\mathcal{L}\mathbf{J}_l)(\mathbf{r}) + (\mathcal{K}\mathbf{M}_l)(\mathbf{r})\right]_{tan} + \frac{1}{2}\hat{\mathbf{n}}_l(\mathbf{r}) \times \mathbf{M}_l(\mathbf{r}) = \left[\mathbf{E}_l^i(\mathbf{r})\right]_{tan} \quad (3.178)$$

and

$$\left[j\omega\epsilon(\mathcal{L}\mathbf{M}_l)(\mathbf{r}) - (\mathcal{K}\mathbf{J}_l)(\mathbf{r})\right]_{tan} - \frac{1}{2}\hat{\mathbf{n}}_l(\mathbf{r}) \times \mathbf{J}_l(\mathbf{r}) = \left[\mathbf{H}_l^i(\mathbf{r})\right]_{tan}. \quad (3.179)$$

For scattering problems, $\mathbf{E}_l^i(\mathbf{r})$ and $\mathbf{H}_l^i(\mathbf{r})$ are zero everywhere except in $R_0$. For antenna and radiation problems, $\mathbf{E}_l^i(\mathbf{r})$ and $\mathbf{H}_l^i(\mathbf{r})$ may be nonzero in one or more regions, depending on where the excitation is applied. We refer to (3.178) as the Electric Field Integral Equation (EFIE), and to (3.179) as the Magnetic Field Integral Equation (MFIE).

### 3.6.2.1 Interior Resonance Problem

Applying (3.178) and (3.179) to dielectrics will yield a result free of spurious solutions [9], however on closed conductors the EFIE and MFIE cannot produce a unique solution for all frequencies [9, 10, 11, 12, 13]. This is because there are homogeneous solutions to these equations that satisfy the boundary conditions with zero incident field. These spurious solutions correspond to interior cavity (resonant) modes of the object itself, and radiate no field outside the object. One method for dealing with this problem is through the application of an extended boundary condition, where the original integral equations are modified or augmented such that the value of the field inside the object is explicitly forced to be zero [7, 14]. Dual-surface electric and magnetic integral equations (DSEFIE, DSMFIE) [15] are an example of such an extended boundary condition. In such a formulation, a second surface is placed just inside the original surface. The appropriate boundary condition for the internal fields on this surface is used to generate an additional integral equation. The combination of this new equation with the original results in a combined equation that produces a unique solution at all frequencies. This method, however, requires additional effort in generating the secondary surface, and the number of currents is doubled, which drives up the demand for CPU time and system memory.

*Combined Field Integral Equation*

To date, the most popular method for handling the resonance problem is a linear combination of the EFIE and a modified MFIE. This enforces the boundary conditions on the electric and magnetic field simultaneously and is free of spurious solutions, as the null spaces of the EFIE and the modified MFIE differ. To create the modified MFIE, we take the cross product of $\hat{\mathbf{n}}_l$ and (3.179), yielding

$$\hat{\mathbf{n}}_l \times \left[ j\omega\epsilon(\mathcal{L}\mathbf{M}_l)(\mathbf{r}) - (\mathcal{K}\mathbf{J}_l)(\mathbf{r}) \right]_{tan} + \frac{1}{2}\mathbf{J}_l(\mathbf{r}) = \hat{\mathbf{n}}_l \times \left[ \mathbf{H}_l^i(\mathbf{r}) \right]_{tan} , \quad (3.180)$$

where there are now $\hat{\mathbf{n}} \times \mathcal{L}$ and $\hat{\mathbf{n}} \times \mathcal{K}$ operators. Herein, we will refer to (3.180) as the nMFIE, which when combined with the EFIE yields the Combined Field Integral Equation (CFIE) given by [9]

$$\alpha\text{EFIE} + (1 - \alpha)\eta_l\text{nMFIE} , \quad (3.181)$$

where $\eta_l = \sqrt{\mu_l/\epsilon_l}$ and $0 \leq \alpha \leq 1$, with $\alpha = 0.5$ commonly used. We apply the CFIE to closed, conducting volumes in this book. For open, thin conductors, we will use the EFIE only ($\alpha = 1$).

*A Note on Nomenclature*

Many other books and journal papers that discuss surface integral equations are concerned with conducting (PEC) objects only. The equations in these references are typically the EFIE (3.178) and the nMFIE of (3.180) with no magnetic currents, except that the nMFIE is often referred to as simply the MFIE. References that treat conducting as well as dielectric objects will more often use the nomenclature outlined in this book.

### 3.6.2.2 Discretization and Testing

We will now us the Method of Moments to convert (3.178), (3.179), and (3.180) into a matrix system using Galerkin-type testing in each region. Expanding the electric and magnetic currents in $R_l$ using a sum of weighted basis functions yields

$$\mathbf{J}_l(\mathbf{r}) = \sum_{n=1}^{N_J^{(l)}} I_n^{J(l)} \mathbf{f}_n(\mathbf{r}) \quad (3.182)$$

and

$$\mathbf{M}_l(\mathbf{r}) = \sum_{n=1}^{N_M^{(l)}} I_n^{M(l)} \mathbf{g}_n(\mathbf{r}) , \quad (3.183)$$

where $N_J^{(l)}$ and $N_M^{(l)}$ are the number of electric and magnetic basis functions in $R_l$, respectively, and the basis functions $\mathbf{f}_n(\mathbf{r})$ and $\mathbf{g}_n(\mathbf{r})$ will remain undefined for now. We next test the EFIE and nMFIE with the electric testing functions $\mathbf{f}_m(\mathbf{r})$, and the MFIE with the magnetic testing functions $\mathbf{g}_m(\mathbf{r})$ as in [9]. This yields in $R_l$ the matrix system

$$
\begin{bmatrix}
Z^{EJ(l)} & Z^{EM(l)} \\
Z^{HJ(l)} & Z^{HM(l)} \\
Z^{nHJ(l)} & Z^{nHM(l)}
\end{bmatrix}
\begin{bmatrix}
I^{J(l)} \\
I^{M(l)}
\end{bmatrix}
=
\begin{bmatrix}
V^{E(l)} \\
V^{H(l)} \\
V^{nH(l)}
\end{bmatrix},
\tag{3.184}
$$

with matrix elements given by

$$
\begin{aligned}
Z_{mn}^{EJ(l)} &= j\omega\mu_l \int_{\mathbf{f}_m} \mathbf{f}_m(\mathbf{r}) \cdot \int_{\mathbf{f}_n} \mathbf{f}_n(\mathbf{r}')\, G(\mathbf{r},\mathbf{r}')\, d\mathbf{r}'\, d\mathbf{r} \\
&\quad + \frac{j}{\omega\epsilon_l} \int_{\mathbf{f}_m} \mathbf{f}_m(\mathbf{r}) \cdot \left[ \nabla\nabla \cdot \int_{\mathbf{f}_n} \mathbf{f}_n(\mathbf{r}')\, G(\mathbf{r},\mathbf{r}')\, d\mathbf{r}' \right] d\mathbf{r},
\end{aligned}
\tag{3.185}
$$

$$
\begin{aligned}
Z_{mn}^{EM(l)} &= \int_{\mathbf{f}_m} \mathbf{f}_m(\mathbf{r}) \cdot \int_{\mathbf{g}_n} \nabla G(\mathbf{r},\mathbf{r}') \times \mathbf{g}_n(\mathbf{r}')\, d\mathbf{r}'\, d\mathbf{r} \\
&\quad + \frac{1}{2} \int_{\mathbf{f}_m,\mathbf{g}_n} \mathbf{f}_m(\mathbf{r}) \cdot \left[ \hat{\mathbf{n}}_l(\mathbf{r}) \times \mathbf{g}_n(\mathbf{r}) \right] d\mathbf{r},
\end{aligned}
\tag{3.186}
$$

$$
\begin{aligned}
Z_{mn}^{HJ(l)} &= - \int_{\mathbf{g}_m} \mathbf{g}_m(\mathbf{r}) \cdot \int_{\mathbf{f}_n} \nabla G(\mathbf{r},\mathbf{r}') \times \mathbf{f}_n(\mathbf{r}')\, d\mathbf{r}'\, d\mathbf{r} \\
&\quad - \frac{1}{2} \int_{\mathbf{g}_m,\mathbf{f}_n} \mathbf{g}_m(\mathbf{r}) \cdot \left[ \hat{\mathbf{n}}_l(\mathbf{r}) \times \mathbf{f}_n(\mathbf{r}) \right] d\mathbf{r},
\end{aligned}
\tag{3.187}
$$

$$
\begin{aligned}
Z_{mn}^{HM(l)} &= j\omega\epsilon_l \int_{\mathbf{g}_m} \mathbf{g}_m(\mathbf{r}) \cdot \int_{\mathbf{g}_n} \mathbf{g}_n(\mathbf{r}')\, G(\mathbf{r},\mathbf{r}')\, d\mathbf{r}'\, d\mathbf{r} \\
&\quad + \frac{j}{\omega\mu_l} \int_{\mathbf{g}_m} \mathbf{g}_m(\mathbf{r}) \cdot \left[ \nabla\nabla \cdot \int_{\mathbf{g}_n} \mathbf{g}_n(\mathbf{r}')\, G(\mathbf{r},\mathbf{r}')\, d\mathbf{r}' \right] d\mathbf{r},
\end{aligned}
\tag{3.188}
$$

$$
\begin{aligned}
Z_{mn}^{nHJ(l)} &= - \int_{\mathbf{f}_m} \mathbf{f}_m(\mathbf{r}) \cdot \left[ \hat{\mathbf{n}}_l(\mathbf{r}) \times \int_{\mathbf{f}_n} \nabla G(\mathbf{r},\mathbf{r}') \times \mathbf{f}_n(\mathbf{r}')\, d\mathbf{r}' \right] d\mathbf{r} \\
&\quad + \frac{1}{2} \int_{\mathbf{f}_m,\mathbf{f}_n} \mathbf{f}_m(\mathbf{r}) \cdot \mathbf{f}_n(\mathbf{r})\, d\mathbf{r},
\end{aligned}
\tag{3.189}
$$

$$
\begin{aligned}
Z_{mn}^{nHM(l)} &= j\omega\epsilon_l \int_{\mathbf{f}_m} \mathbf{f}_m(\mathbf{r}) \cdot \left[ \hat{\mathbf{n}}_l(\mathbf{r}) \times \int_{\mathbf{g}_n} \mathbf{g}_n(\mathbf{r}')\, G(\mathbf{r},\mathbf{r}')\, d\mathbf{r}' \right] d\mathbf{r} \\
&\quad + \frac{j}{\omega\mu_l} \int_{\mathbf{f}_m} \mathbf{f}_m(\mathbf{r}) \cdot \left[ \hat{\mathbf{n}}_l(\mathbf{r}) \times \nabla\nabla \cdot \int_{\mathbf{g}_n} \mathbf{g}_n(\mathbf{r}')\, G(\mathbf{r},\mathbf{r}')\, d\mathbf{r}' \right] d\mathbf{r},
\end{aligned}
\tag{3.190}
$$

where the $\mathcal{K}$ and $\hat{\mathbf{n}} \times \mathcal{K}$ operators were modified following (3.96). The elements of the right-hand side vectors are

$$V_m^{E(l)} = \int_{\mathbf{f}_m} \mathbf{f}_m(\mathbf{r}) \cdot \mathbf{E}_l^i(\mathbf{r}) \, d\mathbf{r} \,, \qquad (3.191)$$

$$V_m^{H(l)} = \int_{\mathbf{f}_m} \mathbf{f}_m(\mathbf{r}) \cdot \mathbf{H}_l^i(\mathbf{r}) \, d\mathbf{r} \,, \qquad (3.192)$$

and

$$V_m^{nH(l)} = \int_{\mathbf{f}_m} \mathbf{f}_m(\mathbf{r}) \cdot \left[ \hat{\mathbf{n}}_l \times \mathbf{H}_l^i(\mathbf{r}) \right] \, d\mathbf{r} \,. \qquad (3.193)$$

Note that all blocks of the matrix in (3.184) are square. If we now combine the matrix equations for all regions, we obtain a block-diagonal system of the form

$$\begin{bmatrix} Z^{(1)} & 0 & \cdots & 0 \\ 0 & Z^{(2)} & \cdots & 0 \\ \vdots & \vdots & \ddots & 0 \\ 0 & 0 & \cdots & Z^{(N)} \end{bmatrix} \begin{bmatrix} I^{(1)} \\ I^{(2)} \\ \vdots \\ I^{(N)} \end{bmatrix} = \begin{bmatrix} V^{(1)} \\ V^{(2)} \\ \vdots \\ V^{(N)} \end{bmatrix} \,, \qquad (3.194)$$

where

$$Z^{(l)} = \begin{bmatrix} Z^{EJ(l)} & Z^{EM(l)} \\ Z^{HJ(l)} & Z^{HM(l)} \\ Z^{nHJ(l)} & Z^{nHM(l)} \end{bmatrix} \,, \qquad (3.195)$$

$$I^{(l)} = \begin{bmatrix} I^{J(l)} \\ I^{M(l)} \end{bmatrix} \,, \qquad (3.196)$$

and

$$V^{(l)} = \begin{bmatrix} V^{E(l)} \\ V^{H(l)} \\ V^{nH(l)} \end{bmatrix} \,. \qquad (3.197)$$

### 3.6.2.3 Modification of Matrix Elements

If the basis and testing functions have a well-behaved divergence, it is possible to re-distribute the vector derivatives in the $\mathcal{L}$ operator so that they do not operate on the Green's function. This is advantageous as they would otherwise compound the $1/r$ singularity. To do so, let us concentrate on the expression having the form

$$\int_{\mathbf{f}_m} \mathbf{f}_m(\mathbf{r}) \cdot \left[ \nabla \nabla \cdot \int_{\mathbf{f}_n} \mathbf{f}_n(\mathbf{r}') G(\mathbf{r}, \mathbf{r}') \, d\mathbf{r}' \right] \, d\mathbf{r} \,. \qquad (3.198)$$

Following Section 3.4.4, this can be rewritten as

$$\int_{\mathbf{f}_m} \mathbf{f}_m(\mathbf{r}) \cdot \nabla S(\mathbf{r}) \, d\mathbf{r} = \int_{\mathbf{f}_m} \mathbf{f}_m(\mathbf{r}) \cdot \left[ \nabla \int_{\mathbf{f}_n} \nabla' \cdot \mathbf{f}(\mathbf{r}') \, G(\mathbf{r}, \mathbf{r}') \, d\mathbf{r}' \right] d\mathbf{r} \, . \quad (3.199)$$

Using the vector identity

$$\mathbf{f}(\mathbf{r}) \cdot \nabla S(\mathbf{r}) = \nabla \cdot \left[ \mathbf{f}(\mathbf{r}) S(\mathbf{r}) \right] - \left[ \nabla \cdot \mathbf{f}(\mathbf{r}) \right] S(\mathbf{r}) \, , \quad (3.200)$$

we can write the above as

$$\int_{\mathbf{f}_m} \mathbf{f}_m(\mathbf{r}) \cdot \nabla S(\mathbf{r}) \, d\mathbf{r} = \int_{\mathbf{f}_m} \nabla \cdot \left[ \mathbf{f}_m(\mathbf{r}) S(\mathbf{r}) \right] d\mathbf{r} - \int_{\mathbf{f}_m} \left[ \nabla \cdot \mathbf{f}_m(\mathbf{r}) \right] S(\mathbf{r}) \, d\mathbf{r} \, . \quad (3.201)$$

We convert the first term on the right-hand side to a surface integral using the divergence theorem

$$\int_V \nabla \cdot \left[ \mathbf{f}_m(\mathbf{r}) S(\mathbf{r}) \right] d\mathbf{r} = \int_S \hat{\mathbf{n}} \cdot \left[ \mathbf{f}_m(\mathbf{r}) S(\mathbf{r}) \right] d\mathbf{r} \, . \quad (3.202)$$

Since the bounding surface can be made large enough that $\mathbf{f}_m(\mathbf{r})$ vanishes, the above goes to zero leaving only the second term, which is

$$\int_{\mathbf{f}_m} \mathbf{f}_m(\mathbf{r}) \cdot \nabla S(\mathbf{r}) \, d\mathbf{r} = - \int_{\mathbf{f}_m} \nabla \cdot \mathbf{f}_m(\mathbf{r}) \int_n \nabla' \cdot \mathbf{f}(\mathbf{r}') \, G(\mathbf{r}, \mathbf{r}') \, d\mathbf{r}' \, d\mathbf{r} \, . \quad (3.203)$$

Substitution of the above into (3.185) yields

$$\begin{aligned} Z_{mn}^{EJ(l)} &= j\omega\mu_l \int_{\mathbf{f}_m} \mathbf{f}_m(\mathbf{r}) \cdot \int_{\mathbf{f}_n} \mathbf{f}_n(\mathbf{r}') \, G(\mathbf{r}, \mathbf{r}') \, d\mathbf{r}' \, d\mathbf{r} \\ &\quad - \frac{j}{\omega\epsilon_l} \int_{\mathbf{f}_m} \nabla \cdot \mathbf{f}_m(\mathbf{r}) \int_{\mathbf{f}_n} \nabla' \cdot \mathbf{f}_n(\mathbf{r}') \, G(\mathbf{r}, \mathbf{r}') \, d\mathbf{r}' \, d\mathbf{r} \, , \end{aligned} \quad (3.204)$$

and a similar substitution can also be made in (3.188).

Using the vector identity $\mathbf{a} \cdot (\mathbf{b} \times \mathbf{c}) = (\mathbf{a} \times \mathbf{b}) \cdot \mathbf{c}$, we can also write the second integral in (3.190) as

$$- \int_{\mathbf{f}_m} \nabla \cdot \left[ \mathbf{f}_m(\mathbf{r}) \times \hat{\mathbf{n}}_l(\mathbf{r}) \right] \int_{\mathbf{g}_n} \nabla' \cdot \mathbf{g}_n(\mathbf{r}') \, G(\mathbf{r}, \mathbf{r}') \, d\mathbf{r}' \, d\mathbf{r} \, . \quad (3.205)$$

Using the divergence theorem, the outermost integral can be converted to a contour integral, which yields

$$- \oint_{\delta\mathbf{f}_m} \hat{\mathbf{c}}_m(\mathbf{r}) \cdot \left[ \mathbf{f}_m(\mathbf{r}) \times \hat{\mathbf{n}}_l(\mathbf{r}) \right] \int_{\mathbf{g}_n} \nabla' \cdot \mathbf{g}_n(\mathbf{r}') \, G(\mathbf{r}, \mathbf{r}') \, d\mathbf{r}' \, d\mathbf{r} \, . \quad (3.206)$$

This integral is now performed along the boundary of the support of $\mathbf{f}_m(\mathbf{r})$, where $\hat{\mathbf{c}}_m(\mathbf{r})$ is the outward-facing normal to that boundary. The choice of (3.205) or (3.206) in the implementation of (3.190) depends on the geometry and testing functions, and we will use both forms in this book.

### 3.6.3   Enforcement of Boundary Conditions

The block-diagonal system in (3.194) was defined region-wise without regard to the boundary conditions on the interfaces between regions. We know from (3.5) and (3.6) that the tangential components of the electric and magnetic fields are discontinuous across a dielectric interface by an amount equal to the magnetic and electric surface currents on that interface, respectively. However, in our integral equation formulation there are no real surface currents on dielectric interfaces; they are fictitious. This implies that at the interface between dielectric regions $R_l$ and $R_m$,

$$\mathbf{J}_l(\mathbf{r}) = -\mathbf{J}_m(\mathbf{r}) \ , \qquad (3.207)$$

$$\mathbf{M}_l(\mathbf{r}) = -\mathbf{M}_m(\mathbf{r}) \ , \qquad (3.208)$$

and therefore, (3.194) contains linearly dependent unknowns. This can be addressed by assigning an opposing orientation to the basis functions on opposite sides of an interface, so that they will have the same coefficient. As these basis functions now represent a single unknown, the corresponding columns of (3.194) can be combined. Note that by doing this, (3.194) becomes overdetermined as it has more rows than columns, and so we must address that problem as well.

#### 3.6.3.1   EFIE-CFIE-PMCHWT Approach

The overdetermined system can be handled in several ways, though the most common approach is the Poggio-Miller-Chang-Harrington-Wu-Tsai (PMCHWT) formulation [16, 17]. In this method, the EFIEs and MFIEs on each side of dielectric interfaces are added together by combining the corresponding rows of the matrix. Doing so, (3.194) becomes square, and can be solved. We can combine this with the treatment for open and closed conductors (EFIE and CFIE, respectively) to yield a unified approach for composite objects that we refer to as the EFIE-CFIE-PMCHWT formulation [9]. This approach can be summarized as follows:

1. On dielectric interfaces we test the EFIE and MFIE, and eliminate linearly dependent unknowns using PMCHWT.

2. On interfaces between a dielectric and a closed conductor, we test the CFIE.

3. On interfaces between a dielectric and an open conductor, we test the EFIE only.

*Treatment at Junctions*

Points where three or more interfaces meet are referred to as *junctions*. Enforcement of the EFIE-CFIE-PMCHWT formulation at junctions requires special consideration for specific problem types. We will consider junctions in more detail in later chapters.

### 3.6.4 Physical Optics Equivalent

Though integral equation methods have existed for a long time, and in principle can treat a wide variety of problems, the size of the problems that can be solved through direct means is always limited by the currently available computing technology. Though the requirements for one- and two-dimensional problems are somewhat reasonable, three-dimensional problems present a far more formidable challenge. Historically, Moment Method solutions have been viable only for 3D problems of small electrical size (5 to 10 wavelengths, or less), though this has been mitigated in recent years by advances in computing power, and compressive methods such as the Adaptive Cross Approximation (ACA) and the Fast Multipole Method, both of which are discussed in later chapters. Larger 3D problems, such as computing radiation patterns of large antennas (such as radio telescopes), or the radar cross section of space vehicles and aircraft, can range upwards of 10 to 100 wavelengths per linear dimension, or even more depending on the waveband of interest. Because these problems are of practical interest, it is often better to have a less accurate answer than no answer at all. Fortunately, in the absence of direct solutions, approximations can be made that make the problem tractable, while yielding fair to modest accuracy in most cases. We will now briefly consider one such approximation still widely used for this purpose.

Consider again the equivalent problem in Figure 3.2c, where the surface currents are specified by (3.149) and (3.150). Let us now assume that the object is electrically large, so that at any point on $S$, the surface is considered to be locally planar and infinite in extent. If the object is perfectly conducting, $\mathbf{M}_1(\mathbf{r}) = 0$ and the scattered magnetic field is equal in amplitude and phase to the incident field. The electric surface current is then

$$\mathbf{J}_1(\mathbf{r}) = 2\hat{\mathbf{n}}_1 \times \mathbf{H}_1^i(\mathbf{r}) \,, \tag{3.209}$$

which is commonly referred to as the physical optics (PO) approximation for conductors, and is illustrated in Figure 3.4. PO also assumes that current only exists on surfaces having a direct line of sight to the source, and in shadowed areas the current is zero. In subsequent chapters, we will compare results using the PO approximation to those computed via the MoM. Though it is reasonably accurate at near-specular angles, PO performs poorly elsewhere. As a result, PO is often supplemented by adding to it secondary effects such as

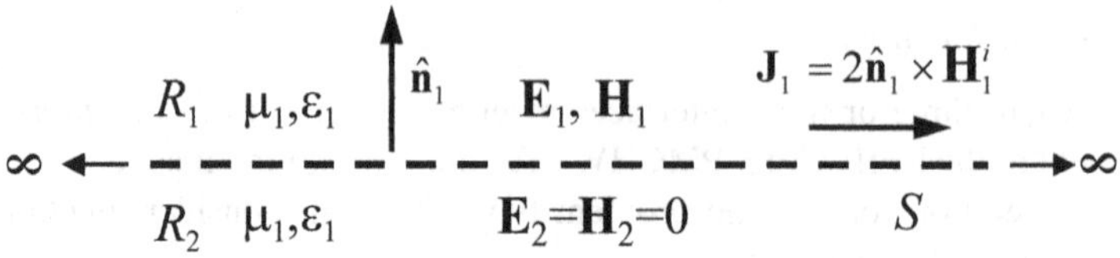

$$R_1 \quad \mu_1, \varepsilon_1 \quad \hat{\mathbf{n}}_1 \quad \mathbf{E}_1, \mathbf{H}_1 \qquad \mathbf{J}_1 = 2\hat{\mathbf{n}}_1 \times \mathbf{H}_1^i$$

$$\infty \longleftarrow - - - - - - - - - - - - - - - - \bullet \longrightarrow \infty$$

$$R_2 \quad \mu_1, \varepsilon_1 \qquad \mathbf{E}_2 = \mathbf{H}_2 = 0 \qquad S$$

**FIGURE 3.4:** Physical Optics Approximation

edge diffraction via the physical theory of diffraction [18, 19] and multiple bounce interactions via shooting and bouncing rays [20]. In Appendix A, we will consider scattering using PO in greater detail.

## References

[1] C. A. Balanis, *Advanced Engineering Electromagnetics*. John Wiley and Sons, 1989.

[2] W. C. Chew, *Waves and Fields in Inhomogeneous Media*. IEEE Press, 1995.

[3] P. M. Morse and H. Feshbach, *Methods of Theoretical Physics*. McGraw-Hill, 1953.

[4] M. Abramowitz and I. Stegun, *Handbook of Mathematical Functions*. National Bureau of Standards, 1966.

[5] M. Tong and W. Chew, "Super-hyper singularity treatment for solving 3D electric field integral equations," *Microw. Opt. Technol. Lett*, vol. 49, pp. 1383–1388, June 2007.

[6] M. Tong and W. Chew, "On the near-interaction elements in integral equation solvers for electromagnetic scattering by three-dimensional thin objects," *IEEE Trans. Antennas Propagat.*, vol. 57, pp. 2500–2506, August 2009.

[7] N. Morita, N. Kumagai, and J. R. Mautz, *Integral Equation Methods for Electromagnetics*. Artech House, 1990.

[8] J. M. Rius, E. Úbeda, and J. Parrón, "On the testing of the magnetic field integral equation with RWG basis functions in the method of moments," *IEEE Trans. Antennas Propagat.*, vol. 49, pp. 1550–1553, November 2001.

[9] P. Ylä-Oijala, M. Taskinen, and J. Sarvas, "Surface integral equation method for general composite metallic and dielectric structures with junctions," *Progress in Electromagnetics Research*, vol. 52, pp. 81–108, 2005.

[10] L. N. Medgyesi-Mitschang and J. M. Putnam, "Electromagnetic scattering from axially inhomogeneous bodies of revolution," *IEEE Trans. Antennas Propagat.*, vol. 32, pp. 796–806, March 1984.

[11] L. N. Medgyesi-Mitschang and J. M. Putnam, "Combined field integral equation formulation for inhomogeneous two- and three-dimensional bodies: The junction problem," *IEEE Trans. Antennas Propagat.*, vol. 39, pp. 667–672, May 1991.

[12] R. Mautz and R. Harrington, "H-field, E-field and combined solutions for bodies of revolution," Tech. Rep. Report RADC-TR-77-109, Rome Air Development Center, Griffiss AFB, N.Y., March 1977.

[13] W. C. Chew, J. M. Jin, E. Michielssen, and J. Song, *Fast and Efficient Algorithms in Computational Electromagnetics*. Artech House, 2001.

[14] A. F. Peterson, S. L. Ray, and R. Mittra, *Computational Methods for Electromagnetics*. IEEE Press, 1998.

[15] R. A. Shora and A. D. Yaghjian, "Dual-surface integral equations in electromagnetic scattering," *IEEE Trans. Antennas Propagat.*, vol. 53, pp. 1706–1709, May 2005.

[16] K. Umashankar, A. Taflove, and S. M. Rao, "Electromagnetic scattering by arbitrary shaped three-dimensional homogeneous lossy dielectric bodies," *IEEE Trans. Antennas Propagat.*, vol. 34, pp. 758–766, June 1986.

[17] S. M. Rao, C. C. Cha, R. L. Cravey, and D. L. Wilkes, "Electromagnetic scattering from arbitrary shaped conducting bodies coated with lossy materials of arbitrary thickness," *IEEE Trans. Antennas Propagat.*, vol. 39, pp. 627–631, May 1991.

[18] K. M. Mitzner, "Incremental length diffraction coefficients," Tech. Rep. AFAL-TR-73-296, Northrop Corporation, Aircraft Division, April 1974.

[19] A. Michaeli, "Equivalent edge currents for arbitrary aspects of observation," *IEEE Trans. Antennas Propagat.*, vol. 32, pp. 252–257, March 1984.

[20] H. Ling, S. W. Lee, and R. Chou, "Shooting and bouncing rays: calculating the RCS of an arbitrarily shaped cavity," *IEEE Trans. Antennas Propagat.*, vol. 37, pp. 194–205, February 1989.

# Chapter 4

## Solution of Matrix Equations

The discretization and testing of integral equations via the Method of Moments results in a linear system which must be solved numerically. Thus, an understanding of how to do this is necessary. Though engineers will likely not have to implement their own linear equation solver, it is useful to understand how such solvers work, what pitfalls exist, and what software libraries exist for helping in this task. In this chapter we will briefly describe common methods for solving linear equations such as Gaussian Elimination, scalar and block LU factorization, and iterative solvers.

## 4.1 Direct Methods

In this section we discuss Gaussian elimination (GE) and LU factorization, methods that allow a direct solution of a linear system via matrix factorization. These algorithms require a compute time of the order $O(N^3)$, where $N$ is the number of unknowns. These methods work very well when $N$ is relatively small, however for larger problems, the factorization time may grow prohibitively large, or computer memory may be simply exhausted. Even though computing power continues to grow year to year, these issues will always present challenges as engineers try to solve larger problems.

### 4.1.1 Gaussian Elimination

Gaussian elimination is a simple method of reducing a matrix to row echelon form through the use of elementary row operations. Once this is done, the unknown vector is obtained through simple back-substitution. To illustrate the method, let us form an augmented matrix from the $N \times N$ matrix equation

$\mathbf{Ax} = \mathbf{b}$, which can be written as

$$
\left[
\begin{array}{ccccc|c}
A_{11} & A_{12} & A_{13} & \cdots & A_{1N} & b_1 \\
A_{21} & A_{22} & A_{23} & \cdots & A_{2N} & b_2 \\
A_{31} & A_{32} & A_{33} & \cdots & A_{3N} & b_3 \\
\vdots & \vdots & \vdots & \ddots & \vdots & \vdots \\
A_{N1} & A_{N2} & A_{N3} & \cdots & A_{NN} & b_N
\end{array}
\right]. \tag{4.1}
$$

We start with the first row and divide every entry by $A_{11}$, leaving a 1 on the diagonal. For every row below this, we then subtract a multiple of this row so that a zero remains in the first column. The result is

$$
\left[
\begin{array}{ccccc|c}
1 & A'_{12} & A'_{13} & \cdots & A'_{1N} & b'_1 \\
0 & A'_{22} & A'_{23} & \cdots & A'_{2N} & b'_2 \\
0 & A'_{32} & A'_{33} & \cdots & A'_{3N} & b'_3 \\
\vdots & \vdots & \vdots & \ddots & \vdots & \vdots \\
0 & A'_{N2} & A'_{N3} & \cdots & A'_{NN} & b'_N
\end{array}
\right]. \tag{4.2}
$$

We repeat this same operation for the second row and all rows below that, until only ones remain on the diagonal and the lower triangle is zero. The original matrix equation now reads

$$
\left[
\begin{array}{ccccc|c}
1 & A'_{12} & A'_{13} & \cdots & A'_{1N} & b'_1 \\
0 & 1 & A'_{23} & \cdots & A'_{2N} & b'_2 \\
0 & 0 & 1 & \cdots & A'_{3N} & b'_3 \\
\vdots & \vdots & \vdots & \ddots & \vdots & \vdots \\
0 & 0 & 0 & \cdots & 1 & b'_N
\end{array}
\right]. \tag{4.3}
$$

The solution to this system is obtained through the following back-substitution operations

$$
x_N = b'_N \tag{4.4}
$$

and

$$
x_i = b'_i - \sum_{k=i+1}^{N} A'_{ik} x_k \qquad i < N. \tag{4.5}
$$

The elimination requires a total of $N^3/3$ operations, and $N^2/2$ for the back-substitution. Note that the row operations leading to (4.3) modify the right-hand side vector as well. This is undesirable when dealing with multiple right-hand sides, as they must all be precomputed and available when performing the elimination. This limitation is avoided by using LU factorization.

#### 4.1.1.1 Pivoting

When performing the row operations for Gaussian elimination, the values of the diagonal entries (the *pivots*) were not addressed. A problem arises when one of these entries is zero, and if the relative magnitude of the pivot is small, round-off errors may become a problem in the subsequent multiplications and subtractions. Because the augmented matrix system is not altered by exchanging any two rows, the current row can be exchanged with any of the ones below it that contain the "strongest" diagonal entry (one of greatest magnitude). This is referred to as *partial pivoting*. An exchange of rows *and* columns in the matrix is called *full pivoting*, and requires recording the permutations of the solution vector as well [1]. Many linear algebra packages employ a pivoting strategy for robustness, with partial pivoting being the most common.

### 4.1.2 LU Factorization

As previously discussed, a factorization that does not modify the right-hand side vector is more advantageous than Gaussian elimination. Therefore, let us consider a factorization of the matrix **A** into lower and upper triangular parts. We can write this as

$$\mathbf{LU} = \mathbf{A} \,, \tag{4.6}$$

where

$$\mathbf{LU} = \begin{bmatrix} L_{11} & 0 & 0 & \cdots & 0 \\ L_{21} & L_{22} & 0 & \cdots & 0 \\ L_{31} & L_{32} & L_{33} & \cdots & 0 \\ \vdots & \vdots & \vdots & \ddots & \vdots \\ L_{N1} & L_{N2} & L_{N3} & \cdots & L_{NN} \end{bmatrix} \begin{bmatrix} U_{11} & U_{12} & U_{13} & \cdots & U_{1N} \\ 0 & U_{22} & U_{23} & \cdots & U_{2N} \\ 0 & 0 & U_{33} & \cdots & U_{3N} \\ \vdots & \vdots & \vdots & \ddots & \vdots \\ 0 & 0 & 0 & \cdots & U_{NN} \end{bmatrix}. \tag{4.7}$$

Using this factorization, we can solve the following equation

$$\mathbf{Ax} = (\mathbf{LU})\mathbf{x} = \mathbf{L}(\mathbf{Ux}) = \mathbf{b} \tag{4.8}$$

by first solving the equation

$$\mathbf{Ly} = \mathbf{b} \,, \tag{4.9}$$

where **y** is obtained through the forward substitution

$$y_1 = \frac{b_1}{L_{11}} \,, \tag{4.10}$$

and

$$y_i = \frac{1}{L_{ii}} \left[ b_i - \sum_{k=1}^{i-1} L_{ik} y_k \right] \qquad i > 1 \,. \tag{4.11}$$

We next solve

$$\mathbf{U x} = \mathbf{y} \,, \tag{4.12}$$

where **x** is obtained through the back-substitution

$$x_N = \frac{y_N}{U_{NN}} \,, \tag{4.13}$$

and

$$x_i = \frac{1}{U_{ii}} \left[ y_i - \sum_{k=i+1}^{N} U_{ik} x_k \right] \quad i < N \,. \tag{4.14}$$

The remaining task is to determine the elements of **L** and **U**. If we carry out the matrix multiplication **LU**, we will have $N^2$ equations for the $N^2 + N$ unknowns comprising $L_{ij}$ and $U_{ij}$. Because the diagonal is represented twice, we can specify $N$ of the unknowns ourselves, so we choose

$$L_{ii} = 1 \quad i = 1, 2, \ldots, N \,. \tag{4.15}$$

Thus,

$$\mathbf{LU} = \begin{bmatrix} 1 & 0 & 0 & \ldots & 0 \\ L_{21} & 1 & 0 & \ldots & 0 \\ L_{31} & L_{32} & 1 & \ldots & 0 \\ \vdots & \vdots & \vdots & \ddots & \vdots \\ L_{N1} & L_{N2} & L_{N3} & \ldots & 1 \end{bmatrix} \begin{bmatrix} U_{11} & U_{12} & U_{13} & \ldots & U_{1N} \\ 0 & U_{22} & U_{23} & \ldots & U_{2N} \\ 0 & 0 & U_{33} & \ldots & U_{3N} \\ \vdots & \vdots & \vdots & \ddots & \vdots \\ 0 & 0 & 0 & \ldots & U_{NN} \end{bmatrix} \,, \tag{4.16}$$

and we can immediately set $U_{11} = A_{11}$. Multiplying **L** by the first column of **U** we obtain

$$\begin{aligned} L_{21} &= \frac{A_{21}}{U_{11}} \,, \\ L_{31} &= \frac{A_{31}}{U_{11}} \,, \\ &\cdots \end{aligned} \tag{4.17}$$

and doing the same with the second column of **U** yields $U_{12} = A_{12}$ and

$$\begin{aligned} U_{22} &= A_{22} - L_{21} U_{12} \,, \\ L_{32} &= \frac{1}{U_{22}} \left[ A_{32} - L_{31} U_{12} \right] \,, \\ L_{42} &= \frac{1}{U_{22}} \left[ A_{42} - L_{41} U_{12} \right] \,, \\ &\cdots \end{aligned} \tag{4.18}$$

which leads to a generalized method known as *Crout's algorithm*. For each column $j = 1, 2, \ldots, N$, we first solve for the elements of **U** on and above the diagonal, i.e.,

$$U_{ij} = A_{ij} - \sum_{k=1}^{i-1} L_{ik} U_{kj} \qquad i = 1, 2, \ldots, j , \qquad (4.19)$$

and then for elements of **L** below the diagonal

$$L_{ij} = \frac{1}{U_{jj}} \left[ A_{ij} - \sum_{k=1}^{j-1} L_{ik} U_{kj} \right] \qquad i = j+1, j+2, \ldots, N . \qquad (4.20)$$

Following this algorithm, the elements needed at each step are already computed when they are needed. Note that because **L** is a unit triangular matrix, its diagonal does not need to be stored explicitly. As a result, **A** can be factored *in place*, requiring no additional storage for **L** and **U**. Therefore, throughout the rest of this book we will denote the matrix [**LU**], or often simply the **LU** matrix, as a matrix where **L** and **U** are stored together this way.

A practical LU factorization algorithm will also incorporate partial pivoting to ensure numerical stability at each step of the process. The operation count for the LU factorization and back-substitution are of the same complexity as Gaussian elimination. Once a factorization is completed, it can then be reused for an arbitrary number of right-hand sides. The factored matrix may also be written to disk and read back in later if needed, negating the need to compute and factor the matrix a second time.

### 4.1.3   Block LU Factorization

The LU factorization described in Section 4.1.2 is a scalar algorithm that operates on individual matrix elements. If implemented as-is, it does not take advantage of CPU vector instructions or efficiently manage the loading of matrix elements to and from system RAM and the CPU cache. If we now partition the matrix **A** into sub-blocks, the scalar operations become block operations, for which optimized kernels can be implemented for various block sizes and CPU architectures.

To illustrate, consider an $N \times N$ matrix **A** partitioned into a $3 \times 3$ block matrix, each having approximately $N/3$ rows and columns. This can be written as

$$\mathbf{A} = \begin{bmatrix} \mathbf{A}_{11} & \mathbf{A}_{12} & \mathbf{A}_{13} \\ \mathbf{A}_{21} & \mathbf{A}_{22} & \mathbf{A}_{23} \\ \mathbf{A}_{31} & \mathbf{A}_{32} & \mathbf{A}_{33} \end{bmatrix} = \begin{bmatrix} \mathbf{L}_{11} & 0 & 0 \\ \mathbf{L}_{21} & \mathbf{L}_{22} & 0 \\ \mathbf{L}_{31} & \mathbf{L}_{32} & \mathbf{L}_{33} \end{bmatrix} \begin{bmatrix} \mathbf{U}_{11} & \mathbf{U}_{12} & \mathbf{U}_{13} \\ 0 & \mathbf{U}_{22} & \mathbf{U}_{23} \\ 0 & 0 & \mathbf{U}_{33} \end{bmatrix} , \qquad (4.21)$$

where **L** is unit lower triangular, and **U** is upper triangular. As the **L** and **U** blocks are stored together, we write (4.21) using the more compact notation

$$[\mathbf{LU}] = \begin{bmatrix} [\mathbf{LU}]_{11} & \mathbf{U}_{12} & \mathbf{U}_{13} \\ \mathbf{L}_{21} & [\mathbf{LU}]_{22} & \mathbf{U}_{23} \\ \mathbf{L}_{31} & \mathbf{L}_{32} & [\mathbf{LU}]_{33} \end{bmatrix}. \tag{4.22}$$

Let us now compute a scalar LU factorization of the entire first column of $\mathbf{A}$, but leave the rest of the matrix unchanged. This results in

$$\begin{bmatrix} [\mathbf{LU}]_{11} & \mathbf{A}_{12} & \mathbf{A}_{13} \\ \mathbf{L}_{21} & \mathbf{A}_{22} & \mathbf{A}_{23} \\ \mathbf{L}_{31} & \mathbf{A}_{32} & \mathbf{A}_{33} \end{bmatrix}. \tag{4.23}$$

We next compute the first trailing sub-row of $\mathbf{U}$, which can be written as

$$\begin{bmatrix} \mathbf{U}_{12} & \mathbf{U}_{13} \end{bmatrix} = \mathbf{L}_{11}^{-1} \begin{bmatrix} \mathbf{A}_{12} & \mathbf{A}_{13} \end{bmatrix}, \tag{4.24}$$

where $\mathbf{L}_{11}^{-1}$ indicates a triangular back-substitution. This results in

$$\begin{bmatrix} [\mathbf{LU}]_{11} & \mathbf{U}_{12} & \mathbf{U}_{13} \\ \mathbf{L}_{21} & \mathbf{A}_{22} & \mathbf{A}_{23} \\ \mathbf{L}_{31} & \mathbf{A}_{32} & \mathbf{A}_{33} \end{bmatrix}. \tag{4.25}$$

We now update the lower-right sub-matrix, yielding

$$\begin{bmatrix} [\mathbf{LU}]_{11} & \mathbf{U}_{12} & \mathbf{U}_{13} \\ \mathbf{L}_{21} & \tilde{\mathbf{A}}_{22} & \tilde{\mathbf{A}}_{23} \\ \mathbf{L}_{31} & \tilde{\mathbf{A}}_{32} & \tilde{\mathbf{A}}_{33} \end{bmatrix}, \tag{4.26}$$

where

$$\tilde{\mathbf{A}} = \begin{bmatrix} \tilde{\mathbf{A}}_{22} & \tilde{\mathbf{A}}_{23} \\ \tilde{\mathbf{A}}_{32} & \tilde{\mathbf{A}}_{33} \end{bmatrix} = \begin{bmatrix} \mathbf{A}_{22} & \mathbf{A}_{23} \\ \mathbf{A}_{32} & \mathbf{A}_{33} \end{bmatrix} - \begin{bmatrix} \mathbf{L}_{21} \\ \mathbf{L}_{31} \end{bmatrix} \begin{bmatrix} \mathbf{U}_{12} & \mathbf{U}_{13} \end{bmatrix}. \tag{4.27}$$

The notation $\tilde{\mathbf{A}}$ indicates that the factorization is not yet complete for that sub-matrix. This process is now repeated for $\tilde{\mathbf{A}}$, where a scalar factorization of the first column yields

$$\begin{bmatrix} [\mathbf{LU}]_{22} & \tilde{\mathbf{A}}_{23} \\ \mathbf{L}_{32} & \tilde{\mathbf{A}}_{33} \end{bmatrix}. \tag{4.28}$$

Updating the upper-right block yields

$$\begin{bmatrix} [\mathbf{LU}]_{22} & \mathbf{U}_{23} \\ \mathbf{L}_{32} & \tilde{\mathbf{A}}_{33} \end{bmatrix}, \tag{4.29}$$

where $\mathbf{U}_{23} = \mathbf{L}_{22}^{-1} \tilde{\mathbf{A}}_{23}$. Updating the trailing sub-matrix $\tilde{\mathbf{A}}_{33}$ yields

$$\begin{bmatrix} [\mathbf{LU}]_{22} & \mathbf{U}_{23} \\ \mathbf{L}_{32} & \tilde{\mathbf{B}}_{33} \end{bmatrix} \tag{4.30}$$

where $\tilde{\mathbf{B}}_{33} = \tilde{\mathbf{A}}_{33} - \mathbf{L}_{32}\mathbf{U}_{23}$. As the trailing sub-matrix comprises a single block, we now simply compute $\text{LU}(\tilde{\mathbf{B}}_{33}) = [\mathbf{LU}]_{33}$, yielding finally the **LU** matrix in 4.22.

The algorithm just described comprises a right-looking block LU factorization [2, 3]. It is so-called because at each step, the trailing submatrix is located to the lower-right. With the exception of the scalar factorization, all other steps are matrix-matrix operations, which can be carried out efficiently using Level 3 BLAS functions (gemm and trsm). A similar right-looking algorithm is used in LAPACK (getrf), which also incorporates row pivoting. The BLAS and LAPACK software libraries are discussed further in Sections 4.3.1 and 4.3.2.

### 4.1.4 Condition Number

Given the matrix system $\mathbf{Ax} = \mathbf{b}$, it is desirable to know the impact on the solution $\mathbf{x}$ given any inaccuracies in the estimate of $\mathbf{b}$. If after computing the singular value decomposition (SVD) of $\mathbf{A}$ we observe a large difference between the largest and smallest singular values, we know that small changes in $\mathbf{b}$ may result in large changes in $\mathbf{x}$. We now define the *condition number* of a matrix, which is

$$\kappa(\mathbf{A}) = \|\mathbf{A}^{-1}\| \, \|\mathbf{A}\| \tag{4.31}$$

where $\| \cdot \|$ is a matrix norm. If we choose the 2-norm, then the condition number is given by

$$\kappa(\mathbf{A}) = \frac{\lambda_{max}(\mathbf{A})}{\lambda_{min}(\mathbf{A})} \tag{4.32}$$

where $\lambda_{min}(\mathbf{A})$ and $\lambda_{max}(\mathbf{A})$ are the minimum and maximum eigenvalues of $\mathbf{A}$, respectively [4]. The condition number is important because many elements of the MOM system matrix and right-hand side vectors are obtained through numerical integration. Small errors in the integration can be amplified by a poorly conditioned system matrix. The condition number is also important to iterative methods such as those in the next section, as it has a direct influence on their rate of convergence.

## 4.2 Iterative Methods

We now discuss iterative solver techniques, which have become increasingly popular in recent years. In these algorithms, the bulk of the compute time is spent computing the matrix-vector products between the system matrix and one or more vectors at each iteration. Though the full matrix must still be stored, the overall compute time is of the order $O(MN^2)$, where $N^2$ is the operations for the matrix-vector product and $M$ is the number of iterations.

Iterative methods do not modify the original matrix, and instead minimize a residual vector at each iteration. At the core of each lies the update operation

$$\mathbf{x}_{k+1} = \mathbf{x}_k + \alpha_k \mathbf{p}_k \,, \tag{4.33}$$

where the new estimate of the solution $\mathbf{x}_{k+1}$ is obtained from the previous solution $\mathbf{x}_k$ and a search direction $\mathbf{p}_k$, where $\alpha_k$ is a constant. The residual at each step is

$$\mathbf{r}_{k+1} = \mathbf{A}\mathbf{x}_{k+1} - \mathbf{b} \,. \tag{4.34}$$

In this section we will briefly discuss several commonly used iterative methods and summarize templates that the reader can use to write routines of their own. Because the iterative solver algorithms comprise simple operations such as simple vector and matrix-vector products, they are easy and quick to implement in software, even if the user is not familiar with their inner workings. Some of these methods involve matrix-vector products involving the transpose or conjugate transpose of the system matrix, whereas some of the others do not. The latter are referred to as transpose-free.

### 4.2.1 Conjugate Gradient

The Conjugate Gradient (CG) algorithm is a method for solving a matrix system where $\mathbf{A}$ is symmetric and positive definite. It is related to the Method of Steepest Descent, and minimizes a quadratic function by generating a sequence of conjugate-direction search vectors that are $\mathbf{A}$-orthogonal at each step [5, 6]. Because the system matrix obtained for integral equation problems via the Moment Method is neither symmetric or positive definite, this algorithm cannot be applied directly. Instead, we can apply it to the following system called the "normal equations"

$$\mathbf{A}^\dagger \mathbf{A}\mathbf{x} = \mathbf{A}^\dagger \mathbf{b} \,, \tag{4.35}$$

where $\mathbf{A}^\dagger$ denotes the conjugate transpose, and $\mathbf{A}^\dagger \mathbf{A}$ is symmetric and positive definite. Pseudocode [6] for the preconditioned CG method for the normal equations is provided in Algorithm 1, where $\mathbf{M}$ is a preconditioner matrix. This method requires two matrix-vector products per iteration, one with $\mathbf{A}$ and one with $\mathbf{A}^\dagger$.

The CG method has been used extensively for electromagnetic field problems, and the relationship between the matrix eigenvalues and its convergence studied [7, 8, 9, 10]. For a matrix system with $N$ unknowns, CG theoretically converges to the exact solution in at most $N$ iterations assuming there are no round-off errors, and the residual error decreases at each step. For many practical radiation and scattering problems, CG performs very well. However, when the system matrix is poorly conditioned, its convergence rate can be very slow and may stagnate with no additional decrease in the residual norm.

### 4.2.2 Biconjugate Gradient

The Biconjugate Gradient (BiCG) method was developed by Lanczos [11] and is applicable to general, nonsymmetric systems. BiCG approaches the problem by generating a pair of bi-orthogonal residual sequences using $\mathbf{A}$ and $\mathbf{A}^T$. Pseudocode [12] for the preconditioned BiCG method is provided in Algorithm 2. This method requires two matrix-vector products per iteration, one with $\mathbf{A}$ and one with $\mathbf{A}^T$. Though the residual error of the BiCG does not necessarily decrease at each step and its convergence may be erratic, it does exhibit good performance for many problems.

### 4.2.3 Conjugate Gradient Squared

The Conjugate Gradient Squared (CGS) algorithm [13] is a method that avoids using the transpose of $\mathbf{A}$ and attempts to obtain a faster rate of convergence than CG and BiCG. The method does converge faster in many cases, though because of its more aggressive formulation it is also more sensitive to residual errors and may diverge quickly if the system is ill-conditioned. Pseudocode [12] for the preconditioned CGS method is provided in Algorithm 3. This method requires two matrix-vector products with $\mathbf{A}$ per iteration.

### 4.2.4 Biconjugate Gradient Stabilized

The Biconjugate Gradient Stabilized (BiCG-Stab) algorithm is similar to CGS but attempts to avoid its irregular convergence patterns. Pseudocode [12] for the preconditioned BiCG-Stab method is provided in Algorithm 4. This method requires two matrix-vector products with $\mathbf{A}$ per iteration. Note that there are two end-condition tests in this method.

### 4.2.5   GMRES

In 1986 Saad and Schultz proposed [14] an algorithm known as the Generalized Minimum Residual (GMRES) method, which can be used for the iterative solution of general, nonsymmetric matrices. In the Conjugate Gradient method, the residuals form an orthogonal basis for the space span $\{r_0, Ar_0, A^2 r_0, \dots\}$. In GMRES, this basis is formed explicitly via the following algorithm:

> $w_i = A v_i$
> **for** $k = 1, 2, \dots, i$ **do**
>      $w_i = w_i - (w_i \cdot v_k) v_k$
> **end for**
> $v_{i+1} = w_i / \|w_i\|$

which is a modified Gram-Schmidt orthogonalization. Applied to the Krylov sequence $\{A^k r_0\}$, this is called the Arnoldi method. The inner product coefficients $w_i \cdot v_k$ and $\|w_i\|$ are stored in an upper Hessenberg matrix. The GMRES iterates are then constructed as

$$x_i = x_0 + y_1 v_1 + y_2 v_2 + \cdots y_i v_i \,, \qquad (4.36)$$

where the coefficients $y_k$ are chosen to minimize the residual norm $\|b - Ax_i\|$. GMRES has the property that this residual norm can be computed without the iterate having been formed. Thus, the expensive action of forming the iterate can be postponed until the residual norm has grown small enough. The drawback to GMRES is that the work and storage required per iteration increase linearly with the iteration count. Unless convergence happens quickly, the cost can quickly become prohibitive. One common way of handling this is a reset of the iteration: after $m$ iterations, the accumulated data are cleared and the intermediate results used as the initial data for the next $m$ iterations. This procedure is then repeated until convergence is achieved. The difficulty now lies in choosing an appropriate value for $m$. If $m$ is too small, the algorithm may be slow to converge, or fail to converge entirely. If $m$ is too large, the required computations and storage may be excessive.

Pseudocode [12] for the preconditioned, restarted GMRES($m$) method is provided in Algorithm 5. GMRES is employed often to solve electromagnetic problems due to its good convergence properties, robust performance, and the fact it is transpose-free. We will use it along with the Fast Multipole Method in Chapter 11.

---

**Algorithm 1** Preconditioned Conjugate Gradient (CGNR) Method

---

Make an initial guess for $\mathbf{x}_0$

$\mathbf{r}_0 = \mathbf{b} - \mathbf{A}\mathbf{x}_0$

$\tilde{\mathbf{r}}_0 = \mathbf{A}^\dagger \mathbf{r}_0$

Solve $\mathbf{M}\mathbf{z}_0 = \tilde{\mathbf{r}}_0$

$\mathbf{p}_0 = \mathbf{z}_0$

**for** $i = 1, 2, ...,$ until convergence **do**

$\quad \mathbf{w}_{i-1} = \mathbf{A}\mathbf{p}_{i-1}$

$\quad \alpha_{i-1} = (\mathbf{z}_{i-1}^* \cdot \tilde{\mathbf{r}}_{i-1})/\|\mathbf{w}_{i-1}\|_2^2$

$\quad \mathbf{x}_i = \mathbf{x}_{i-1} + \alpha_{i-1}\mathbf{p}_{i-1}$

$\quad \mathbf{r}_i = \mathbf{r}_{i-1} - \alpha_{i-1}\mathbf{w}_{i-1}$

$\quad$ Convergence check

$\quad \tilde{\mathbf{r}}_i = \mathbf{A}^\dagger \mathbf{r}_i$

$\quad$ Solve $\mathbf{M}\mathbf{z}_i = \tilde{\mathbf{r}}_i$

$\quad \beta_{i-1} = (\mathbf{z}_i^* \cdot \tilde{\mathbf{r}}_i)/(\mathbf{z}_{i-1}^* \cdot \tilde{\mathbf{r}}_{i-1})$

$\quad \mathbf{p}_i = \mathbf{z}_i + \beta_{i-1}\mathbf{p}_{i-1}$

**end for**

---

---

**Algorithm 2** Preconditioned Biconjugate Gradient (BiCG) Method

---

Make an initial guess for $\mathbf{x}_0$
$\mathbf{r}_0 = \mathbf{b} - \mathbf{A}\mathbf{x}_0$
$\tilde{\mathbf{r}}_0 = \mathbf{r}_0$
**for** $i = 1, 2, ...$, until convergence **do**
    Solve $\mathbf{M}\mathbf{z}_{i-1} = \mathbf{r}_{i-1}$
    Solve $\mathbf{M}^T\tilde{\mathbf{z}}_{i-1} = \tilde{\mathbf{r}}_{i-1}$
    $\rho_{i-1} = \mathbf{z}_{i-1} \cdot \tilde{\mathbf{r}}_{i-1}$
    **if** $\rho_{i-1} = 0$ **then**
        method fails
    **end if**
    **if** $i = 1$ **then**
        $\mathbf{p}_i = \mathbf{z}_{i-1}$
        $\tilde{\mathbf{p}}_i = \tilde{\mathbf{z}}_{i-1}$
    **else**
        $\beta_{i-i} = \rho_{i-1}/\rho_{i-2}$
        $\mathbf{p}_i = \mathbf{z}_{i-1} + \beta_{i-1}\mathbf{p}_{i-1}$
        $\tilde{\mathbf{p}}_i = \tilde{\mathbf{z}}_{i-1} + \beta_{i-1}\tilde{\mathbf{p}}_{i-1}$
    **end if**
    $\mathbf{q}_i = \mathbf{A}\mathbf{p}_i$
    $\tilde{\mathbf{q}}_i = \mathbf{A}^T\tilde{\mathbf{p}}_i$
    $\alpha_i = \rho_{i-1}/(\tilde{\mathbf{p}}_i \cdot \mathbf{q}_i)$
    $\mathbf{x}_i = \mathbf{x}_{i-1} + \alpha_i\mathbf{p}_i$
    $\mathbf{r}_i = \mathbf{r}_{i-1} - \alpha_i\mathbf{q}_i$
    $\tilde{\mathbf{r}}_i = \tilde{\mathbf{r}}_{i-1} - \alpha_i\tilde{\mathbf{q}}_i$
    Convergence check
**end for**

---

---

**Algorithm 3** Preconditioned Conjugate Gradient Squared (CGS) Method

---

Make an initial guess for $\mathbf{x}_0$
$\mathbf{r}_0 = \mathbf{b} - \mathbf{A}\mathbf{x}_0$
$\tilde{\mathbf{r}} = \mathbf{r}_0$
**for** $i = 1, 2, ...$, until convergence **do**
    $\rho_{i-1} = \tilde{\mathbf{r}} \cdot \mathbf{r}_{i-1}$
    **if** $\rho_{i-1} = 0$ **then**
        method fails
    **end if**
    **if** $i = 1$ **then**
        $\mathbf{u}_1 = \mathbf{r}_0$
        $\mathbf{p}_1 = \mathbf{u}_1$
    **else**
        $\beta_{i-i} = \rho_{i-1}/\rho_{i-2}$
        $\mathbf{u}_i = \mathbf{r}_{i-1} + \beta_{i-1}\mathbf{q}_{i-1}$
        $\mathbf{p}_i = \mathbf{u}_i + \beta_{i-1}(\mathbf{q}_{i-1} + \beta_{i-1}\mathbf{p}_{i-1})$
    **end if**
    Solve $\mathbf{M}\hat{\mathbf{p}} = \mathbf{p}_i$
    $\hat{\mathbf{v}} = \mathbf{A}\hat{\mathbf{p}}$
    $\alpha_i = \rho_{i-1}/(\tilde{\mathbf{r}} \cdot \hat{\mathbf{v}})$
    $\mathbf{q}_i = \mathbf{u}_i - \alpha_i\hat{\mathbf{v}}$
    Solve $\mathbf{M}\hat{\mathbf{u}} = \mathbf{u}_i + \mathbf{q}_i$
    $\mathbf{x}_i = \mathbf{x}_{i-1} + \alpha_i\hat{\mathbf{u}}$
    $\hat{\mathbf{q}} = \mathbf{A}\hat{\mathbf{u}}$
    $\mathbf{r}_i = \mathbf{r}_{i-1} - \alpha_i\hat{\mathbf{q}}_i$
    Convergence check
**end for**

---

---

**Algorithm 4** Preconditioned Biconjugate Gradient Stabilized (BiCG-Stab) Method

---

Make an initial guess for $\mathbf{x}_0$
$\mathbf{r}_0 = \mathbf{b} - \mathbf{A}\mathbf{x}_0$
$\tilde{\mathbf{r}} = \mathbf{r}_0$
**for** $i = 1, 2, ...,$ until convergence **do**
    $\rho_{i-1} = \tilde{\mathbf{r}} \cdot \mathbf{r}_{i-1}$
    **if** $\rho_{i-1} = 0$ **then**
        method fails
    **end if**
    **if** $i = 1$ **then**
        $\mathbf{p}_1 = \mathbf{r}_0$
    **else**
        $\beta_{i-i} = (\rho_{i-1}/\rho_{i-2})(\alpha_{i-1}/\omega_{i-1})$
        $\mathbf{p}_i = \mathbf{r}_{i-1} + \beta_{i-1}(\mathbf{p}_{i-1} - \omega_{i-1}\mathbf{v}_{i-1})$
    **end if**
    Solve $\mathbf{M}\hat{\mathbf{p}} = \mathbf{p}_i$
    $\mathbf{v}_i = \mathbf{A}\hat{\mathbf{p}}$
    $\alpha_i = \rho_{i-1}/(\tilde{\mathbf{r}} \cdot \mathbf{v}_i)$
    $\mathbf{s} = \mathbf{r}_{i-1} - \alpha_i\mathbf{v}_i$
    Check norm of $s$, if small enough set $\mathbf{x}_i = \mathbf{x}_{i-1} + \alpha_i\hat{\mathbf{p}}$, stop
    Solve $\mathbf{M}\hat{\mathbf{s}} = \mathbf{s}$
    $\mathbf{t} = \mathbf{A}\hat{\mathbf{s}}$
    $\omega_i = (\mathbf{t} \cdot \mathbf{s})/(\mathbf{t} \cdot \mathbf{t})$
    $\mathbf{x}_i = \mathbf{x}_{i-1} + \alpha_i\hat{\mathbf{p}} + \omega_i\hat{\mathbf{s}}$
    $\mathbf{r}_i = \mathbf{s} - \omega_i\mathbf{t}$
    Convergence check
    For continuation, it is necessary that $\omega_i \neq 0$
**end for**

---

---

**Algorithm 5** Preconditioned GMRES($m$) Method

---

Make an initial guess for $\mathbf{x}_0$
**for** $j = 1, 2, ...,$ **do**
   Solve $\mathbf{Mr} = \mathbf{b} - \mathbf{Ax}_0$
   $\mathbf{v}_1 = \mathbf{r}/||\mathbf{r}||_2$
   $\mathbf{s} := ||\mathbf{r}||_2\mathbf{e}_1$
   **for** $i = 1, ..., m$ **do**
      Solve $\mathbf{Mw} = \mathbf{Av}_i$
      **for** $k = 1, ..., i$ **do**
         $h_{k,i} = \mathbf{w} \cdot \mathbf{v}_k$
         $\mathbf{w} = \mathbf{w} - h_{k,i}\mathbf{v}_k$
      **end for**
      $h_{i+1,i} = ||\mathbf{w}||_2$
      $\mathbf{v}_{i+1} = \mathbf{w}/h_{i+1,i}$
      Apply $J_1, \ldots, J_i - 1$ on $(h_{1,i}, \ldots, h_{i+1,i})$
      Construct $J_i$ acting on the $i$th and $(i + 1)$st component
      of $h_{.,i}$ such that $(i + 1)$st component of $J_i h_{.,i}$ is 0
      $\mathbf{s} := J_i\mathbf{s}$
      **if** $\mathbf{s}(i + 1)$ small enough **then**
         UPDATE($\tilde{\mathbf{x}}, i$)
         stop
      **end if**
   **end for**
   UPDATE($\tilde{\mathbf{x}}, m$)
**end for**

**procedure** UPDATE($\tilde{\mathbf{x}}, i$)
   Solve $\mathbf{Hy} = \tilde{\mathbf{s}}$
   Where the upper $i \times i$ triangular part of $\mathbf{H}$ has $h_{i,j}$ as
   its elements, and $\tilde{\mathbf{s}}$ represents the first $i$ components of $\mathbf{s}$.
   $\tilde{\mathbf{x}} = \mathbf{x}_0 + y_1\mathbf{v}_1 + y_2\mathbf{v}_2 + \cdots y_i\mathbf{v}_i$
   $\mathbf{s}(i + 1) = ||\mathbf{b} - \mathbf{A}\tilde{\mathbf{x}}||_2$
   **if** $\tilde{\mathbf{x}}$ is accurate enough **then**
      stop
   **else**
      $\mathbf{x}_0 = \tilde{\mathbf{x}}$
   **end if**
**end procedure**

---

### 4.2.6   Stopping Criteria

When using an iterative method, we need to know when to terminate the iteration. Since we do not know the solution vector $\mathbf{x}$, we cannot compute the error vector

$$\mathbf{e}_i = \mathbf{x} - \mathbf{x}_i \qquad (4.37)$$

but instead compute the residual norm $N_i$ at iteration $i$, which is

$$N_i = \frac{\|\mathbf{r}_n\|}{\|\mathbf{b}\|} = \frac{\|\mathbf{A}\mathbf{x}_i - \mathbf{b}\|}{\|\mathbf{b}\|} . \qquad (4.38)$$

We terminate the iteration once the residual norm is below a certain value, such as $10^{-3}$. Note that the residual norm is only an indirect measure of the amount of error, because [4]

$$\frac{\|\mathbf{e}_n\|}{\|\mathbf{e}_0\|} \leq \kappa(\mathbf{A}) \frac{\|\mathbf{r}_n\|}{\|\mathbf{r}_0\|} , \qquad (4.39)$$

and if $\mathbf{A}$ is badly conditioned, $\mathbf{x}_i$ may not be a very good estimate of the solution for a given $N_i$. In practice, the iteration is stopped when the residual norm goes below a predefined value, or a maximum number of iterations is reached.

### 4.2.7   Preconditioning

Previous sections have pointed out the relationship between increasing condition number of a matrix and slow convergence of iterative solvers. In some cases, the solution may converge but only after many iterations and a long compute time. In others, the solution may be unobtainable because the iteration stagnates or even diverges. In both cases, though especially in the latter, a method of improving the convergence behavior of the solver is desirable. To investigate such a method, let us modify the original linear system to create the new system

$$\mathbf{M}^{-1}\mathbf{A}\mathbf{x} = \mathbf{M}^{-1}\mathbf{b} , \qquad (4.40)$$

where $\mathbf{M}$ is a *preconditioner matrix*. If $\mathbf{M}$ resembles $\mathbf{A}$ in some way, then the solution of the above matrix system remains the same, however the eigenvalues of $\mathbf{M}^{-1}\mathbf{A}$ may be more attractive and allow better performance in the iterative solver. Algorithms 1-5 include a preconditioning step, requiring the solution of one or more auxiliary linear systems of the form $\mathbf{M}\mathbf{z} = \mathbf{r}$ at each iteration.

The problem we now face is finding the value of $\mathbf{M}^{-1}$. Obviously, if $\mathbf{M}^{-1}$ is exactly equal to $\mathbf{A}^{-1}$ we have not improved anything, as this requires a complete factorization of $\mathbf{A}$. The key is determining a value of $\mathbf{M}^{-1}$ that has a reasonable setup time and memory demand, and that does not impose an unacceptable overhead when solving the auxiliary problem at each iteration. In some cases, the application of a preconditioner is absolutely necessary as the original system does not converge at all. In other cases, the hope is that the improved rate of convergence will result in fewer iterations and an overall savings in compute time. This is particularly attractive when the calculations involve multiple right-hand sides, as the preconditioner setup time can be amortized quickly. Several effective preconditioning methods are discussed at length in [6] for general matrices. We present versions of some of these in Section 11.5, adapted specifically for use with the Fast Multiple Method.

## 4.3 Software for Linear Systems

There are many off-the-shelf software libraries available for working with and solving linear systems of equations. Fortunately, many of these are well written, time-tested, robust and freely available. This means that the programmer does not have to worry about writing their own routines, and can instead simply install the libraries and follow the provided API. In this section we will briefly discuss some of the software most widely used for matrix algebra problems.

### 4.3.1 BLAS

The BLAS (Basic Linear Algebra Subprograms) comprises a set of subroutines that act as building blocks for more complex vector and matrix operations [15]. BLAS comprises three levels of functions: BLAS level 1 for scalar, vector and vector-vector operations, BLAS level 2 for matrix-vector operations, and BLAS level 3 for matrix-matrix operations. Separate routines exist for real and complex arithmetic, and in single and double precision. The BLAS source code is freely available from the Netlib BLAS repository[1], and may be compiled and included in commercial applications at no cost. This version comprises the original FORTRAN "reference implementation", which is the standard from which other versions are derived. There are several highly optimized versions of BLAS available from various vendors. For Intel CPUs, the most popular are the Intel Math Kernel Library (MKL) [16] and OpenBLAS [17], both of which are distributed for free as of 2020. These libraries are written to take advantage of CPU vector instructions, and carefully manage the use of the CPU cache and the slower system RAM. These libraries also possess their own internal thread-level parallelism, and can take advantage of multiple processors without the user having to implement a multithreaded program themselves.

There also exist BLAS libraries for systems having one or more graph-

---

[1] http://www.netlib.org/blas

ics processing units (GPUs). These units are massively parallel, having thousands of small processing cores, and are capable of orders of magnitude higher performance in BLAS operations than on a single CPU. For users with NVIDIA GPUs, the cuBLAS library [18] is a GPU-accelerated implementation of BLAS made specifically for NVIDIA GPUs, and uses the NVIDIA CUDA API programming model [19]. Another GPU-accelerated implementation of BLAS is MAGMA [20], which provides implementations for CUDA as well as OpenCL, which is an open, parallel programming standard supported by AMD GPUs.

### 4.3.2   LAPACK

LAPACK is a library for solving dense linear systems, least-squares solutions of linear systems of equations, eigenvalue problems, and singular value problems [21]. It is also available from the Netlib LAPACK repository[2], and may be used freely in commercial applications. LAPACK routines use BLAS functions for low-level work, and a BLAS implementation must accompany an installation of LAPACK. Separate versions of most LAPACK routines are available for real and complex arithmetic, and in single and double precision. As with BLAS, there are also highly optimized versions of LAPACK available. The Intel Math Kernel Library (MKL) also contains an implementation of LAPACK for CPUs. Similarly, MAGMA also implements LAPACK functions for GPUs.

### 4.3.3   MATLAB

MATLAB is a popular numerical computing environment and programming language commercially available from The Mathworks, Inc. It was originally developed to aid in solving matrix problems and many of its matrix functions operate via calls to BLAS and LAPACK. The MATLAB scripting language is easy to learn and fairly complex simulations can be written in it. Though its scripts are interpreted and less efficient than compiled code, MATLAB is easy to learn, accurate, and can be used to solve many of the examples in this book. It is also excellent at generating plots, and was used to generate all the plots in this book.

---

[2]http://www.netlib.org/lapack

# References

[1] W. H. Press, B. P. Flannery, S. A. Teukolsky, and W. T. Vetterling, *Numerical Recipes in C : The Art of Scientific Computing*. Cambridge University Press, 1992.

[2] J. W. Demmel, N. J. Higham, and R. S. Schreiber, "Stability of block LU factorization," *Numer. Lin. Algebra Applic*, vol. 2, pp. 173–190, 1995.

[3] J. Dongarra, S. Hammarlin, and D. Walker, "Key concepts for parallel out-of-core LU factorization," *Computers and Mathematics With Applications*, vol. 35, pp. 13–31, April 1998.

[4] A. F. Peterson, S. L. Ray, and R. Mittra, *Computational Methods for Electromagnetics*. IEEE Press, 1998.

[5] M. Hestenes and E. Steifel, "Methods of conjugate gradients for solving linear systems," *J. Res. Nat. Bur. Stand.*, vol. 49, pp. 409–435, 1952.

[6] Y. Saad, *Iterative Methods for Sparse Linear Systems*. PWS, 1st ed., 1996.

[7] T. K. Sarkar, K. R. Siarkiewicz, and R. F. Stratton, "Survey of numerical methods for solution of large systems of linear equations for electromagnetic field problems," *IEEE Trans. Antennas Propagat.*, vol. 29, pp. 847–856, November 1981.

[8] T. K. Sarkar, "The conjugate gradient method as applied to electromagnetic field problems," *IEEE Antennas Propagat. Soc. Newsletter*, pp. 5–14, August 1986.

[9] A. F. Peterson and R. Mittra, "Convergence of the conjugate gradient method when applied to matrix equations representing electromagnetic scattering problems," *IEEE Trans. Antennas Propagat.*, vol. 34, pp. 1447–1454, December 1986.

[10] A. F. Peterson, C. F. Smith, and R. Mittra, "Eigenvalues of the moment-method matrix and their effect on the convergence of the conjugate gradient algorithm," *IEEE Trans. Antennas Propagat.*, vol. 36, pp. 1177–1179, August 1988.

[11] C. Lanczos, "Solution of systems of linear equations by minimized iterations," *J. Res. Nat. Bur. Standards*, vol. 49, pp. 33–53, 1952.

[12] R. Barrett, M. Berry, T. F. Chan, J. Demmel, J. Donato, J. Dongarra, V. Eijkhout, R. Pozo, C. Romine, and H. V. der Vorst, *Templates for the Solution of Linear Systems: Building Blocks for Iterative Methods.* SIAM, second ed., 1994.

[13] P. Sonneveld, "CGS, a fast Lanczos-type solver for nonsymmetric linear systems," *SIAM J. Sci. Statist. Comput.*, vol. 10, pp. 36–52, 1989.

[14] Y. Saad and M. H. Schultz, "GMRES: a generalized minimal residual algorithm for solving nonsymmetric linear systems," *SIAM J. Sci. Stat. Comput.*, vol. 7, pp. 856–869, July 1986.

[15] L. S. Blackford, J. Demmel, J. Dongarra, I. Duff, S. Hammarling, G. Henry, M. Heroux, L. Kaufman, A. Lumsdaine, A. Petitet, R. Pozo, K. Remington, and R. C. Whaley, "An updated set of Basic Linear Algebra Subprograms (BLAS)," *ACM Transactions on Mathematical Software*, vol. 28, pp. 135–151, June 2002.

[16] Intel Math Kernel Library (MKL), software.intel.com/mkl.

[17] OpenBLAS, An optimized BLAS library, www.openblas.net.

[18] cuBLAS, docs.nvidia.com/cuda/cublas/index.html.

[19] CUDA Toolkit Documentation, docs.nvidia.com/cuda.

[20] J. Dongarra, M. Gates, A. Haidar, J. Kurzak, P. Luszczek, S. Tomov, and I. Yamazaki, "Accelerating numerical dense linear algebra calculations with gpus," *Numerical Computations with GPUs*, pp. 1–26, 2014.

[21] E. Anderson, Z. Bai, C. Bischof, S. Blackford, J. Demmel, J. Dongarra, J. Du Croz, A. Greenbaum, S. Hammarling, A. McKenney, and D. Sorensen, *LAPACK Users' Guide.* Philadelphia, PA: Society for Industrial and Applied Mathematics, third ed., 1999.

# Chapter 5

# Thin Wires

In this chapter we will use the Method of Moments to analyze the radiation and scattering by thin, conducting wires. This is an area of great practical interest, as many realistic antennas can be modeled by wires whose radius is much smaller than their length and the operating wavelength. We will first derive a thin wire approximation to the magnetic vector potential, and consider the well-known Hallén and Pocklington integral equations for thin, straight wires. Next, we will discuss modeling of the feed system at an antenna's terminals, apply the EFIE (3.178) to thin wires of more general shape, and compare the effectiveness of each model through a few simple examples. We will then present several examples with geometries having more general, realistic configurations.

## 5.1 Thin Wire Approximation

Consider a perfectly conducting long, $\hat{\mathbf{z}}$-oriented thin wire of length $L$ whose radius $a$ is much less than $L$ and $\lambda$. An incident electric field $\mathbf{E}^i(\mathbf{r})$ excites on this wire a surface current $\mathbf{J}(\mathbf{r})$. As the wire is very thin, we assume that $\mathbf{J}(\mathbf{r})$ is a filamentary electric current $I_z(\mathbf{r})$ which can be written as

$$\mathbf{J}(\mathbf{r}) = \frac{I_z(z)}{2\pi a}\hat{\mathbf{z}} , \qquad (5.1)$$

where there is no dependence on wire azimuthal angle $\phi$. We also assume that the current goes to zero at the ends of a wire without any current flow onto the wire end-caps. In cylindrical coordinates, we can write the corresponding magnetic vector potential $A_z$ as

$$A_z(\rho, \phi, z) = \mu \int_{-L/2}^{L/2} \int_0^{2\pi} \frac{I_z(z')}{2\pi} \frac{e^{-jkr}}{4\pi r} \, d\phi' \, dz' , \qquad (5.2)$$

where

$$r = |\mathbf{r} - \mathbf{r}'| = \sqrt{(z - z')^2 + |\boldsymbol{\rho} - \boldsymbol{\rho}'|^2} . \qquad (5.3)$$

Using the fact that $\rho' = a$ allows us to write

$$|\boldsymbol{\rho} - \boldsymbol{\rho}'| = \rho^2 + a^2 - 2\boldsymbol{\rho} \cdot \boldsymbol{\rho}' = \rho^2 + a^2 - 2\rho a \cos(\phi' - \phi) \, . \tag{5.4}$$

Since the above is a function of $\phi' - \phi$, the result is cylindrically symmetric. Therefore, we can simply replace $\phi' - \phi$ with $\phi'$ and write

$$A_z(\rho, z) = \mu \int_{-L/2}^{L/2} \frac{I_z(z')}{2\pi} \int_0^{2\pi} \frac{e^{-jkr}}{4\pi r} \, d\phi' \, dz' \, , \tag{5.5}$$

where

$$r = \sqrt{(z - z')^2 + \rho^2 + a^2 - 2\rho a \cos \phi'} \tag{5.6}$$

and the integral

$$\int_0^{2\pi} \frac{e^{-jkr}}{4\pi r} \, d\phi' \tag{5.7}$$

is referred to as the cylindrical wire kernel in the literature [1, 2]. If we assume $a$ to be very small, we can approximate $r$ as

$$r = \sqrt{(z - z')^2 + \rho^2} \, , \tag{5.8}$$

and the innermost integral is no longer a function of $\phi'$, resulting in

$$A_z(\rho, z) = \mu \int_{-L/2}^{L/2} I_z(z') \frac{e^{-jkr}}{4\pi r} \, dz' \, . \tag{5.9}$$

The above is often referred to as a *thin wire approximation* with reduced kernel. The original surface integral has now been effectively replaced by a line integral along the axis of the wire. In cases where the dimensions of the problem invalidate the assumptions of the reduced kernel, the cylindrical wire kernel should be evaluated by more exact means, such as those found in [3]. The total radiated field is obtained via (3.73), and is

$$-j\omega \left[ 1 + \frac{1}{k^2} \frac{\partial^2}{\partial z^2} \right] A_z = E_z^s \, . \tag{5.10}$$

By enforcing the boundary condition of zero tangential electric fields on the surface of the wire, we can now write the above in terms of the incident field $E_z^i$

$$j\omega \left[ 1 + \frac{1}{k^2} \frac{\partial^2}{\partial z^2} \right] A_z = E_z^i \, . \tag{5.11}$$

When solving the thin wire equations in this chapter, we will assume the testing points to be located on the axis of the wire and the source points to be on the surface. Also note that (5.11) comprises a simplified version of (3.178) where $\mathbf{M} = 0$. There are two common forms by which (5.11) is commonly written.

The first retains the differential operator outside the integral, and can be written as

$$j\omega\mu\left[1 + \frac{1}{k^2}\frac{\partial^2}{\partial z^2}\right]\int_{-L/2}^{L/2} I_z(z')\frac{e^{-jkr}}{4\pi r}\,dz' = E_z^i(z) \qquad (5.12)$$

This is called *Hallén's integral equation* [4]. We can also move the differential operator under the integral sign, yielding

$$j\omega\mu\int_{-L/2}^{L/2} I_z(z')\left[1 + \frac{1}{k^2}\frac{\partial^2}{\partial z^2}\right]\frac{e^{-jkr}}{4\pi r}\,dz' = E_z^i(z) \qquad (5.13)$$

which is called *Pocklington's integral equation* [5]. Pocklington's equation is not as well behaved as Hallén's, as the differential operator is acting on the Green's function. As we will see, results obtained using it typically exhibit slower convergence and less accuracy than those obtained from Hallén's.

---

## 5.2   Thin Wire Excitations

In antenna problems, the unknowns of interest are most often the input impedance at a particular feed location and the resulting radiation pattern, directivity, and gain. The easiest way to compute these is to consider the antenna in its transmitting mode, which requires a reasonable model of the feed system at the input terminals. In practical situations, the antenna might be fed by an open-wire transmission line, or by a coaxial line through a ground plane. These various feed systems impact the antenna impedance characteristics in different ways. With this in mind, we need a way to model the field introduced by the feed system without having to model the system itself. In this section we consider two common feed methods used in thin wire problems, the delta-gap source and the magnetic frill. The delta-gap source treats the feed as if the field due to the feedline exists only in the gap between the antenna terminals, with a value of zero outside (no fringing effects). This method typically produces less accurate results for input impedance, though it still performs well in computing radiation patterns. The magnetic frill models the feed as a coaxial line that terminates in a monopole over a ground plane. Use of the frill results in more accurate input impedance values at the expense of increased computations in computing the right-hand side vector elements.

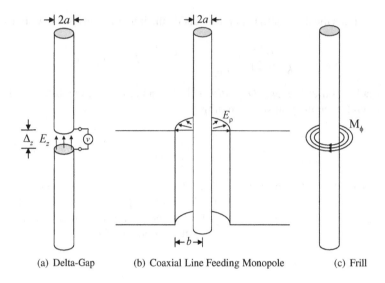

(a) Delta-Gap      (b) Coaxial Line Feeding Monopole      (c) Frill

**FIGURE 5.1:** Thin Wire Excitation Models

### 5.2.1  Delta-Gap Source

The delta-gap source model assumes that the impressed electric field in the thin gap between the antenna terminals can be expressed as

$$\mathbf{E}^i = \frac{V_o}{\Delta_z}\hat{\mathbf{z}}\,, \tag{5.14}$$

where $\Delta_z$ is the width of the gap. This is illustrated in Figure 5.1a. In our numerical simulation, we will assume that this field exists on a single wire segment and is zero on all others. The resulting excitation vector will have nonzero elements only for basis functions having support on that segment.

### 5.2.2  Magnetic Frill

In Figure 5.1b, we depict a coaxial line feeding a monopole antenna over an infinite ground plane. If we assume the field distribution in the aperture to be purely TEM, we can use the method of images and replace the ground plane and aperture with the magnetic frill shown in Figure 5.1c. With an aperture electric field given by

$$\mathbf{E}(\rho) = \frac{1}{2\rho \log(b/a)}\hat{\boldsymbol{\rho}}\,, \tag{5.15}$$

the equivalent magnetic current density is

$$\mathbf{M}(\rho) = -2\,\hat{\mathbf{n}} \times \mathbf{E}(\rho) = \frac{-1}{\rho \log(b/a)}\hat{\boldsymbol{\phi}} \qquad a \le \rho \le b\,. \tag{5.16}$$

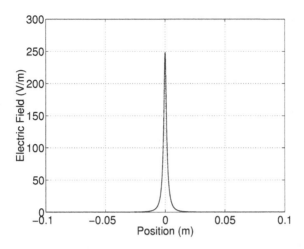

**FIGURE 5.2:** Field Intensity Due To Frill

This current generates an electric field along the wire. For a frill centered at the origin, the field intensity on the axis of the wire ($\rho = 0$) is [6]

$$E_z^i(z) = \frac{1}{2 \log(b/a)} \left( \frac{e^{-jkR_1}}{R_1} - \frac{e^{-jkR_2}}{R_2} \right), \tag{5.17}$$

where

$$R_1 = \sqrt{z^2 + a^2} \tag{5.18}$$

and

$$R_2 = \sqrt{z^2 + b^2}. \tag{5.19}$$

The $\hat{\mathbf{z}}$-directed field in (5.17) is the incident field used for the right-hand side vector. Note that in using this model, we are effectively modeling a dipole with the feed system of a monopole. A true monopole will have an input impedance only half that of the dipole as it only radiates into the half space above the ground. A plot of the axis electric field intensity due to a frill with $a = 1$ mm, $b = 5$ mm and $\lambda = 1$ m is shown in Figure 5.2. Given the sharp drop-off of the field near $z = 0$, when computing the right-hand side vector elements special attention should be paid to the numerical integration so that maximum accuracy is obtained.

## 5.2.3 Plane Wave

The tangential electric field on a filamentary wire can be written as

$$E_{tan}(\mathbf{r}) = \hat{\mathbf{t}}(\mathbf{r}) \cdot \mathbf{E}(\mathbf{r}), \tag{5.20}$$

where $\hat{t}(\mathbf{r})$ is the vector tangent to the wire at $\mathbf{r}$. For a $\hat{z}$-oriented wire illuminated by a $\hat{\theta}$-oriented plane wave of unit amplitude incident from direction $\hat{\mathbf{r}}^i$, (5.20) becomes

$$E_{tan}(\mathbf{r}) = \hat{z} \cdot \hat{\theta} \, e^{-jk\mathbf{r}\cdot\hat{\mathbf{r}}^i} = \sin\theta^i e^{jkz\cos\theta^i} . \qquad (5.21)$$

Because the thin-wire approximation assumes no azimuthally excited currents, an $\hat{\phi}$-polarized incident wave will induce no current on the wire.

---

## 5.3   Hallén's Equation

We will now consider the solution to Hallén's equation (5.12). If we consider it in the following form

$$j\omega\left[1 + \frac{1}{k^2}\frac{\partial^2}{\partial z^2}\right]A_z(z) = E_z^i(z) , \qquad (5.22)$$

we see that it is an inhomogeneous scalar Helmholtz equation for $A_z(z)$, which can be solved by the Green's function method. The general solution to the homogeneous equation

$$\left[1 + \frac{1}{k^2}\frac{\partial^2}{\partial z^2}\right]A_z(z) = 0 \qquad (5.23)$$

is

$$A_z(z) = C_1 e^{jkz} + C_2 e^{-jkz} . \qquad (5.24)$$

To obtain a particular solution we must obtain the Green's function $G(z)$ which satisfies the equation

$$j\omega\left[1 + \frac{1}{k^2}\frac{\partial^2}{\partial z^2}\right]G(z) = \delta(z) . \qquad (5.25)$$

Once $G(z)$ is known, the solution for $A_z(z)$ can be obtained by

$$A_z(z) = C_1 e^{jkz} + C_2 e^{-jkz} + \int_{-L/2}^{L/2} G(z, z') \, E_z^i(z') \, dz' . \qquad (5.26)$$

To obtain the Green's function, let us use the trial function

$$G(z) = C \sin(k|z|) , \qquad (5.27)$$

which is continuous at $z = 0$ with discontinuous derivative at $z = 0$ as is required for the Green's function [7]. To determine the constant C, we integrate (5.25) from $-\epsilon$ to $\epsilon$, yielding

$$j\omega C \int_{-\epsilon}^{\epsilon} \sin(k|z|)\, dz - \frac{jw}{k} C[\cos(-kz)]\big|_{-\epsilon}^{0} + \frac{jw}{k} C[\cos(kz)]\big|_{0}^{\epsilon} = 1 \,. \quad (5.28)$$

If we allow $\epsilon$ to approach zero, the first term goes to zero and after evaluating the remaining terms we get

$$C = -\frac{jk}{2\omega} = -\frac{j\mu}{2\eta} \,. \quad (5.29)$$

Thus,

$$G(z) = -\frac{j\mu}{2\eta} \sin(k|z|) \,, \quad (5.30)$$

and the solution for $A_z(z)$ is

$$A_z(z) = C_1 e^{jkz} + C_2 e^{-jkz} - \frac{j\mu}{2\eta} \int_{-L/2}^{L/2} \sin(k|z - z'|)\, E_z^i(z')\, dz' \,. \quad (5.31)$$

Substituting the original expression for $A_z$ on the left-hand side of the above yields

$$\int_{-L/2}^{L/2} I_z(z') \frac{e^{-jkr}}{4\pi r}\, dz' = C_1 e^{jkz} + C_2 e^{-jkz}$$

$$- \frac{j}{2\eta} \int_{-L/2}^{L/2} \sin(k|z - z'|)\, E_z^i(z')\, dz' \,. \quad (5.32)$$

By similar reasoning, a second solution for the Green's function is found to be

$$G_z(z) = \frac{\mu}{2\eta} e^{-jk|z|} \,, \quad (5.33)$$

leading to a second expression for $A_z$, which is

$$\int_{-L/2}^{L/2} I_z(z') \frac{e^{-jkr}}{4\pi r}\, dz' = C_1 e^{jkz} + C_2 e^{-jkz} + \frac{1}{2\eta} \int_{-L/2}^{L/2} e^{-jk|z-z'|}\, E_z^i(z')\, dz' \,. \quad (5.34)$$

Note also that the two homogeneous terms in the above can also be written as

$$C_1 e^{jkz} + C_2 e^{-jkz} = D_1 \cos(kz) + D_2 \sin(kz) \,, \quad (5.35)$$

where $D_1$ and $D_2$ are complex. The form we use will depend on the symmetry of the problem.

### 5.3.1  Symmetric Problems

We will first solve Hallén's equation for symmetric problems, such as the induced current and input impedance of a center-fed dipole antenna. We first expand the left-hand side of (5.32) using $N$ weighted basis functions, yielding

$$\sum_{n=1}^{N} a_n \int_{f_n} f_n(z') \frac{e^{-jkr}}{4\pi r} \, dz' \, . \tag{5.36}$$

Applying the Method of Moments and testing (5.32) using $N$ testing functions $f_m$ yields matrix elements $Z_{mn}$ given by

$$Z_{mn} = \int_{f_m} f_m(z) \int_{f_n} f_n(z') \frac{e^{-jkr}}{4\pi r} \, dz' \, dz \, . \tag{5.37}$$

The choice of right-hand side depends on the symmetry of the problem being considered. Since we are feeding the antenna at its center, we expect that the induced current will be symmetric. Therefore, we will retain the right-hand side of (5.32) but rewrite the homogeneous terms following (5.35), yielding a right-hand side given by

$$D_1 \cos(kz) + D_2 \sin(kz) - \frac{j}{2\eta} \int_{-L/2}^{L/2} \sin(k|z - z'|) \, E_z^i(z') \, dz' \, . \tag{5.38}$$

As we expect the solution to be symmetric, we set $D_2$ in the above to zero. Applying the MOM to the right-hand side yields

$$D_1 \int_{f_m} f_m(z) \cos(kz) \, dz - \frac{j}{2\eta} \int_{f_m} f_m(z) \int_{-L/2}^{L/2} \sin(k|z-z'|) \, E_z^i(z') \, dz' \, dz \, , \tag{5.39}$$

which comprises a linear system of the form

$$\mathbf{Z}\mathbf{a} = D_1 \mathbf{s} + \mathbf{b} \, . \tag{5.40}$$

To obtain the solution vector $\mathbf{a}$, we must determine the constant $D_1$. This can be done by enforcing the boundary conditions at the ends of the wire, which are

$$I_z(-L/2) = I_z(L/2) = 0 \, . \tag{5.41}$$

We can do this this in our discretized version by forcing the basis function coefficients at the wire ends to be zero [8]. This can be expressed vectorially as $\mathbf{u}^T \mathbf{a} = 0$, where $\mathbf{u}^T = [1, 0, \ldots, 0, 1]$. Solving (5.40) for $\mathbf{a}$ we obtain

$$\mathbf{a} = D_1 \mathbf{Z}^{-1} \mathbf{s} + \mathbf{Z}^{-1} \mathbf{b} \, , \tag{5.42}$$

and multiplying both sides by $\mathbf{u}^T$ results in

$$\mathbf{u}^T \mathbf{a} = D_1 \mathbf{u}^T \mathbf{Z}^{-1} \mathbf{s} + \mathbf{u}^T \mathbf{Z}^{-1} \mathbf{b} = 0 . \tag{5.43}$$

Finally, solving for $D_1$ yields

$$D_1 = -\frac{\mathbf{u}^T \mathbf{Z}^{-1} \mathbf{b}}{\mathbf{u}^T \mathbf{Z}^{-1} \mathbf{s}} . \tag{5.44}$$

The right-hand side vector can now be assembled from $\mathbf{s}$, $\mathbf{b}$, and $D_1$.

### 5.3.1.1 Solution Using Pulse Functions and Point Matching

To solve the symmetric Hallén's equation with pulse basis functions and point matching, we subdivide the wire into $N$ equally spaced segments of length $L/N$. The matrix elements of (5.37) are then

$$Z_{mn} = \int_{z_n - \Delta_z/2}^{z_n + \Delta_z/2} \frac{e^{-jkR}}{4\pi R} dz' \tag{5.45}$$

where the matching is done at the center of the testing segment ($z = z_m$), and $R = \sqrt{(z_m - z')^2 + a^2}$. The non-self terms are computed from (5.45) using an $M$-point numerical quadrature.

*Self Terms*

When the source and testing segments overlap, we use a small-argument approximation to the Green's function to write

$$Z_{mm} = \int_{-\Delta_z/2}^{\Delta_z/2} \frac{e^{-jkR}}{4\pi R} dz' \approx \int_{-\Delta_z/2}^{\Delta_z/2} \frac{1 - jkR}{4\pi R} dz' \tag{5.46}$$

which evaluates to [9] (200.01)

$$Z_{mm} = \frac{1}{4\pi} \log \left[ \frac{\sqrt{1 + 4a^2/\Delta_z^2} + 1}{\sqrt{1 + 4a^2/\Delta_z^2} - 1} \right] - \frac{jk\Delta_z}{4\pi} \tag{5.47}$$

We assume that the approximation in (5.47) is valid when the segment is small compared to the wavelength.

*Excitation*

The elements of the right-hand side vectors $\mathbf{s}$ and $\mathbf{b}$ are

$$s_m = \cos(kz_m) \tag{5.48}$$

and

$$b_m = -\frac{j}{2\eta} \int_{-L/2}^{L/2} \sin\left(k|z_m - z'|\right) E_z^i(z')\, dz' \,, \tag{5.49}$$

where we have not yet specified the incident field $E_z^i(z')$. If we use a delta-gap source located at the center of the wire, $E_z^i(z') = \delta(z')$, and (5.49) simplifies to

$$b_m = -\frac{j}{2\eta} \sin(k|z_m|) \,. \tag{5.50}$$

If we use the magnetic frill of (5.17), the convolution of (5.49) must be performed numerically for each $z_m$. Because of the fast drop-off in the frill amplitude, each integral should be carried out carefully to ensure its accuracy. In such cases, an adaptive quadrature technique is recommended.

### 5.3.2 Asymmetric Problems

When the feedpoint is not at the center of the antenna, or the wire is excited by an incident wave, we can no longer assume the solution to be symmetric. In this case, the elements of the system matrix remain the same as in (5.37), however we will use the more general right-hand side of (5.34). Testing the right-hand side with testing function $f_m$ yields

$$C_1 \int_{f_m} f_m(z)e^{jkz}\, dz + C_2 \int_{f_m} f_m(z)e^{-jkz}\, dz$$
$$+\frac{1}{2\eta} \int_{f_m} f_m(z) \int_{-L/2}^{L/2} e^{-jk|z_m - z'|}\, E_z^i(z')\, dz'\, dz \,. \tag{5.51}$$

The resulting matrix equation can be written as

$$\mathbf{Z}\mathbf{a} = C_1\mathbf{s}_1 + C_2\mathbf{s}_2 + \mathbf{b} = [\mathbf{s}_1\mathbf{s}_2][C_1 C_2]^T + \mathbf{b} = \mathbf{SC} + \mathbf{b} \,, \tag{5.52}$$

where $\mathbf{S}$ is a $N \times 2$ matrix, and $\mathbf{C}$ is a $2 \times 1$ matrix. Solving for $\mathbf{a}$ then yields

$$\mathbf{a} = \mathbf{Z}^{-1}\left[\mathbf{SC} + \mathbf{b}\right] \tag{5.53}$$

To obtain $\mathbf{a}$, we must also solve for the constants $C_1$ and $C_2$, and as in the symmetric case, we will enforce the boundary conditions at the end of the wire. To do so, let us define the $2 \times N$ matrix

$$\mathbf{U} = \begin{bmatrix} \mathbf{u}_1^T \\ \mathbf{u}_2^T \end{bmatrix}, \tag{5.54}$$

where the row vectors $\mathbf{u}_1^T$ and $\mathbf{u}_2^T$ are

$$\mathbf{u}_1^T = [1, 0, \ldots, 0, 0] \tag{5.55}$$

and

$$\mathbf{u}_2^T = [0, 0, \dots, 0, 1] \, . \tag{5.56}$$

Multiplying both sides of (5.53) by $\mathbf{U}$ yields

$$\mathbf{Ua} = \mathbf{U}\mathbf{Z}^{-1}[\mathbf{SC} + \mathbf{b}] = 0 \, , \tag{5.57}$$

and solving for $\mathbf{C}$ results in

$$\mathbf{C} = -[\mathbf{U}\mathbf{Z}^{-1}\mathbf{S}]^{-1}\mathbf{U}\mathbf{Z}^{-1}\mathbf{b} \, . \tag{5.58}$$

#### 5.3.2.1 Solution Using Pulse Functions and Point Matching

The solution of the asymmetric Hallén's equation by point matching is similar to the symmetric case, as the system matrix elements remain the same. The only difference lies in the right-hand side vector elements.

*Excitation*

Using point matching, the right-hand side vector elements in (5.52) are

$$s_{1,m} = e^{jkz_m} \, , \tag{5.59}$$

$$s_{2,m} = e^{-jkz_m} \, , \tag{5.60}$$

and

$$b_m = -\frac{j}{2\eta} \int_{-L/2}^{L/2} e^{-jk|z_m - z'|} E_z^i(z') \, dz' \, , \tag{5.61}$$

where we have not yet specified the incident field $E_z^i(z')$. If we place a delta-gap source within a wire segment centered at $z_c$, the field $E_z^i(z') = \delta(z_c)$ and (5.61) becomes

$$b_m = -\frac{j}{2\eta} e^{-jk|z_m - z_c|} \, . \tag{5.62}$$

If instead the incident field comprises a magnetic frill or other incident field, (5.61) must be computed numerically for each $z_m$.

---

## 5.4 Pocklington's Equation

Pocklington's Equation (5.13) is relatively straightforward to solve using the Method of Moments, as the differential operator acts only on the Green's function in the innermost integral. Expanding the current into a sum of $N$

weighted basis functions, and testing with $N$ testing functions, we obtain a linear system with matrix elements given by

$$Z_{mn} = \frac{j\omega\mu}{4\pi} \int_{f_m} f_m(z) \int_{f_n} f_n(z') \left[1 + \frac{1}{k^2} \frac{\partial^2}{\partial z^2}\right] \frac{e^{-jkr}}{r} \, dz' \, dz \qquad (5.63)$$

and excitation vector elements given by

$$b_m = \int_{f_m} f_m(z) \, E_z^i(z) \, dz . \qquad (5.64)$$

### 5.4.1 Solution Using Pulse Functions and Point Matching

We again subdivide the wire into $N$ equally spaced segments of length $L/N$. Using pulse basis functions point matching at $z_m$, the first part of (5.63) is

$$\frac{j\omega\mu}{4\pi} \int_{z_n - \Delta_z/2}^{z_n + \Delta_z/2} \frac{e^{-jkR}}{R} \, dz' , \qquad (5.65)$$

where

$$R = \sqrt{(z_m - z')^2 + a^2} . \qquad (5.66)$$

We note that (5.65) has the same form as (5.45), and can be computed the same way. The second part of (5.63) can be written as

$$\frac{j}{4\pi\omega\epsilon} \int_{f_m} f_m(z) \int_{f_n} f_n(z') \frac{\partial^2}{\partial z^2} \frac{e^{-jkr}}{r} \, dz' \, dz . \qquad (5.67)$$

Evaluating the first partial derivative in the integrand yields

$$\frac{\partial}{\partial z} \frac{e^{-jkr}}{r} = -(z - z') \frac{1 + jkr}{r^3} e^{-jkr} , \qquad (5.68)$$

and inserting the above into (5.67) yields

$$\frac{j}{4\pi\omega\epsilon} \int_{f_m} f_m(z) \int_{f_n} f_n(z') \frac{\partial}{\partial z} \left[ -(z - z') \frac{1 + jkr}{r^3} e^{-jkr} \right] dz' . \qquad (5.69)$$

Using pulse basis functions point matching at $z_m$, the above evaluates to

$$\frac{j}{4\pi\omega\epsilon} \left[ (z_m - z') \frac{1 + jkR}{R^3} e^{-jkR} \right] \Big|_{z'=z_n - \Delta_z/2}^{z'=z_n + \Delta_z/2} , \qquad (5.70)$$

which allows us to write the matrix elements as

$$Z_{mn} = \frac{j\omega\mu}{4\pi} \int_{z_n - \Delta_z/2}^{z_n + \Delta_z/2} \frac{e^{-jkR}}{R} \, dz'$$

$$+ \frac{j}{4\pi\omega\epsilon} \left[ (z_m - z') \frac{1 + jkR}{R^3} e^{-jkR} \right] \Big|_{z'=z_n - \Delta_z/2}^{z'=z_n + \Delta_z/2} . \qquad (5.71)$$

The second term is analytic and can be used as-is in computing each matrix element. Because of the strongly singular $1/R^3$ term, we will find that our results converge less quickly than those obtained using Hallén's equation for the same problem. The corresponding right-hand side vector elements are

$$b_m = E_z^i(z_m) \, . \tag{5.72}$$

## 5.5 Thin Wires of Arbitrary Shape

The formulations considered thus far treated wires that are perfectly straight. Because realistic antennas have curves, bends, and wire-to-wire junctions, we need to develop a thin wire treatment that takes on a more general form. We will do so in this section.

### 5.5.1 Method of Moments Discretization

Consider a curved thin wire, illustrated in Figure 5.3a. We subdivide the curve into $N$ segments with $N+1$ endpoints as shown in Figure 5.3b. Though the segments need not be of equal length, they should be small enough to reflect the curvature of the wire, and should support the required number of basis functions per wavelength. The tangent vector on the curve is now a piecewise continuous function, as is illustrated with the arrows. As the wire supports only electric currents, we will enforce the EFIE (3.178). The filamentary current $\mathbf{J}(\mathbf{r})$ is expanded using electric basis functions $\mathbf{f}(\mathbf{r})$ whose vector components are tangent to the segment(s) on which it resides. Testing the EFIE with these

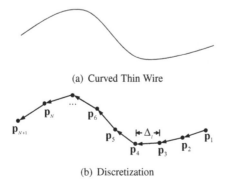

(a) Curved Thin Wire

(b) Discretization

**FIGURE 5.3:** Curved Thin Wire Discretization

same functions yields the matrix elements (3.185) for the thin wire case, which
are

$$Z_{mn}^{EJ} = j\omega\mu \int_{\mathbf{f}_m} \mathbf{f}_m(\mathbf{r}) \cdot \int_{\mathbf{f}_n} \mathbf{f}_n(\mathbf{r}') \, G(\mathbf{r}, \mathbf{r}') \, d\mathbf{r}' \, d\mathbf{r}$$

$$- \frac{j}{\omega\epsilon} \int_{\mathbf{f}_m} \nabla \cdot \mathbf{f}_m(\mathbf{r}) \int_{\mathbf{f}_n} \nabla' \cdot \mathbf{f}_n(\mathbf{r}') \, G(\mathbf{r}, \mathbf{r}') \, d\mathbf{r}' \, d\mathbf{r} \,, \qquad (5.73)$$

where we have re-distributed the differential operators following (3.204). The
right-hand side vector elements (3.191) are

$$V_m^E = \int_{\mathbf{f}_m} \mathbf{f}_m(\mathbf{r}) \cdot \mathbf{E}^i(\mathbf{r}) \, d\mathbf{r} \,. \qquad (5.74)$$

We will refer to the approach in this section as a thin-wire EFIE. Similar approaches are found in the literature.

### 5.5.2  Solution Using Triangle Basis and Testing Functions

We will use triangle functions to solve the thin-wire EFIE. As the current must go to zero on the ends of the wire, we assign triangle functions
everywhere except at the wire endpoints. Also, as the basis functions have
vector components, they must be oriented properly. Consider a triangle function which has support on segments $s+$ and $s-$ whose endpoints comprise
$(\mathbf{p}_{k-1}, \mathbf{p}_k, \mathbf{p}_{k+1})$, as shown in Figure 5.4. The vector components are $\mathbf{t}^+$ on
$s+$, and $\mathbf{t}^-$ on $s-$. Although the vector orientations are shown going right to
left, it is equally valid to orient them from left to right; the only requirement is
that the orientations satisfy Kirchoff's law at $p_k$.

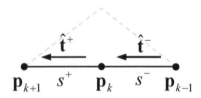

**FIGURE 5.4:** Vector Triangle Function Orientation

### 5.5.2.1 Non-Self Terms

For segments that do not overlap, the matrix elements (5.73) are computed using an $M$-point numerical quadrature formula, yielding

$$Z_{mn}^{EJ} = \frac{1}{4\pi} \sum_{p=1}^{M} \sum_{q=1}^{M} w_p(\mathbf{r}_p) w_q(\mathbf{r}_q') \left[ j\omega\mu\, \mathbf{f}_m(\mathbf{r}_p)\cdot\mathbf{f}_n(\mathbf{r}_q') \pm \frac{j}{\omega\epsilon\Delta_p\Delta_q} \right] \frac{e^{-jkR_{pq}}}{R_{pq}} ,$$
(5.75)

where $\mathbf{r}_p$ and $\mathbf{r}_q$ are quadrature points along the wire axis, and

$$R_{pq} = |\mathbf{r}_p - \mathbf{r}_q'| .$$
(5.76)

The sums in (5.75) are performed over all segments where the source and testing functions reside. For a triangle function that flows into or out of its anchor point on a segment of length $\Delta_l$, its divergence is $1/\Delta_l$ or $-1/\Delta_l$ on that segment, respectively. The sign of the second term then depends on the orientation of the triangle functions on the source and testing segments. Note that it is more efficient to compute the matrix elements by looping over pairs of segments instead of pairs of basis and testing functions.

### 5.5.2.2 Self Terms

For overlapping segments, we will evaluate the innermost integral analytically and the outermost integral numerically. To calculate the first term on the right in (5.73), consider the following innermost integral with the triangle function of positive slope $x'/\Delta_l$

$$S_1(x) = \int_0^{\Delta_l} \frac{x'}{\Delta_l} \frac{e^{-jkr}}{r} \, dx' ,$$
(5.77)

where $r = \sqrt{(x-x')^2 + a^2}$. Using the small-argument approximation to the Green's function this becomes

$$\int_0^{\Delta_l} \frac{x'}{\Delta_l} \frac{e^{-jkr}}{r} \, dx' \approx \int_0^{\Delta_l} \frac{x'}{\Delta_l} \frac{1 - jkr}{r} \, dx' ,$$
(5.78)

which is

$$\int_0^{\Delta_l} \frac{x'}{\Delta_l} \frac{1 - jkr}{r} \, dx' = \frac{1}{\Delta_l} \int_0^{\Delta_l} \frac{x'}{r} \, dx' - \frac{jk\Delta_l}{2} .$$
(5.79)

The first term on the right then evaluates to [10]

$$\frac{1}{\Delta_l}\sqrt{a^2 + (x - \Delta_l)^2} - \frac{1}{\Delta_l}\sqrt{a^2 + x^2} + \frac{x}{\Delta_l} \log\left[ \frac{x + \sqrt{a^2 + x^2}}{x - \Delta_l + \sqrt{a^2 + (x - \Delta_l)^2}} \right] ,$$

resulting in

$$S_1(x) = \frac{1}{\Delta_l}\sqrt{a^2 + (x - \Delta_l)^2} - \frac{1}{\Delta_l}\sqrt{a^2 + x^2}$$

$$+ \frac{x}{\Delta_l}\log\left[\frac{x + \sqrt{a^2 + x^2}}{x - \Delta_l + \sqrt{a^2 + (x - \Delta_l)^2}}\right] - \frac{jk\Delta_l}{2}. \quad (5.80)$$

The integral involving the triangle of positive slope is sufficient to obtain all the self terms because of symmetry. We will compute the second term on the right in (5.73) the same way as the first term. The innermost integral is

$$S_2(x) = \pm\frac{1}{\Delta_l^2}\int_0^{\Delta_l}\frac{1 - jkr}{r}\,dx', \quad (5.81)$$

where we have again used the small-argument approximation to the Green's function, and the sign of the integral depends on whether the derivatives of the source and testing functions are of opposing sign. The above evaluates to [10]

$$S_2(x) = \pm\frac{1}{\Delta_l^2}\left[\log\left[\frac{x + \sqrt{a^2 + x^2}}{x - \Delta_l + \sqrt{a^2 + (x - \Delta_l)^2}}\right] - jk\Delta_l\right]. \quad (5.82)$$

The self-term contributions to (5.73) can then be obtained via numerical quadrature as

$$\frac{1}{4\pi}\sum_{p=1}^{M}w_p(x_p)\left[j\omega\mu f_m(x_p)\,S_1(x_p) - \frac{j}{\omega\epsilon}S_2(x_p)\right]. \quad (5.83)$$

### 5.5.3 Solution Using Sinusoidal Basis and Testing Functions

The solution using sinusoidal basis and testing functions is practically the same as with triangle functions. The difference lies in the self terms, which we compute using numerical outer and analytic inner integrations as before.

#### 5.5.3.1 Self Terms

To calculate the first term on the right in (5.73), consider the innermost integral having a sinusoid function of positive slope, given by

$$S_1(x_p) = \frac{1}{\sin(k\Delta_l)}\int_0^{\Delta_l}\sin(kx')\frac{e^{-jkr}}{r}\,dx'. \quad (5.84)$$

Again using the small-argument approximation to the Green's function this becomes

$$\int_0^{\Delta_l}\sin(kx')\frac{e^{-jkr}}{r}\,dx' \approx \int_0^{\Delta_l}\sin(kx')\left[\frac{1 - jkr}{r}\right]dx', \quad (5.85)$$

which is

$$\int_0^{\Delta_l} \sin(kx') \left[ \frac{1 - jkr}{r} \right] dx' = \int_0^{\Delta_l} \frac{\sin(kx')}{r} dx' + j \left[ \cos(k\Delta_l) - 1 \right] .$$
(5.86)

The first term on the right is not tractable in its present form, so we will approximate the sine by its third-order polynomial approximation. This yields

$$\int_0^{\Delta_l} \frac{\sin(kx')}{r} dx' \approx \int_0^{\Delta_l} \frac{kx' - (kx')^3/6}{\sqrt{(x' - x_p)^2 + a^2}} dx' .$$
(5.87)

Evaluating the above results in [10]

$$S_1(x_p) = C \frac{kx_p}{12} \Bigg( \left[ -3a^2 k^2 + 2x_p^2 k^2 - 12 \right] \log \left[ x_p - x' + \sqrt{a^2 + (x_p - x')^2} \right]$$
$$- \frac{k}{36} \left[ -4a^2 k^2 + k^2 \left( 11 x_p^2 + 5x' x_p - 2(x')^2 \right) - 36 \right] \cdot$$
$$\sqrt{a^2 + (x_p - x')^2} \Bigg) \Bigg|_0^{\Delta_l} + jC \left[ \cos(k\Delta_l) - 1 \right] ,$$
(5.88)

where $C = 1/\sin(k\Delta_l)$. This result is lengthy but can be evaluated numerically in a straightforward manner. We compute the second term on the right of (5.73) the same way. Under the small-argument approximation to the Green's function, the innermost integral in this case is

$$S_2(x_p) = \frac{k}{\sin(k\Delta_l)} \int_0^{\Delta_l} \frac{\cos(kx')}{r} dx' - jk .$$
(5.89)

We approximate the cosine by the second-order approximation, yielding the integral

$$\int_0^{\Delta_l} \frac{\cos(kx')}{r} dx' \approx \int_0^{\Delta_l} \frac{1 - (kx')^2/2}{\sqrt{(x' - x_p)^2 + a^2}} dx' .$$
(5.90)

The above evaluates to [10]

$$S_2(x_p) = -\frac{C}{4} \Bigg( k^2 \left[ 3x_p + x' \right] \sqrt{a^2 + (x_p - x')^2} + (a^2 k^2 - 2x_p^2 k^2 + 4)$$
$$\log \left[ x_p - x' + \sqrt{a^2 + (x_p - x')^2} \right] \Bigg) \Bigg|_0^{\Delta_l} - jk .$$
(5.91)

The contributions to the self terms can then be computed via numerical quadrature as

$$\frac{1}{4\pi} \sum_{p=1}^{M} w_p(x_p) \left[ j\omega\mu \, s_m(x_p) S_1(x_p) - \frac{j}{\omega\epsilon} s_m'(x_p) S_2(x_p) \right] ,$$
(5.92)

where $s_m(x)$ is the sinusoidal testing function.

### 5.5.4   Lumped and Distributed Impedances

In some cases it may be necessary to modify the impedance characteristics of an antenna by loading it at one or more points. Feedpoint matching is one of the most common adjustments, and is done by placing an impedance in parallel at the feedpoint, or by inserting coils or capacitors at one or more points along the wire. To model these lumped impedances in the thin wire EFIE, we must take into account the boundary conditions on the segments where the loads exist. Given a complex impedance $Z_l$ on a segment of length $\Delta_l$, the boundary condition on this segment is

$$E_{tan}^i + E_{tan}^s = E_{load} = \frac{V_l}{\Delta_l} = \frac{Z_l I_s}{\Delta_l} \, . \tag{5.93}$$

The right-hand side of the EFIE now becomes

$$\frac{Z_l I_l}{\Delta_l} - E_{tan}^i \, , \tag{5.94}$$

where the current $I_l$ on the segment is supported by one or more basis functions. If we use triangle functions in the thin wire EFIE, the testing operation with the basis function $f_m$ creates a "self term" on the right-hand side of the form

$$b_m' = \frac{Z_l}{2} \, , \tag{5.95}$$

which can then be moved to the left-hand side and subtracted from the diagonal element $Z_{mm}^{EJ}$. Any basis function having support on this segment will have its self term modified in this manner.

## 5.6 Examples

In this section we consider several thin wire antenna problems. We first compare the relative performance of the Hallén, Pocklington, and thin-wire EFIE models in computing the input impedance of a straight wire. We then consider a half-wavelength dipole, circular loop, folded dipole, and a multi-element Yagi, where we use the thin-wire EFIE approach from Section 5.5.

### 5.6.1 Comparison of Thin Wire Models

For this comparison, we will compute the input impedance and induced current distribution on center-fed dipole antennas. To judge the effectiveness of each method, we compare the rates of convergence versus the number of wire segments used in the discretization. We use a delta-gap source of 1V and an odd number of wire segments so the voltage source is positioned exactly at the center of the antenna. For numerical integration, a 5-point Gaussian quadrature is used on each segment. Each dipole has a radius of $10^{-3}\lambda$, and in computing the induced current distributions, the amplitude of the applied voltage is unity.

#### 5.6.1.1 Input Impedance

First, we will compare the convergence of the input impedance versus the number of wire segments used. We first consider a dipole of length $\lambda/2$, where the resistance and reactance are plotted in Figures 5.5a and 5.5b, respectively. The results from the thin wire EFIE models are seen to converge very quickly, whereas those from Hallén's and Pocklington's equation converge to similar values, but much more slowly. This is not surprising, as triangle and sinusoidal functions are better at modeling the current distribution than pulses, and they enforce the boundary conditions more accurately than with point matching. Similar comparisons for a $3\lambda/2$ dipole are made in Figures 5.6a and 5.6b, respectively. The results from the thin wire EFIE models again converge quickly, whereas Pocklington's Equation performs quite poorly at low segment counts. This clearly demonstrates the limitations of pulse basis functions and point matching, as well as the $1/R^3$ singularity in the integrand. This singularity would be compounded even further if the radius of the wire were made smaller.

We next set the number of segments per wavelength to 25, and vary the length of the dipole from 0.1 to $3\lambda$. The results are shown in Figure 5.7. The thin-wire EFIE results agree very well at almost every data point, suggesting that triangles and sinusoids have similar performance characteristics for thin wire problems. The impedances obtained from Hallén's and Pocklington's Equation do not agree very well at this level of discretization, as expected.

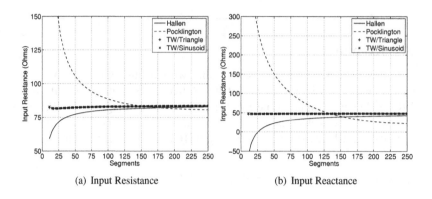

(a) Input Resistance                    (b) Input Reactance

**FIGURE 5.5:** Input Impedance of a $\lambda/2$ Dipole

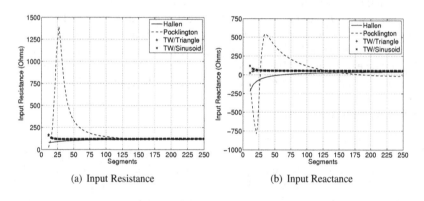

(a) Input Resistance                    (b) Input Reactance

**FIGURE 5.6:** Input Impedance of a $3\lambda/2$ Dipole

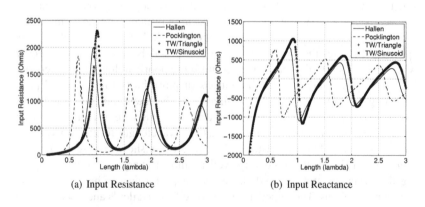

(a) Input Resistance                    (b) Input Reactance

**FIGURE 5.7:** Input Impedance versus Dipole Length

### 5.6.1.2   Induced Current Distribution

We next compare the currents induced on a center-fed dipole computed using each thin wire model. For the comparison, we plot the current at the center of each wire segment. Note that the solution via Hallén's equation will yield exactly zero current in the first and last segments, as this was enforced explicitly. In Figures 5.8a and 5.8b are shown the currents induced on a $\lambda/2$ dipole computed using 31 and 81 wire segments, respectively. The thin-wire EFIE models show almost no change between the two figures indicating excellent convergence, whereas the currents obtained from Hallén's and Pocklington's equation vary significantly. Results for a $2\lambda$ dipole are shown in Figures 5.8c and 5.8d for 81 and 181 wire segments, respectively. Hallén's equation does somewhat better at the lower segment count than does Pocklington's equation, though they are both fairly good at the higher count. The convergence of the thin-wire EFIE models is again excellent.

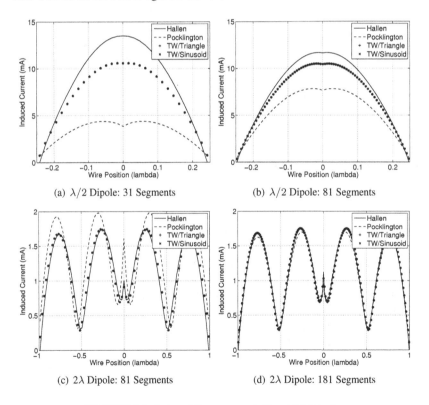

(a) $\lambda/2$ Dipole: 31 Segments          (b) $\lambda/2$ Dipole: 81 Segments

(c) $2\lambda$ Dipole: 81 Segments          (d) $2\lambda$ Dipole: 181 Segments

**FIGURE 5.8:** Induced Currents on $\lambda/2$ and $2\lambda$ Dipoles

### 5.6.2　Half-Wavelength Dipole

It is known that a dipole in free space has its first resonance at a length just under $\lambda/2$. Using the thin-wire EFIE, a delta-gap source of 1V and 35 wire segments, we find via iteration that a dipole of radius $10^{-3}\lambda$ and length of approximately $0.475\lambda$ has a purely resistive input impedance of 72 $\Omega$. This antenna could be fed directly with a 50 $\Omega$ or 75 $\Omega$ coaxial cable (RG-58 or RG-59, respectively) and have a reasonably good standing wave ratio (SWR).

We now set the feedpoint at the origin and align the antenna with the $z$ axis. Using the induced current from the delta-gap source, we compute the near field in a $20\lambda \times 20\lambda$ region of the $xz$ plane using (3.108). The real part of the electric field (in V/m) is plotted in Figure 5.9. We note that the near field region is small in extent, and the field amplitude decays quickly along the $z$ axis.

Now we will compare the computed field pattern to that of an ideal dipole from antenna theory. We recall that the radiated electric field of an infinitesimal, $\hat{\mathbf{z}}$-oriented electric current has $\hat{\mathbf{r}}$ and $\hat{\boldsymbol{\theta}}$ components given by

$$E_r(r,\theta) = -\frac{k^2\eta I\Delta_l}{2\pi}\left[\frac{1}{(jkr)^2} + \frac{1}{(jkr)^3}\right]\cos\theta\, e^{-jkr} \qquad (5.96)$$

and

$$E_\theta(r,\theta) = -\frac{k^2\eta I\Delta_l}{2\pi}\left[\frac{1}{jkr} + \frac{1}{(jkr)^2} + \frac{1}{(jkr)^3}\right]\sin\theta\, e^{-jkr}, \qquad (5.97)$$

where $I$ is the current amplitude and $\Delta_l$ is its length, where $\Delta_l \ll \lambda$. If we know $I(z)$, we can then integrate over the length of the dipole to determine the total near field. For the dipole, we will use an ideal sinusoidal distribution model for current, where

$$I(z) = I_0 \sin\left[k(L/2 - z)\right]. \qquad (5.98)$$

For $I_0$, we use the numerically computed feedpoint current, and $L = 0.475\lambda$ as before.

In Figure 5.10 is plotted the real part of the near electric field (in V/m) in the same region, computed using (5.96–5.98). To the eye, the results are nearly indistinguishable. In Figure 5.11, we compare the computed currents to those of (5.98), where the currents have slightly different amplitude envelopes but are otherwise very similar.

Finally, we sample the computed near field pattern in the $xz$ plane at a fixed distance of $10\lambda$ from the antenna for $0 \leq \theta \leq 2\pi$. To this we compare the normalized far electric field of an ideal $\lambda/2$ dipole, which is [11]

$$E_\theta(\theta) = \frac{\cos\left(\frac{\pi}{2}\cos\theta\right)}{\sin\theta}. \qquad (5.99)$$

The normalized amplitudes are compared in Figure 5.12, where the agreement is quite good.

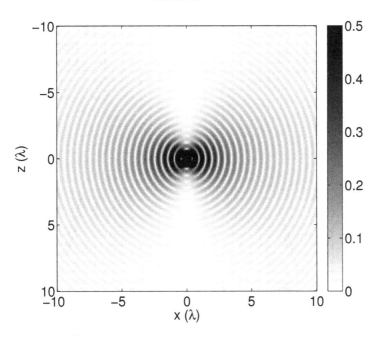

**FIGURE 5.9:** Near Field ($|\text{Re}(\mathbf{E})|$, V/m) of a $0.475\lambda$ Dipole (MoM)

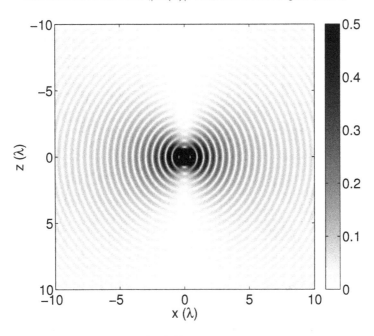

**FIGURE 5.10:** Near Field ($|\text{Re}(\mathbf{E})|$, V/m) of a $0.475\lambda$ Dipole (Equation)

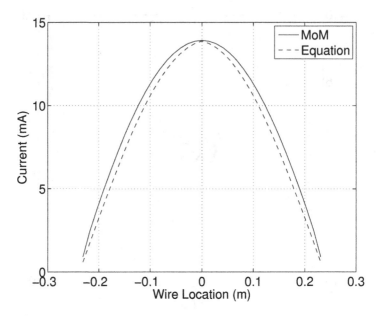

**FIGURE 5.11:** Comparison of Current Amplitudes on the 0.475λ Dipole

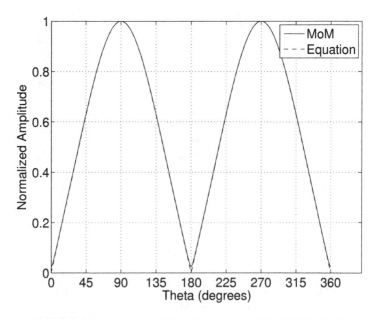

**FIGURE 5.12:** Normalized Far Field Patterns of the 0.475λ Dipole

### 5.6.3 Circular Loop Antenna

We next consider the input impedance of a circular loop antenna in the $xy$ plane. For a wire of radius $10^{-4}\lambda$, we vary the circumference of the loop up to $2.5\lambda$. At each step, we use approximately 10 segments per wavelength, with no fewer than 35 segments at the smallest length. This ensures that the model retains the shape of a circle at the lowest frequency. The feedpoint of the antenna is placed at the point where the $+x$ axis intersects the loop, and a delta-gap excitation is used. The real and imaginary parts of the input impedance are shown in Figure 5.13. For lengths under $0.5\lambda$, we see that the loop has a high inductance and virtually no radiation resistance. At approximately $\lambda/2$, the loop becomes an open circuit. We find first and second resonances at approximately $1.03\lambda$ and $2.05\lambda$, where the resistance is about 140 $\Omega$ and 182 $\Omega$, respectively.

Using the induced current on the $2.05\lambda$ loop antenna, we compute the near field in a $20\lambda \times 20\lambda$ region of the $xy$ plane via (3.108). The real part of the electric field (in V/m) is plotted in Figure 5.14. The corresponding normalized field amplitude at a distance of $10\lambda$ is shown versus $\phi$ in Figure 5.15. Four nulls are visible, and a slight asymmetry in the pattern is seen. Similar plots of the near field in the $xz$ plane are shown in Figures 5.16 and 5.17, where the pattern is significantly different. To better visualize the overall radiation pattern, we compute the normalized radiated field for all $\theta$ and $\phi$ on the unit sphere at a distance of $10\lambda$, which is shown in Figure 5.18.

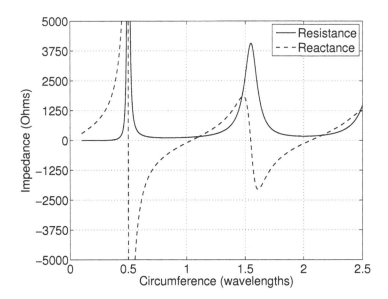

**FIGURE 5.13:** Circular Loop Input Impedance

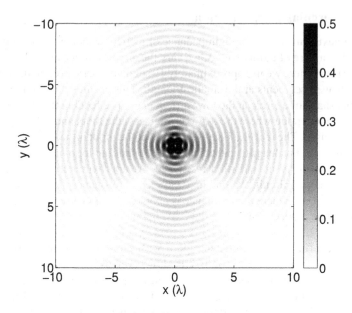

**FIGURE 5.14:** Near Field ($|\text{Re}(\mathbf{E})|$, V/m) of the Circular Loop ($xy$ plane)

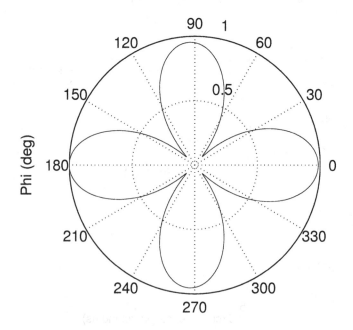

**FIGURE 5.15:** Radiated Field Pattern of the Circular Loop ($xy$ plane)

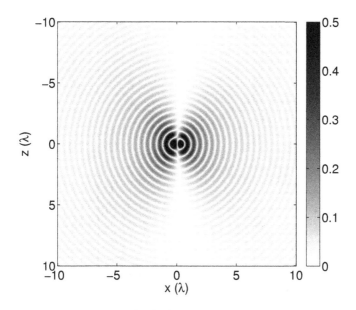

**FIGURE 5.16:** Near Field ($|\mathrm{Re}(\mathbf{E})|$, V/m) of the Circular Loop ($xz$ plane)

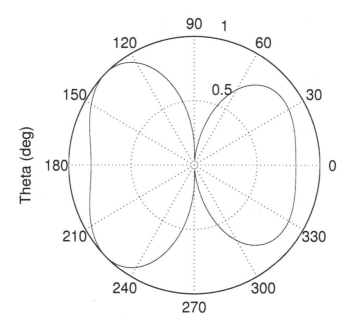

**FIGURE 5.17:** Radiated Field Pattern of the Circular Loop ($xz$ plane)

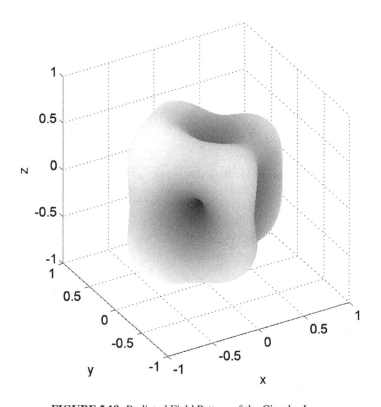

**FIGURE 5.18:** Radiated Field Pattern of the Circular Loop

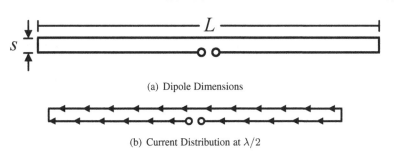

(a) Dipole Dimensions

(b) Current Distribution at $\lambda/2$

**FIGURE 5.19:** Folded Dipole Antenna

## 5.6.4 Folded Dipole Antenna

The folded dipole is an antenna commonly used in HF and VHF systems. It gets this name at it is a dipole that is essentially "folded over on itself," comprising a square loop that is very short in one dimension. This is illustrated in Figure 5.19a, where the loop has a length of $L$ and width of $s$, and a feedpoint in the center of one of the long sides. At DC, this antenna is a short circuit, but as its length increases the current distribution changes, and when $L \approx \lambda/2$ it assumes the characteristics of Figure 5.19b. In this case, the currents on the secondary side of the loop take on the same magnitude and phase as those on the primary side. As $s$ is very small, the far radiated field remains much the same, however the current is divided evenly between the two halves. Therefore, the input impedance of the folded dipole is four times that of an ordinary dipole at its first resonance, or approximately 288 $\Omega$. Folded dipoles are often used because of this property, as they can be matched to open-wire feed lines having a much higher impedance than coaxial cable.

In Figure 5.20 is plotted the input impedance of a folded dipole for $L \leq 1\lambda$, where $s = L/20$ and the radius of the wire is $10^{-4}\lambda$. We use 30 segments for each long conductor and 5 segments for the short ends, and a delta-gap source is used. We observe that when the dipole is electrically small it is similar to the circular loop, having virtually no radiation resistance, and at a length of about $0.3\lambda$ it is an open circuit. The first resonance occurs at a length of about $0.46\lambda$ where the input resistance is 284 $\Omega$, close to the expected value.

We next place the $0.46\lambda$ folded dipole at the origin, oriented in the $xz$ plane. Using a delta-gap source of 1V, we then compute the radiated near field in the $xz$ plane, where the real part of the electric field (in V/m) is plotted in Figure 5.21. It is similar, but not identical, to the near field pattern of the $0.475\lambda$ dipole observed in Figure 5.9.

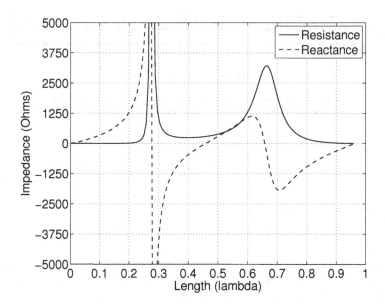

**FIGURE 5.20:** Folded Dipole Input Impedance

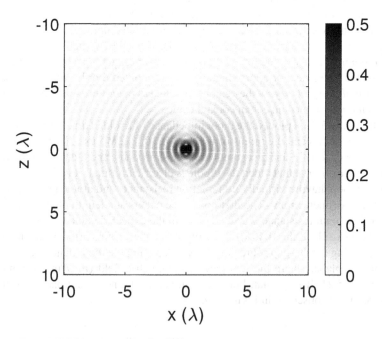

**FIGURE 5.21:** Near Field ($|\mathrm{Re}(\mathbf{E})|$, V/m) of the Folded Dipole ($xz$ plane)

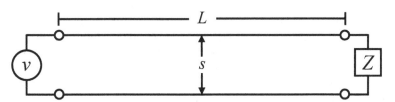

**FIGURE 5.22:** Two-Wire Transmission Line

## 5.6.5 Two-Wire Transmission Line

As transmission lines are implicit in virtually all antenna problems, we will next examine the behavior of a common two-wire transmission line illustrated in Figure 5.22, Two-wire line (also called "ladder line") is typically constructed of two multistranded copper wires, which are held a precise distance apart by a vinyl sheath or flat plastic ribbon. In two-wire line, the ribbon often has rectangular holes cut into it at regular intervals for weight reduction, giving it the appearance of a ladder. For the purposes of simulation, we will ignore the presence of a sheath or ribbon, and treat the two-wire line in an identical manner to the folded dipole of Section 5.6.4, except that the feedpoint and load are placed at the short ends. We will analyze a two-wire line 2.5 meters long, constructed of wires 1 mm in radius having a separation distance of 3 cm. We use 30 segments for the main wires and 5 segments for each end.

Our first task is to determine the characteristic impedance of the line. The theoretical value for a two-wire line is obtained from transmission line theory [8] and is

$$Z_o = \sqrt{L_0/C_0} \tag{5.100}$$

where $L_0$ is the inductance per unit length

$$L_O = \frac{\mu_0}{\pi} \cosh^{-1}\left(\frac{s}{2a}\right), \tag{5.101}$$

and $C_0$ is the capacitance per unit length

$$C_O = (\pi\epsilon_0)/\cosh^{-1}\left(\frac{s}{2a}\right), \tag{5.102}$$

where $s$ is the distance between wire centers, and $a$ is the radius of each wire. Using (5.100–5.102), we find that the impedance of our line is approximately $Z_o = 407\ \Omega$. The true impedance can be determined numerically as

$$Z_o = \sqrt{Z_{oc}Z_{sc}}, \tag{5.103}$$

where $Z_{oc}$ and $Z_{sc}$ are the input impedances with an opened and shorted load end, respectively. We must be careful, however, that we do not take the measurement at a frequency where the length of the line would result in an extremely small or large impedance at the input end, as numerical inaccuracy would result. We therefore make the measurement at a frequency of 15 MHz, where the length of the line is approximately $\lambda/8$. Using the thin-wire EFIE with a delta-gap source, we find the impedance to be $Z_o = 405 \, \Omega$, very close to the theoretical value.

We next compare the computed input impedance to that of an ideal lossless transmission line, which is [8]

$$Z_{in} = Z_o \frac{Z_l + j Z_o \tan(k_0 l)}{Z_o + j Z_l \tan(k_0 l)} \, , \qquad (5.104)$$

where $Z_l$ is the load impedance, and $Z_{in}$ the impedance measured at a distance $l$ from the load. For this comparison, we use a load with a pure resistance of 200 $\Omega$. The results are plotted in Figure 5.23 for frequencies between 100 kHz and 100 MHz. The comparison is quite good. At lower frequencies, the input impedance tends toward the DC value, as expected.

Two-wire lines are only useful at frequencies where they are not effective radiators, which is typically satisfied when $s \ll \lambda$. Let us retain the 200 $\Omega$ load, and place the line in the $xy$ plane with the feedpoint on the $-x$ axis. We then compute the real part of the radiated near electric field (in V/m) at 15 MHz in the $xz$ plane, which is shown in Figure 5.24.

We note that even though the load is not matched and there are standing waves on the line, the fields are concentrated in a small, tight region surrounding the line, with virtually no field radiated. Increasing the frequency to 500 MHz, we see in Figure 5.25 that the near field is no longer confined close to the line. Increasing the frequency further to 1 GHz, we observe in Figure 5.26 that the line is now clearly behaving as an antenna, making it unsuitable as a transmission line at this frequency.

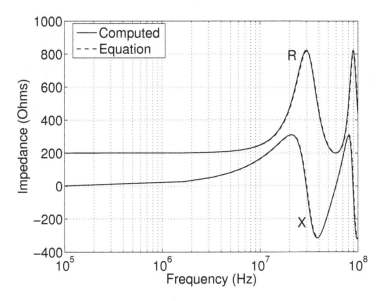

**FIGURE 5.23:** Two-Wire Line Input Impedance

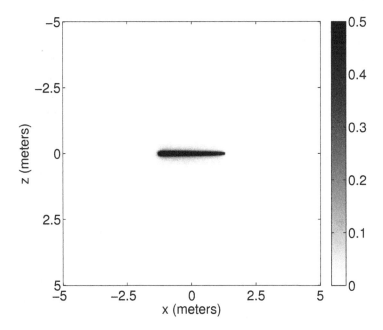

**FIGURE 5.24:** Near Field ($|\mathrm{Re}(\mathbf{E})|$, V/m) of the Two-Wire Line at 15 MHz ($xz$ plane)

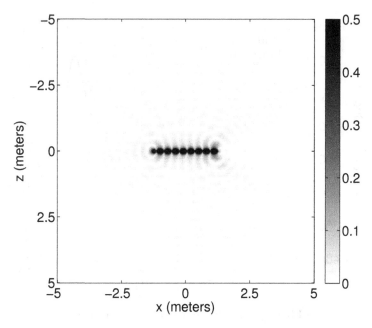

**FIGURE 5.25:** Near Field ($|\mathrm{Re}(\mathbf{E})|$, V/m) of the Two-Wire Line at 500 MHz ($xz$ plane)

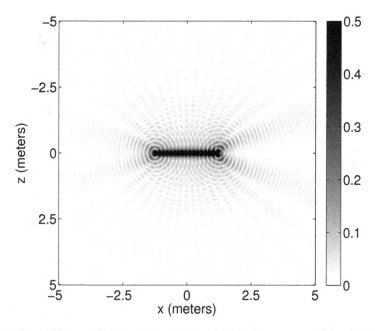

**FIGURE 5.26:** Near Field ($|\mathrm{Re}(\mathbf{E})|$, V/m) of the Two-Wire Line at 1 GHz ($xz$ plane)

### 5.6.6   Yagi Antenna for 146 MHz

We now examine a more practical antenna: a Yagi antenna for the 2-meter amateur radio band (144–148 MHz). The dimensions and locations for the antenna elements are from [12] and are summarized in Table 5.1. The diameter for each element is 6.35 mm (1/4 inch). Our thin-wire EFIE model comprises separate straight wires in the $xz$ plane, as shown in Figure 5.27. Each wire is assigned 25 segments. The feedpoint with delta-gap source is at the center of the *driven element*, noted with a dot in Figure 5.27. Though a real Yagi antenna would use a piece of metal or plastic tubing called a *boom* to affix the antenna elements, the boom does not typically have much current flow on it and we omit it from our analysis.

Most Yagi antennas have a single element placed behind the driven element called a *reflector*. This element reduces the amount of power radiated along the $-x$ axis in Figure 5.27 (the "back" side). The elements to the right of the driven element are called *directors*, and they act to focus or *direct* most of the power along the $+x$ axis (the "front" side). Though these elements are not connected to the feedline, they interact with the other elements via the near field, and are often referred to as *parasitic* elements.

The input resistance and reactance of the Yagi antenna over the band are shown in Figures 5.28a and 5.28b, respectively. The antenna has a feedpoint resistance ranging from 18 to 33 $\Omega$ over the frequency band, and a capacitive reactance, which is typical of Yagi antennas. This antenna can be matched to a 50 $\Omega$ coaxial feed using a lumped inductive element at the feedpoint (a coil) or

**TABLE 5.1:** 2 Meter Yagi Antenna Parameters

| Element | REF | DE | D1 | D2 | D3 | D4 | D5 | D6 | D7 |
|---|---|---|---|---|---|---|---|---|---|
| Length (mm) | 1038 | 955 | 956 | 932 | 916 | 906 | 897 | 891 | 887 |
| $x$ location (mm) | 0 | 312 | 447 | 699 | 1050 | 1482 | 1986 | 2553 | 3168 |

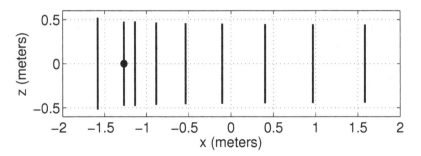

**FIGURE 5.27:** Yagi Antenna For the 2 Meter Amateur Radio Band

a tuning stub such as the hairpin match [13] and a 2:1 balun. We will match the antenna at about 146.5 MHz, where its reactance is approximately $-j18.8\ \Omega$. To achieve the match we follow Section 5.28, and place an inductance of about 20 nH at the feedpoint, which will tune out the capacitive reactance. The reactance after matching is shown in Figure 5.28c, and the SWR on a $Z_0 = 50\ \Omega$ line with 2:1 balun is shown in Figure 5.28d. The SWR is computed as [8]

$$\text{SWR} = \frac{1 + |\Gamma_L|}{1 - |\Gamma_L|}, \tag{5.105}$$

where $\Gamma_L$ is the reflection coefficient at the load, given by

$$\Gamma_L = \frac{Z_L - Z_0}{Z_L + Z_0}. \tag{5.106}$$

To account for the 2:1 balun, we use double the computed feedpoint impedance for the load impedance $Z_L$ in (5.106). We next compute the front-to-back ratio, which comprises the ratio of the forward-broadside to rear-broadside radiated power. In the far field, the power density is computed using the electric field **E** as

$$P = \frac{1}{2\eta_0}|\mathbf{E}|^2. \tag{5.107}$$

The results are plotted in dB in Figure 5.28e. Finally, we compute the forward power gain (relative to isotropically radiated power), which comprises the ratio of the broadside radiated power density to the average radiated power density $P_{avg}$. This average is computed as [8]

$$P_{avg} = \frac{P_{in}}{4\pi} = \frac{1}{4\pi}\frac{1}{R_{in}}. \tag{5.108}$$

where for a delta-gap source of 1V, $P_{in} = V^2/R_{in} = 1/R_{in}$. The results are plotted in dB in Figure 5.28f, where the forward gain remains fairly constant over the operating band.

We next consider the radiated near field in the $xy$ plane. The real part of the total radiated near electric field in a $40 \times 40$ meter region around the antenna is shown in Figure 5.29. The $\hat{\boldsymbol{\theta}}$-polarized far radiated field pattern is shown versus $\phi$ in Figure 5.30. A similar plot of the near field in the $xz$ plane is shown in Figure 5.31. The corresponding $\hat{\boldsymbol{\theta}}$-polarized far radiated field pattern is shown versus $\theta$ in Figure 5.32. The normalized total far field pattern on the unit sphere for all $\theta$ and $\phi$ is shown in Figure 5.33.

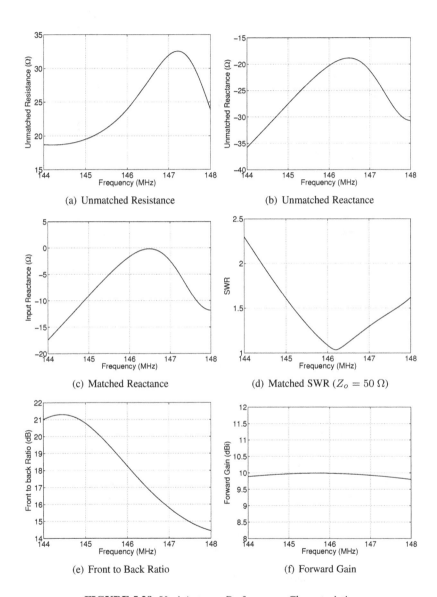

(a) Unmatched Resistance

(b) Unmatched Reactance

(c) Matched Reactance

(d) Matched SWR ($Z_o = 50\ \Omega$)

(e) Front to Back Ratio

(f) Forward Gain

**FIGURE 5.28:** Yagi Antenna Performance Characteristics

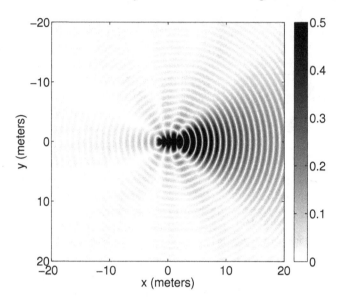

**FIGURE 5.29:** Near Field ($|\text{Re}(\mathbf{E})|$, V/m) of Yagi Antenna ($xy$ plane)

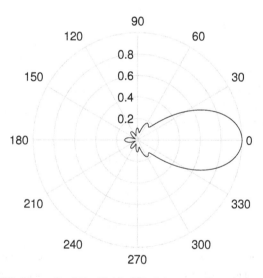

**FIGURE 5.30:** Normalized Far Field of Yagi Antenna ($0 \leq \phi \leq 2\pi$, $xy$ plane)

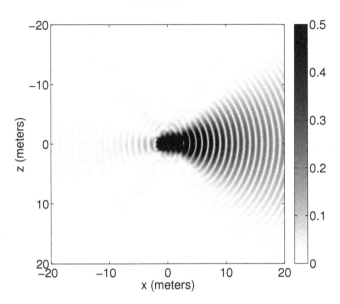

**FIGURE 5.31:** Near Field ($|\mathrm{Re}(\mathbf{E})|$, V/m) of Yagi Antenna ($xz$ plane)

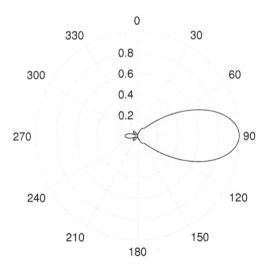

**FIGURE 5.32:** Normalized Far Field of Yagi Antenna ($0 \leq \theta \leq 2\pi$, $xz$ plane)

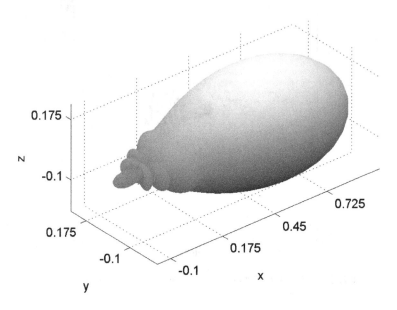

**FIGURE 5.33:** 3D Far Field Pattern of Yagi Antenna

---

## References

[1] R. W. P. King, "The linear antenna – Eighty years of progress," *Proc. IEEE*, vol. 55, pp. 2–26, January 1967.

[2] W. Wang, "The exact kernel for cylindrical antenna," *IEEE Trans. Antennas Propagat.*, vol. 39, pp. 434–435, April 1991.

[3] D. R. Wilton and N. J. Champagne, "Evaluation and integration of the thin wire kernel," *IEEE Trans. Antennas Propagat.*, vol. 54, pp. 1200–1206, April 2006.

[4] E. Hallén, "Theoretical investigations into the transmitting and receiving qualities of antennae," *Nova Acta (Uppsala)*, vol. 11, pp. 1–44, 1938.

[5] H. C. Pocklington, "Electrical oscillations in wires," *Proc. Camb. Phil. Soc.*, vol. 9, 1897.

[6] W. Stutzman and G. Thiele, *Antenna Theory and Design*. John Wiley and Sons, 1981.

[7] C. A. Balanis, *Advanced Engineering Electromagnetics*. John Wiley and Sons, 1989.

[8] S. J. Orfanidis, "Electromagnetic waves and antennas." Electronic version: ece.rutgers.edu/ orfanidi/ewa/.

[9] H. B. Dwight, *Tables of Integrals and Other Mathematical Data*. The Macmillan Company, 1961.

[10] *The Integrator*. Wolfram Research, integrals.wolfram.com.

[11] C. A. Balanis, *Antenna Theory: Analysis and Design*. John Wiley and Sons, second ed., 1997.

[12] *The ARRL Antenna Book*. The Amateur Radio Relay League, 17th ed., 1994.

[13] W. Orr, *Radio Handbook*. Sams, 1992.

# Chapter 6

# Two-Dimensional Problems

In this chapter we will investigate scattering problems in the context of two-dimensional structures. The discussion and examples here will serve to illustrate the approach outlined in Section 3.6.3.1, practical computation of the matrix elements (3.185–3.190), and the choice of basis and testing functions. We will first consider conducting objects, and discuss the implementation of the $\mathcal{L}$ and $\hat{\mathbf{n}} \times \mathcal{K}$ operators in the EFIE and nMFIE. We will then consider dielectric objects and the $\mathcal{K}$ operator in the EFIE/MFIE, and the $\hat{\mathbf{n}} \times \mathcal{L}$ operator in the nMFIE. Because general two-dimensional problems can be split into TM- and TE-polarized sub-problems, we will consider each of these individually.

## 6.1 Conducting Objects

First we will consider conducting objects and the solution of the EFIE and nMFIE in TM and TE polarization. As the $\mathcal{L}$ operator can be used on thin, open structures, we will first apply the EFIE to a thin, conducting strip of finite length, and develop the needed expressions for the self terms. We will then generalize the EFIE to a cylindrical contour of arbitrary shape. Because the $\hat{\mathbf{n}} \times \mathcal{K}$ operator is valid only on closed surfaces, we will consider the nMFIE on closed cylindrical geometries only.

### 6.1.1 EFIE: TM Polarization

Consider a thin, perfectly conducting strip illuminated by a TM-polarized plane wave as illustrated in Figure 6.1a. The electric field has unit amplitude and is given by

$$\mathbf{E}^i(\boldsymbol{\rho}) = e^{jk\hat{\boldsymbol{\rho}}_i \cdot \boldsymbol{\rho}} \, \hat{\mathbf{z}} = e^{jk(x\cos\phi^i + y\sin\phi^i)} \, \hat{\mathbf{z}} \,, \qquad (6.1)$$

where $\hat{\boldsymbol{\rho}}^i = \cos\phi^i \hat{\mathbf{x}} + \sin\phi^i \hat{\mathbf{y}}$. In the absence of magnetic current, the EFIE (3.178) is

$$j\omega\mu\big[(\mathcal{L}\mathbf{J}_l)(\mathbf{r})\big]_{tan} = \big[\mathbf{E}_l^i(\mathbf{r})\big]_{tan} \,. \qquad (6.2)$$

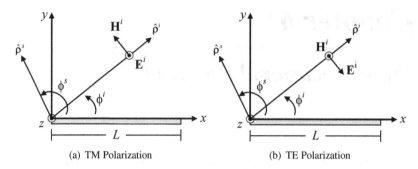

(a) TM Polarization                    (b) TE Polarization

**FIGURE 6.1:** Thin Strip

On the thin strip, this can be written as

$$j\omega\mu\left[1 + \frac{1}{k^2}\nabla\nabla\cdot\right]\int_0^L \mathbf{J}(x')\,G(\boldsymbol{\rho}, \boldsymbol{\rho}')\,dx' = e^{jkx\cos\phi^i}\,, \qquad (6.3)$$

where the integration is only in $x'$ as the strip is assumed to be very thin. We note that since the TM-polarized field has only a $\hat{\mathbf{z}}$ component, so will the electric current, therefore $\nabla' \cdot \mathbf{J}(x') = 0$. Using this fact along with (3.49) and (3.79), the above becomes

$$\frac{\omega\mu}{4}\int_0^L \mathbf{J}(x')\,H_0^{(2)}\left(k|x - x'|\right)\,dx' = e^{jkx\cos\phi^i}\hat{\mathbf{z}}\,. \qquad (6.4)$$

The scattered far electric field (3.142) can be computed as

$$\mathbf{E}^s(\boldsymbol{\rho}) = -\omega\mu\sqrt{\frac{j}{8\pi k}}\,\frac{e^{-jk\rho}}{\sqrt{\rho}}\int_0^L \mathbf{J}(x')\,e^{jkx'\cos\phi^s}\,dx'\,. \qquad (6.5)$$

The corresponding Physical Optics current on the strip for the TM case is

$$\mathbf{J}(\boldsymbol{\rho}) = \frac{2}{\eta}\,\hat{\mathbf{y}}\times\left[-\hat{\boldsymbol{\rho}}^i\times\mathbf{E}^i(\boldsymbol{\rho})\right] = \frac{2}{\eta}\sin\phi^i e^{jkx\cos\phi^i}\hat{\mathbf{z}}\,. \qquad (6.6)$$

#### 6.1.1.1   Solution Using Pulse Functions

We will first solve this problem using pulse basis functions and point matching for testing. Dividing the strip into $N$ segments of width $\Delta_x = L/N$ yields $N$ basis functions. Point matching at the center of the testing segment $x_m$ yields matrix elements given by

$$Z_{mn} = \frac{\omega\mu}{4}\int_{x_n}^{x_{n+1}} H_0^{(2)}\left(kP\right)\,dx'\,, \qquad (6.7)$$

where $P = |x_m - x'|$, and right-hand side vector elements

$$V_m = e^{jkx_m \cos \phi^i} \tag{6.8}$$

Note that the system matrix has a symmetric Toeplitz structure due to the choice of basis functions and the linear geometry of the strip.

*Non-Self Terms*

For non-overlapping segments separated by sufficient distance, we apply a centroidal approximation to (6.7), which yields

$$Z_{mn} = \frac{\omega\mu\Delta_x}{4} H_0^{(2)}(kP) \,, \tag{6.9}$$

where $x_m$ and $x_n$ are at the center of the field and source points, respectively.

*Self Terms*

For overlapping segments, let us consider the small argument approximation to the Hankel function, which is

$$H_0^{(2)}(kP) \approx 1 - j\frac{2}{\pi} \log \left( \frac{\gamma kP}{2} \right) \qquad x \to 0 \,, \tag{6.10}$$

where $\gamma = 1.781$. We note that the second term is singular when $x' = x_m$, so it must be treated carefully. If we rewrite (6.7) as

$$Z_{mm} = \frac{\omega\mu}{4} \int_0^{\Delta_x} \left[ H_0^{(2)}(kP) + j\frac{2}{\pi} \log P \right] dx' - \frac{\omega\mu}{4} j\frac{2}{\pi} \int_0^{\Delta_x} \log P \, dx', \tag{6.11}$$

the first integral on the right-hand side is bounded and can be integrated numerically, and the second term can evaluated analytically. To evaluate the second term, we add an offset $a$ that moves the testing point off the $x$ axis, yielding

$$I_s(x_m) = \int_0^{\Delta_x} \log P \, dx' = \int_0^{\Delta_x} \log \sqrt{(x' - x_m)^2 + a^2} \, dx' \,. \tag{6.12}$$

This integral evaluates to

$$I_s(x_m) = -\frac{1}{2}(x_m - x') \log(a^2 + (x_m - x')^2 - 2) - a \tan^{-1} \left( \frac{x_m - x'}{a} \right) \Big|_0^{\Delta_x} \,. \tag{6.13}$$

Allowing $a \to 0$, this becomes

$$I_s(x_m) = -\frac{1}{2}(x_m - x') \log \left[ (x_m - x')^2 - 2 \right] \Big|_0^{\Delta_x} \,, \tag{6.14}$$

and after substituting $x_m = \Delta_x/2$, and doing some simplification, we get

$$I_s(x_m) = \Delta_x \log(\Delta_x/2 - 1) \, . \tag{6.15}$$

*Near Terms*

When the segments are adjacent but not overlapping ($|m - n| = 1$), the integrand can still exhibit singular behavior near the endpoints which may lead to inaccuracies. In this case, the approach used for the self terms can be used as well, except that $x_m$ is no longer within the limits of integration. Fortunately, (6.14) is valid for all $x_m$ and can be used as-is.

We note that that due to the linear geometry of the strip, we can use (6.11) to compute all the matrix elements. However, evaluation of near terms can be time consuming for more complicated problems, particularly in the three-dimensional case (Chapter 8). Thus, we restrict our use of near term expressions only to distances where they are needed for maximum accuracy.

*Solution*

Consider a strip of width $3\lambda$ centered at the origin, divided into 45 segments (15 segments/$\lambda$). In Figure 6.2a, we plot the induced surface current amplitude at broadside incidence ($\phi^i = \pi/2$). The current is singular near the edges, and the magnitude falls off quickly with distance from each edge. To this we compare the PO current of (6.6), which matches the EFIE fairly well near the center of the strip, but does poorly at the edges. In Figure 6.2b we plot the 2D RCS for incident angles ranging from 0 to 180 degrees. PO compares well to the EFIE at angles near broadside and within the first main lobe, but performs poorly beyond that, as edge diffraction becomes dominant and is omitted by PO.

Using (3.120), we next compute the scattered near field in a $20\lambda \times 20\lambda$ region around the strip. The real component of the electric field is shown (in V/m) for $\phi^i = \pi/4$ and $\phi^i = \pi/2$ in Figures 6.3a and 6.3b, respectively.

### 6.1.1.2  Solution Using Triangle Functions

We will next solve this problem using triangle basis and testing functions. Because the current has only a $\hat{\mathbf{z}}$-component, we will omit the vector notation and work with their scalar components, which we refer to as $f_m^z$ and $f_n^z$, respectively. The strip is again divided into $N$ segments of width $\Delta_x = L/N$. As we know the current is nonzero at the ends of the strip, we use $N - 2$ whole triangle functions and a half-triangle function at each end of the strip, yielding $N$ unknowns. The matrix elements are

$$Z_{mn} = \frac{\omega \mu}{4} \int_{f_m^z} f_m^z(x) \int_{f_n^z} f_n^z(x') \, H_0^{(2)}(kP) \, dx' \, dx \, , \tag{6.16}$$

where $P = |x - x'|$. To evaluate (6.16) numerically, we will use an $M$-point quadrature rule over the source and testing triangles, which yields

$$Z_{mn} = \frac{\omega\mu}{4} \sum_{p=1}^{M} w(x_p) f_m^z(x_p) \sum_{q=1}^{M} w(x_q) f_n^z(x_q) H_0^{(2)}\left(k|x_p - x_q|\right) , \quad (6.17)$$

where $x_q$ and $x_p$ are the source and testing locations, respectively. The corresponding right-hand side vector elements are

$$V_m = \sum_{p=1}^{M} w(x_p) f_m^z(x_p) e^{jkx_p \cos \phi^i} . \quad (6.18)$$

*Self Terms*

When segments overlap, a singularity occurs in the Hankel function. To address this, we will evaluate the innermost integral in (6.16) analytically and the outermost integral numerically. For the innermost integral, we will use the same approach we used with pulse functions. Referring to (6.11), the inner part of (6.16) can be written similarly as

$$I(x_p) = \int_0^{\Delta_x} f_n(x') \left[ H_0^{(2)}(kP) + j\frac{2}{\pi} \log P \right] dx' - j\frac{2}{\pi} \int_0^{\Delta_x} f_n(x') \log P \, dx' . \quad (6.19)$$

where $x_p$ is the testing point and $P = |x_p - x'|$. As before, the first term is computed numerically, whereas the second term is evaluated analytically.

To evaluate the second term, we note that on a segment where $0 \le x \le \Delta_x$, there are two half-triangle functions, $f_1(x) = 1 - x'/\Delta_x$ (negative slope), and $f_2(x) = x'/\Delta_x$ (positive slope). We will first compute the analytic integral involving $f_2(x)$, and following (6.12), it can be written as

$$I_S(x_p) = \frac{1}{\Delta_x} \int_0^{\Delta_x} x' \log \sqrt{(x' - x_p)^2 + a^2)} \, dx' . \quad (6.20)$$

This evaluates to

$$I_S(x_p) = \frac{1}{4\Delta_x} \left( \left[ a^2 - x_p^2 + (x')^2 \right] \log \left[ a^2 + (x_p - x')^2 \right] - \right.$$
$$\left. 4ax_p \tan^{-1}\left( \frac{x_p - x'}{a} \right) + 3x_p^2 - 2x_p x' - (x')^2 \right) \Bigg|_0^{\Delta_x} \quad (6.21)$$

Allowing $a \to 0$, this becomes

$$I_s(x_p) = \frac{1}{4\Delta_x} \left( \left[ -x_p^2 + (x')^2 \right] \log \left[ (x_p - x')^2 \right] + 3x_m^2 - 2x_p x' - (x')^2 \right) \Bigg|_0^{\Delta_x} . \quad (6.22)$$

We can use (6.22) to compute the integral for any value of $x_p$. Although it is lengthy, it is straightforward to implement. The corresponding integral involving $f_1(x)$ is very similar, involving integrals of the form $x \log P$ and $\log P$. Its result can be easily assembled using scaled versions of (6.15) and (6.22).

*Solution*

We again consider the $3\lambda$ strip. On each segment, a 5-point Gaussian quadrature is used, and we again use 45 segments. In Figure 6.2a is shown the computed surface current. The current is somewhat higher at the edges than with pulse functions, but the difference is minimal. In Figure 6.2b is shown the 2D RCS, which is nearly identical to that of the pulse functions. This suggests that pulse functions may be sufficient to characterize the RCS for this problem.

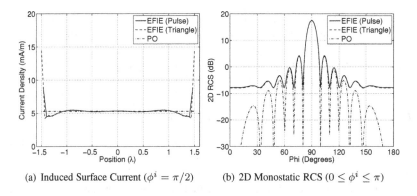

(a) Induced Surface Current ($\phi^i = \pi/2$)    (b) 2D Monostatic RCS ($0 \le \phi^i \le \pi$)

**FIGURE 6.2:** $3\lambda$ Strip: Currents and RCS, TM Polarization

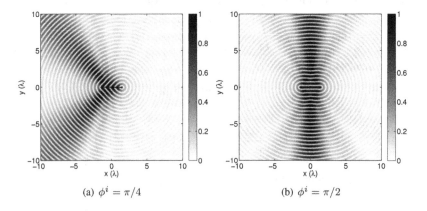

(a) $\phi^i = \pi/4$    (b) $\phi^i = \pi/2$

**FIGURE 6.3:** Scattered Near Field ($|\mathrm{Re}(\mathbf{E})|$, V/m) of $3\lambda$ Strip

## 6.1.2 Generalized EFIE: TM Polarization

Following (6.3) and (6.4), the EFIE (3.178) for a conducting, two-dimensional contour $C$ of arbitrary shape can be written as

$$\frac{\omega\mu}{4} \int_C \mathbf{J}(\boldsymbol{\rho}')\, H_0^{(2)}\left(k|\boldsymbol{\rho} - \boldsymbol{\rho}'|\right)\, d\boldsymbol{\rho}' = \mathbf{E}^i(\boldsymbol{\rho})\,. \tag{6.23}$$

### 6.1.2.1 MoM Discretization

The contour $C$ is divided into a number of linear segments, and the current $\mathbf{J}(\boldsymbol{\rho})$ is expanded using $\hat{\mathbf{z}}$-oriented electric basis functions $\mathbf{f}^z$. Testing with these same functions yields the matrix elements (3.185) for the TM case, given by

$$Z_{mn}^{EJ} = \mathrm{L}^{zz}(\mathbf{f}_m^z, \mathbf{f}_n^z)\,, \tag{6.24}$$

where

$$\mathrm{L}^{zz}(\mathbf{f}_m^z, \mathbf{f}_n^z) = \frac{\omega\mu}{4} \int_{\mathbf{f}_m^z} \mathbf{f}_m^z(\boldsymbol{\rho}) \cdot \int_{\mathbf{f}_n^z} \mathbf{f}_n^z(\boldsymbol{\rho}')\, H_0^{(2)}\left(k|\boldsymbol{\rho} - \boldsymbol{\rho}'|\right)\, d\boldsymbol{\rho}'\, d\boldsymbol{\rho}\,. \tag{6.25}$$

The right-hand side vector elements (3.191) are

$$V_m^{E,TM} = \int_{\mathbf{f}_m^z} \mathbf{f}_m^z(\boldsymbol{\rho}) \cdot \mathbf{E}_z^i(\boldsymbol{\rho})\, d\boldsymbol{\rho}\,. \tag{6.26}$$

### 6.1.2.2 Solution Using Triangle Functions

Using triangle basis and testing functions, (6.25) can be evaluated on non-overlapping segments using an $M$-point numerical quadrature, yielding

$$\mathrm{L}^{zz}(\mathbf{f}_m^z, \mathbf{f}_n^z) = \frac{\omega\mu}{4} \sum_{p=1}^{M} w(\boldsymbol{\rho}_p) f_m^z(\boldsymbol{\rho}_p) \sum_{q=1}^{M} w(\boldsymbol{\rho}_q) f_n^z(\boldsymbol{\rho}_q)\, H_0^{(2)}\left(k|\boldsymbol{\rho}_p - \boldsymbol{\rho}_q|\right)\,, \tag{6.27}$$

where we have again omitted the vector notation for convenience. Self-terms can be computed using the procedure outlined in Section 6.1.1.2. When generating the matrix elements, the right-hand side of (6.27) should be evaluated inside a loop over segment pairs and the resulting values placed into the corresponding matrix locations. As each segment supports at most two triangles, (6.27) will contribute to a maximum of four matrix elements. The right-hand side vector elements (6.26) are

$$V_m^{E,TM} = \sum_{p=1}^{M} w(\boldsymbol{\rho}_p) f_m^z(\boldsymbol{\rho}_p) E_z^i(\boldsymbol{\rho}_p)\,. \tag{6.28}$$

For a plane wave having an electric field of unit amplitude, (6.28) becomes

$$V_m^{E,TM} = \sum_{p=1}^{M} w(x_p) f_m^z(x_p) e^{jk(x_p \cos \phi_i + y_p \sin \phi_i)} . \tag{6.29}$$

### 6.1.3   EFIE: TE Polarization

We next consider the thin strip for TE polarization, as illustrated in Figure 6.1b, where an incident electric field of unit amplitude is now given by

$$\mathbf{E}^i(\boldsymbol{\rho}) = (\sin \phi^i \hat{\mathbf{x}} - \cos \phi^i \hat{\mathbf{y}}) e^{jk(x \cos \phi^i + y \sin \phi^i)} \tag{6.30}$$

As the incident field has $\hat{\mathbf{x}}$ and $\hat{\mathbf{y}}$ components, the induced electric currents will as well. However, since $\hat{\mathbf{y}}$-directed currents (normal to the strip) cannot exist in this case, the current has only an $\hat{\mathbf{x}}$ component. Therefore, (3.178) can be written as

$$\frac{\omega \mu}{4} \int_0^L J_x(x') \left[ 1 + \frac{1}{k^2} \frac{\partial^2}{\partial x^2} \right] H_0^{(2)}(k\rho) \, dx' = \sin \phi^i e^{jkx \cos \phi^i} , \tag{6.31}$$

where $\rho = \sqrt{(x - x')^2 + (y - y')^2}$. Let us now evaluate the expression

$$\left[ 1 + \frac{1}{k^2} \frac{\partial^2}{\partial x^2} \right] H_0^{(2)}(k\rho) . \tag{6.32}$$

To begin, we note that [1]

$$\frac{d}{du} H_n^{(2)}(u) = -H_{n+1}^{(2)}(u) + \frac{n}{u} H_n^{(2)}(u) , \tag{6.33}$$

and

$$\frac{\partial}{\partial x} H_n^{(2)}(k\rho) = \frac{d}{du} \left[ H_n^{(2)}(u) \right] \frac{\partial u}{\partial x} , \tag{6.34}$$

where $u = k\rho$. Therefore,

$$\frac{\partial u}{\partial x} = \frac{\partial (k\rho)}{\partial x} = \frac{k(x - x')}{\sqrt{(x - x')^2 + (y - y')^2}} , \tag{6.35}$$

and so

$$\frac{\partial}{\partial x} H_0^{(2)}(k\rho) = -H_1^{(2)}(k\rho) \frac{k(x - x')}{\sqrt{(x - x')^2 + (y - y')^2}} . \tag{6.36}$$

Differentiating a second time yields

$$\frac{\partial^2}{\partial x^2} H_0^{(2)}(k\rho) = -\frac{k(x-x')}{\sqrt{(x-x')^2+(y-y')^2}}\left[\frac{\partial}{\partial x}H_1^{(2)}(k\rho)\right]$$
$$-H_1^{(2)}(k\rho)\left[\frac{\partial}{\partial x}\frac{k(x-x')}{\sqrt{(x-x')^2+(y-y')^2}}\right], \qquad (6.37)$$

which is

$$\frac{\partial^2}{\partial x^2} H_0^{(2)}(k\rho) = -\frac{k^2(x-x')^2}{(x-x')^2+(y-y')^2}\left[-H_2^{(2)}(k\rho)+\frac{1}{k\rho}H_1^{(2)}(k\rho)\right]$$
$$-H_1^{(2)}(k\rho)\left[\frac{k}{(x-x')^2+(y-y')^2}-\frac{k^2(x-x')^2}{[(x-x')^2+(y-y')^2]^{3/2}}\right].$$
$$\qquad (6.38)$$

The second and fourth term cancel, leaving

$$\frac{\partial^2}{\partial x^2} H_0^{(2)}(k\rho) = \frac{k^2(x-x')^2}{(x-x')^2+(y-y')^2}H_2^{(2)}(k\rho)$$
$$-\frac{k}{(x-x')^2+(y-y')^2}H_1^{(2)}(k\rho). \qquad (6.39)$$

Using the recurrence relationship [1]

$$\frac{2}{k\rho}H_1^{(2)}(k\rho) = H_0^{(2)}(k\rho)+H_2^{(2)}(k\rho), \qquad (6.40)$$

we can then write

$$\left[1+\frac{1}{k^2}\frac{\partial^2}{\partial x^2}\right]H_0^{(2)}(k\rho) = \frac{1}{2}\left[\frac{2(x-x')^2}{(x-x')^2+(y-y')^2}-1\right]H_2^{(2)}(k\rho)+\frac{1}{2}H_0^{(2)}(k\rho).$$
$$\qquad (6.41)$$

We note from the geometry of Figure 6.1b that

$$\cos\theta = \frac{(x-x')}{\sqrt{(x-x')^2+(y-y')^2}}, \qquad (6.42)$$

where the angle $\theta$ is measured in the right-hand sense from the positive x-axis. Using this, the previous expression can be written as

$$\left[1+\frac{1}{k^2}\frac{\partial^2}{\partial x^2}\right]H_0^{(2)}(k\rho) = \frac{1}{2}\left[\cos(2\theta)H_2^{(2)}(k\rho)+H_0^{(2)}(k\rho)\right], \qquad (6.43)$$

and we can now write (6.31) as

$$\frac{\omega\mu}{8}\int_0^L J_x(x')\left[\cos(2\theta)H_2^{(2)}(k\rho)+H_0^{(2)}(k\rho)\right]dx' = \sin\phi^i e^{jkx\cos\phi^i}.$$
$$\qquad (6.44)$$

Note that for observations on the strip, $\rho = |x - x'|$, and for any point $x'$ where $x' \neq x$, $\theta$ is either 0 or $\pi$, thus $\cos(2\theta) = 1$. In these cases, (6.40) can be used to write (6.44) as

$$\frac{\omega\mu}{4} \int_0^L J_x(x') \frac{H_1^{(2)}(k\rho)}{k\rho}\, dx' = \sin\phi^i e^{jkx \cos\phi^i}\,. \tag{6.45}$$

Note that for any matrix elements where $x = x'$, we will need to use (6.44) along with (6.42).

The corresponding PO current on the strip for the TE case is

$$\mathbf{J}(\rho) = 2\,\hat{\mathbf{y}} \times \mathbf{H}^i(\rho) = 2\,\hat{\mathbf{y}} \times \frac{\hat{\mathbf{z}}}{\eta} = \frac{2}{\eta} e^{jkx \cos\phi^i}\hat{\mathbf{x}}\,, \tag{6.46}$$

which at broadside incidence is equivalent in magnitude to the PO current for the TM case.

### 6.1.3.1  Pulse Function Solution

We will approach this problem numerically as we did for the TM case. We subdivide the strip into $N$ segments with $N$ pulse basis functions, and point match at the center of each segment. For segments that are not "close" ($|m - n| > 2$), we will compute the matrix elements using a centroidal approximation of (6.45), which is

$$Z_{mn} = \frac{\omega\mu\Delta_x}{4} \frac{H_1^{(2)}\big(k|x_m - x_n|\big)}{k|x_m - x_n|}\,. \tag{6.47}$$

The right-hand side elements are

$$b_m = \sin\phi^i e^{jkx_m \cos\phi^i}\,. \tag{6.48}$$

*Near Terms*

When segments are adjacent or separated by one segment ($|m - n| = 1, 2$), we will treat these as near terms and perform the integration analytically. The integral for these matrix elements is

$$Z_{mn} = \frac{\omega\mu}{4} \int_{-\Delta_x/2}^{\Delta_x/2} \frac{H_1^{(2)}\big(k|x_m - x_n| - x'\big)}{k\big(|x_m - x_n| - x'\big)}\, dx' \qquad |m - n| = 1, 2\,. \tag{6.49}$$

Using the small-argument approximation to $H_1^{(2)}(k\rho)$ [1], which is

$$H_1^{(2)}(k\rho) \approx \frac{k\rho}{2} + \frac{2j}{\pi k\rho} \qquad r \to 0\,, \tag{6.50}$$

this integral can be written as

$$Z_{mn} = \frac{\omega\mu}{8} \int_{-\Delta_x/2}^{\Delta_x/2} \left[ 1 + \frac{4j}{\pi k^2 (|x_m - x_n| - x')^2} \right] dx', \tag{6.51}$$

which evaluates to

$$Z_{mn} = \frac{\omega\mu\Delta_x}{8} \left[ 1 + \frac{4j}{\pi k^2} \frac{1}{|x_m - x_n|^2 - \Delta_x^2/4} \right]. \tag{6.52}$$

*Self Terms*

For the self terms, we will need to work with (6.44). For the part involving $H_0^{(2)}(k\rho)$, we can use (6.15), but for the part involving $H_2^{(2)}(k\rho)$, we must be careful as it has a non-integrable singularity. To overcome this problem, we will do the integration and then apply a limiting argument as the observation point approaches the strip. The integral to be evaluated is

$$I = \int_{-\Delta_x/2}^{\Delta_x/2} \cos(2\theta) H_2^{(2)}(k\rho) \, dx'. \tag{6.53}$$

First we will employ the small-argument approximation to the Hankel function $H_2^{(2)}(k\rho)$ [1], which is

$$H_2^{(2)}(k\rho) \approx \frac{(k\rho)^2}{8} + \frac{4j}{\pi(k\rho)^2} \qquad r \to 0. \tag{6.54}$$

We next fix the observation point at $(x, y) = (0, a)$, where $a$ is very small. The distance $r$ is then

$$r = \sqrt{(x')^2 + a^2}. \tag{6.55}$$

and we can write the integral as

$$I = \frac{k^2}{4} \int_0^{\Delta_x/2} ((x')^2 + a^2) \left[ \frac{2(x')^2}{(x')^2 + a^2} - 1 \right] dx'$$
$$+ \frac{8j}{\pi k^2} \int_0^{\Delta_x/2} \frac{1}{(x')^2 + a^2} \left[ \frac{2x'^2}{(x')^2 + a^2} - 1 \right] dx', \tag{6.56}$$

where we have reintroduced the expression for $\cos(2\theta)$ from (6.42). The above is then

$$I = \frac{k^2}{4} \int_0^{\Delta_x/2} ((x')^2 - a^2) \, dx' + \frac{8j}{\pi k^2} \int_0^{\Delta_x/2} \left[ \frac{2(x')^2}{((x')^2 + a^2)^2} - \frac{1}{(x')^2 + a^2} \right] dx'. \tag{6.57}$$

The first integral is evaluated easily and is

$$I_1 = \frac{k^2}{4}\left[\frac{\Delta_x^3}{24} - \frac{a^2\Delta_x}{2}\right]. \tag{6.58}$$

The second integral is obtained by way of [2] (120.1 and 122.2) and is

$$I_2 = \frac{8j}{\pi k^2}\left[\frac{1}{a}\tan^{-1}\left(\frac{\Delta_x}{2a}\right) - \frac{\Delta_x/2}{a^2 + (\Delta_x/2)^2} - \frac{1}{a}\tan^{-1}\left(\frac{\Delta_x}{2a}\right)\right], \tag{6.59}$$

which reduces to

$$I_2 = \frac{-8j}{\pi k^2}\left[\frac{\Delta_x/2}{a^2 + (\Delta_x/2)^2}\right]. \tag{6.60}$$

If we now take the limit as the observation point approaches the strip ($a \to 0$), the results are

$$I_1 = k^2\frac{\Delta_x^3}{96} \tag{6.61}$$

and

$$I_2 = \frac{-16j}{\pi k^2\Delta_x}, \tag{6.62}$$

and the resulting self term is

$$Z_{mm} = \frac{\omega\mu}{8}\left[\Delta_x + k^2\frac{\Delta_x^3}{96} - \frac{j\Delta_x}{\pi}\left(2\log\left[\frac{\gamma k\Delta_x}{4e}\right] + \frac{16}{(k\Delta_x)^2}\right)\right]. \tag{6.63}$$

*Solution*

In Figure 6.4a is shown the amplitude of the computed surface current on the $3\lambda$ strip at broadside incidence, using 45 segments. The current tends toward zero near the ends of the strip as expected. In Figure 6.4b we compare the 2D RCS obtained from the EFIE versus that of PO. Again, PO does well near normal incidence, but poorly beyond that. Next, we compute the scattered near field in a $20\lambda \times 20\lambda$ region around the strip. The real component of the electric field is shown (in V/m) for $\phi^i = \pi/4$ and $\phi^i = \pi/2$ in Figures 6.5a and 6.5b, respectively.

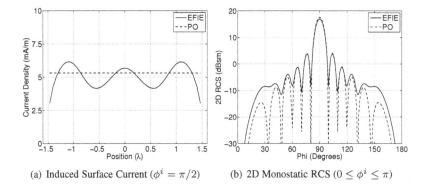

(a) Induced Surface Current ($\phi^i = \pi/2$)  (b) 2D Monostatic RCS ($0 \leq \phi^i \leq \pi$)

**FIGURE 6.4:** Strip with Pulse Functions: TE Polarization

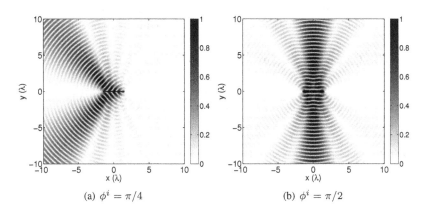

(a) $\phi^i = \pi/4$  (b) $\phi^i = \pi/2$

**FIGURE 6.5:** $3\lambda$ Strip: Scattered TE Near Field (V/m)

### 6.1.4 Generalized EFIE: TE Polarization

Extending (3.178) to a conducting, two-dimensional contour $C$ of arbitrary shape yields

$$\frac{\omega\mu}{4}\,\hat{\mathbf{t}}(\rho) \cdot \int_C \left[1 + \frac{1}{k^2}\nabla\nabla\cdot\right] \mathbf{J}(\rho')H_0^{(2)}\big(k|\rho - \rho'|\big)\,d\rho' = \hat{\mathbf{t}}(\rho)\cdot\mathbf{E}^i(\rho) \quad (6.64)$$

where $\hat{\mathbf{t}}(\mathbf{r})$ and $\mathbf{J}(\rho)$ are everywhere tangent to $C$.

#### 6.1.4.1 MoM Discretization

The contour $C$ is divided into a number of linear segments, and the current $\mathbf{J}(\rho)$ expanded using vector basis functions that are tangent to each segment. Testing with those same functions yields the matrix elements (3.185) for the

TE case, which are

$$Z_{mn}^{EJ} = L^{tt}(\mathbf{f}_m^t, \mathbf{f}_n^t) \tag{6.65}$$

where

$$
\begin{aligned}
L^{tt}(\mathbf{f}_m^t, \mathbf{f}_n^t) =& \frac{\omega\mu}{4} \int_{\mathbf{f}_m^t} \mathbf{f}_m^t(\boldsymbol{\rho}) \cdot \int_{\mathbf{f}_n^t} \mathbf{f}_n^t \boldsymbol{t}\boldsymbol{\rho}') \, H_0^{(2)}\left(k|\boldsymbol{\rho} - \boldsymbol{\rho}'|\right) \, d\boldsymbol{\rho}' \, d\boldsymbol{\rho} \\
&+ \frac{1}{4\omega\epsilon} \int_{\mathbf{f}_m^t} \mathbf{f}_m^t(\boldsymbol{\rho}) \cdot \int_{\mathbf{f}_n^t} \nabla\nabla \cdot \left[\mathbf{f}_n^t(\boldsymbol{\rho}') \, H_0^{(2)}\left(k|\boldsymbol{\rho} - \boldsymbol{\rho}'|\right)\right] \, d\boldsymbol{\rho}' \, d\boldsymbol{\rho}
\end{aligned}
\tag{6.66}
$$

The analysis of the strip in Section 6.1.3 was complicated by the differential operators acting on the Green's function, which will make the analysis of complex geometries even more difficult. Assuming that the basis and testing functions have a well-behaved divergence, we can re-distribute the differential operators following (3.204). Doing so allows us to rewrite (6.66) as

$$
\begin{aligned}
L^{tt}(\mathbf{f}_m^t, \mathbf{f}_n^t) =& \frac{\omega\mu}{4} \int_{\mathbf{f}_m^t} \mathbf{f}_m^t(\boldsymbol{\rho}) \cdot \int_{\mathbf{f}_n^t} \mathbf{f}_n^t \boldsymbol{t}\boldsymbol{\rho}') \, H_0^{(2)}\left(k|\boldsymbol{\rho} - \boldsymbol{\rho}'|\right) \, d\boldsymbol{\rho}' \, d\boldsymbol{\rho} \\
&+ \frac{1}{4\omega\epsilon} \int_{\mathbf{f}_m^t} \nabla \cdot \mathbf{f}_m^t(\boldsymbol{\rho}) \int_{\mathbf{f}_n^t} \nabla' \cdot \mathbf{f}_n^t(\boldsymbol{\rho}') \, H_0^{(2)}\left(k|\boldsymbol{\rho} - \boldsymbol{\rho}'|\right) \, d\boldsymbol{\rho}' \, d\boldsymbol{\rho}
\end{aligned}
\tag{6.67}
$$

The right-hand side vector elements (3.191) are

$$V_m^{E,TE} = \int_{\mathbf{f}_m^t} \mathbf{f}_m^t(\boldsymbol{\rho}) \cdot \mathbf{E}^i(\boldsymbol{\rho}) \, d\boldsymbol{\rho} \tag{6.68}$$

### 6.1.4.2   Solution Using Triangle Functions

The geometry in the TE case is similar to the thin-wire problem of Section 5.5.2, and so the orientation of the basis and testing functions follows that of Figure 5.4. We can evaluate (6.67) on non-overlapping segments using an $M$-point numerical quadrature, yielding

$$L^{tt}(\mathbf{f}_m^t, \mathbf{f}_n^t) = \sum_{p=1}^{M} \sum_{q=1}^{M} w(\boldsymbol{\rho}_p) w(\boldsymbol{\rho}_q) \left[\frac{\omega\mu}{4} \mathbf{f}_m^t(\boldsymbol{\rho}_p) \cdot \mathbf{f}_n^t(\boldsymbol{\rho}_q) \pm \frac{1}{4\omega\epsilon\Delta_l^2}\right] H_0^{(2)}\left(k|\boldsymbol{\rho}_p - \boldsymbol{\rho}_q|\right) \tag{6.69}$$

where we have used the fact that the divergence of each triangle function is

$$\nabla \cdot \mathbf{f}^t(\boldsymbol{\rho}) = \pm\frac{1}{\Delta_l} \tag{6.70}$$

The sign of this term depends on the orientation of the basis and testing functions on each segment. The right-hand side of (6.69) should be evaluated inside a loop over segment pairs and the resulting values placed into the corresponding matrix locations. On overlapping segments, the first integral in (6.67) can be computed following Section (6.1.1.2), and the second integral has the form

$$\int_0^{\Delta_x} \int_0^{\Delta_x} H_0^{(2)}\left(k|x - x'|\right) dx' \, dx \, . \tag{6.71}$$

We will evaluate the outer integral in (6.71) numerically, and the innermost integral following the solution to (6.11). The right-hand side vector elements are

$$V_m^{E,TE} = \sum_{p=1}^{M} w(\boldsymbol{\rho}_p) \, \mathbf{f}_m^t(\boldsymbol{\rho}_p) \cdot \mathbf{E}^i(\boldsymbol{\rho}_p) \, . \tag{6.72}$$

For a plane wave having an electric field of unit amplitude, (6.72) becomes

$$V_m^{E,TE} = \sum_{p=1}^{M} w(\boldsymbol{\rho}_p) \, \mathbf{f}_m^t(\boldsymbol{\rho}_p) \cdot (\sin \phi^i \hat{\mathbf{x}} - \cos \phi^i \hat{\mathbf{y}}) e^{jk(x_p \cos \phi^i + y_p \sin \phi^i)} \, .$$

$$\tag{6.73}$$

### 6.1.5 nMFIE: TM Polarization

Consider the closed, conducting cylindrical contour $C$ of Figure 6.6. Given the TM-polarized incident field of (6.1), the corresponding magnetic field is

$$\mathbf{H}^i(\boldsymbol{\rho}) = -\frac{1}{\eta} \, \hat{\boldsymbol{\rho}}^i \times \mathbf{E}^i(\boldsymbol{\rho}) = \frac{1}{\eta} \left(-\sin \phi^i \hat{\mathbf{x}} + \cos \phi^i \hat{\mathbf{y}}\right) . e^{jk(x \cos \phi^i + y \sin \phi^i)} \, .$$

$$\tag{6.74}$$

Using (3.49) and (3.94), we can write the nMFIE (3.180) as

$$\frac{1}{2}\mathbf{J}(\boldsymbol{\rho}) - \frac{j}{4} \, \hat{\mathbf{n}}(\boldsymbol{\rho}) \times \int_{C'} \mathbf{J}(\boldsymbol{\rho}') \times \nabla H_0^{(2)}(k\rho) \, d\boldsymbol{\rho}' = \hat{\mathbf{n}}(\boldsymbol{\rho}) \times \mathbf{H}^i(\boldsymbol{\rho}) \, , \tag{6.75}$$

where we have used the notation $C'$ to indicate the integration is not valid for $\boldsymbol{\rho} = \boldsymbol{\rho}'$. Using the fact that

$$\nabla H_0^{(2)}(k\rho) = -k \left[\frac{x - x'}{R}\hat{\mathbf{x}} + \frac{y - y'}{R}\hat{\mathbf{y}}\right] H_1^{(2)}(k\rho) = -k \, \hat{\mathbf{R}}(\boldsymbol{\rho}, \boldsymbol{\rho}') H_1^{(2)}(k\rho) \, ,$$

$$\tag{6.76}$$

where

$$\hat{\mathbf{R}}(\boldsymbol{\rho}, \boldsymbol{\rho}') = \frac{x - x'}{R}\hat{\mathbf{x}} + \frac{y - y'}{R}\hat{\mathbf{y}} = \frac{\boldsymbol{\rho} - \boldsymbol{\rho}'}{|\boldsymbol{\rho} - \boldsymbol{\rho}'|} \, , \tag{6.77}$$

we can write (6.75) as

$$\frac{1}{2}\mathbf{J}(\boldsymbol{\rho}) + \frac{jk}{4}\hat{\mathbf{n}}(\boldsymbol{\rho}) \times \int_{C'} \mathbf{J}(\boldsymbol{\rho}') \times \hat{\mathbf{R}}(\boldsymbol{\rho}, \boldsymbol{\rho}') \, H_1^{(2)}(k\rho) \, d\boldsymbol{\rho}' = \hat{\mathbf{n}}(\boldsymbol{\rho}) \times \mathbf{H}^i(\boldsymbol{\rho}) \, .$$

$$\tag{6.78}$$

The TM CFIE is formed by combining the TM EFIE (6.23) and the TM nMFIE (6.78) following (3.181).

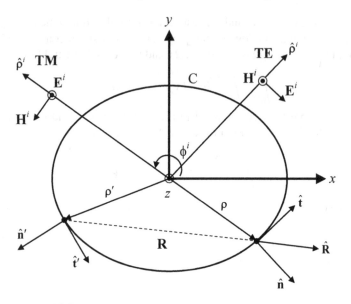

**FIGURE 6.6:** 2D Contour: TM and TE Polarization

## MoM Discretization

The contour $C$ is divided into a number of linear segments, and the current $\mathbf{J}(\boldsymbol{\rho})$ is expanded using $\hat{\mathbf{z}}$-oriented electric basis functions. Testing with these same functions yields the matrix elements (3.189) for the TM case, which are

$$Z_{mn}^{nHJ} = nK^{zz}(\mathbf{f}_m^z, \mathbf{f}_n^z) , \tag{6.79}$$

where

$$nK^{zz}(\mathbf{f}_m^z, \mathbf{f}_n^z) = \frac{1}{2} \int_{\mathbf{f}_m^z = \mathbf{f}_n^z} \mathbf{f}_m^z(\boldsymbol{\rho}) \cdot \mathbf{f}_n^z(\boldsymbol{\rho}) \, d\boldsymbol{\rho} + \frac{jk}{4} \int_{\mathbf{f}_m^z} \mathbf{f}_m^z(\boldsymbol{\rho}) \cdot$$
$$\hat{\mathbf{n}}(\boldsymbol{\rho}) \times \int_{\mathbf{f}_n^z} \mathbf{f}_n^z(\boldsymbol{\rho}') \times \hat{\mathbf{R}}(\boldsymbol{\rho}, \boldsymbol{\rho}') \, H_1^{(2)}(k\rho) \, d\boldsymbol{\rho}' \, d\boldsymbol{\rho} . \tag{6.80}$$

The first term in (6.80) is evaluated over boundary segments where $\mathbf{f}_m^z$ and $\mathbf{f}_n^z$ overlap, and the second where they do not. If we use the vector identity

$$\mathbf{A} \times \mathbf{B} \times \mathbf{C} = (\mathbf{A} \cdot \mathbf{C})\mathbf{B} - (\mathbf{A} \cdot \mathbf{B})\mathbf{C} , \tag{6.81}$$

and the fact that $\hat{\mathbf{n}}(\boldsymbol{\rho}) \cdot \mathbf{f}_n^z(\boldsymbol{\rho}') = 0$, we can write

$$\hat{\mathbf{n}}(\boldsymbol{\rho}) \times \mathbf{f}_n^z(\boldsymbol{\rho}') \times \hat{\mathbf{R}}(\boldsymbol{\rho}, \boldsymbol{\rho}') = \mathbf{f}_n^z(\boldsymbol{\rho}') \Big[ \mathbf{n}(\boldsymbol{\rho}) \cdot \hat{\mathbf{R}}(\boldsymbol{\rho}, \boldsymbol{\rho}') \Big] . \tag{6.82}$$

This allows us to write (6.80) as

$$
\mathrm{nK}^{zz}(\mathbf{f}_m^z, \mathbf{f}_n^z) = \frac{1}{2} \int_{\mathbf{f}_m^z = \mathbf{f}_n^z} \mathbf{f}_m^z(\boldsymbol{\rho}) \cdot \mathbf{f}_n^z(\boldsymbol{\rho}) \, d\boldsymbol{\rho} + \frac{jk}{4} \int_{\mathbf{f}_m^z} \mathbf{f}_m^z(\boldsymbol{\rho}) \cdot
$$

$$
\int_{\mathbf{f}_n^z} \mathbf{f}_n^z(\boldsymbol{\rho}') \left[ \hat{\mathbf{n}}(\boldsymbol{\rho}) \cdot \hat{\mathbf{R}}(\boldsymbol{\rho}, \boldsymbol{\rho}') \right] H_1^{(2)}(k\rho) \, d\boldsymbol{\rho}' \, d\boldsymbol{\rho} \, . \qquad (6.83)
$$

The right-hand side vector elements (3.193) are

$$
V_m^{nH,TM} = \int_{\mathbf{f}_m^z} \mathbf{f}_m^z(\boldsymbol{\rho}) \cdot \left[ \hat{\mathbf{n}}(\boldsymbol{\rho}) \times \mathbf{H}^i(\boldsymbol{\rho}) \right] d\boldsymbol{\rho} \, . \qquad (6.84)
$$

### 6.1.5.1 Solution Using Triangle Functions

Using triangle basis and testing functions, (6.83) is computed using an $M$-point numerical quadrature yielding

$$
\mathrm{nK}^{zz}(\mathbf{f}_m^z, \mathbf{f}_n^z) = \frac{1}{2} \sum_{p=1}^{M} w(\boldsymbol{\rho}_p) f_m^z(\boldsymbol{\rho}_p) f_n^z(\boldsymbol{\rho}_p) + \frac{jk}{4} \sum_{p=1}^{M} w(\boldsymbol{\rho}_p) f_m^z(\boldsymbol{\rho}_p) \cdot
$$

$$
\sum_{q=1}^{M} w(\boldsymbol{\rho}_q) f_n^z(\boldsymbol{\rho}_q)(\hat{\mathbf{n}}_p \cdot \hat{\mathbf{R}}_{pq}) H_1^{(2)}(k|\boldsymbol{\rho}_p - \boldsymbol{\rho}_q|) \, , \qquad (6.85)
$$

where we have again omitted the vector notation for $\mathbf{f}_m^z$ and $\mathbf{f}_n^z$. The right-hand side elements are

$$
V_m^{nH,TM} = \sum_{p=1}^{M} w(\boldsymbol{\rho}_p) \, \mathbf{f}_m^z(\boldsymbol{\rho}_p) \cdot \left[ \hat{\mathbf{n}}(\boldsymbol{\rho}_p) \times \mathbf{H}^i(\boldsymbol{\rho}_p) \right] . \qquad (6.86)
$$

For a plane wave having an electric field of unit amplitude, (6.86) becomes

$$
V_m^{nH,TM} = \frac{1}{\eta} \sum_{p=1}^{M} w(\boldsymbol{\rho}_p) \, \mathbf{f}_m^z(\boldsymbol{\rho}_p) \cdot \left[ \hat{\mathbf{n}}(\boldsymbol{\rho}_p) \times (-\sin\phi^i \hat{\mathbf{x}} \cos\phi^i \hat{\mathbf{y}}) \right] \cdot \qquad (6.87)
$$

$$
e^{jk(x_p \cos\phi^i + y_p \sin\phi^i)} \, . \qquad (6.88)
$$

### 6.1.6 nMFIE: TE Polarization

Consider the same contour as before, with a TE-polarized incident field as shown in Figure 6.6. Given the electric field of (6.30), the corresponding magnetic field is

$$
\mathbf{H}^i(\boldsymbol{\rho}) = -\frac{1}{\eta} \, \hat{\boldsymbol{\rho}}^i \times \mathbf{E}^i(\boldsymbol{\rho}) = \frac{1}{\eta} \, e^{jk(x\cos\phi^i + y\sin\phi^i)} \, \hat{\mathbf{z}} \, . \qquad (6.89)
$$

We know from the discussion in Section 6.1.3 that the induced current $\mathbf{J}(\rho)$ only has components along the vector tangent to $C$. The nMFIE for the TE case is very similar to that of the TM case (6.78), with some small differences that will appear in the MoM discretization. The TE CFIE can be created by combining the TE EFIE (6.64) and the TE nMFIE (6.78) following (3.181).

*MoM Discretization*

We again divide the boundary $C$ into $N$ segments, and the current $\mathbf{J}(\rho)$ is expanded using electric basis functions that are tangent to $C$. Testing with these same functions yields the matrix elements (3.189) for the TE case, which are

$$Z_{mn}^{nHJ} = \text{n}K^{tt}(\mathbf{f}_m^t, \mathbf{f}_n^t) , \tag{6.90}$$

where

$$\text{n}K^{tt}(\mathbf{f}_m^t, \mathbf{f}_n^t) = \frac{1}{2} \int_{\mathbf{f}_m^t = \mathbf{f}_n^t} \mathbf{f}_m^t(\rho) \cdot \mathbf{f}_n^t(\rho) \, d\rho + \frac{jk}{4} \int_{\mathbf{f}_m^t} \mathbf{f}_m^t(\rho) \cdot$$

$$\hat{\mathbf{n}}(\rho) \times \int_{\mathbf{f}_n^t} \mathbf{f}_n^t(\rho') \times \hat{\mathbf{R}}(\rho, \rho') \, H_1^{(2)}(k\rho) \, d\rho' \, d\rho . \tag{6.91}$$

The first term in (6.91) is evaluated over boundary segments where $\mathbf{a}^t$ and $\mathbf{b}^t$ overlap, and the second where they do not. The right-hand side vector elements (3.193) are

$$V_m^{nH,TE} = \int_{\mathbf{f}_m^t} \mathbf{f}_m^t(\rho) \cdot \left[ \hat{\mathbf{n}}(\rho) \times \mathbf{H}^i(\rho) \right] d\rho . \tag{6.92}$$

### 6.1.6.1  Solution Using Triangle Functions

The triangle basis and testing functions are oriented in the same manner as in Section 6.1.4.2. Applying an $M$-point numerical quadrature to (6.91) yields

$$\text{n}K^{tt}(\mathbf{f}_m^t, \mathbf{f}_n^t) = \frac{1}{2} \sum_{p=1}^{M} w(\rho_p) \, \mathbf{f}_m^t(\rho_p) \cdot \mathbf{f}_n^t(\rho_p) + \frac{jk}{4} \sum_{p=1}^{M} w(\rho_p) \, \mathbf{f}_m^t(\rho_p) \cdot$$

$$\left[ \hat{\mathbf{n}}(\rho_p) \times \sum_{q=1}^{M} w(\rho_q) \, \mathbf{f}_n^t(\rho_q) \times \hat{\mathbf{R}}_{pq} \, H_1^{(2)}(k|\rho_p - \rho_q|) \right], \tag{6.93}$$

and right-hand side vector elements are

$$V_m^{nH,TE} = \sum_{p=1}^{M} w(\rho_p) \, \mathbf{f}_m^t(\rho_p) \cdot \left[ \hat{\mathbf{n}}(\rho_p) \times \mathbf{H}^i(\rho_p) \right] . \tag{6.94}$$

For a plane wave having an electric field of unit amplitude, (6.94) becomes

$$V_m^{nH,TE} = \frac{1}{\eta} \sum_{p=1}^{M} w(\rho_p) \, \mathbf{f}_m^t(\rho_p) \cdot \left[ \hat{\mathbf{n}}(\rho_p) \times \hat{\mathbf{z}} \right] e^{jk(x_p \cos \phi^i + y_p \sin \phi^i)} . \tag{6.95}$$

### 6.1.7 Examples

We will now use the generalized EFIE and nMFIE to analyze the scattering of plane waves by a two-dimensional conducting circular cylinder for TM and TE polarizations. The integral equation solution uses the MoM discretizations formulated via triangle basis and testing functions, and we use $\alpha = 0.5$ for the CFIE.

#### 6.1.7.1 Conducting Cylinder: TM Polarization

For TM polarization, the exact solution for the scattered electric field of a conducting cylinder is [3]

$$E_z^s(\rho, \phi^s) = \sum_{n=0}^{\infty} j^n c_n A_n H_n^{(2)}(k_0 \rho) \cos(n\phi^s), \qquad (6.96)$$

where $|E_z^i| = 1$ for convenience, $\phi^s$ is the bistatic scattering angle, and $c_n = 1$ for $n = 0$, and $c_n = 2$ otherwise. For the conducting cylinder, the coefficients $A_n$ are

$$A_n = -\frac{J_n(k_0 a)}{H_n^{(2)}(k_0 a)}, \qquad (6.97)$$

where $a$ is the radius of the cylinder. The scattered far electric field is

$$E_z^s(\rho, \phi^s) = \sqrt{\frac{2}{\pi}} \frac{e^{-j(k_0 \rho - \pi/4)}}{\sqrt{k_0 \rho}} \sum_{n=0}^{\infty} (-1)^n c_n A_n \cos n\phi^s. \qquad (6.98)$$

The induced electric current $J_z(\phi^a)$ is

$$J_z(\phi^a) = \frac{2}{\pi \eta_0 k_0 a} \sum_{n=0}^{\infty} \frac{(j)^n c_n \cos n\phi^a}{H_n^{(2)}(k_0 a)}, \qquad (6.99)$$

where $\phi^a$ is the azimuthal angle on the surface of the cylinder. Using (6.74), the PO current is

$$J_z^{PO}(\phi^a) = \frac{2}{\eta}(\cos \phi^a \cos \phi^i + \sin \phi^a \sin \phi^i)e^{jk(x \cos \phi^i + y \sin \phi^i)}. \qquad (6.100)$$

In the high-frequency limit ($k_o \to \infty$), the radar cross section in the backscattering direction ($\phi^s = 0$) approaches $\pi a$, and in the forward-scattering direction ($\phi^s = \pi$) it approaches $4k_o a$ [3].

*Results*

We first compute the radar cross section of a conducting cylinder of a radius of 1 meter, for frequencies up to 800 MHz. For the MoM solution, 10 segments per $\lambda$ at 800 MHz were used, yielding 167 segments of equal length. For the modal solution, we terminate the summation at $N = 40$. The backscattering results are summarized in Figures 6.7a–6.7c using EFIE, nMFIE and CFIE, respectively. The EFIE does fairly well, at least within the range of frequencies considered, whereas the nMFIE shows many errors due to internal resonances. The CFIE is free of errors, as expected. The RCS in each case is seen to approach the expected high-frequency value of $\pi a$ m$^2$, which is approximately 4.97 dBsm in this case. The forward-scattering results are shown in Figures 6.7d–6.7f. The nMFIE again shows some errors but they are not quite as apparent as they were in the backscattering case. The RCS in each case tends toward the expected high-frequency value of $4k_o$ m$^2$, which at 800 MHz is approximately 18.26 dBsm.

We next compute the bistatic RCS for $0 \leq \phi^s \leq 2\pi$ and $\phi^i = 0$ at 800 MHz. The CFIE results are compared to the exact solution in Figure 6.8a, where the agreement is excellent. The results obtained using the PO current of (6.100) are shown in Figure 6.8b, where the agreement is good near the specular angles but fair to poor elsewhere.

We then compute the electric current $J_z(\phi^a)$ across all azimuthal angles for $\phi^i = 0$ at 800 MHz. The real and imaginary components of the currents obtained via the CFIE and PO are compared to the exact solution (6.99) in Figures 6.9a–6.9d, and the amplitudes compared in Figures 6.9e and 6.9f. The agreement between the modal solution and the CFIE is very good. The PO current compares fairly well in the illuminated region, however in the shadow region ($\pi/2 \leq \phi^a \leq 3\pi/2$) the PO current is zero.

Finally, we compute the scattered near field in a $10 \times 10$ meter region around the cylinder at 800 MHz for $\phi^i = 0$. The real component of the electric field (in V/m) obtained using the CFIE is shown in Figure 6.10a. The modal solution is indistinguishable to the eye and is not shown. The solution obtained using physical optics is shown in Figure 6.10b.

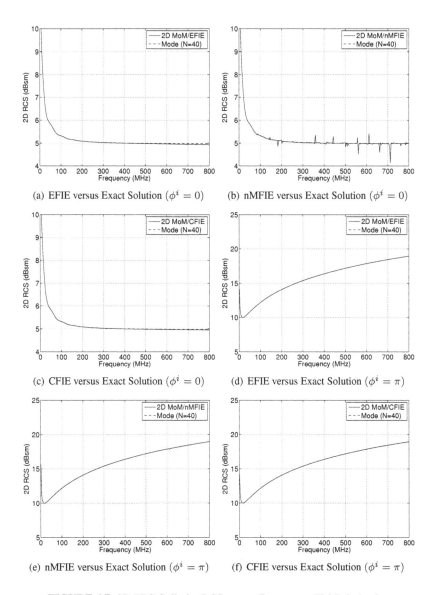

(a) EFIE versus Exact Solution ($\phi^i = 0$)

(b) nMFIE versus Exact Solution ($\phi^i = 0$)

(c) CFIE versus Exact Solution ($\phi^i = 0$)

(d) EFIE versus Exact Solution ($\phi^i = \pi$)

(e) nMFIE versus Exact Solution ($\phi^i = \pi$)

(f) CFIE versus Exact Solution ($\phi^i = \pi$)

**FIGURE 6.7:** 2D PEC Cylinder RCS versus Frequency: TM Polarization

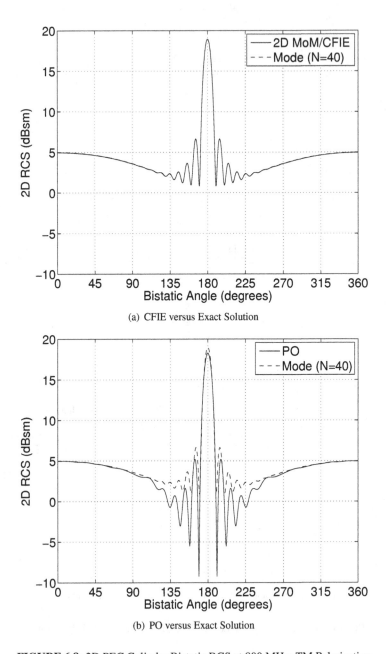

(a)  CFIE versus Exact Solution

(b)  PO versus Exact Solution

**FIGURE 6.8:** 2D PEC Cylinder Bistatic RCS at 800 MHz: TM Polarization

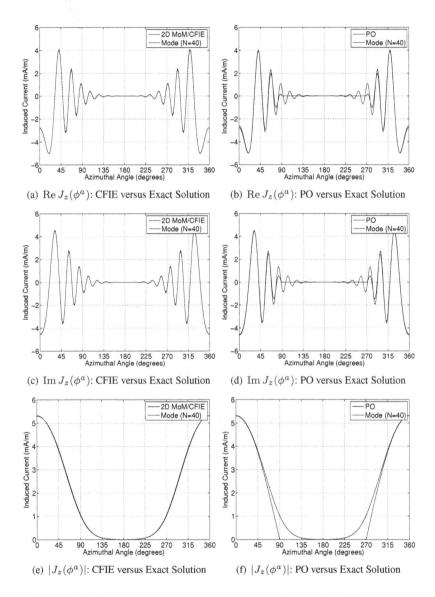

(a) Re $J_z(\phi^a)$: CFIE versus Exact Solution

(b) Re $J_z(\phi^a)$: PO versus Exact Solution

(c) Im $J_z(\phi^a)$: CFIE versus Exact Solution

(d) Im $J_z(\phi^a)$: PO versus Exact Solution

(e) $|J_z(\phi^a)|$: CFIE versus Exact Solution

(f) $|J_z(\phi^a)|$: PO versus Exact Solution

**FIGURE 6.9:** 2D PEC Cylinder Current versus Azimuthal Angle: TM Polarization

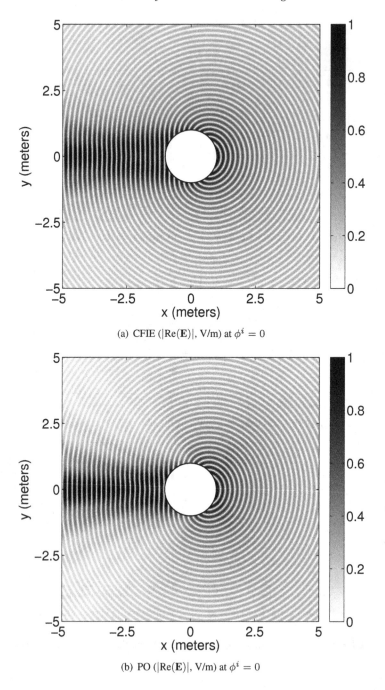

(a) CFIE ($|\text{Re}(\mathbf{E})|$, V/m) at $\phi^i = 0$

(b) PO ($|\text{Re}(\mathbf{E})|$, V/m) at $\phi^i = 0$

**FIGURE 6.10:** 2D PEC Cylinder Scattered Near Field at 800 MHz: TM Polarization

### 6.1.7.2 Conducting Cylinder: TE Polarization

For TE polarization, with a $\hat{\mathbf{z}}$-directed incident magnetic field, the exact solution for the scattered magnetic field is [3]

$$H_z^s(\rho, \phi^s) = \frac{1}{\eta_0} \sum_{n=0}^{\infty} j^n c_n B_n H_n^{(2)}(k_0 \rho) \cos(n\phi^s), \qquad (6.101)$$

where $|E_\phi^i| = 1$ for convenience. For the conducting cylinder, the coefficients $B_n$ are

$$B_n = -\frac{J_n'(k_0 a)}{H_n'^{(2)}(k_0 a)}. \qquad (6.102)$$

The scattered far magnetic field is

$$H_z^s(\rho, \phi^s) = \frac{1}{\eta_0} \sqrt{\frac{2}{\pi}} \frac{e^{-j(k_0\rho - \pi/4)}}{\sqrt{k_0\rho}} \sum_{n=0}^{\infty} (-1)^n c_n B_n \cos n\phi^s. \qquad (6.103)$$

The induced azimuthal electric current is

$$\mathbf{J}(\phi^a) = \frac{2j}{\eta_0 \pi k_0 a} \sum_{n=0}^{\infty} \frac{(j)^n c_n \cos n\phi^a}{H_n'^{(2)}(k_0 a)} \hat{\phi}, \qquad (6.104)$$

where $\phi^a$ is the azimuthal angle on the surface of the cylinder. The recurrence relationship [1]

$$Z_n'(ka) = \frac{n}{ka} Z_n(ka) - Z_{n+1}(ka) \qquad (6.105)$$

can be used to compute the derivatives, where $Z_n$ represents either $J_n$ or $H_n$. Using (6.89), the PO current is

$$\mathbf{J}^{PO}(\phi^a) = \frac{2}{\eta}(\sin\phi^a \hat{\mathbf{x}} - \cos\phi^a \hat{\mathbf{y}})e^{jk(x\cos\phi^i + y\sin\phi^i)}. \qquad (6.106)$$

*Results*

For TE polarization, we use the same conducting cylinder and numerical parameters as in the TM case. The backscatter results are summarized in Figures 6.11a–6.11c using EFIE, nMFIE, and CFIE, respectively. The nMFIE again shows errors due to internal resonances, and the CFIE is free of errors, as expected. The RCS in each case again approaches the high-frequency limit of approximately 4.97 dBsm. The forward-scattering results are shown in Figures 6.11d–6.11f. The nMFIE again shows some errors but they are again not as apparent as in the backscattering direction. The RCS in each case approaches the expected high-frequency limit of approximately 18.26 dBsm at 800 MHz. We next compute the bistatic RCS for $0 \le \phi^s \le 2\pi$ and $\phi^i = 0$ at 800

MHz. The CFIE results are compared to the exact solution in Figure 6.12a, and the agreement is again excellent. The results obtained using the PO current of (6.106) are shown in Figure 6.12b. The agreement is fair near the specular angles, but poor elsewhere.

We then we compute the electric current $J_z(\phi^a)$ across all azimuthal angles for $\phi^i = 0$ at 800 MHz. The real and imaginary components of the currents obtained via CFIE and PO are compared to the exact solution (6.99) in Figures 6.13a–6.13d, and the amplitudes compared in Figures 6.13e and 6.13f. The agreement between the modal solution and the CFIE is fairly good, however the PO current does not agree very well at all, as its amplitude comprises a step function in the TE case.

Finally, we compute the scattered near field in a $10 \times 10$ meter region about the cylinder at 800 MHz for $\phi^i = 0$. The amplitude of the real part of the electric field (in V/m) obtained using the CFIE is shown in Figure 6.14a. The modal solution is indistinguishable to the eye and is not shown. The corresponding solution obtained using Physical Optics is shown in Figure 6.14b.

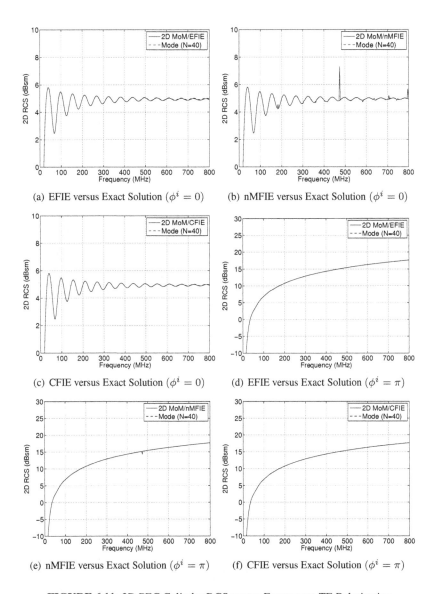

(a) EFIE versus Exact Solution $(\phi^i = 0)$

(b) nMFIE versus Exact Solution $(\phi^i = 0)$

(c) CFIE versus Exact Solution $(\phi^i = 0)$

(d) EFIE versus Exact Solution $(\phi^i = \pi)$

(e) nMFIE versus Exact Solution $(\phi^i = \pi)$

(f) CFIE versus Exact Solution $(\phi^i = \pi)$

**FIGURE 6.11:** 2D PEC Cylinder RCS versus Frequency: TE Polarization

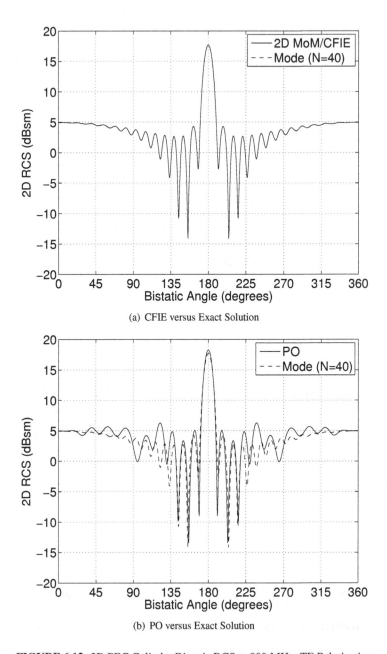

(a) CFIE versus Exact Solution

(b) PO versus Exact Solution

**FIGURE 6.12:** 2D PEC Cylinder Bistatic RCS at 800 MHz: TE Polarization

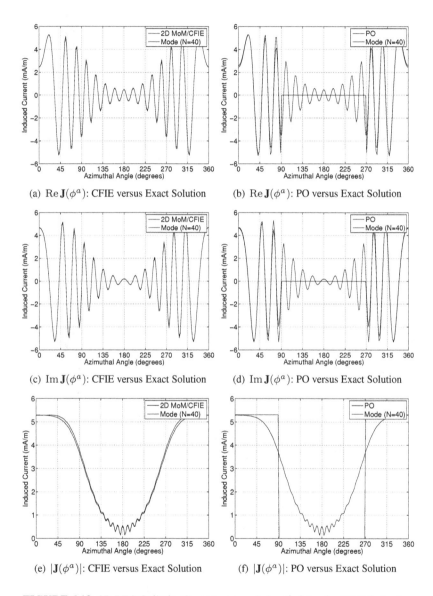

(a) $\operatorname{Re}\mathbf{J}(\phi^a)$: CFIE versus Exact Solution

(b) $\operatorname{Re}\mathbf{J}(\phi^a)$: PO versus Exact Solution

(c) $\operatorname{Im}\mathbf{J}(\phi^a)$: CFIE versus Exact Solution

(d) $\operatorname{Im}\mathbf{J}(\phi^a)$: PO versus Exact Solution

(e) $|\mathbf{J}(\phi^a)|$: CFIE versus Exact Solution

(f) $|\mathbf{J}(\phi^a)|$: PO versus Exact Solution

**FIGURE 6.13:** 2D PEC Cylinder Current versus Azimuthal Angle: TE Polarization

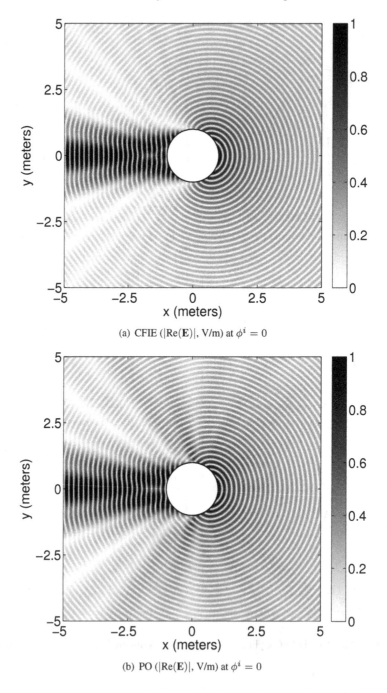

(a) CFIE ($|\text{Re}(\mathbf{E})|$, V/m) at $\phi^i = 0$

(b) PO ($|\text{Re}(\mathbf{E})|$, V/m) at $\phi^i = 0$

**FIGURE 6.14:** 2D PEC Cylinder Scattered Near Field at 800 MHz: TE Polarization

## 6.2 Dielectric and Composite Objects

In this section, we build upon the work from Section 6.1 to create a framework for treating two-dimensional dielectric and composite dielectric/conducting objects (without junctions). We will use the expressions previously derived for the EFIE and nMFIE that involved the $\mathcal{L}$ and $\hat{\mathbf{n}} \times \mathcal{K}$ operators, and develop new expressions involving the $\mathcal{K}$ and $\hat{\mathbf{n}} \times \mathcal{L}$ operators. As before, we will consider each integral equation in TM and TE polarization separately, and we will present generalized MoM discretizations exclusively in terms of triangle basis and testing functions. We will then present several examples.

### 6.2.1 Basis Function Orientation

As we begin to study multi-region problems in more detail, the orientation of the basis functions on the various dielectric interfaces must be considered carefully. As there are $\hat{\mathbf{z}}$-directed electric currents and tangentially directed magnetic currents in the TM case, and vice-versa for the TE case, we will treat the orientations of both currents in a self-consistent manner. Consider a triangle function having support on segments $s+$ and $s-$, whose endpoints comprise $(\mathbf{p}_{k-1}, \mathbf{p}_k, \mathbf{p}_{k+1})$. The segments reside on the interface $S_{12}$ between dielectric regions $R_1$ and $R_2$, as shown in Figure 6.15. First, we define the unit vectors

$$\hat{\mathbf{t}}^{s+} = \frac{\mathbf{p}_{k+1} - \mathbf{p}_k}{|\mathbf{p}_{k+1} - \mathbf{p}_k|} \tag{6.107}$$

and

$$\hat{\mathbf{t}}^{s-} = \frac{\mathbf{p}_k - \mathbf{p}_{k-1}}{|\mathbf{p}_k - \mathbf{p}_{k-1}|} . \tag{6.108}$$

We know from Section 3.6.3 that the currents on each side of the interface must

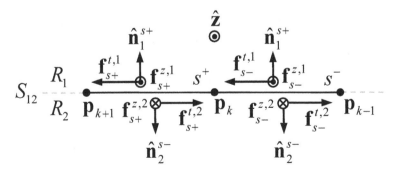

**FIGURE 6.15:** Basis Function Orientation for 2D Problems

flow in opposite directions. Therefore, let us define $R_1$ as the "positive" side of the interface. On the positive side, the vectors normal to each segment are

$$\hat{\mathbf{n}}_1^{s+} = \hat{\mathbf{t}}^{s+} \times \hat{\mathbf{z}} \tag{6.109}$$

and

$$\hat{\mathbf{n}}_1^{s-} = \hat{\mathbf{t}}^{s-} \times \hat{\mathbf{z}} . \tag{6.110}$$

As we are still free to orient the basis functions, we begin in $R_1$, and make the following assignments for the tangential basis function:

$$\hat{\mathbf{f}}_{s+}^{t,1} = \hat{\mathbf{t}}^{s+} \tag{6.111}$$

and

$$\hat{\mathbf{f}}_{s-}^{t,1} = \hat{\mathbf{t}}^{s-} . \tag{6.112}$$

and the following assignments for the $\hat{\mathbf{z}}$-oriented basis function:

$$\hat{\mathbf{f}}_{s+}^{z,1} = \hat{\mathbf{z}} \tag{6.113}$$

and

$$\hat{\mathbf{f}}_{s-}^{z,1} = \hat{\mathbf{z}} . \tag{6.114}$$

If we now move to the opposite or "negative" side of the interface in $R_2$, we need only reverse the direction of the normal vector and the basis functions in $R_1$, as shown in Figure 6.15. We refer to this as a right-handed boundary orientation. Note that all currents have opposite orientations on each side of the interface, and for tangentially oriented currents, Kirchoff's law is satisfied at $p_k$. We will use this orientation again when discussing bodies of revolution in Chapter 7.

### 6.2.2    EFIE: TM Polarization

The EFIE (3.179) for TM polarization has an $\mathcal{L}$ operator acting on the $\hat{\mathbf{z}}$-directed electric current $\mathbf{J}(\boldsymbol{\rho})$, and a $\mathcal{K}$ operator acting on the tangential magnetic current $\mathbf{M}(\boldsymbol{\rho})$. In region $R_l$, these expressions are

$$\frac{\omega \mu_l}{4} \hat{\mathbf{t}}(\boldsymbol{\rho}) \cdot \int_C \left[ 1 + \frac{1}{k_l^2} \nabla \nabla \cdot \right] \mathbf{J}(\boldsymbol{\rho}') H_0^{(2)}\left(k_l |\boldsymbol{\rho} - \boldsymbol{\rho}'|\right) d\boldsymbol{\rho}' , \tag{6.115}$$

and

$$\frac{1}{2}\hat{\mathbf{n}}_l(\boldsymbol{\rho}) \times \mathbf{M}(\boldsymbol{\rho}) - \frac{jk_l}{4} \int_{C'} \mathbf{M}(\boldsymbol{\rho}') \times \hat{\mathbf{R}}(\boldsymbol{\rho}, \boldsymbol{\rho}') \, H_1^{(2)}\left(k_l \rho\right) d\boldsymbol{\rho}' . \tag{6.116}$$

### 6.2.2.1 MoM Discretization

Testing (6.115) and (6.116) with the $\hat{z}$-oriented electric basis functions yields the matrix elements (3.185) and (3.186) for the TM case, which are

$$Z_{mn}^{EJ(l),TM} = L^{zz}(\mathbf{f}_m^z, \mathbf{f}_n^z) \tag{6.117}$$

and

$$Z_{mn}^{EM(l),TM} = K^{zt}(\mathbf{f}_m^z, \mathbf{g}_n^t) . \tag{6.118}$$

The function $L^{zz}$ was summarized in (6.25). The function $K^{zt}$ is

$$K^{zt}(\mathbf{f}_m^z, \mathbf{g}_n^t) = \frac{1}{2} \int_{\mathbf{f}_m^z = \mathbf{g}_n^t} \mathbf{f}_m^z(\rho) \cdot \left[ \hat{\mathbf{n}}_l(\rho) \times \mathbf{g}_n^t(\rho) \right] d\rho$$
$$- \frac{jk_l}{4} \int_{\mathbf{f}_m^z} \mathbf{f}_m^z(\rho) \cdot \int_{\mathbf{g}_n^t} \mathbf{g}_n^t(\rho') \times \hat{\mathbf{R}}(\rho, \rho') H_1^{(2)}(k_l\rho) \, d\rho' \, d\rho . \tag{6.119}$$

where the first term is evaluated over boundary segments where $\mathbf{f}_m^z$ and $\mathbf{g}_n^t$ overlap, and the second where they do not.

## 6.2.3 MFIE: TM Polarization

The MFIE (3.179) for TM-polarization has an $\mathcal{L}$ operator acting on the tangential magnetic current $\mathbf{M}(\rho)$, and a $\mathcal{K}$ operator acting on the $\hat{z}$-directed electric current $\mathbf{J}(\rho)$. In region $R_l$, these expressions are

$$\frac{\omega \epsilon_l}{4} \, \hat{\mathbf{t}}(\rho) \cdot \int_C \left[ 1 + \frac{1}{k_l^2} \nabla \nabla \cdot \right] \mathbf{M}(\rho') H_0^{(2)} \left( k_l |\rho - \rho'| \right) d\rho' \tag{6.120}$$

and

$$-\frac{1}{2} \hat{\mathbf{n}}_l(\rho) \times \mathbf{J}(\rho) + \frac{jk_l}{4} \int_{C'} \mathbf{J}(\rho') \times \hat{\mathbf{R}}(\rho, o\rho') H_1^{(2)}(k_l\rho) \, d\rho' . \tag{6.121}$$

### 6.2.3.1 MoM Discretization

Testing (6.120) and (6.121) with the tangential magnetic basis functions yields the matrix elements (3.187) and (3.188), which are

$$Z_{mn}^{HM(l),TM} = \frac{\epsilon_l}{\mu_l} L^{tt}(\mathbf{g}_m^t, \mathbf{g}_n^t) \tag{6.122}$$

and

$$Z_{mn}^{HJ(l),TM} = -K^{tz}(\mathbf{g}_m^t, \mathbf{f}_n^z) . \tag{6.123}$$

The function $\mathbf{L}^{tt}$ was summarized in (6.67). The function $\mathbf{K}^{tz}$ is identical to (6.119) except that $t$ and $z$ are interchanged. Therefore, it can be written as

$$
\mathbf{K}^{tz}(\mathbf{g}_m^t, \mathbf{f}_n^z) = \frac{1}{2} \int_{\mathbf{g}_m^t = \mathbf{f}_n^z} \mathbf{g}_m^t(\rho) \cdot \left[ \hat{\mathbf{n}}_l(\rho) \times \mathbf{f}_n^z(\rho) \right] d\rho
$$

$$
- \frac{jk_l}{4} \int_{\mathbf{g}_m^t} \mathbf{g}_m^t(\rho) \cdot \int_{\mathbf{f}_n^z} \mathbf{f}_n^z(\rho') \times \hat{\mathbf{R}}(\rho, \rho') \, H_1^{(2)}(k_l \rho) \, d\rho' \, d\rho \, .
$$

$$(6.124)$$

### 6.2.4 nMFIE: TM Polarization

The nMFIE (3.180) for TM polarization has an $\hat{\mathbf{n}} \times \mathcal{L}$ operator acting on the tangential magnetic current $\mathbf{M}(\rho)$, and an $\hat{\mathbf{n}} \times \mathcal{K}$ operator acting on the $\hat{\mathbf{z}}$-directed electric current $\mathbf{J}(\rho)$. In region $R_l$, these expressions are

$$
\frac{1}{2} \mathbf{J}(\rho) + \frac{jk_l}{4} \hat{\mathbf{n}}(\rho) \times \int_{C'} \mathbf{J}(\rho') \times \hat{\mathbf{R}}(\rho, \rho') \, H_1^{(2)}(k_l \rho) \, d\rho' \tag{6.125}
$$

and

$$
\frac{\omega \epsilon_l}{4} \hat{\mathbf{n}}_l(\rho) \times \int_C \left[ 1 + \frac{1}{k_l^2} \nabla \nabla \cdot \right] \mathbf{M}(\rho') H_0^{(2)}(k_l |\rho - \rho'|) \, d\rho' \, . \tag{6.126}
$$

#### 6.2.4.1 MoM Discretization

Testing (6.125) and (6.126) with the $\hat{\mathbf{z}}$-oriented electric basis functions yields the matrix elements (3.189) and (3.190) for the TM case, which are

$$
Z_{mn}^{nHJ(l),TM} = n\mathbf{K}^{zz}(\mathbf{f}_m^z, \mathbf{f}_n^z) \tag{6.127}
$$

and

$$
Z_{mn}^{nHM(l),TM} = \frac{\epsilon_l}{\mu_l} n\mathbf{L}^{zt}(\mathbf{f}_m^z, \mathbf{g}_n^t) \, . \tag{6.128}
$$

The function $n\mathbf{K}^{zz}$ was summarized in (6.80). The function $n\mathbf{L}^{zt}$ is

$$
n\mathbf{L}^{zt}(\mathbf{f}_m^z, \mathbf{g}_n^t) = \frac{\omega \mu_l}{4} \int_{\mathbf{f}_m^z} \mathbf{f}_m^z(\rho) \cdot \hat{\mathbf{n}}_l(\rho) \times \int_{\mathbf{g}_n^t} \mathbf{g}_n^t(\rho') \, H_0^{(2)}(k_l |\rho - \rho'|) \, d\rho' \, d\rho
$$

$$
- \frac{1}{4\omega \epsilon_l} \int_{\mathbf{f}_m^z} \nabla \cdot \left[ \mathbf{f}_m^z(\rho) \times \hat{\mathbf{n}}_l(\rho) \right] \int_{\mathbf{g}_n^t} \nabla' \cdot \mathbf{g}_n^t(\rho') \, H_0^{(2)}(k_l |\rho - \rho'|) \, d\rho' \, d\rho \, .
$$

$$(6.129)$$

where we have used (3.205). Note that the term $\mathbf{f}_m^z(\rho) \times \hat{\mathbf{n}}_l(\rho)$ results in a tangentially oriented vector on each segment.

## 6.2.5 EFIE: TE Polarization

The operators in the EFIE for TE polarization are the dual of those from the MFIE in the TM case, where the electric and magnetic currents are exchanged. Therefore, we will simply summarize the MoM discretization.

### 6.2.5.1 MoM Discretization

The matrix elements (3.185) and (3.186) for the TE case are

$$Z_{mn}^{EJ(l),TE} = L^{tt}(\mathbf{f}_m^t, \mathbf{f}_n^t) \tag{6.130}$$

and

$$Z_{mn}^{EM(l),TE} = K^{tz}(\mathbf{f}_m^t, \mathbf{g}_n^z) . \tag{6.131}$$

## 6.2.6 MFIE: TE Polarization

Similarly, the operators in the MFIE for TE polarization are the dual of those from the EFIE in the TM case, with the electric and magnetic currents exchanged. Thus, we will simply summarize the MoM discretization.

### 6.2.6.1 MoM Discretization

The matrix elements (3.187) and (3.188) for the TE case are

$$Z_{mn}^{HM(l),TE} = \frac{\epsilon_l}{\mu_l} L^{zz}(\mathbf{g}_m^z, \mathbf{g}_n^z) \tag{6.132}$$

and

$$Z_{mn}^{HJ(l),TE} = -K^{zt}(\mathbf{g}_m^z, \mathbf{f}_n^t) . \tag{6.133}$$

## 6.2.7 nMFIE: TE Polarization

The expressions for the nMFIE for TE-polarization are functionally identical to those of the nMFIE for TM-polarization, except that $t$ and $z$ are interchanged. Thus, we will simply summarize the MoM discretization.

### 6.2.7.1 MoM Discretization

The matrix elements (3.189) and (3.190) for the TE case are

$$Z_{mn}^{nHJ(l),TE} = nK^{tt}(\mathbf{f}_m^t, \mathbf{f}_n^t) \tag{6.134}$$

and

$$Z_{mn}^{nHM(l),TE} = \frac{\epsilon_l}{\mu_l} nL^{tz}(\mathbf{f}_m^t, \mathbf{g}_n^z) . \tag{6.135}$$

The function $\mathrm{n}K^{tt}$ was summarized in (6.91). The function $\mathrm{n}L^{tz}$ is

$$\mathrm{n}L^{tz}(\mathbf{f}_m^t, \mathbf{g}_n^z) = \frac{\omega\mu_l}{4} \int_{\mathbf{f}_m^t} \mathbf{f}_m^t(\boldsymbol{\rho}) \cdot \int_{\mathbf{g}_n^z} \hat{\mathbf{n}}_l(\boldsymbol{\rho}) \times \mathbf{g}_n^z(\boldsymbol{\rho}') \, H_0^{(2)}(k_l|\boldsymbol{\rho} - \boldsymbol{\rho}'|) \, d\rho' \, d\rho \,,$$

(6.136)

where the second term on the right in (6.129) goes to zero, since

$$\nabla \cdot \left[ \mathbf{f}_m^t(\boldsymbol{\rho}) \times \hat{\mathbf{n}}_l(\boldsymbol{\rho}) \right] = 0 \,.$$

(6.137)

### 6.2.8 Numerical Stability

Consider an MoM system matrix that has been constructed from the EFIE and MFIE. In very general terms, it can be represented via the block structure

$$Z = \left[ \begin{array}{cc} L(\mathbf{f}, \mathbf{f}) & K(\mathbf{f}, \mathbf{g}) \\ -K(\mathbf{g}, \mathbf{f}) & \frac{\epsilon}{\mu} L(\mathbf{g}, \mathbf{g}) \end{array} \right] \,.$$

(6.138)

Note that the matrix block in the lower right is scaled down by $1/\eta^2$ versus the block in the upper left. This disparity in amplitude tends to cause numerical stability issues in the linear system, due to limitations in the floating point representation. As a remedy, we will use the approach from [4], where the magnetic basis function is everywhere scaled by $\eta_0$. That is,

$$\mathbf{g}^{t,z} = \eta_0 \mathbf{f}^{t,z} \,.$$

(6.139)

This modification works very well in practice, and we will incorporate this scale factor into all the magnetic basis functions throughout the remainder of this book.

### 6.2.9 Examples

In this section we apply the expressions developed in this section to the scattering of plane waves by a homogeneous dielectric cylinder and a coated, conducting cylinder. Lossless as well as lossy dielectrics will be considered.

#### 6.2.9.1 Dielectric Cylinder

Consider a homogeneous dielectric circular cylinder of radius $a$ with dielectric parameters $(\epsilon_1, \mu_1)$. We divide the problem into an interior region $R_1$ with the aforementioned parameters, and an external free space region $R_0$ with material parameters $(\epsilon_0, \mu_0)$, which are separated by the interface $S_{01}$. Using the PMCHWT approach, we compute the EFIE and MFIE on $S_0$ in $R_0$ and $S_1$ in $R_1$. We then sum the resulting EFIEs and MFIEs on each side of the interface by combining the columns and rows for the basis and testing functions on $S_{01}$.

### 6.2.9.2 Dielectric Cylinder: TM Polarization

For a dielectric cylinder, the series coefficients $A_n$ for TM polarization are [3]

$$A_n = -\frac{(k_1/\mu_1)J_n(k_0a)J_n'(k_1a) - (k_0/\mu_0)J_n'(k_0a)J_n'(k_1a)}{(k_1/\mu_1)H_n^{(2)}(k_0a)J_n'(k_1a) - (k_0/\mu_0)H_n'^{(2)}(k_0a)J_n'(k_1a)},$$
(6.140)

where the scattered field is given by (6.96).

*MoM Linear System*

For the TM case, the MoM linear system can be written as

$$\begin{bmatrix} Z^{zz,TM} & Z^{zt,TM} \\ Z^{tz,TM} & Z^{tt,TM} \end{bmatrix} \begin{bmatrix} I^J \\ I^M \end{bmatrix} = \begin{bmatrix} V^{z,TM} \\ V^{t,TM} \end{bmatrix},$$
(6.141)

where the blocks of the MoM system matrix are

$$Z^{zz,TM} = \mathbf{L}^{zz}\left(\mathbf{f}_{S_0}^z, \mathbf{f}_{S_0}^z, R_0\right) + \mathbf{L}^{zz}\left(\mathbf{f}_{S_1}^z, \mathbf{f}_{S_1}^z, R_1\right),$$
(6.142)

$$Z^{tz,TM} = \mathbf{K}^{zt}\left(\mathbf{f}_{S_0}^z, \mathbf{g}_{S_0}^t, R_0\right) + \mathbf{K}^{zt}\left(\mathbf{f}_{S_1}^z, \mathbf{g}_{S_1}^t, R_1\right),$$
(6.143)

$$Z^{zt,TM} = -\mathbf{K}^{tz}\left(\mathbf{g}_{S_0}^t, \mathbf{f}_{S_0}^z, R_0\right) - \mathbf{K}^{tz}\left(\mathbf{g}_{S_1}^t, \mathbf{f}_{S_1}^z, R_1\right),$$
(6.144)

$$Z^{tt,TM} = \frac{\epsilon_0}{\mu_0}\mathbf{L}^{tt}\left(\mathbf{g}_{S_0}^t, \mathbf{g}_{S_0}^t, R_0\right) + \frac{\epsilon_1}{\mu_1}\mathbf{L}^{tt}\left(\mathbf{g}_{S_1}^t, \mathbf{g}_{S_1}^t, R_1\right).$$
(6.145)

The notation $\mathrm{X}\left(\mathbf{a}_{S_l}, \mathbf{b}_{S_m}, R_l\right)$ indicates that X is evaluated in $R_l$ for the testing and basis functions $\mathbf{a}$ and $\mathbf{b}$ that reside on the boundaries $S_l$ and $S_m$, respectively. The right-hand side vectors are

$$V^{z,TM} = V^{E,TM}\left(\mathbf{f}_{S_0}^z, R_0\right)$$
(6.146)

and

$$V^{t,TM} = V^{H,TM}\left(\mathbf{g}_{S_0}^t, R_0\right),$$
(6.147)

and the solution vectors are

$$I^J = I_{S_{01}}^{J,z}$$
(6.148)

and

$$I^M = I_{S_{01}}^{M,t}.$$
(6.149)

*Results*

We will consider cylinders of radius $a = 5$, 10, and 20 meters, first with $\epsilon_1 = 2.56$ (lossless) and then with $\epsilon_1 = 2.56 - j.102$ (lossy). Setting $k_0 = 1$, we compute the bistatic RCS for $0 \leq \phi^s \leq 2\pi$ with $\phi^i = 0$. For the MoM we use 10 segments per free space wavelength on the boundary, with a minimum of 64 segments, and for the modal solution $N = 40$. The MoM results are compared to the modal solution in Figure 6.16. The agreement is excellent in all cases. We next compute the RCS in the backscattering and forward-scattering directions for $0 < k_0 a \leq 40$. The MoM uses the same parameters as before, and for the modal solution $N = 60$. For aesthetic purposes, the RCS has been normalized with respect to $\pi a$. The agreement is seen to be very good.

To determine the near fields in $R_0$ and $R_1$, we can use (3.120) and (3.122) and integrate the MoM currents on $S_0$ for points in $R_0$, and the currents on $S_1$ for points in $R_1$. In Figure 6.19a is shown the scattered near electric field in $R_0$ (in V/m) for a lossless dielectric cylinder with radius $1\lambda$ and $\epsilon_1 = 2.56$. In Figure 6.19a are shown the total fields, where in $R_0$ the field is the sum of the incident and scattered electric field, and in $R_1$ it is the transmitted field. We see that the total fields are continuous across the dielectric boundary.

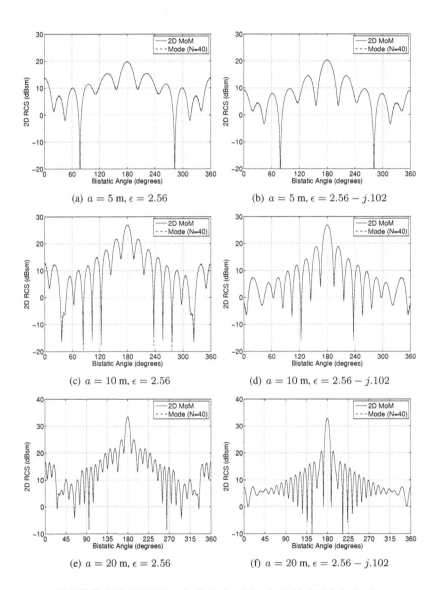

**FIGURE 6.16:** 2D Dielectric Cylinder: Bistatic RCS, TM Polarization

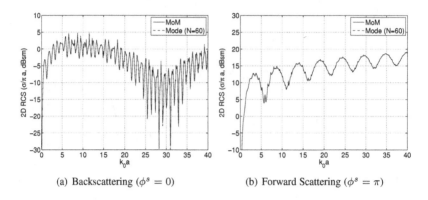

(a) Backscattering ($\phi^s = 0$)   (b) Forward Scattering ($\phi^s = \pi$)

**FIGURE 6.17:** 2D Lossless Dielectric Cylinder: RCS versus Radius, TM Polarization

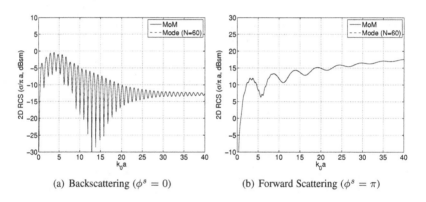

(a) Backscattering ($\phi^s = 0$)   (b) Forward Scattering ($\phi^s = \pi$)

**FIGURE 6.18:** 2D Lossy Dielectric Cylinder: RCS versus Radius, TM Polarization

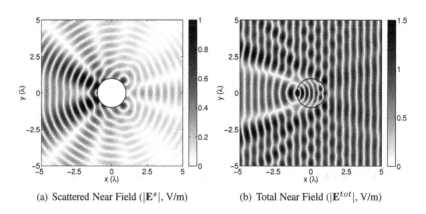

(a) Scattered Near Field ($|\mathbf{E}^s|$, V/m)   (b) Total Near Field ($|\mathbf{E}^{tot}|$, V/m)

**FIGURE 6.19:** 2D Dielectric Cylinder: TM Near Fields for $\phi^i = 0$

### 6.2.9.3 Dielectric Cylinder: TE Polarization

For the dielectric cylinder, the coefficients $B_n$ for TE polarization are [3]

$$B_n = -\frac{(k_1/\epsilon_1)J_n(k_0a)J_n'(k_1a) - (k_0/\epsilon_0)J_n'(k_0a)J_n(k_1a)}{(k_1/\epsilon_1)H_n^{(2)}(k_0a)J_n'(k_1a) - (k_0/\epsilon_0)H_n'^{(2)}(k_0a)J_n(k_1a)} , \quad (6.150)$$

where the scattered field is given by (6.101).

*MoM Linear System*

For the TE case, the MoM linear system can be written as

$$\begin{bmatrix} Z^{tt,TE} & Z^{tz,TE} \\ Z^{zt,TE} & Z^{zz,TE} \end{bmatrix} \begin{bmatrix} I^J \\ I^M \end{bmatrix} = \begin{bmatrix} V^{t,TE} \\ V^{z,TE} \end{bmatrix}, \quad (6.151)$$

where the blocks of the system matrix are

$$Z^{tt,TE} = \mathbf{L}^{tt}\left(\mathbf{f}_{S_0}^t, \mathbf{f}_{S_0}^t, R_0\right) + \mathbf{L}^{tt}\left(\mathbf{f}_{S_1}^t, \mathbf{f}_{S_1}^t, R_1\right), \quad (6.152)$$

$$Z^{tz,TE} = \mathbf{K}^{tz}\left(\mathbf{f}_{S_0}^t, \mathbf{g}_{S_0}^z, R_0\right) + \mathbf{K}^{tz}\left(\mathbf{f}_{S_1}^t, \mathbf{g}_{S_1}^z, R_1\right), \quad (6.153)$$

$$Z^{zt,TE} = -\mathbf{K}^{zt}\left(\mathbf{g}_{S_0}^z, \mathbf{f}_{S_0}^t, R_0\right) - \mathbf{K}^{zt}\left(\mathbf{g}_{S_1}^z, \mathbf{f}_{S_1}^t, R_1\right), \quad (6.154)$$

$$Z^{zz,TE} = \frac{\epsilon_0}{\mu_0}\mathbf{L}^{zz}\left(\mathbf{g}_{S_0}^z, \mathbf{g}_{S_0}^z, R_0\right) + \frac{\epsilon_1}{\mu_1}\mathbf{L}^{zz}\left(\mathbf{g}_{S_1}^z, \mathbf{g}_{S_1}^z, R_1\right). \quad (6.155)$$

The right-hand side vectors are

$$V^{t,TE} = V^{E,TE}\left(\mathbf{f}_{S_0}^t, R_0\right) \quad (6.156)$$

and

$$V^{z,TE} = V^{H,TE}\left(\mathbf{g}_{S_0}^z, R_0\right), \quad (6.157)$$

and the solution vectors are

$$I^J = I_{S_{01}}^{J,t} \quad (6.158)$$

and

$$I^M = I_{S_{01}}^{M,z}. \quad (6.159)$$

*Results*

We consider the same examples presented in Section 6.2.9.2, but for TE polarization. The discretization and other parameters remain identical. The results are shown in Figures 6.20–6.23. The comparisons between the MoM and modal solutions are again very good.

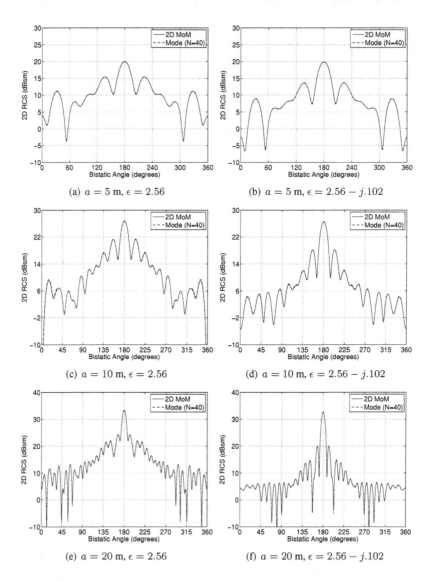

**FIGURE 6.20:** 2D Dielectric Cylinder: Bistatic RCS, TE Polarization

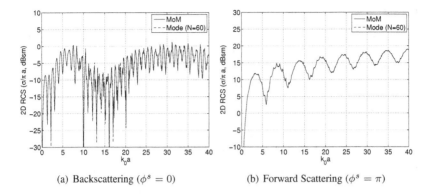

(a) Backscattering ($\phi^s = 0$)

(b) Forward Scattering ($\phi^s = \pi$)

**FIGURE 6.21:** 2D Lossless Dielectric Cylinder: RCS versus Radius, TE Polarization

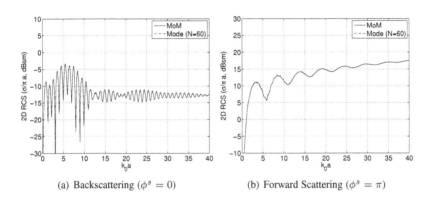

(a) Backscattering ($\phi^s = 0$)

(b) Forward Scattering ($\phi^s = \pi$)

**FIGURE 6.22:** 2D Lossy Dielectric Cylinder: RCS versus Radius, TE Polarization

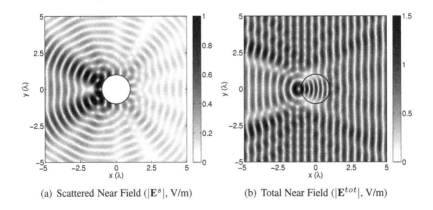

(a) Scattered Near Field ($|\mathbf{E}^s|$, V/m)

(b) Total Near Field ($|\mathbf{E}^{tot}|$, V/m)

**FIGURE 6.23:** 2D Dielectric Cylinder: TE Near Fields For $\phi^i = 0$

### 6.2.9.4   Coated Cylinder

We next consider a conducting cylinder of radius $a_1$ with a dielectric coating of thickness $d$. This problem has three regions: the PEC region $R_2$, the dielectric region $R_1$, and the free space region $R_0$. There are two interfaces: the conducting interface $S_{12}$ with radius $a_1$, and the dielectric interface $S_{01}$ with radius $a_0 = a_1 + d$. In this case, we test the EFIE and MFIE on $S_{01}$, and the CFIE on $S_{12}$.

### 6.2.9.5   Coated Cylinder: TM Polarization

For the coated cylinder, the coefficients $A_n$ for TM polarization are [3]

$$A_n = -\frac{J_n(k_0 a_0) - iZ_n J_n'(k_0 a_0)}{H_n^{(2)}(k_0 a_0) - iZ_n H_n'^{(2)}(k_0 a_0)}, \tag{6.160}$$

where

$$iZ_n = \frac{k_0 \mu_1}{k_1 \mu_0} \left[ \frac{J_n(k_1 a_0) H_n^{(2)}(k_1 a_1) - H_n^{(2)}(k_1 a_0) J_n(k_1 a_1)}{J_n'(k_1 a_0) H_n^{(2)}(k_1 a_1) - H_n'^{(2)}(k_1 a_0) J_n(k_1 a_1)} \right]. \tag{6.161}$$

*MoM Linear System*

For the TM case, the MoM linear system is given by (6.141), where the sub-matrix blocks are

$$Z^{zz,TM} = \begin{bmatrix} \mathbf{L}^{zz}\left(\mathbf{f}_{S_{01}}^z, \mathbf{f}_{S_{01}}^z, R_0\right) + \mathbf{L}^{zz}\left(\mathbf{f}_{S_{10}}^z, \mathbf{f}_{S_{10}}^z, R_1\right) & \mathbf{L}^{zz}\left(\mathbf{f}_{S_{10}}^z, \mathbf{f}_{S_{12}}^z, R_1\right) \\ \alpha \mathbf{L}^{zz}\left(\mathbf{f}_{S_{12}}^z, \mathbf{f}_{S_{10}}^z, R_1\right) & \alpha \mathbf{L}^{zz}\left(\mathbf{f}_{S_{12}}^z, \mathbf{f}_{S_{12}}^z, R_1\right) \end{bmatrix}$$

$$+ \begin{bmatrix} 0 & 0 \\ (1-\alpha)\eta_1 n \mathbf{K}^{zz}\left(\mathbf{f}_{S_{12}}^z, \mathbf{f}_{S_{10}}^z, R_1\right) & (1-\alpha)\eta_1 n \mathbf{K}^{zz}\left(\mathbf{f}_{S_{12}}^z, \mathbf{f}_{S_{12}}^z, R_1\right) \end{bmatrix}, \tag{6.162}$$

$$Z^{zt,TM} = \begin{bmatrix} \mathbf{K}^{zt}\left(\mathbf{f}_{S_{01}}^z, \mathbf{g}_{S_{01}}^t, R_0\right) + \mathbf{K}^{zt}\left(\mathbf{f}_{S_{10}}^z, \mathbf{g}_{S_{10}}^t, R_1\right) \\ \alpha \mathbf{K}^{zt}\left(\mathbf{f}_{S_{12}}^z, \mathbf{g}_{S_{10}}^t, R_1\right) \end{bmatrix}$$

$$+ \begin{bmatrix} 0 \\ (1-\alpha)\frac{1}{\eta_1} n \mathbf{L}^{zt}\left(\mathbf{f}_{S_{12}}^z, \mathbf{g}_{S_{10}}^t, R_1\right) \end{bmatrix}, \tag{6.163}$$

$$Z^{tz,TM} = \left[ -\mathbf{K}^{tz}\left(\mathbf{g}_{S_{01}}^t, \mathbf{f}_{S_{01}}^z, R_0\right) - \mathbf{K}^{tz}\left(\mathbf{g}_{S_{10}}^t, \mathbf{f}_{S_{10}}^z, R_1\right) \mid -\mathbf{K}^{tz}\left(\mathbf{g}_{S_{10}}^t, \mathbf{f}_{S_{12}}^z, R_1\right) \right], \tag{6.164}$$

and

$$Z^{tt,TM} = \left[ \frac{\epsilon_0}{\mu_0} \mathbf{L}^{tt}\left(\mathbf{g}_{S_{01}}^t, \mathbf{g}_{S_{01}}^t, R_0\right) + \frac{\epsilon_1}{\mu_1} \mathbf{L}^{tt}\left(\mathbf{g}_{S_{10}}^t, \mathbf{g}_{S_{10}}^t, R_1\right) \right]. \tag{6.165}$$

The right-hand side vectors are

$$V^{z,TM} = \begin{bmatrix} V^{E,TM}(\mathbf{f}^z_{S_{01}}, R_0) \\ 0 \end{bmatrix} \tag{6.166}$$

and

$$V^{t,TM} = \begin{bmatrix} V^{H,TM}(\mathbf{g}^t_{S_{01}}, R_0) \\ 0 \end{bmatrix}, \tag{6.167}$$

and the solution vectors are

$$I^J = \begin{bmatrix} I^{J,z}_{S_{01}} \\ I^{J,z}_{S_{12}} \end{bmatrix}, \qquad I^{M,t} = \begin{bmatrix} I^{M,t}_{S_{01}} \\ 0 \end{bmatrix}. \tag{6.168}$$

*Results*

We will analyze a coated cylinder where $a_0 = 1.1a_1$. Setting $k_0 = 1$, we compute the RCS in the backscattering and forward-scattering directions for $0 < k_0 a \le 40$. For the MoM solution, both interfaces are discretized using 10 points per $\lambda_0$ with a minimum of 64 segments on each, and for the modal solution, $N = 60$. The MoM results are compared to the modal solution for a lossless ($\epsilon_1 = 2.56$) and lossy ($\epsilon_1 = 2.56 - j.102$) coating in Figures 6.24 and 6.25, respectively. The RCS has again been normalized with respect to $\pi a$. The agreement is seen to be fairly good, except that in the backscattering direction, the MoM results are about $\frac{1}{2}$ dB below the modal results. The reason for this difference is discussed further in Section 6.2.9.7.

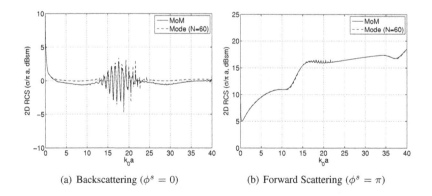

(a) Backscattering ($\phi^s = 0$)  (b) Forward Scattering ($\phi^s = \pi$)

**FIGURE 6.24:** 2D PEC Cylinder w/Lossless Coating: RCS versus $k_0 a$, TM Polarization

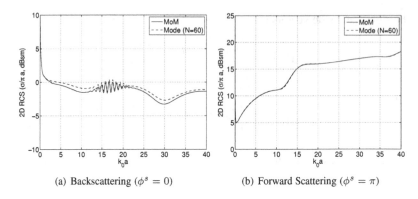

(a) Backscattering ($\phi^s = 0$)          (b) Forward Scattering ($\phi^s = \pi$)

**FIGURE 6.25:** 2D PEC Cylinder w/Lossy Coating: RCS versus $k_0 a$, TM Polarization

### 6.2.9.6    Coated Cylinder: TE Polarization

For the coated cylinder, the coefficients $B_n$ for TE polarization are [3]

$$B_n = -\frac{J_n(k_0 a_0) - iY_n J_n'(k_0 a_0)}{H_n^{(2)}(k_0 a_0) - iY_n H_n'^{(2)}(k_0 a_0)} , \qquad (6.169)$$

where

$$iY_n = \frac{k_0 \epsilon_1}{k_1 \epsilon_0}\left[\frac{J_n(k_1 a_0)H_n'^{(2)}(k_1 a_1) - H_n^{(2)}(k_1 a_0)J_n'(k_1 a_1)}{J_n'(k_1 a_0)H_n'^{(2)}(k_1 a_1) - H_n'^{(2)}(k_1 a_0)J_n'(k_1 a_1)}\right]. \qquad (6.170)$$

*MoM Linear System*

For the TE case, the MoM linear system is given by (6.151), where the sub-matrix blocks are

$$Z^{tt,TE} = \begin{bmatrix} L^{tt}\left(\mathbf{f}_{S_{01}}^t, \mathbf{f}_{S_{01}}^t, R_0\right) + L^{tt}\left(\mathbf{f}_{S_{10}}^t, \mathbf{f}_{S_{10}}^t, R_1\right) & L^{tt}\left(\mathbf{f}_{S_{10}}^t, \mathbf{f}_{S_{12}}^t, R_1\right) \\ \alpha L^{tt}\left(\mathbf{f}_{S_{12}}^t, \mathbf{f}_{S_{10}}^t, R_1\right) & \alpha L^{tt}\left(\mathbf{f}_{S_{12}}^t, \mathbf{f}_{S_{12}}^t, R_1\right) \end{bmatrix}$$
$$+ \begin{bmatrix} 0 & 0 \\ (1-\alpha)\eta_1 n K^{tt}\left(\mathbf{f}_{S_{12}}^t, \mathbf{f}_{S_{10}}^t, R_1\right) & (1-\alpha)\eta_1 n K^{tt}\left(\mathbf{f}_{S_{12}}^t, \mathbf{f}_{S_{12}}^t, R_1\right) \end{bmatrix}, \qquad (6.171)$$

$$Z^{tz,TE} = \begin{bmatrix} K^{tz}\left(\mathbf{f}_{S_{01}}^t, \mathbf{g}_{S_{01}}^z, R_0\right) + K^{tz}\left(\mathbf{f}_{S_{10}}^t, \mathbf{g}_{S_{10}}^z, R_1\right) \\ \alpha K^{tz}\left(\mathbf{f}_{S_{12}}^t, \mathbf{g}_{S_{10}}^z, R_1\right) \end{bmatrix}$$
$$+ \begin{bmatrix} 0 \\ (1-\alpha)\frac{1}{\eta_1} n L^{tz}\left(\mathbf{f}_{S_{12}}^t, \mathbf{g}_{S_{10}}^z, R_1\right) \end{bmatrix}, \qquad (6.172)$$

$$Z^{zt,TE} = \left[ -K^{zt}\left(\mathbf{g}^z_{S_{01}}, \mathbf{f}^t_{S_{01}}, R_0\right) - K^{zt}\left(\mathbf{g}^z_{S_{10}}, \mathbf{f}^t_{S_{10}}, R_1\right) \mid -K^{zt}\left(\mathbf{g}^z_{S_{10}}, \mathbf{f}^t_{S_{12}}, R_1\right) \right],$$
(6.173)

and

$$Z^{zz,TE} = \left[ \tfrac{\epsilon_0}{\mu_0} L^{zz}\left(\mathbf{g}^z_{S_{01}}, \mathbf{g}^z_{S_{01}}, R_0\right) + \tfrac{\epsilon_1}{\mu_1} L^{zz}\left(\mathbf{g}^z_{S_{10}}, \mathbf{g}^z_{S_{10}}, R_1\right) \right].$$
(6.174)

The right-hand side vectors are

$$V^{t,TE} = \begin{bmatrix} V^{E,TE}\left(\mathbf{f}^t_{S_{01}}, R_0\right) \\ 0 \end{bmatrix}$$
(6.175)

and

$$V^{z,TE} = \begin{bmatrix} V^{H,TE}\left(\mathbf{g}^z_{S_{01}}, R_0\right) \\ 0 \end{bmatrix},$$
(6.176)

and the solution vectors are

$$I^J = \begin{bmatrix} I^{J,t}_{S_{01}} \\ I^{J,t}_{S_{12}} \end{bmatrix}$$
(6.177)

and

$$I^{M,t} = \begin{bmatrix} I^{M,z}_{S_{01}} \\ 0 \end{bmatrix}.$$
(6.178)

*Results*

We consider again the coated cylinder of Section 6.2.9.5. The MoM results are compared to the modal solution for the lossless ($\epsilon_1 = 2.56$) and lossy ($\epsilon_1 = 2.56 - j.102$) coatings in Figures 6.26 and 6.27, respectively. The amplitude of the MoM results in the backscattering direction is again seen to be about $\frac{1}{2}$ dB below the modal solution.

### 6.2.9.7 Effect of Number of Segments per Wavelength on Accuracy

The MoM results in Sections 6.2.9.5 and 6.2.9.6 used 10 segments per $\lambda_0$ on both interfaces. While this number is sufficient on $S_{10}$, it is not on $S_{12}$, since $\lambda_1 < \lambda_0$. For $\epsilon_1 = 2.56$, $\lambda_1 = \lambda_0/\sqrt{\epsilon_1} = 0.625\lambda_0$. This suggests that $S_{12}$ should have almost twice the number of segments as does $S_{10}$. To demonstrate, consider a coated cylinder with $\epsilon_1 = 2.56$, $k_0 a_1 = 20$ and $k_0 a_0 = 22$. The bistatic RCS obtained using 10 and 20 segments per $\lambda_0$ on $S_{12}$ is shown in Figures 6.28a and 6.28b, respectively, for TE polarization. We see that the difference is eliminated by increasing to 20 segments per $\lambda_0$.

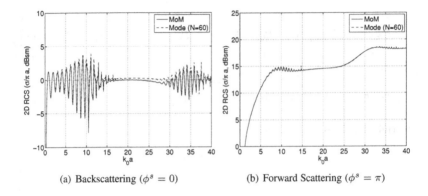

(a) Backscattering ($\phi^s = 0$)    (b) Forward Scattering ($\phi^s = \pi$)

**FIGURE 6.26:** 2D PEC Cylinder w/Lossless Coating: RCS versus $k_0 a$, TE Pol

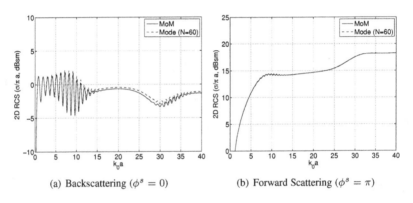

(a) Backscattering ($\phi^s = 0$)    (b) Forward Scattering ($\phi^s = \pi$)

**FIGURE 6.27:** 2D PEC Cylinder w/Lossy Coating: RCS versus $k_0 a$, TE Polarization

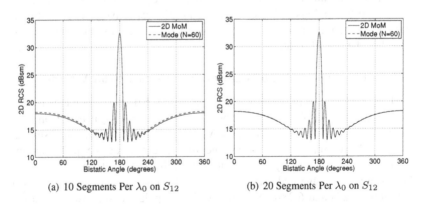

(a) 10 Segments Per $\lambda_0$ on $S_{12}$    (b) 20 Segments Per $\lambda_0$ on $S_{12}$

**FIGURE 6.28:** 2D PEC Cylinder with Lossless Coating: Bistatic RCS, TE Polarization

# References

[1] M. Abramowitz and I. Stegun, *Handbook of Mathematical Functions*. National Bureau of Standards, 1966.

[2] H. B. Dwight, *Tables of Integrals and Other Mathematical Data*. The Macmillan Company, 1961.

[3] G. T. Ruck, ed., *Radar Cross Section Handbook*. Plenum Press, 1970.

[4] L. N. Medgyesi-Mitschang and J. M. Putnam, "Electromagnetic scattering from axially inhomogeneous bodies of revolution," *IEEE Trans. Antennas Propagat.*, vol. 32, pp. 796–806, March 1984.

# Chapter 7

## Bodies of Revolution

In this chapter we will use the Method of Moments to solve surface integral equation problems involving bodies of revolution (BoRs), which comprise rotationally symmetric objects such as spheres, ellipsoids, and finite cylinders. This is of great practical interest, as such objects are commonly used to benchmark new computational electromagnetics techniques and codes, and to calibrate instruments that measure RCS in a static range. Furthermore, many realistic objects such as the conic reentry vehicle (RV) can be modeled as a body of revolution.

In Chapter 6, we solved the EFIE, MFIE, and nMFIE by formulating the common functional expressions (L, nL, K, and nK), and then assembling the MoM system matrix in a systematic way using those functions. In this chapter, we will use the same methodology to treat BoRs of general shape and configuration. We will first discuss modeling of the surface and the surface current expansion, and then derive the functional expressions required in each integral equation. We will also consider the issues with treating the intersection (junction) between three or more dielectric/conducting regions.

## 7.1 BoR Surface Description

Consider a body of revolution such as the cone-sphere in Figure 7.1a. Such an object can be completely described by a bounding curve and an axis of symmetry, where a rotation of the bounding curve by $2\pi$ about the symmetry axis sweeps through all points comprising the surface. Therefore, let us define a BoR in the cylindrical coordinate system as shown in Figure 7.1b. In this system, the axis of symmetry is the $\hat{\mathbf{z}}$ axis, and any point $\mathbf{r}$ on the surface can be identified by its cylindrical components $(\rho, \phi, z)$, or by its corresponding Cartesian components, given by

$$\mathbf{r} = \rho \cos \phi \hat{\mathbf{x}} + \rho \sin \phi \hat{\mathbf{y}} + z\hat{\mathbf{z}} .  \tag{7.1}$$

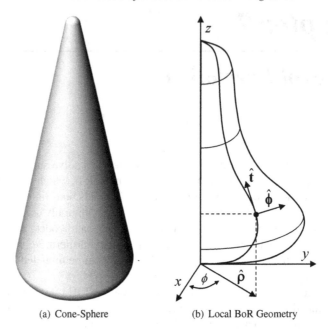

(a) Cone-Sphere                    (b) Local BoR Geometry

**FIGURE 7.1:** Body of Revolution (BoR)

The distance $r$ between any two points on the surface can be written as

$$r = |\mathbf{r} - \mathbf{r}'| = \sqrt{(\rho - \rho')^2 + (z - z')^2 + 2\rho\rho'[1 - \cos(\phi - \phi')]} . \quad (7.2)$$

We next define the longitudinal vector $\hat{\mathbf{t}}(\mathbf{r})$, which along with the cylindrical vector $\hat{\boldsymbol{\phi}}(\mathbf{r})$ is everywhere tangent to the surface. Given the surface normal $\hat{\mathbf{n}}(\mathbf{r})$, their relationship is defined as

$$\hat{\mathbf{n}}(\mathbf{r}) = \hat{\boldsymbol{\phi}}(\mathbf{r}) \times \hat{\mathbf{t}}(\mathbf{r}) . \quad (7.3)$$

As $\hat{\mathbf{t}}(\mathbf{r})$ and $\hat{\boldsymbol{\phi}}(\mathbf{r})$ are everywhere normal to each other, they comprise an orthogonal basis set that we will use to expand the surface currents.

## 7.2  Expansion of Surface Currents

Due to the rotational symmetry, we will describe the surface currents $\mathbf{J}$ and $\mathbf{M}$ using $\hat{\mathbf{t}}$- and $\hat{\boldsymbol{\phi}}$-oriented basis functions that are local in the longitudinal dimension and a Fourier series in the azimuthal dimension [1]. In a region

$R_l$ with bounding surface $S_l$, the electric current is expanded using $N$ basis functions of each type ($2N$ total basis functions), yielding [2]

$$\mathbf{J}_l(\mathbf{r}) = \sum_{\alpha=-\infty}^{\infty} \sum_{n=1}^{N} \left[ I_{\alpha n}^{J,t} \mathbf{f}_{\alpha n}^{t}(\mathbf{r}) - I_{\alpha n}^{J,\phi} \mathbf{f}_{\alpha n}^{\phi}(\mathbf{r}) \right], \qquad (7.4)$$

where $a_{\alpha n}^{t}$ and $a_{\alpha n}^{\phi}$ are the longitudinal and azimuthal expansion coefficients of basis function $n$ for Fourier mode $\alpha$. The functions $\mathbf{f}_{\alpha n}^{t,\phi}(\mathbf{r})$ are

$$\mathbf{f}_{\alpha n}^{t}(\mathbf{r}) = f_n(t) \, e^{j\alpha\phi} \, \hat{\mathbf{t}}(\mathbf{r}) \qquad (7.5)$$

and

$$\mathbf{f}_{\alpha n}^{\phi}(\mathbf{r}) = f_n(t) \, e^{j\alpha\phi} \, \hat{\boldsymbol{\phi}}(\mathbf{r}) , \qquad (7.6)$$

where for the one-dimensional function $f_n(t)$ we will use triangle functions (Section 2.3.2). For convenience, we will treat the integrations in the $t$ dimension in a generic sense in the derivations that follow, and present a more detailed discussion in Section 7.7. Similarly, the magnetic current in $R_l$ is

$$\mathbf{M}_l(\mathbf{r}) = \sum_{\alpha=-\infty}^{\infty} \sum_{n=1}^{N} \left[ I_{\alpha n}^{M,t} \mathbf{g}_{\alpha n}^{t}(\mathbf{r}) - I_{\alpha n}^{M,\phi} \mathbf{g}_{\alpha n}^{\phi}(\mathbf{r}) \right] \qquad (7.7)$$

where the functions $\mathbf{g}_{\alpha n}^{t,\phi}(\mathbf{r})$ are

$$\mathbf{g}_{\alpha n}^{t}(\mathbf{r}) = \eta_0 \, g_n(t) \, e^{j\alpha\phi} \, \hat{\mathbf{t}}(\mathbf{r}) \qquad (7.8)$$

and

$$\mathbf{g}_{\alpha n}^{\phi}(\mathbf{r}) = \eta_0 \, g_n(t) \, e^{j\alpha\phi} \, \hat{\boldsymbol{\phi}}(\mathbf{r}) . \qquad (7.9)$$

As it will simplify several expressions later, $f_n(t)$ and $g_n(t)$ are defined as

$$f_n(t) = g_n(t) = \frac{T_n(t)}{\rho(t)} , \qquad (7.10)$$

where $T_n(t)$ is the triangle function. The minus sign on the second term in (7.4) and (7.7) was chosen as it will yield symmetries in several terms in the sections that follow.

## 7.3 EFIE

The EFIE (3.178) contains $\mathcal{L}$ and $\mathcal{K}$ operators, which were discretized and tested via the MoM in (3.185) and (3.186). To implement these on a BoR, we require expressions for the testing functions. Following the approach in [1], we will use the complex conjugate of the basis functions as the electric testing functions. Taking the conjugate of the $\hat{t}$ and $\hat{\phi}$ components from (7.5) and (7.6) yields

$$\mathbf{f}_{\beta m}^{t}(\mathbf{r}) = f_n(t)\, e^{-j\beta\phi}\, \hat{t}(\mathbf{r}) \tag{7.11}$$

and

$$\mathbf{f}_{\beta m}^{\phi}(\mathbf{r}) = f_n(t)\, e^{-j\beta\phi}\, \hat{\phi}(\mathbf{r}) . \tag{7.12}$$

### 7.3.1 $\mathcal{L}$ Operator

Substitution of the basis (7.5, 7.6) and testing (7.11, 7.12) functions into (3.185) yields a matrix of the form

$$Z^{EJ(l)} = \begin{bmatrix} \mathrm{L}(\mathbf{f}_{\beta}^{t}, \mathbf{f}_{\alpha}^{t}) & \mathrm{L}(\mathbf{f}_{\beta}^{t}, \mathbf{f}_{\alpha}^{\phi}) \\ \mathrm{L}(\mathbf{f}_{\beta}^{\phi}, \mathbf{f}_{\alpha}^{t}) & \mathrm{L}(\mathbf{f}_{\beta}^{\phi}, \mathbf{f}_{\alpha}^{\phi}) \end{bmatrix} = \begin{bmatrix} \mathrm{L}^{tt} & \mathrm{L}^{t\phi} \\ \mathrm{L}^{\phi t} & \mathrm{L}^{\phi\phi} \end{bmatrix}, \tag{7.13}$$

where in region $R_l$, the elements of each sub-matrix block are given by

$$\mathrm{L}_{mn}^{**} = j\omega\mu_l \int_{\mathbf{f}_{\beta m}^{t,\phi}} \int_{\mathbf{f}_{\alpha n}^{t,\phi}} \mathbf{f}_{\beta m}^{t,\phi}(\mathbf{r}) \cdot \pm \mathbf{f}_{\alpha n}^{t,\phi}(\mathbf{r}') \frac{e^{-jk_l r}}{4\pi r}\, d\mathbf{r}'\, d\mathbf{r}$$
$$- \frac{j}{\omega\epsilon_l} \int_{\mathbf{f}_{\beta m}^{t,\phi}} \int_{\mathbf{f}_{\alpha n}^{t,\phi}} \left[\nabla \cdot \mathbf{f}_{\beta m}^{t,\phi}(\mathbf{r})\right]\left[\nabla' \cdot \pm \mathbf{f}_{\alpha n}^{t,\phi}(\mathbf{r}')\right] \frac{e^{-jk_l r}}{4\pi r}\, d\mathbf{r}'\, d\mathbf{r}, \tag{7.14}$$

and we have used (3.204) to re-distribute the differential operators. The $\hat{t}$- and $\hat{\phi}$-oriented basis functions (7.14) have positive and negative signs, respectively, following the convention in (7.4).

#### 7.3.1.1  L Matrix Elements

To compute the sub-matrix elements in (7.14) we must first determine the divergences of the basis and testing functions. These are

$$\nabla' \cdot \mathbf{f}_{\alpha n}^{t}(\mathbf{r}') = \frac{1}{\rho(t')}\frac{\partial}{\partial t'}\left[\rho(t')f_n(t')\right]e^{j\alpha\phi'} = \frac{1}{\rho(t')}T_n'(t')e^{j\alpha\phi'}, \tag{7.15}$$

$$\nabla \cdot \mathbf{f}_{\beta m}^{t}(\mathbf{r}) = \frac{1}{\rho(t)}\frac{\partial}{\partial t}\left[\rho(t)f_m(t)\right]e^{-j\beta\phi} = \frac{1}{\rho(t)}T_m'(t)e^{-j\beta\phi}, \tag{7.16}$$

$$\nabla' \cdot \mathbf{f}_{\alpha n}^{\phi}(\mathbf{r}') = \frac{f_n(t')}{\rho(t')}\frac{\partial}{\partial \phi'}e^{j\alpha\phi'} = \frac{j\alpha}{\rho(t')^2}T_n(t')e^{j\alpha\phi'}, \tag{7.17}$$

$$\nabla \cdot \mathbf{f}_{\beta m}^{\phi}(\mathbf{r}) = \frac{f_m(t)}{\rho(t)}\frac{\partial}{\partial \phi}e^{-j\beta\phi} = -\frac{j\beta}{\rho(t)^2}T_m(t)e^{-j\beta\phi}. \tag{7.18}$$

Using (7.15–7.18) and noting that the differential surface element in cylindrical coordinates is $ds = \rho \, d\phi \, dt$, the sub-matrix elements can be written as

$$
\mathrm{L}_{mn}^{tt} = \int_{T_m} \int_{T_n} \int_0^{2\pi} \int_0^{2\pi} \left[ j\omega\mu_l T_m(t) T_n(t') \, \hat{\mathbf{t}}(\mathbf{r}) \cdot \hat{\mathbf{t}}(\mathbf{r}') \right.
$$
$$
\left. - \frac{j}{\omega\epsilon_l} T_m'(t) T_n'(t') \right] e^{j(\alpha\phi' - \beta\phi)} \frac{e^{-jk_l r}}{4\pi r} \, d\phi' \, d\phi \, dt' \, dt \,, \tag{7.19}
$$

$$
\mathrm{L}_{mn}^{t\phi} = - \int_{T_m} \int_{T_n} \int_0^{2\pi} \int_0^{2\pi} \left[ j\omega\mu_l T_m(t) T_n(t') \, \hat{\mathbf{t}}(\mathbf{r}) \cdot \hat{\boldsymbol{\phi}}(\mathbf{r}') \right.
$$
$$
\left. + \frac{\alpha}{\omega\epsilon_l \rho(t')} T_m'(t) T_n(t') \right] e^{j(\alpha\phi' - \beta\phi)} \frac{e^{-jk_l r}}{4\pi r} \, d\phi' \, d\phi \, dt' \, dt \,, \tag{7.20}
$$

$$
\mathrm{L}_{mn}^{\phi t} = \int_{T_m} \int_{T_n} \int_0^{2\pi} \int_0^{2\pi} \left[ j\omega\mu_l T_m(t) T_n(t') \, \hat{\boldsymbol{\phi}}(\mathbf{r}) \cdot \hat{\mathbf{t}}(\mathbf{r}') \right.
$$
$$
\left. - \frac{\beta}{\omega\epsilon_l \rho(t)} T_m(t) T_n'(t') \right] e^{j(\alpha\phi' - \beta\phi)} \frac{e^{-jk_l r}}{4\pi r} \, d\phi' \, d\phi \, dt' \, dt \,, \tag{7.21}
$$

$$
\mathrm{L}_{mn}^{\phi\phi} = - \int_{T_m} \int_{T_n} \int_0^{2\pi} \int_0^{2\pi} \left[ j\omega\mu_l T_m(t) T_n(t') \, \hat{\boldsymbol{\phi}}(\mathbf{r}) \cdot \hat{\boldsymbol{\phi}}(\mathbf{r}') \right.
$$
$$
\left. - \frac{j\alpha\beta}{\omega\epsilon_l \rho(t)\rho(t')} T_m(t) T_n(t') \right] e^{j(\alpha\phi' - \beta\phi)} \frac{e^{-jk_l r}}{4\pi r} \, d\phi' \, d\phi \, dt' \, dt \,. \tag{7.22}
$$

We note that because integrals are periodic functions of $\phi'$ with period $2\pi$, $\phi'$ can be replaced by $\phi' + \phi$ without altering their values [1]. Making this substitution allows us to write

$$
\int_0^{2\pi} e^{j(\alpha\phi' - \beta\phi)} \, d\phi = e^{j\alpha\phi'} \int_0^{2\pi} e^{j(\alpha - \beta)\phi} \, d\phi \,, \tag{7.23}
$$

and noting that

$$
\int_0^{2\pi} e^{j(\alpha - \beta)\phi} \, d\phi = \begin{cases} 0 & \alpha \neq \beta \\ 2\pi & \alpha = \beta \end{cases} \,, \tag{7.24}
$$

only those integrals where $\alpha = \beta$ will remain. Additionally, we observe that for a BoR,

$$
\frac{e^{jkr(\rho, -\phi, z)}}{r(\rho, -\phi, z)} = \frac{e^{jkr(\rho, \phi, z)}}{r(\rho, \phi, z)} \qquad 0 \leq \phi \leq \pi \,. \tag{7.25}
$$

Let us now define $\hat{\mathbf{t}}(\mathbf{r})$ such that

$$
\hat{\mathbf{t}}(\mathbf{r}) = \sin\gamma \hat{\boldsymbol{\rho}} + \cos\gamma \hat{\mathbf{z}} \,, \tag{7.26}
$$

which allows us to write the following dot products:

$$\hat{\mathbf{t}}(\mathbf{r}) \cdot \hat{\mathbf{t}}(\mathbf{r}') = \sin\gamma \sin\gamma' \cos(\phi' - \phi) + \cos\gamma \cos\gamma' , \tag{7.27}$$

$$\hat{\mathbf{t}}(\mathbf{r}) \cdot \hat{\boldsymbol{\phi}}(\mathbf{r}') = -\sin\gamma \sin(\phi' - \phi) , \tag{7.28}$$

$$\hat{\boldsymbol{\phi}}(\mathbf{r}) \cdot \hat{\mathbf{t}}(\mathbf{r}') = \sin\gamma' \sin(\phi' - \phi) , \tag{7.29}$$

$$\hat{\boldsymbol{\phi}}(\mathbf{r}) \cdot \hat{\boldsymbol{\phi}}(\mathbf{r}') = \cos(\phi' - \phi) . \tag{7.30}$$

Taking into account (7.27–7.30), and using the orthogonality relationships that exist between the trigonometric terms in the various integrands, the sub-matrix elements are:

$$L_{mn}^{tt} = \int_{T_m} \int_{T_n} \left[ j\omega\mu_l T_m(t) T_n(t') \left( \sin\gamma \sin\gamma' G_2 + \cos\gamma \cos\gamma' G_1 \right) \right.$$
$$\left. - \frac{j}{\omega\epsilon_l} T_m'(t) T_n'(t') G_1 \right] dt' \, dt , \tag{7.31}$$

$$L_{mn}^{t\phi} = -\int_{T_m} \int_{T_n} \left[ \omega\mu_l T_m(t) T_n(t') \sin\gamma G_3 + \frac{\alpha}{\omega\epsilon_l \rho(t')} T_m'(t) T_n(t') G_1 \right] dt' \, dt , \tag{7.32}$$

$$L_{mn}^{\phi t} = -\int_{T_m} \int_{T_n} \left[ \omega\mu_l T_m(t) T_n(t') \sin\gamma' G_3 + \frac{\alpha}{\omega\epsilon_l \rho(t)} T_m(t) T_n'(t') G_1 \right] dt' \, dt , \tag{7.33}$$

$$L_{mn}^{\phi\phi} = -\int_{T_m} \int_{T_n} \left[ j\omega\mu_l T_m(t) T_n(t') G_2 - \frac{j\alpha^2}{\omega\epsilon_l \rho(t)\rho(t')} T_m(t) T_n(t') G_1 \right] dt' \, dt . \tag{7.34}$$

where the integrals $G_1$, $G_2$ and $G_3$ are

$$G_1 = \int_0^\pi \cos(\alpha\phi') \frac{e^{-jk_l r}}{r} d\phi' , \tag{7.35}$$

$$G_2 = \int_0^\pi \cos(\phi') \cos(\alpha\phi') \frac{e^{-jk_l r}}{r} d\phi' , \tag{7.36}$$

$$G_3 = \int_0^\pi \sin(\phi') \sin(\alpha\phi') \frac{e^{-jk_l r}}{r} d\phi' . \tag{7.37}$$

These integrals are evaluated by a one-dimensional Gaussian quadrature. When $\mathbf{r} \neq \mathbf{r}'$, $r$ can be obtained via

$$r = |\mathbf{r} - \mathbf{r}'| = \sqrt{(\rho - \rho')^2 + (z - z')^2 + 2\rho\rho'[1 - \cos(\phi')]} . \tag{7.38}$$

When $\mathbf{r} = \mathbf{r}'$, a modification of (7.38) for the above is needed, which is discussed in Section 7.7.

Inspecting the $L$ sub-matrix elements, we find that the relationship between positive and negative modes is

$$\begin{bmatrix} L^{tt} & L^{t\phi} \\ L^{\phi t} & L^{\phi\phi} \end{bmatrix}_{-\alpha} = \begin{bmatrix} L^{tt} & -L^{t\phi} \\ -L^{\phi t} & L^{\phi\phi} \end{bmatrix}_{\alpha} . \tag{7.39}$$

These properties survive inversion.

## 7.3.2 $\mathcal{K}$ Operator

Substitution of the magnetic basis and electric testing functions into (3.186) yields a matrix of the form

$$Z^{EM(l)} = \begin{bmatrix} \mathbf{K}(\mathbf{f}_\beta^t, \mathbf{g}_\alpha^t) & \mathbf{K}(\mathbf{f}_\beta^t, \mathbf{g}_\alpha^\phi) \\ \mathbf{K}(\mathbf{f}_\beta^\phi, \mathbf{g}_\alpha^t) & \mathbf{K}(\mathbf{f}_\beta^\phi, \mathbf{g}_\alpha^\phi) \end{bmatrix} = \begin{bmatrix} K^{tt} & K^{t\phi} \\ K^{\phi t} & K^{\phi\phi} \end{bmatrix}, \tag{7.40}$$

where in Region $R_l$, the elements of each sub-matrix block are given by

$$K_{mn}^{**} = \frac{1}{2} \int_{\mathbf{f}_{\beta m}^{t,\phi}, \mathbf{g}_{\alpha m}^{t,\phi}} \mathbf{f}_{\beta m}^{t,\phi}(\mathbf{r}) \cdot \left[\hat{\mathbf{n}}_l(\mathbf{r}) \times \pm \mathbf{g}_{\alpha m}^{t,\phi}(\mathbf{r})\right] d\mathbf{r}$$
$$- \int_{\mathbf{f}_{\beta m}^{t,\phi}} \mathbf{f}_{\beta m}^{t,\phi}(\mathbf{r}) \cdot \int_{\mathbf{g}_{\alpha m}^{t,\phi}} \left[(\mathbf{r} - \mathbf{r}') \times \pm \mathbf{g}_{\alpha m}^{t,\phi}(\mathbf{r}')\right] \frac{1 + jk_l r}{4\pi r^3} e^{-jk_l r} \, d\mathbf{r}' d\mathbf{r} . \tag{7.41}$$

### 7.3.2.1 K Matrix Elements

To evaluate the sub-matrix elements in (7.41), let us first make the following definitions

$$\mathbf{r} = \rho\hat{\boldsymbol{\rho}} + z\hat{\mathbf{z}}, \tag{7.42}$$

$$\mathbf{r}' = \rho' \cos(\phi' - \phi)\hat{\boldsymbol{\rho}} + \rho' \sin(\phi' - \phi)\hat{\boldsymbol{\phi}} + z'\hat{\mathbf{z}}, \tag{7.43}$$

$$\hat{\mathbf{t}}(\mathbf{r}') = \sin\gamma'\left[\cos(\phi' - \phi)\hat{\boldsymbol{\rho}} + \sin(\phi' - \phi)\hat{\boldsymbol{\phi}}\right] + \cos\gamma'\hat{\mathbf{z}}, \tag{7.44}$$

$$\hat{\boldsymbol{\phi}}(\mathbf{r}') = -\sin(\phi' - \phi)\hat{\boldsymbol{\rho}} + \cos(\phi' - \phi)\hat{\boldsymbol{\phi}}, \tag{7.45}$$

and then compute the following cross products

$$(\mathbf{r} - \mathbf{r}') \times \hat{\mathbf{t}}(\mathbf{r}') = \sin(\phi' - \phi)\left[(z' - z)\sin\gamma' - \rho'\cos\gamma'\right]\hat{\boldsymbol{\rho}}$$
$$- \left[(\rho - \rho'\cos(\phi' - \phi))\cos\gamma' + (z' - z)\sin\gamma'\cos(\phi' - \phi)\right]\hat{\boldsymbol{\phi}}$$
$$+ \rho\sin\gamma'\sin(\phi' - \phi)\hat{\mathbf{z}}, \tag{7.46}$$

and

$$(\mathbf{r} - \mathbf{r}') \times \hat{\boldsymbol{\phi}}(\mathbf{r}') = (z' - z)\cos(\phi' - \phi)\hat{\boldsymbol{\rho}} + (z' - z)\sin(\phi' - \phi)\hat{\boldsymbol{\phi}}$$
$$+ \left[\rho\cos(\phi' - \phi) - \rho'\right]\hat{\mathbf{z}}. \tag{7.47}$$

These definitions then allow us to write

$$\hat{t}(\mathbf{r}) \cdot \left[ (\mathbf{r} - \mathbf{r}') \times \hat{t}(\mathbf{r}') \right] = \left[ (z' - z) \sin\gamma \sin\gamma' - \rho' \cos\gamma' \sin\gamma \right.$$
$$\left. + \rho \cos\gamma \sin\gamma' \right] \sin(\phi' - \phi) , \tag{7.48}$$

$$\hat{t}(\mathbf{r}) \cdot \left[ (\mathbf{r} - \mathbf{r}') \times \hat{\phi}(\mathbf{r}') \right] = \left[ (z' - z) \sin\gamma \cos(\phi' - \phi) \right.$$
$$\left. + \left[ \rho \cos(\phi' - \phi) - \rho' \right] \cos\gamma \right], \tag{7.49}$$

$$\hat{\phi}(\mathbf{r}) \cdot \left[ (\mathbf{r} - \mathbf{r}') \times \hat{t}(\mathbf{r}') \right] = - \left[ (z' - z) \sin\gamma' \cos(\phi' - \phi) \right.$$
$$\left. + \left[ \rho - \rho' \cos(\phi' - \phi) \right] \cos\gamma' \right], \tag{7.50}$$

$$\hat{\phi}(\mathbf{r}) \cdot \left[ (\mathbf{r} - \mathbf{r}') \times \hat{\phi}(\mathbf{r}') \right] = (z' - z) \sin(\phi' - \phi) . \tag{7.51}$$

Using (7.48–7.51) as well as the periodic properties of the integrands outlined in Section 7.3.1.1, the system matrix can be re-written as

$$\begin{bmatrix} \mathbf{K}^{tt} & \mathbf{K}^{t\phi} + \tilde{\mathbf{K}}^{t\phi} \\ \mathbf{K}^{\phi t} + \tilde{\mathbf{K}}^{\phi t} & \mathbf{K}^{\phi\phi} \end{bmatrix} . \tag{7.52}$$

The sub-matrix elements for mode $\alpha$ are

$$\mathbf{K}^{tt}_{mn} = j\eta_0 \int_{T_m} \int_{T_n} T_m(t) \, T_n(t') \left( \left[ z(t) - z(t') \right] \sin\gamma(t) \sin\gamma(t') \right.$$
$$\left. + \sin\gamma(t) \cos\gamma(t')\rho(t') - \sin\gamma(t') \cos\gamma(t)\rho(t) \right) G_6 \, dt' \, dt , \tag{7.53}$$

$$\mathbf{K}^{t\phi}_{mn} = - \eta_0 \int_{T_m} \int_{T_n} T_m(t) \, T_n(t') \cdot \left( \left[ z(t) - z(t') \right] \sin\gamma(t) G_5 \right.$$
$$\left. + \cos\gamma(t) \left[ \rho(t') G_4 - \rho(t) G_5 \right] \right) dt' \, dt , \tag{7.54}$$

$$\mathbf{K}^{\phi t}_{mn} = - \eta_0 \int_{T_m} \int_{T_n} T_m(t) \, T_n(t') \cdot \left( \left[ z(t) - z(t') \right] \sin\gamma(t') G_5 \right.$$
$$\left. + \cos\gamma(t') \left[ \rho(t') G_5 - \rho(t) G_4 \right] \right) dt' \, dt , \tag{7.55}$$

$$\mathbf{K}^{\phi\phi}_{mn} = - j\eta_0 \int_{T_m} \int_{T_n} T_m(t) \, T_n(t') \left[ z(t) - z(t') \right] G_6 \, dt' \, dt , \tag{7.56}$$

where the integrations are performed on non-overlapping segments only. For overlapping segments, they are

$$\tilde{\mathbf{K}}^{t\phi}_{mn} = \tilde{\mathbf{K}}^{\phi t}_{mn} = \tilde{\mathbf{K}}_{mn} = -\pi\eta_0 \int_{T_m, T_n} \frac{T_m(t) T_n(t)}{\rho(t)} \, dt . \tag{7.57}$$

The integrals $G_4$, $G_5$, and $G_6$ are

$$G_4 = \int_0^\pi \cos(\alpha\phi') \frac{1 + jk_l r}{r^3} e^{-jk_l r} d\phi' , \tag{7.58}$$

$$G_5 = \int_0^\pi \cos(\phi') \cos(\alpha\phi') \frac{1 + jk_l r}{r^3} e^{-jk_l r} d\phi' , \tag{7.59}$$

$$G_6 = \int_0^\pi \sin(\phi') \sin(\alpha\phi') \frac{1 + jk_l r}{r^3} e^{-jk_l r} d\phi' , \tag{7.60}$$

where $r$ was defined in (7.38).

Inspecting the K sub-matrix elements, we find that the relationship between positive and negative modes is

$$\begin{bmatrix} \mathbf{K}^{tt} & \mathbf{K}^{t\phi} \\ \mathbf{K}^{\phi t} & \mathbf{K}^{\phi\phi} \end{bmatrix}_{-\alpha} = \begin{bmatrix} -\mathbf{K}^{tt} & \mathbf{K}^{t\phi} \\ \mathbf{K}^{\phi t} & -\mathbf{K}^{\phi\phi} \end{bmatrix}_\alpha . \tag{7.61}$$

These properties survive inversion.

### 7.3.3 Excitation

Right-hand side vector elements for the EFIE can be obtained by substituting the testing functions $\mathbf{f}_{\alpha m}^{t,\phi}(\mathbf{r})$ into (3.191). This yields a column vector of the form

$$V^{E(l)} = \begin{bmatrix} V^{E(l),t} \\ V^{E(l),\phi} \end{bmatrix} . \tag{7.62}$$

The individual elements are given by

$$V_m^{E(l),(t,\phi)} = \int_{T_m} \int_0^{2\pi} \mathbf{f}_{\alpha m}^{t,\phi}(\mathbf{r}) \cdot \mathbf{E}^i(\mathbf{r}) \, \rho \, d\phi \, dt , \tag{7.63}$$

where $\mathbf{E}^i(\mathbf{r})$ is the impressed electric field. In general, the above integral must be evaluated numerically, however for plane waves it can be evaluated in the $\phi$ dimension analytically.

### 7.3.3.1 Plane Wave Excitation

An incident plane wave in the free space region $R_0$ has a phase component at a point $\mathbf{r}$ given by

$$e^{jk_0 \hat{\mathbf{r}}^i \cdot \mathbf{r}} = e^{jk_0 z \cos\theta^i} e^{jk_0 \sin\theta^i (x \cos\phi^i + y \sin\phi^i)} , \tag{7.64}$$

where $(\theta^i, \phi^i)$ are the incident spherical angles, and $\hat{\mathbf{r}}^i$ the direction of incidence. In cylindrical coordinates, (7.64) can be written as

$$e^{jk_0 z \cos\theta^i} e^{jk_0 \sin\theta^i (x \cos\phi^i + y \sin\phi^i)} = e^{jk_0 z \cos\theta^i} e^{jk_0 \sin\theta^i \rho \cos(\phi - \phi^i)} . \tag{7.65}$$

To obtain a complete scattering matrix, we must consider $\hat{\theta}^i$- and $\hat{\phi}^i$-polarized incident waves, so we will treat each polarization separately. Using the fact that

$$\hat{\theta}^i = \cos\theta^i \cos\phi^i \hat{\mathbf{x}} + \cos\theta^i \sin\phi^i \hat{\mathbf{y}} - \sin\theta^i \hat{\mathbf{z}}, \tag{7.66}$$

and

$$\hat{\phi}^i = -\sin\phi^i \hat{\mathbf{x}} + \cos\phi^i \hat{\mathbf{y}}, \tag{7.67}$$

we take the dot products with the $\hat{\mathbf{t}}$ and $\hat{\phi}$ component vectors to yield

$$\hat{\mathbf{t}} \cdot \hat{\theta}^i = \cos\theta^i \sin\gamma \cos(\phi - \phi^i) - \sin\theta^i \cos\gamma, \tag{7.68}$$

$$\hat{\phi} \cdot \hat{\theta}^i = -\cos\theta^i \sin(\phi - \phi^i), \tag{7.69}$$

$$\hat{\mathbf{t}} \cdot \hat{\phi}^i = \sin\gamma \sin(\phi - \phi^i), \tag{7.70}$$

$$\hat{\phi} \cdot \hat{\phi}^i = \cos(\phi - \phi^i). \tag{7.71}$$

Testing the $\hat{\theta}^i$-polarized electric field with the $\hat{\mathbf{t}}$ and $\hat{\phi}$-oriented electric testing functions yields a right-hand side vector of the form

$$V^{E,\theta^i} = \begin{bmatrix} V^{E,t\theta^i} \\ V^{E,\phi\theta^i} \end{bmatrix}, \tag{7.72}$$

with elements given by

$$V_m^{E,t\theta^i} = \int_{T_m} T_m(t) e^{jk_0 z(t) \cos\theta^i} \int_0^{2\pi} \left[ \cos\theta^i \sin\gamma(t) \cos\phi - \sin\theta^i \cos\gamma(t) \right] \cdot$$
$$e^{jk_0 \sin\theta^i \rho(t) \cos\phi - j\alpha\phi} \, d\phi \, dt \tag{7.73}$$

and

$$V_m^{E,\phi\theta^i} = -\int_{T_m} T_m(t) e^{jk_0 z(t) \cos\theta^i} \int_0^{2\pi} \cos\theta^i \sin\phi \cdot$$
$$e^{jk_0 \sin\theta^i \rho(t) \cos\phi - j\alpha\phi} \, d\phi \, dt. \tag{7.74}$$

Similarly, testing the $\hat{\phi}^i$-polarized electric field yields a right-hand side vector of the form

$$V^{E,\phi^i} = \begin{bmatrix} V^{E,t\phi^i} \\ V^{E,\phi\phi^i} \end{bmatrix}, \tag{7.75}$$

with elements given by

$$V_m^{E,t\phi^i} = \int_{T_m} T_m(t) e^{jk_0 z(t) \cos\theta^i} \int_0^{2\pi} \sin\gamma(t) \sin\phi \, e^{jk_0 \sin\theta^i \rho(t) \cos\phi - j\alpha\phi} \, d\phi \, dt \tag{7.76}$$

and

$$V_m^{E,\phi\phi^i} = \int_{T_m} T_m(t)e^{jk_0 z(t)\cos\theta^i} \int_0^{2\pi} \cos\phi\, e^{jk_0 \sin\theta^i \rho(t)\cos\phi - j\alpha\phi}\, d\phi\, dt\,. \tag{7.77}$$

Due to the rotational symmetry, we have set $\phi^i$ to zero for simplicity. If we need to solve bistatic scattering problems, we can simply replace $\phi^s$ by $\phi^s - \phi^i$ when computing the scattered field in Section 7.5.2.1.

Let us now consider the innermost integral of $V_m^{E,t\theta^i}$ in (7.73), which is

$$I = \cos\theta^i \sin\gamma(t) \int_0^{2\pi} \cos\phi e^{jk_0 \sin\theta^i \rho(t)\cos\phi - j\alpha\phi}\, d\phi$$

$$- \sin\theta^i \cos\gamma(t) \int_0^{2\pi} e^{jk_0 \sin\theta^i \rho(t)\cos\phi - j\alpha\phi}\, d\phi\,. \tag{7.78}$$

Using the identity [3]

$$2\pi j^\alpha J_\alpha(z) = \int_0^{2\pi} e^{jz\cos\phi}\cos(\alpha\phi)\,, \tag{7.79}$$

the second integral in (7.3.3.1) is

$$\int_0^{2\pi} e^{jk_0 \sin\theta^i \rho(t)\cos\phi - j\alpha\phi}\, d\phi = 2\pi j^\alpha J_\alpha(x)\,, \tag{7.80}$$

where $x = k_0 \sin\theta^i \rho(t)$. We can then write (7.3.3.1) as

$$I = \cos\theta^i \sin\gamma(t) \int_0^{2\pi} \cos\phi e^{jk_0 \sin\theta^i \rho_p \cos\phi - j\alpha\phi}\, d\phi$$

$$- 2\pi j^\alpha J_\alpha(x)\sin\theta^i \cos\gamma(t)\,. \tag{7.81}$$

To evaluate the first integral in (7.3.3.1), we substitute $\cos\phi = e^{j\phi} - j\sin\phi$ to yield

$$\int_0^{2\pi} \cos\phi e^{jk \sin\theta^i \rho(t)\cos\phi - j\alpha\phi}\, d\phi = \int_0^{2\pi} (e^{j\phi} - j\sin\phi)e^{jk \sin\theta^i \rho(t)\cos\phi - j\alpha\phi}\, d\phi\,. \tag{7.82}$$

The first term is evaluated as before, yielding

$$2\pi j^{\alpha-1} J_{\alpha-1}(x) - j \int_0^{2\pi} \sin\phi e^{jk \sin\theta^i \rho(t)\cos\phi - j\alpha\phi}\, d\phi\,, \tag{7.83}$$

and the second term can be integrated by parts, yielding

$$2\pi j^{\alpha-1} J_{\alpha-1}(x) + j\left[\frac{2\pi\alpha}{x} j^\alpha J_\alpha(x)\right]\,. \tag{7.84}$$

Using the recurrence relationship

$$\frac{2\alpha}{x} J_\alpha(x) = J_{\alpha+1}(x) + J_{\alpha-1}(x) , \tag{7.85}$$

we can write (7.84) as

$$\pi j^\alpha j \Big[ J_{\alpha+1}(x) - J_{\alpha-1}(x) \Big] , \tag{7.86}$$

and $V_m^{E,t\theta^i}$ finally becomes

$$V_m^{E,t\theta^i} = \pi j^\alpha \int_{T_m} T_m(t) j^{kz(t)\cos\theta^i} \Big[ j \cos\theta^i \sin\gamma(t) \big[ J_{\alpha+1}(x) - J_{\alpha-1}(x) \big]$$

$$-2\sin\theta^i \cos\gamma(t) J_\alpha(x) \Big] \, dt . \tag{7.87}$$

The other elements are obtained similarly, yielding

$$V_m^{E,\phi\theta^i} = \pi j^\alpha \int_{T_m} T_m(t) e^{jkz(t)\cos\theta^i} \cos\theta^i \Big[ J_{\alpha+1}(x) + J_{\alpha-1}(x) \Big] \, dt , \tag{7.88}$$

$$V_m^{E,t\phi^i} = -\pi j^\alpha \int_{T_m} T_m(t) e^{jkz(t)\cos\theta^i} \sin\gamma(t) \Big[ J_{\alpha+1}(x) + J_{\alpha-1}(x) \Big] \, dt , \tag{7.89}$$

and

$$V_m^{E,\phi\phi^i} = \pi j^{\alpha+1} \int_{T_m} T_m(t) e^{jkz(t)\cos\theta^i} \Big[ J_{\alpha+1}(x) - J_{\alpha-1}(x) \Big] \, dt . \tag{7.90}$$

Inspecting (7.3.3.1–7.90), we find that the relationship between positive and negative modes is

$$\begin{bmatrix} V^{E,t\theta^i} \\ V^{E,\phi\theta^i} \end{bmatrix}_{-\alpha} = \begin{bmatrix} V^{E,t\theta^i} \\ -V^{E,\phi\theta^i} \end{bmatrix}_{\alpha} \tag{7.91}$$

and

$$\begin{bmatrix} V^{E,t\phi^i} \\ V^{E,\phi\phi^i} \end{bmatrix}_{-\alpha} = \begin{bmatrix} -V^{E,t\phi^i} \\ V^{E,\phi\phi^i} \end{bmatrix}_{\alpha} . \tag{7.92}$$

## 7.4 MFIE

The MFIE (3.179) also has $\mathcal{K}$ and $\mathcal{L}$ operators, which were discretized and tested via the MoM in (3.187) and (3.188). For the testing functions, we use the complex conjugates of the magnetic basis functions (7.8) and (7.9). Substitution of the electric basis and magnetic testing functions into (3.187) yields in region $R_l$ a matrix of the form

$$
Z_{mn}^{HJ(l)} = - \begin{bmatrix} K(\mathbf{g}_\beta^t, \mathbf{f}_\alpha^t) & K(\mathbf{g}_\beta^t, \mathbf{f}_\alpha^\phi) \\ K(\mathbf{g}_\beta^\phi, \mathbf{f}_\alpha^t) & K(\mathbf{g}_\beta^\phi, \mathbf{f}_\alpha^\phi) \end{bmatrix} = - \begin{bmatrix} K^{tt} & K^{t\phi} \\ K^{\phi t} & K^{\phi\phi} \end{bmatrix}, \qquad (7.93)
$$

and substitution of the magnetic basis and testing functions into (3.188) yields

$$
Z_{mn}^{HM(l)} = \frac{\epsilon_l}{\mu_l} \begin{bmatrix} L(\mathbf{g}_\beta^t, \mathbf{g}_\alpha^t) & L(\mathbf{g}_\beta^t, \mathbf{g}_\alpha^\phi) \\ L(\mathbf{g}_\beta^\phi, \mathbf{g}_\alpha^t) & L(\mathbf{g}_\beta^\phi, \mathbf{g}_\alpha^\phi) \end{bmatrix} = \eta_0^2 \frac{\epsilon_l}{\mu_l} \begin{bmatrix} L^{tt} & L^{t\phi} \\ L^{\phi t} & L^{\phi\phi} \end{bmatrix}, \qquad (7.94)
$$

where the expressions for the L and K sub-matrix elements were derived previously in Sections 7.3.1.1 and 7.3.2.1, respectively.

### 7.4.1 Excitation

Right-hand side vector elements for the MFIE can be obtained by substituting the testing functions $\mathbf{g}_{\alpha m}^{t,\phi}(\mathbf{r})$ into (3.192). This yields a column vector of length $2N$ of the form

$$
V^{H(l)} = \begin{bmatrix} V^{H(l),t} \\ V^{H(l),\phi} \end{bmatrix}. \qquad (7.95)
$$

The individual elements are given by

$$
V_m^{H(l),(t,\phi)} = \int_{T_m} \int_0^{2\pi} \mathbf{g}_{\alpha m}^{t,\phi}(\mathbf{r}) \cdot \mathbf{H}^i(\mathbf{r}) \, \rho \, d\phi \, dt , \qquad (7.96)
$$

where $\mathbf{H}^i(\mathbf{r})$ is the impressed magnetic field. As with the EFIE, this can be evaluated analytically in the $\phi$ dimension for incident plane waves.

#### 7.4.1.1 Plane Wave Excitation

Given planar, $\hat{\theta}^i$- and $\hat{\phi}^i$-polarized incident electric fields, the corresponding magnetic fields are $-\hat{\phi}^i$- and $\hat{\theta}^i$-polarized, respectively. Thus, we can re-use the components of $V^{E(l)}$, yielding MFIE right-hand side vectors given by

$$
V^{H,\theta^i} = \begin{bmatrix} V^{H,t\theta^i} \\ V^{H,\phi\theta^i} \end{bmatrix} = \begin{bmatrix} -V^{E,t\phi^i} \\ -V^{E,\phi\phi^i} \end{bmatrix}, \qquad (7.97)
$$

and

$$V^{H,\phi^i} = \begin{bmatrix} V^{H,t\phi^i} \\ V^{H,\phi\phi^i} \end{bmatrix} = \begin{bmatrix} V^{E,t\theta^i} \\ V^{E,\phi\theta^i} \end{bmatrix}. \tag{7.98}$$

## 7.5   Solution

Assembling the sub-matrix blocks for the EFIE and MFIE, the system matrix (3.184) for a general BoR has the form

$$\begin{bmatrix} Z^{EJ,tt} & Z^{EJ,t\phi} & Z^{EM,tt} & Z^{EM,t\phi} \\ Z^{EJ,\phi t} & Z^{EJ,\phi\phi} & Z^{EM,\phi t} & Z^{EM,\phi\phi} \\ Z^{HJ,tt} & Z^{HJ,t\phi} & Z^{HM,tt} & Z^{HM,t\phi} \\ Z^{HJ,\phi t} & Z^{HJ,\phi\phi} & Z^{HM,\phi t} & Z^{HM,\phi\phi} \end{bmatrix} \begin{bmatrix} I^{J,t} \\ I^{J,\phi} \\ I^{M,t} \\ I^{M,\phi} \end{bmatrix} = \begin{bmatrix} V^{E,t} \\ V^{E,\phi} \\ V^{H,t} \\ V^{H,\phi} \end{bmatrix}, \tag{7.99}$$

where the nMFIE is omitted for now.

### 7.5.1   Plane Wave Solution

Substitution of the matrix and vector sub-blocks for $\hat{\theta}^i$-polarization into (7.99) yields as the solution for mode $\alpha$

$$\begin{bmatrix} I^{J,t\theta^i} \\ I^{J,\phi\theta^i} \\ I^{M,t\theta^i} \\ I^{M,\phi\theta^i} \end{bmatrix}_\alpha = \begin{bmatrix} L_E^{tt} & L_E^{t\phi} & K_E^{tt} & K_E^{t\phi} \\ L_E^{\phi t} & L_E^{\phi\phi} & K_E^{\phi t} & K_E^{\phi\phi} \\ K_H^{tt} & K_H^{t\phi} & L_H^{tt} & L_H^{t\phi} \\ K_H^{\phi t} & K_H^{\phi\phi} & L_H^{\phi t} & L_H^{\phi\phi} \end{bmatrix}_\alpha^{-1} \begin{bmatrix} V^{E,t\theta^i} \\ V^{E,\phi\theta^i} \\ -V^{E,t\phi^i} \\ -V^{E,\phi\phi^i} \end{bmatrix}_\alpha. \tag{7.100}$$

Using (7.39) and (7.61), together with (7.91) and (7.92), the solution (7.100) for negative modes can be written as

$$\begin{bmatrix} I^{J,t\theta^i} \\ I^{J,\phi\theta^i} \\ I^{M,t\theta^i} \\ I^{M,\phi\theta^i} \end{bmatrix}_{-\alpha} = \begin{bmatrix} L_E^{tt} & -L_E^{t\phi} & -K_E^{tt} & K_E^{t\phi} \\ -L_E^{\phi t} & L_E^{\phi\phi} & K_E^{\phi t} & -K_E^{\phi\phi} \\ -K_H^{tt} & K_H^{t\phi} & L_H^{tt} & -L_H^{t\phi} \\ K_H^{\phi t} & -K_H^{\phi\phi} & -L_H^{\phi t} & L_H^{\phi\phi} \end{bmatrix}_\alpha^{-1} \begin{bmatrix} V^{E,t\theta^i} \\ -V^{E,\phi\theta^i} \\ V^{E,t\phi^i} \\ -V^{E,\phi\phi^i} \end{bmatrix}_\alpha. \tag{7.101}$$

Therefore,

$$\begin{bmatrix} I^{J,t\theta^i} \\ I^{J,\phi\theta^i} \\ I^{M,t\theta^i} \\ I^{M,\phi\theta^i} \end{bmatrix}_{-\alpha} = \begin{bmatrix} I^{J,t\theta^i} \\ -I^{J,\phi\theta^i} \\ -I^{M,t\theta^i} \\ I^{M,\phi\theta^i} \end{bmatrix}_\alpha, \tag{7.102}$$

where only $I^{J,t\theta^i}$ and $I^{M,\phi\theta^i}$ are nonzero for $\alpha = 0$. Similarly, for $\hat{\phi}^i$-polarization, the solution to (7.99) for mode $\alpha$ is

$$
\begin{bmatrix}
I^{J,t\phi^i} \\
I^{J,\phi\phi^i} \\
I^{M,t\phi^i} \\
I^{M,\phi\phi^i}
\end{bmatrix}_\alpha
=
\begin{bmatrix}
L_E^{tt} & L_E^{t\phi} & K_E^{tt} & K_E^{t\phi} \\
L_E^{\phi t} & L_E^{\phi\phi} & K_E^{\phi t} & K_E^{\phi\phi} \\
K_H^{tt} & K_H^{t\phi} & L_H^{tt} & L_H^{t\phi} \\
K_H^{\phi t} & K_H^{\phi\phi} & L_H^{\phi t} & L_H^{\phi\phi}
\end{bmatrix}_\alpha^{-1}
\begin{bmatrix}
V^{E,t\phi^i} \\
V^{E,\phi\phi^i} \\
V^{E,t\theta^i} \\
V^{E,\phi\theta^i}
\end{bmatrix}_\alpha ,
\tag{7.103}
$$

and for negative modes it is

$$
\begin{bmatrix}
I^{J,t\phi^i} \\
I^{J,\phi\phi^i} \\
I^{M,t\phi^i} \\
I^{M,\phi\phi^i}
\end{bmatrix}_{-\alpha}
=
\begin{bmatrix}
L_E^{tt} & -L_E^{t\phi} & -K_E^{tt} & K_E^{t\phi} \\
-L_E^{\phi t} & L_E^{\phi\phi} & K_E^{\phi t} & -K_E^{\phi\phi} \\
-K_H^{tt} & K_H^{t\phi} & L_H^{tt} & -L_H^{t\phi} \\
K_H^{\phi t} & -K_H^{\phi\phi} & -L_H^{\phi t} & L_H^{\phi\phi}
\end{bmatrix}_\alpha^{-1}
\begin{bmatrix}
-V^{E,t\phi^i} \\
V^{E,\phi\phi^i} \\
V^{E,t\theta^i} \\
-V^{E,\phi\theta^i}
\end{bmatrix}_\alpha .
\tag{7.104}
$$

Therefore,

$$
\begin{bmatrix}
I^{J,t\phi^i} \\
I^{J,\phi\phi^i} \\
I^{M,t\phi^i} \\
I^{M,\phi\phi^i}
\end{bmatrix}_{-\alpha}
=
\begin{bmatrix}
-I^{J,t\phi^i} \\
I^{J,\phi\phi^i} \\
I^{M,t\phi^i} \\
-I^{M,\phi\phi^i}
\end{bmatrix}_\alpha ,
\tag{7.105}
$$

where only $I^{J,\phi\phi^i}$ and $I^{M,t\phi^i}$ are nonzero for $\alpha = 0$.

### 7.5.1.1   Currents

For $\hat{\theta}^i$-polarization, the electric and magnetic currents for basis function $m$ can be written as

$$
\mathbf{J}_m^{\theta^i}(\mathbf{r}) = \sum_{\alpha=-\infty}^{\infty} \left[ I_{\alpha m}^{J,t\theta^i} f_m(t)\hat{\mathbf{t}}(\mathbf{r}) - I_{\alpha m}^{J,\phi\theta^i} f_m(t)\hat{\phi}(\mathbf{r}) \right] e^{j\alpha\phi} ,
\tag{7.106}
$$

and

$$
\mathbf{M}_m^{\theta^i}(\mathbf{r}) = \sum_{\alpha=-\infty}^{\infty} \left[ I_{\alpha m}^{M,t\theta^i} g_m(t)\hat{\mathbf{t}}(\mathbf{r}) - I_{\alpha m}^{M,\phi\theta^i} g_m(t)\hat{\phi}(\mathbf{r}) \right] e^{j\alpha\phi} .
\tag{7.107}
$$

Using (7.102), these currents can be re-written using only positive modes as

$$
\mathbf{J}_m^{\theta^i}(\mathbf{r}) = I_{0m}^{J,t\theta^i} f_m(t)\hat{\mathbf{t}}(\mathbf{r})
$$
$$
+ 2\sum_{\alpha=1}^{\infty} \left[ I_{\alpha m}^{J,t\theta^i} f_m(t)\cos(\alpha\phi)\hat{\mathbf{t}}(\mathbf{r}) - jI_{\alpha m}^{J,\phi\theta^i} f_m(t)\sin(\alpha\phi)\hat{\phi}(\mathbf{r}) \right],
$$
$$
\tag{7.108}
$$

and

$$
\mathbf{M}_m^{\theta^i}(\mathbf{r}) = - I_{0m}^{M,\phi\theta^i} g_m(t) \hat{\phi}(\mathbf{r})
$$
$$
+ 2 \sum_{\alpha=1}^{\infty} \left[ j I_{\alpha m}^{M,t\theta^i} g_m(t) \sin(\alpha\phi) \hat{\mathbf{t}}(\mathbf{r}) - I_{\alpha m}^{M,\phi\theta^i} g_m(t) \cos(\alpha\phi) \hat{\phi}(\mathbf{r}) \right].
$$
$$(7.109)$$

Similarly, for $\hat{\phi}^i$-polarization, the currents are

$$
\mathbf{J}_m^{\phi^i}(\mathbf{r}) = - I_{0m}^{J,\phi\phi^i} f_m(t) \hat{\phi}(\mathbf{r})
$$
$$
+ 2 \sum_{\alpha=1}^{\infty} \left[ j I_{\alpha m}^{J,t\phi^i} f_m(t) \sin(\alpha\phi) \hat{\mathbf{t}}(\mathbf{r}) - I_{\alpha m}^{J,\phi\phi^i} f_m(t) \cos(\alpha\phi) \hat{\phi}(\mathbf{r}) \right],
$$
$$(7.110)$$

and

$$
\mathbf{M}_m^{\phi^i}(\mathbf{r}) = I_{0m}^{M,t\phi^i} g_m(t) \hat{\mathbf{t}}(\mathbf{r})
$$
$$
+ 2 \sum_{\alpha=1}^{\infty} \left[ I_{\alpha m}^{M,t\phi^i} g_m(t) \cos(\alpha\phi) \hat{\mathbf{t}}(\mathbf{r}) - j I_{\alpha m}^{M,\phi\phi^i} g_m(t) \sin(\alpha\phi) \hat{\phi}(\mathbf{r}) \right].
$$
$$(7.111)$$

### 7.5.2 Scattered Field

The scattered near electric field in Region $R_l$ can be obtained by substituting (7.4) and (7.7) into (3.108) and (3.112), and performing the integration over all basis functions in $R_l$. For general problems, the near field integrals must be performed numerically in the $t$ and $\phi$ dimensions for each mode.

#### 7.5.2.1 Scattered Far Fields

Using (7.108–7.111), the scattered far electric fields in the exterior free space region $R_0$ are obtained using (3.133), with the components of $\mathbf{A}(\mathbf{r})$ and $\mathbf{F}(\mathbf{r})$ given by

$$
A_{**}(\mathbf{r}) = \mu_0 \frac{e^{-jk_0 r}}{4\pi r} \sum_{m=1}^{N} \int_{t_m} (\hat{\boldsymbol{\theta}}^s, \hat{\boldsymbol{\phi}}^s) \cdot \int_0^{2\pi} \mathbf{J}_m^{\theta^i,\phi^i}(\mathbf{r}) e^{jk_0 \hat{\mathbf{r}}^s \cdot \mathbf{r}'} \rho \, d\phi \, dt \quad (7.112)
$$

and

$$
F_{**}(\mathbf{r}) = \epsilon_0 \frac{e^{-jk_0 r}}{4\pi r} \sum_{m=1}^{N} \int_{t_m} (\hat{\boldsymbol{\theta}}^s, \hat{\boldsymbol{\phi}}^s) \cdot \int_0^{2\pi} \mathbf{M}_m^{\theta^i,\phi^i}(\mathbf{r}) e^{jk_0 \hat{\mathbf{r}}^s \cdot \mathbf{r}'} \rho \, d\phi \, dt \, ,
$$
$$(7.113)$$

where $(\theta^s, \phi^s)$ are the spherical scattering angles, and $\hat{\mathbf{r}}^s$ the direction of scattering. The integration is performed over all basis functions in $R_0$. As before, the integrations versus $\phi$ in (7.112) and (7.113) can be performed analytically.

Let us focus on the innermost integral in (7.112) for basis function $m$, which has the form

$$(\hat{\theta}^s, \hat{\phi}^s) \cdot \int_0^{2\pi} \mathbf{J}_m^{\theta^i, \phi^i}(\mathbf{r}) e^{jk_0 \hat{\mathbf{r}}^s \cdot \mathbf{r}'} \rho \, d\phi . \tag{7.114}$$

The exponential term can be written as

$$e^{jk_0 \hat{\mathbf{r}}^s \cdot \mathbf{r}'} = e^{jk_0 z \cos \theta^s} e^{jk_0 \sin \theta^s \rho \cos(\phi - \phi^s)} . \tag{7.115}$$

Applying the dot products from before, the $(\theta^s, \theta^i)$ term for modes $\alpha > 0$ is

$$I_{\alpha m}^{J, t\theta^i} \int_0^{2\pi} e^{jk_0 \sin \theta^s \rho(t) \cos \phi} \cos(\alpha\phi + \alpha\phi^s) A(\theta^s, \phi, t) \, d\phi$$

$$+ I_{\alpha m}^{J, \phi\theta^i} \int_0^{2\pi} j e^{jk_0 \sin \theta^s \rho(t) \cos \phi} \sin(\alpha\phi + \alpha\phi^s) \sin \phi \cos \theta^s \, d\phi , \tag{7.116}$$

where

$$A(\theta^s, \phi, t) = \cos \theta^s \sin \gamma(t) \cos \phi - \sin \theta^s \cos \gamma(t) , \tag{7.117}$$

and we have made the substitution $\phi = \phi + \phi^s$ due to the $2\pi$ periodicity. Now consider the first integral in (7.116). Using the identity

$$\cos(\alpha\phi + \alpha\phi^s) = \cos(\alpha\phi) \cos(\alpha\phi^s) - \sin(\alpha\phi) \sin(\alpha\phi^s) , \tag{7.118}$$

this integral becomes

$$I_1 = \cos(\alpha\phi^s) \int_0^{2\pi} e^{jk_0 \sin \theta^s \rho(t) \cos \phi} \cos(\alpha\phi) A(\theta^s, \phi, t) \, d\phi$$

$$- \sin(\alpha\phi^s) \int_0^{2\pi} e^{jk_0 \sin \theta^s \rho(t) \cos \phi} \sin(\alpha\phi) A(\theta^s, \phi, t) \, d\phi . \tag{7.119}$$

The second term goes to zero by inspection. Using (7.79), the integral can then be written as

$$I_1 = \pi j^\alpha \left[ \cos \theta^s \sin \gamma(t) j \left[ J_{\alpha+1}(x) - J_{\alpha-1}(x) \right] - 2 \sin \theta^s \cos \gamma(t) J_\alpha(x) \right] \cos(\alpha\phi^s) \tag{7.120}$$

where $J_\alpha(x) = J_\alpha(k_0 \rho_p \sin \theta^s)$. Using the identity

$$\sin(\alpha\phi + \alpha\phi^s) = \sin(\alpha\phi) \cos(\alpha\phi^s) + \cos(\alpha\phi) \sin(\alpha\phi^s) \tag{7.121}$$

the second integral in (7.116) becomes

$$
\begin{aligned}
I_2 = \cos(\alpha\phi^s)\cos\theta^s \int_0^{2\pi} j e^{jk_0\sin\theta^s\rho(t)\cos\phi}\sin(\alpha\phi)\sin\phi \, d\phi \\
+ \sin(\alpha\phi^s)\cos\theta^s \int_0^{2\pi} j e^{jk_0\sin\theta^s\rho(t)\cos\phi}\cos(\alpha\phi)\sin\phi \, d\phi \, .
\end{aligned} \tag{7.122}
$$

The second term again is zero by inspection. The first term can be integrated by parts yielding

$$
I_2 = \cos(\alpha\phi^s)\cos\theta^s \pi j^\alpha \frac{2\alpha}{x} J_\alpha(x) \, , \tag{7.123}
$$

and again using (7.85) this becomes

$$
I_2 = \pi j^\alpha \left[ J_{\alpha+1}(x) + J_{\alpha-1}(x) \right] \cos\theta^s \cos(\alpha\phi^s) \, . \tag{7.124}
$$

Combining the results, we see that the integrals have the same form as (7.3.3.1) and (7.88), where $\theta^i$ is replaced by $\theta^s$. Thus, $A_{\theta\theta}$ can be written as

$$
A_{\theta\theta}(\mathbf{r}) = \frac{C_A}{2} I_{0m}^{J,t\theta^i} V_0^{E,t\theta^s} + C_A \sum_{\alpha=1}^\infty \left[ I_{\alpha m}^{J,t\theta^i} V_\alpha^{E,t\theta^s} + I_{\alpha m}^{J,\phi\theta^i} V_\alpha^{E,\phi\theta^s} \right] \cos(\alpha\phi^s) \, . \tag{7.125}
$$

The other terms follow similarly, and are

$$
A_{\phi\theta}(\mathbf{r}) = -jC_A \sum_{\alpha=1}^\infty \left[ I_{\alpha n}^{J,t\theta^i} V_\alpha^{E,t\phi^s} + I_{\alpha n}^{J,\phi\theta^i} V_\alpha^{E,\phi\phi^s} \right] \sin(\alpha\phi^s) \, , \tag{7.126}
$$

$$
A_{\theta\phi}(\mathbf{r}) = jC_A \sum_{\alpha=1}^\infty \left[ I_{\alpha n}^{J,t\phi^i} V_\alpha^{E,t\theta^s} + I_{\alpha n}^{J,\phi\phi^i} V_\alpha^{E,\phi\theta^s} \right] \sin(\alpha\phi^s) \, , \tag{7.127}
$$

and

$$
A_{\phi\phi}(\mathbf{r}) = -\frac{C_A}{2} I_{0m}^{J,\phi\phi^i} V_0^{E,\phi\phi^s} - C_A \sum_{\alpha=1}^\infty \left[ I_{\alpha n}^{J,t\phi^i} V_\alpha^{E,t\phi^s} + I_{\alpha n}^{J,\phi\phi^i} V_\alpha^{E,\phi\phi^s} \right] \cos(\alpha\phi^s) \, , \tag{7.128}
$$

where

$$
C_A = \mu_0 \frac{e^{-jk_0 r}}{2\pi r} \, . \tag{7.129}
$$

Similarly, the components of $\mathbf{F}(\mathbf{r})$ are

$$
F_{\theta\theta}(\mathbf{r}) = jC_F \sum_{\alpha=1}^\infty \left[ I_{\alpha n}^{M,t\theta^i} V_\alpha^{E,t\theta^s} + I_{\alpha n}^{M,\phi\theta^i} V_\alpha^{E,\phi\theta^s} \right] \sin(\alpha\phi^s) \, , \tag{7.130}
$$

$$F_{\phi\theta}(\mathbf{r}) = -\frac{C_F}{2}I_{0m}^{M,\phi\theta^i}V_0^{E,\phi\phi^s} - C_F\sum_{\alpha=1}^{\infty}\left[I_{\alpha n}^{M,t\theta^i}V_\alpha^{E,t\phi^s} + I_{\alpha n}^{M,\phi\theta^i}V_\alpha^{E,\phi\phi^s}\right]\cos(\alpha\phi^s),$$

$$(7.131)$$

$$F_{\theta\phi}(\mathbf{r}) = \frac{C_F}{2}I_{0m}^{M,t\phi^i}V_0^{E,t\theta^s} + C_F\sum_{\alpha=1}^{\infty}\left[I_{\alpha m}^{M,t\phi^i}V_\alpha^{E,t\theta^s} + I_{\alpha m}^{M,\phi\phi^i}V_\alpha^{E,\phi\theta^s}\right]\cos(\alpha\phi^s),$$

$$(7.132)$$

and

$$F_{\phi\phi}(\mathbf{r}) = -jC_F\sum_{\alpha=1}^{\infty}\left[I_{\alpha n}^{M,t\phi^i}V_\alpha^{E,t\phi^s} + I_{\alpha n}^{M,\phi\phi^i}V_\alpha^{E,\phi\phi^s}\right]\sin(\alpha\phi^s),\quad(7.133)$$

where

$$C_F = \epsilon_0\eta_0\frac{e^{-jk_0 r}}{2\pi r}.\qquad(7.134)$$

Note that the right-hand side vectors $V^{E,\theta^i}$ and $V^{E,\phi^i}$ contain no references to the incident or observation angles $\phi^i$ or $\phi^s$. For monostatic calculations, these vectors can be computed once per right-hand side and used to obtain the currents as well as the scattered fields.

## 7.6 nMFIE

The nMFIE (3.180) contains the $\hat{\mathbf{n}}\times\mathcal{L}$ and $\hat{\mathbf{n}}\times\mathcal{K}$ operators, which were discretized and tested via the MoM in (3.189) and (3.190). For the testing functions, we will use the electric testing functions in (7.11) and (7.12).

### 7.6.1 $\hat{\mathbf{n}}\times\mathcal{L}$ Operator

Substitution of the basis (7.8, 7.9) and testing (7.11, 7.12) functions into (3.190) yields a $2N\times 2N$ matrix of the form

$$Z^{nHM(l)} = \begin{bmatrix} \mathrm{nL}(\mathbf{f}_\beta^t,\mathbf{g}_\alpha^t) & \mathrm{nL}(\mathbf{f}_\beta^t,\mathbf{g}_\alpha^\phi) \\ \mathrm{nL}(\mathbf{f}_\beta^\phi,\mathbf{g}_\alpha^t) & \mathrm{nL}(\mathbf{f}_\beta^\phi,\mathbf{g}_\alpha^\phi) \end{bmatrix} = \begin{bmatrix} \mathrm{nL}^{tt} & \mathrm{nL}^{t\phi} \\ \mathrm{nL}^{\phi t} & \mathrm{nL}^{\phi\phi} \end{bmatrix}.\qquad(7.135)$$

The elements of each sub-matrix block are given by

$$\mathrm{nL}_{mn}^{**} = j\omega\epsilon_l\int_{\mathbf{f}_{\beta m}^{t,\phi}}\int_{\mathbf{g}_{\alpha n}^{t,\phi}}\left[\mathbf{f}_{\beta m}^{t,\phi}(\mathbf{r})\times\hat{\mathbf{n}}_l(\mathbf{r})\right]\cdot\int_{\mathbf{g}_{\alpha n}^{t,\phi}}\pm\mathbf{g}_{\alpha n}^{t,\phi}(\mathbf{r}')\frac{e^{-jk_l r}}{4\pi r}\,d\mathbf{r}'\right]d\mathbf{r}$$

$$-\frac{j}{\omega\mu_l}\int_{\mathbf{f}_{\beta m}^{t,\phi}}\nabla\cdot\left[\mathbf{f}_{\beta m}^{t,\phi}(\mathbf{r})\times\hat{\mathbf{n}}_l(\mathbf{r})\right]\int_{\mathbf{g}_{\alpha n}^{t,\phi}}\left[\nabla'\cdot\pm\mathbf{g}_{\alpha n}^{t,\phi}(\mathbf{r}')\right]\frac{e^{-jk_l r}}{4\pi r}\,d\mathbf{r}'\,d\mathbf{r},$$

$$(7.136)$$

where we have used (3.205) to re-distribute the cross products and differential operators.

#### 7.6.1.1 nL Matrix Elements

Noting that

$$\hat{\mathbf{t}}(\mathbf{r}) \times \hat{\mathbf{n}}_l(\mathbf{r}) = \hat{\boldsymbol{\phi}}(\mathbf{r}) \qquad (7.137)$$

and

$$\hat{\boldsymbol{\phi}}(\mathbf{r}) \times \hat{\mathbf{n}}_l(\mathbf{r}) = -\hat{\mathbf{t}}(\mathbf{r}) \,, \qquad (7.138)$$

we see that (7.136) has the same form as (7.14). This allows us to write the nL sub-matrix elements in terms of the L sub-matrix elements as

$$\begin{bmatrix} \mathrm{nL}^{tt} & \mathrm{nL}^{t\phi} \\ \mathrm{nL}^{\phi t} & \mathrm{nL}^{\phi\phi} \end{bmatrix} = \eta_0 \frac{\epsilon_l}{\mu_l} \begin{bmatrix} \mathrm{L}^{\phi t} & \mathrm{L}^{\phi\phi} \\ -\mathrm{L}^{tt} & -\mathrm{L}^{t\phi} \end{bmatrix} . \qquad (7.139)$$

Inspecting (7.139), we find that the relationship between positive and negative modes is

$$\begin{bmatrix} \mathrm{nL}^{tt} & \mathrm{nL}^{t\phi} \\ \mathrm{nL}^{\phi t} & \mathrm{nL}^{\phi\phi} \end{bmatrix}_{-\alpha} = \begin{bmatrix} -\mathrm{nL}^{tt} & \mathrm{nL}^{t\phi} \\ \mathrm{nL}^{\phi t} & -\mathrm{nL}^{\phi\phi} \end{bmatrix}_{\alpha} , \qquad (7.140)$$

which are the same properties found in the K matrix.

### 7.6.2  $\hat{\mathbf{n}} \times \mathcal{K}$ Operator

Substitution of the basis (7.5, 7.6) and testing (7.11, 7.12) functions into (3.189) yields a matrix of the form

$$Z^{nHJ(l)} = \begin{bmatrix} \mathrm{nK}(\mathbf{f}_\beta^t, \mathbf{f}_\alpha^t) & \mathrm{nK}(\mathbf{f}_\beta^t, \mathbf{f}_\alpha^\phi) \\ \mathrm{nK}(\mathbf{f}_\beta^\phi, \mathbf{f}_\alpha^t) & \mathrm{nK}(\mathbf{f}_\beta^\phi, \mathbf{f}_\alpha^\phi) \end{bmatrix} = \begin{bmatrix} \mathrm{nK}^{tt} & \mathrm{nK}^{t\phi} \\ \mathrm{nK}^{\phi t} & \mathrm{nK}^{\phi\phi} \end{bmatrix} , \qquad (7.141)$$

where in Region $R_l$, the elements of each sub-matrix block are given by

$$\mathrm{nK}_{mn}^{**} = \frac{1}{2} \int_{\mathbf{f}_{\beta m}^{t,\phi}, \mathbf{f}_{\alpha n}^{t,\phi}} \mathbf{f}_{\beta m}^{t,\phi}(\mathbf{r}) \cdot \pm \mathbf{f}_{\alpha n}^{t,\phi}(\mathbf{r}) \, d\mathbf{r}$$

$$+ \int_{\mathbf{f}_m^{t,\phi}} \mathbf{f}_{\beta m}^{t,\phi}(\mathbf{r}) \cdot \hat{\mathbf{n}}(\mathbf{r}) \times \int_{\mathbf{f}_{\alpha n}^{t,\phi}} \left[ (\mathbf{r} - \mathbf{r}') \times \pm \mathbf{f}_{\alpha n}^{t,\phi}(\mathbf{r}') \right] \frac{1 + jk_l r}{4\pi r^3} e^{-jk_l r} \, d\mathbf{r}' \, d\mathbf{r} \,. \qquad (7.142)$$

#### 7.6.2.1 nK Matrix Elements

In evaluating the sub-matrix elements in (7.142), we again use (7.42–7.47) and note that

$$\hat{\mathbf{t}}(\mathbf{r}) \cdot \hat{\mathbf{n}}(\mathbf{r}) \times \left[ (\mathbf{r} - \mathbf{r}') \times \mathbf{f}(\mathbf{r}') \right] = \hat{\boldsymbol{\phi}}(\mathbf{r}) \cdot \left[ (\mathbf{r} - \mathbf{r}') \times \mathbf{f}(\mathbf{r}') \right] \qquad (7.143)$$

and

$$\hat{\phi}(\mathbf{r}) \cdot \hat{\mathbf{n}}(\mathbf{r}) \times \left[ (\mathbf{r} - \mathbf{r}') \times \mathbf{f}(\mathbf{r}') \right] = -\hat{\mathbf{t}}(\mathbf{r}) \cdot \left[ (\mathbf{r} - \mathbf{r}') \times \mathbf{f}(\mathbf{r}') \right], \quad (7.144)$$

which have the same form as (7.48–7.51). Thus, we can write the nK sub-matrix elements in terms of the K sub-matrix elements as

$$\begin{bmatrix} nK^{tt} & nK^{t\phi} \\ nK^{\phi t} & nK^{\phi\phi} \end{bmatrix} = \frac{1}{\eta_0} \begin{bmatrix} -K^{\phi t} - \tilde{K} & -K^{\phi\phi} \\ K^{tt} & K^{t\phi} + \tilde{K} \end{bmatrix}. \quad (7.145)$$

Inspecting (7.145), we find that the relationship between positive and negative modes is

$$\begin{bmatrix} nK^{tt} & nK^{t\phi} \\ nK^{\phi t} & nK^{\phi\phi} \end{bmatrix}_{-\alpha} = \begin{bmatrix} nK^{tt} & -nK^{t\phi} \\ -nK^{\phi t} & nK^{\phi\phi} \end{bmatrix}_{\alpha}, \quad (7.146)$$

which are the same properties found in the L matrix.

### 7.6.3 Excitation

Right-hand side vector elements for the nMFIE can be obtained by substituting the testing functions $\mathbf{f}_{\alpha m}^{t,\phi}(\mathbf{r})$ into (3.193). This yields a column vector of the form

$$V^{nH(l)} = \begin{bmatrix} V^{nH(l),t} \\ V^{nH(l),\phi} \end{bmatrix}. \quad (7.147)$$

The individual elements are given by

$$V_m^{nH(l),(t,\phi)} = \int_{T_m} \int_0^{2\pi} \mathbf{f}_{\alpha m}^{t,\phi}(\mathbf{r}) \cdot \left[ \hat{\mathbf{n}}_l(\mathbf{r}) \times \mathbf{H}^i(\mathbf{r}) \right] \rho \, d\phi \, dt \,, \quad (7.148)$$

where $\mathbf{H}^i(\mathbf{r})$ is the impressed magnetic field. As with the EFIE and MFIE, this can be evaluated analytically in the $\phi$ dimension for incident plane waves.

#### 7.6.3.1 Plane Wave Excitation

While the components of the incident magnetic field are the same as in Section 7.4.1.1, we must now take into account the presence of $\hat{\mathbf{n}}(\mathbf{r})$ in (7.148). Making the following substitutions

$$\hat{\mathbf{t}}(\mathbf{r}) = \sin\gamma\cos\phi\hat{\mathbf{x}} + \sin\gamma\sin\phi\hat{\mathbf{y}} + \cos\gamma\hat{\mathbf{z}} \,, \quad (7.149)$$

$$\hat{\phi}(\mathbf{r}) = -\sin\phi\hat{\mathbf{x}} + \cos\phi\hat{\mathbf{y}} \,, \quad (7.150)$$

$$\hat{\mathbf{n}}(\mathbf{r}) = \cos\gamma\cos\phi\hat{\mathbf{x}} + \cos\gamma\sin\phi\hat{\mathbf{y}} - \sin\gamma\hat{\mathbf{z}} \,, \quad (7.151)$$

and using (7.66) and (7.67), we find that

$$\hat{\mathbf{t}}(\mathbf{r}) \cdot \left[ \hat{\mathbf{n}}(\mathbf{r}) \times \hat{\boldsymbol{\theta}}^i \right] = -\cos\theta^i \sin\phi \,, \quad (7.152)$$

$$\hat{\phi}(\mathbf{r}) \cdot \left[ \hat{\mathbf{n}}(\mathbf{r}) \times \hat{\boldsymbol{\theta}}^i \right] = \cos\gamma\sin\theta^i - \sin\gamma\cos\theta^i\cos\phi \,, \quad (7.153)$$

$$\hat{t}(\mathbf{r}) \cdot \left[\hat{n}(\mathbf{r}) \times -\hat{\phi}^i\right] = -\cos\phi, \tag{7.154}$$

$$\hat{\phi}(\mathbf{r}) \cdot \left[\hat{n}(\mathbf{r}) \times -\hat{\phi}^i\right] = \sin\gamma\sin\phi, \tag{7.155}$$

where we have again set $\phi^i = 0$ for convenience. We immediately note that these results are identical to those in (7.68–7.71), and so the elements of $V^{nH(l)}$ can be written in terms of $V^{E(l)}$. In particular,

$$V^{nH,\theta^i} = \begin{bmatrix} V^{nH,t\theta^i} \\ V^{nH,\phi\theta^i} \end{bmatrix} = \frac{1}{\eta_0} \begin{bmatrix} -V^{E,\phi\phi^i} \\ V^{E,t\phi^i} \end{bmatrix}, \tag{7.156}$$

and

$$V^{nH,\phi^i} = \begin{bmatrix} V^{nH,t\phi^i} \\ V^{nH,\phi\phi^i} \end{bmatrix} = \frac{1}{\eta_0} \begin{bmatrix} V^{E,\phi\theta^i} \\ -V^{E,t\theta^i} \end{bmatrix}. \tag{7.157}$$

#### 7.6.3.2 Plane Wave Solution

Referring again to (7.3.3.1–7.90), we find that the relationship between positive and negative modes is

$$\begin{bmatrix} V^{nH,t\theta^i} \\ V^{nH,\phi\theta^i} \end{bmatrix}_{-\alpha} = \begin{bmatrix} V^{nH,t\theta^i} \\ -V^{nH,\phi\theta^i} \end{bmatrix}_{\alpha}, \tag{7.158}$$

and

$$\begin{bmatrix} V^{nH,t\phi^i} \\ V^{nH,\phi\phi^i} \end{bmatrix}_{-\alpha} = \begin{bmatrix} -V^{nH,t\phi^i} \\ V^{nH,\phi\phi^i} \end{bmatrix}_{\alpha}, \tag{7.159}$$

which are the same properties observed in (7.91) and (7.92). Noting that the nL and nK matrix have the same behavior as the K and L matrix for negative modes, as well as the EFIE and nMFIE right-hand side vectors, the EFIE and nMFIE can be combined in (7.99) to form the CFIE (3.181) with no additional modifications.

---

## 7.7  Numerical Discretization

A common way to describe a BoR is a piecewise linear approximation of the bounding curve in the $xz$ plane ($\phi = 0$). The curve is divided into $N$ segments with $N + 1$ unique endpoints, where the segments need not be of equal length. Each segment, when rotated 360 degrees about $\hat{z}$, sweeps out an annulus on the surface. To define the longitudinal and normal vectors on each

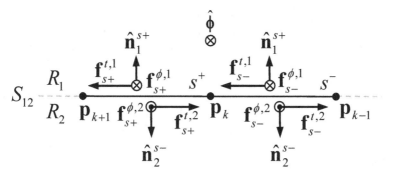

**FIGURE 7.2:** Basis Function Orientation for BoR Problems

segment, we again use a right-handed rule (see Section 6.15). Consider the triangle function with support on segments $s+$ and $s-$, whose endpoints comprise $(\mathbf{p}_{k-1}, \mathbf{p}_k, \mathbf{p}_{k+1})$. These segments reside on the interface $S_{12}$ between dielectric regions $R_1$ and $R_2$, as shown in Figure 7.2. Let us define the unit vectors

$$\hat{\mathbf{t}}^{s+} = \frac{\mathbf{p}_{k+1} - \mathbf{p}_k}{|\mathbf{p}_{k+1} - \mathbf{p}_k|} \, , \tag{7.160}$$

$$\hat{\mathbf{t}}^{s-} = \frac{\mathbf{p}_k - \mathbf{p}_{k-1}}{|\mathbf{p}_k - \mathbf{p}_{k-1}|} \, . \tag{7.161}$$

Following Section 6.2.1, we define $R_1$ as the "positive" side of the interface. On this side, the vectors normal to each segment are then

$$\hat{\mathbf{n}}_1^{s+} = \hat{\boldsymbol{\phi}} \times \hat{\mathbf{t}}^{s+} \, , \tag{7.162}$$

$$\hat{\mathbf{n}}_1^{s-} = \hat{\boldsymbol{\phi}} \times \hat{\mathbf{t}}^{s-} \, . \tag{7.163}$$

We then define the longitudinal basis vectors on each segment in $R_1$ as

$$\hat{\mathbf{f}}_{s+}^{t,1} = \hat{\mathbf{t}}^{s+} \, , \tag{7.164}$$

$$\hat{\mathbf{f}}_{s-}^{t,1} = \hat{\mathbf{t}}^{s-} \, , \tag{7.165}$$

and the azimuthal basis vectors as

$$\hat{\mathbf{f}}_{s+}^{\phi,1} = \hat{\boldsymbol{\phi}} \, , \tag{7.166}$$

$$\hat{\mathbf{f}}_{s-}^{\phi,1} = \hat{\boldsymbol{\phi}} \, . \tag{7.167}$$

On the opposite or "negative" side of the interface in $R_2$, the direction of the normal vectors and the basis functions in $R_1$ are then reversed, as shown in

Figure 7.2. All currents have opposite orientations on each side of the interface, and for the longitudinally oriented currents, Kirchoff's law is satisfied at $p_k$. To facilitate this orientation, the $(x, z)$ locations of the segment endpoints are typically stored in a *boundary* or *points* structure, such that the longitudinal vectors are piecewise continuous between segments.

Other treatments [1, 2, 4] use triangle functions that span $2d$ boundary segments with $d$ segments per half-triangle. Triangles are assigned a constant value on each segment, and integrations in the $t$ dimension are evaluated using a centroidal approximation. Therein, the use of two segments per half-triangle is suggested, though this choice was likely due to the computer limitations of the time. Our treatment here uses triangle functions as defined in Section 2.3.2, with triangles assigned to no more than two segments, and Gaussian quadratures on segments in the $t$ dimension. This assignment is illustrated on a simple, closed ellipse in Figure 7.3a.

When computing the integrals (7.35–7.37) and (7.58–7.60), special treatment is needed when $\mathbf{r} = \mathbf{r}'$. This occurs on overlapping segments when $t = t'$, as illustrated in Figure 7.3b. In this case, and alternative expression [1] for $r$ should be used when computing the integral versus $\phi'$, given by

$$r = \sqrt{(\Delta_s/4)^2 + 2\rho^2[1 - \cos\phi']}. \tag{7.168}$$

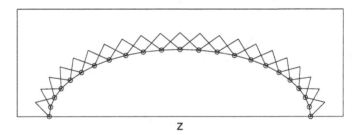

z

(a) Assignment of Triangle Functions

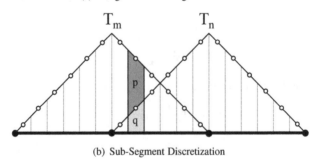

(b) Sub-Segment Discretization

**FIGURE 7.3:** Triangle Functions on a BoR

## 7.8 Notes on Software Implementation

In this section we will discuss software implementation of the MoM/BoR technique.

### 7.8.1 Geometry Processing and Basis Function Assignment

Our approach for the description and processing of BoR geometry is to specify the segments on each material interface as a separate list of one or more boundary segments. Segments that are too long versus the operating wavelength can be subdivided automatically prior to processing. Segments are then read in, endpoints analyzed, and coincident endpoints identified as junctions and combined. The vector orientation of surface normals at each junction is done in a 2D sense, similar to the approach found in Section 8.7.1.3. The assignment and orientation of basis functions at endpoints and junctions is discussed in more detail in Section 7.10.

### 7.8.2 Parallelization

The MoM/BoR formulation produces a system matrix of relatively modest size. Therefore, it is convenient to allocate separate matrix storage space for each compute thread, and to allow each thread to work independently from the others. Each thread then performs the matrix fill, factorization, and calculation of the scattered fields across all excitations and scattering angles for a unique mode. Threads can select mode numbers based on their thread index and size of the thread pool, or simply from modes not already processed. This scheme is equally well suited for shared memory and distributed memory (MPI) systems.

### 7.8.3 Convergence

In practice, the calculation for a given frequency would be halted at a maximum mode number. However, as the amplitude of the scattered field tends to decrease versus mode number, it is likely that the calculation may converge to an accurate solution fairly quickly. One way to determine convergence is to apply a threshold to the amplitude of the radiated/scattered field for each scattering angle. If the amplitudes across all angles fall below this threshold, computations for that frequency are considered as converged. This method is especially useful in monostatic RCS computations due to the many right-hand side operations that can be skipped, increasing the efficiency of the code for larger problems.

## 7.9  Examples

In this section we will use the MoM/BoR approach developed so far in this chapter to compute the RCS of several test articles. We will first consider a series of conducting, dielectric, and coated spheres, then a set of benchmark targets that were constructed and measured by the Electromagnetic Code Consortium (EMCC) [5]. We will then look at more complicated configurations, such as conducting and partially conducting (coated) reentry vehicles (RVs). The numerical results in this section were computed using a software called *Galaxy*, which is part of the *lucernhammer* suite of radar cross section codes. *Galaxy* is written in the C++ programming language, and is parallelized using threads. Unless otherwise specified, the BoR treatment employs 5 subsegments per half-triangle for integration in the $t$ dimension, PMCHWT is applied on dielectrics, and the CFIE is used on closed conductors with $\alpha = 0.5$.

### 7.9.1  Spheres

Spheres are excellent benchmark geometries for computational electromagnetic codes, as analytic solutions for their scattering of plane waves are known. Let us assume that there is a plane wave traveling in the $-\hat{\mathbf{z}}$ direction with an $\hat{\mathbf{x}}$-oriented electric field of unit amplitude. The exact value for the scattered far electric field is [6]

$$\mathbf{E}^s(\theta^s, \phi^s) = \frac{e^{-jk_0 r}}{k_0 r} \left[ \cos \phi^s S_1(\theta^s) \hat{\boldsymbol{\theta}} - \sin \phi^s S_2(\theta^s) \hat{\boldsymbol{\phi}} \right] , \qquad (7.169)$$

where

$$S_1(\theta) = \sum_{n=1}^{\infty} j^{n+1} \left[ A_n \frac{P_n^1(\cos \theta)}{\sin \theta} - j B_n \frac{d}{d\theta} P_n^1(\cos \theta) \right] , \qquad (7.170)$$

$$S_2(\theta) = \sum_{n=1}^{\infty} j^{n+1} \left[ A_n \frac{d}{d\theta} P_n^1(\cos \theta) - j B_n \frac{P_n^1(\cos \theta)}{\sin \theta} \right] , \qquad (7.171)$$

and $P_n^1(\cos \theta)$ is the associated Legendre polynomial of degree 1 and order $n$. In practice, the sums in (7.170) and (7.171) are truncated after a finite number of modes, and the recurrence relationships in [3] used to compute the associated Legendre polynomials of higher order. The coefficients $A_n$ and $B_n$ differ depending on whether the sphere is conducting, or fully or partially dielectric. We will consider examples of each in this section.

### 7.9.1.1 Conducting Sphere

For a conducting sphere of radius $a$, the coefficients $A_n$ and $B_n$ are

$$A_n = -j^n \frac{2n+1}{n(n+1)} \frac{j_n(k_0 a)}{h_n^{(2)}(k_0 a)} , \tag{7.172}$$

and

$$B_n = j^{n+1} \frac{2n+1}{n(n+1)} \frac{\left[k_0 a\, j_n(k_0 a)\right]'}{\left[k_0 a\, h_n^{(2)}(k_0 a)\right]'} , \tag{7.173}$$

where $j_n(k_0 a)$ and $h_n^{(2)}(k_0 a)$ are the spherical Bessel function of the first kind, and spherical Hankel function of the second kind, of order $n$, and the primes indicate differentiation with respect to $k_0 a$ [6]. Spherical Bessel functions can be written as

$$f_n(x) = \sqrt{\frac{\pi}{2x}} F_{n+1/2}(x) , \tag{7.174}$$

where $f_n(x)$ is a spherical Bessel function of the first or second kind ($j_n(x)$ and $y_n(x)$, respectively), and $F_n(x)$ is a regular Bessel function of the first or second kind ($J_n(x)$ and $Y_n(x)$, respectively). The spherical Hankel function is

$$h_n^{(2)}(x) = j_n(x) - j y_n(x) . \tag{7.175}$$

The derivatives of the spherical Bessel functions are

$$\frac{d}{dx}\left[x f_n(x)\right] = \sqrt{\frac{\pi}{2x}}\left[x F_{n-1/2}(x) - n F_{n+1/2}(x)\right] . \tag{7.176}$$

*MoM System Matrix*

The conducting sphere comprises a single bounding curve $S_{01}$ between the free space region $R_0$ and the interior conducting region $R_1$. Applying the CFIE yields a linear system of the form

$$\left[\alpha_c Z^{EJ} + (1 - \alpha_c) Z^{nHJ}\right]\left[I^J\right] = \left[\alpha_c V^E + (1 - \alpha_c) V^{nH}\right], \tag{7.177}$$

where

$$Z^{EJ} = \begin{bmatrix} \mathbf{L}\left(\mathbf{f}_{S_{01}}^t, \mathbf{f}_{S_{01}}^t, R_0\right) & \mathbf{L}\left(\mathbf{f}_{S_{01}}^t, \mathbf{f}_{S_{01}}^\phi, R_0\right) \\ \mathbf{L}\left(\mathbf{f}_{S_{01}}^\phi, \mathbf{f}_{S_{01}}^t, R_0\right) & \mathbf{L}\left(\mathbf{f}_{S_{01}}^\phi, \mathbf{f}_{S_{01}}^\phi, R_0\right) \end{bmatrix}, \tag{7.178}$$

and

$$Z^{nHJ} = \begin{bmatrix} \mathbf{nK}\left(\mathbf{f}_{S_{01}}^t, \mathbf{f}_{S_{01}}^t, R_0\right) & \mathbf{nK}\left(\mathbf{f}_{S_{01}}^t, \mathbf{f}_{S_{01}}^\phi, R_0\right) \\ \mathbf{nK}\left(\mathbf{f}_{S_{01}}^\phi, \mathbf{f}_{S_{01}}^t, R_0\right) & \mathbf{nK}\left(\mathbf{f}_{S_{01}}^\phi, \mathbf{f}_{S_{01}}^\phi, R_0\right) \end{bmatrix}. \tag{7.179}$$

The notation $\mathrm{X}\left(\mathbf{a}_{S_l}, \mathbf{b}_{S_m}, R_l\right)$ indicates that X is evaluated in $R_l$ for the testing and basis functions **a** and **b** that reside on the boundaries $S_l$ and $S_m$, respectively. The right-hand side vectors are

$$V^E = \begin{bmatrix} V^E(\mathbf{f}_{S_0}^t, R_0) \\ V^E(\mathbf{f}_{S_0}^\phi, R_0) \end{bmatrix}, \tag{7.180}$$

and

$$V^{nH} = \begin{bmatrix} V^{nH}(\mathbf{f}_{S_0}^t, R_0) \\ V^{nH}(\mathbf{f}_{S_0}^\phi, R_0) \end{bmatrix}, \tag{7.181}$$

and the solution vector is

$$I^J = \begin{bmatrix} I_{S_{01}}^{J,t} \\ I_{S_{01}}^{J,\phi} \end{bmatrix}. \tag{7.182}$$

*Monostatic RCS versus Frequency*

We first compute the monostatic RCS of a conducting sphere of radius 0.5 meters for frequencies between 10 MHz and 2 GHz. For the Mie terms (7.170) and (7.171) we truncate the summation after 40 modes, and for the BoR bounding curve, we use 180 segments of equal length. In Figures 7.4a and 7.4b we compare the Mie series RCS to that of the EFIE and nMFIE, respectively. The comparisons are good at lower frequencies, though discrepancies begin to appear as frequency increases, particularly in the nMFIE results. These correspond to the internal resonances of the discretized BoR. In Figure 7.5a are plotted the results obtained using the CFIE, where the comparison is very good and the resonance problems have been eliminated, as expected.

*Range Profile*

We next compare the *range profiles* of the sphere using the backscattered field from the CFIE and Mie series. The range profile is a time domain representation of the electromagnetic signature and can be used to localize the effective scattering centers of a target in the down-range dimension. Given an array $S$ of complex-valued backscattered electric field samples over a range of evenly spaced frequencies $f_{min}$ to $f_{max}$, the range profile $R$ is

$$R = \text{IFFT}(S). \tag{7.183}$$

In radar signal processing, this is often referred to as *pulse compression* or *matched filtering*. More specifically, (7.183) is very similar to what is known as *stretch processing*, or "one-pass" pulse compression [7]. The range resolution of the time-domain waveform is

$$\Delta_R = \frac{c}{2B}, \tag{7.184}$$

where the operating bandwidth $B$ is

$$B = f_{max} - f_{min} .$$ (7.185)

In order to reduce sidelobes in the range domain, a window function is usually applied to the frequency domain samples prior to performing the FFT. A commonly used window function is the *Hamming window* [8], which comprises the $N$ element array $W$ with elements given by

$$W_{k+1} = 0.54 - 0.46 \cos\left[\frac{2\pi k}{N-1}\right] \qquad k = 0, \dots, N-1 .$$ (7.186)

This windowing operation decreases the sidelobes at the expense of decreased range resolution. In particular, the Hamming window yields a first sidelobe level of approximately -40 dB, at the expense of increasing the half-power width of the main lobe by around 40%.

Figure 7.5b depicts the monostatic range profiles of the CFIE and Mie series computed from the complex-valued scattered field sampled from 1 to 2 GHz. A total of 32 samples were used, and the 1 GHz bandwidth yields in a range resolution of approximately 15 cm, which is decreased by the application of a Hamming window. Positive range in this figure represents the down-range direction. Noticeable immediately at a range of -0.5 m is the bright specular reflection from the sphere. Located slightly further down-range, at an amplitude of around -28 dB, is the well-known creeping wave response. This results from electromagnetic energy that originates at the shadow boundary and travels along the surface on the shadowed side of the sphere, returning eventually to the direction of incidence. The results are virtually indistinguishable.

*Bistatic RCS*

The bistatic RCS of the sphere is another commonly used benchmark. We retain the same geometry and simulation parameters as before, and consider the bistatic RCS at 0.3, 1.0 and 2.0 GHz, for bistatic angles between 0 and 180 degrees. The results from the CFIE are compared to the Mie series in Figures 7.6a–7.6f. The comparison is fairly good at 0.3 GHz, and at 1.0 and 2.0 GHz, the results are indistinguishable.

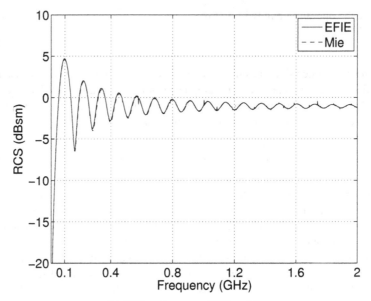

(a) RCS vs. Frequency: EFIE vs. Mie

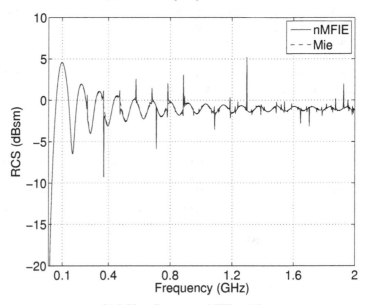

(b) RCS vs. Frequency: nMFIE vs. Mie

**FIGURE 7.4:** Conducting Sphere: EFIE and nMFIE vs. Mie

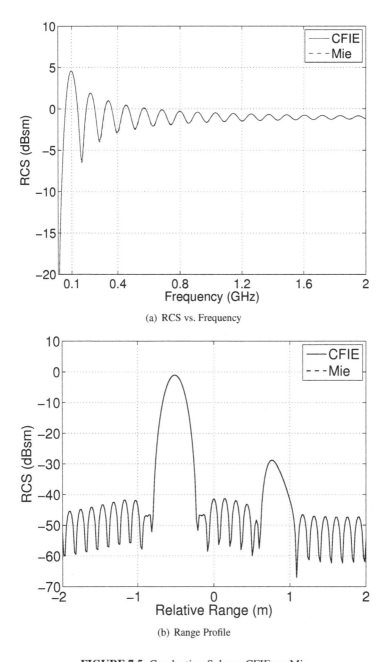

(a) RCS vs. Frequency

(b) Range Profile

**FIGURE 7.5:** Conducting Sphere: CFIE vs. Mie

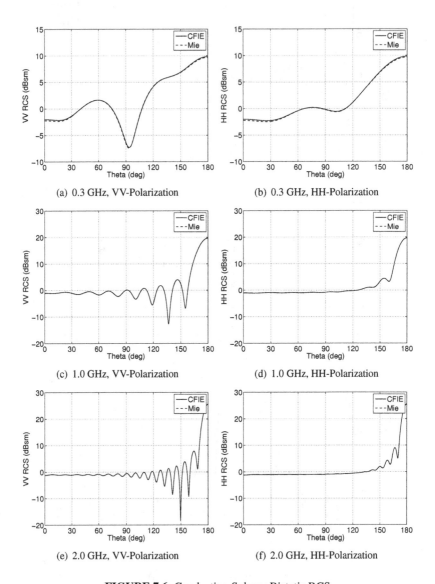

(a) 0.3 GHz, VV-Polarization

(b) 0.3 GHz, HH-Polarization

(c) 1.0 GHz, VV-Polarization

(d) 1.0 GHz, HH-Polarization

(e) 2.0 GHz, VV-Polarization

(f) 2.0 GHz, HH-Polarization

**FIGURE 7.6:** Conducting Sphere: Bistatic RCS

## 7.9.1.2 Stratified Sphere

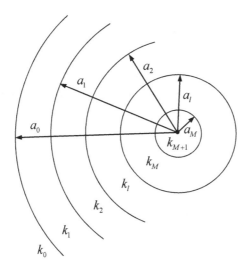

**FIGURE 7.7:** Geometry of an $M$-Layer Stratified Sphere

We next consider the radially stratified geometry of Figure 7.7, where the core may or may not be conducting. The Mie series coefficients [1] for this configuration are [6]

$$A_n = -j^n \frac{2n+1}{n(n+1)} \frac{k_0 a_0 j_n(k_0 a_0) + j Z_n(k_0 a_0)\left[k_0 a_0 j_n(k_0 a_0)\right]'}{k_0 a_0 h_n^{(2)}(k_0 a_0) + j Z_n(k_0 a_0)\left[k_0 a_0 h_n^{(2)}(k_0 a_0)\right]'} \tag{7.187}$$

and

$$B_n = j^{n+1} \frac{2n+1}{n(n+1)} \frac{k_0 a_0 j_n(k_0 a_0) + j Y_n(k_0 a_0)\left[k_0 a_0 j_n(k_0 a_0)\right]'}{k_0 a_0 h_n^{(2)}(k_0 a_0) + j Y_n(k_0 a_0)\left[k_0 a_0 h_n^{(2)}(k_0 a_0)\right]'} , \tag{7.188}$$

where $a_0$ is the outermost radius of the sphere. $Z_n(k_0 a_0)$ and $Y_n(k_0 a_0)$ are called the modal surface impedances and admittance, respectively, which differ depending on the composition of the sphere. We consider the general case in this section, which can be applied to dielectric as well as coated spheres with many layers. Calculation of these coefficients is as follows: First, we define the surface impedance at the interface of radius $a_l$ as $Z_n^l$. The impedance $Z_n^M$ between the core and the first layer is then determined, and from $Z_n^M$, the impedance $Z_n^{M-1}$ at the second interface from the center is found. This process

---

[1] There is an error in the leading coefficient on the right-hand side of (3.4-2) in [6], which has been corrected here.

is repeated until $Z_n^0$ at the outer surface is obtained. Then,

$$Z_n(k_0 a_0) = \frac{j}{\eta_0} Z_n^0 \qquad (7.189)$$

and

$$Y_n(k_0 a_0) = j\eta_0 Y_n^0 , \qquad (7.190)$$

where the admittance $Y_n^0$ is found from $Y_n^M$ using a similar procedure. For the iterative procedure, we first make the following definitions

$$P_n(x) = \frac{x j_n(x)}{[x j_n(x)]'} , \qquad (7.191)$$

$$Q_n(x) = \frac{x h_n^{(2)}(x)}{[x h_n^{(2)}(x)]'} , \qquad (7.192)$$

$$U_n^l = \frac{[k_{l+1} a_{l+1} j_n(k_{l+1} a_{l+1})]'}{[k_{l+1} a_l j_n(k_{l+1} a_l)]'} \frac{[k_{l+1} a_l h_n^{(2)}(k_{l+1} a_l)]'}{[k_{l+1} a_{l+1} h_n^{(2)}(k_{l+1} a_{l+1})]'} , \qquad (7.193)$$

$$V_n^l = \frac{[k_{l+1} a_{l+1} j_n(k_{l+1} a_{l+1})]}{[k_{l+1} a_l j_n(k_{l+1} a_l)]} \frac{[k_{l+1} a_l h_n^{(2)}(k_{l+1} a_l)]}{[k_{l+1} a_{l+1} h_n^{(2)}(k_{l+1} a_{l+1})]} , \qquad (7.194)$$

$$Z_n^M = \eta_{M+1} P_n(k_{M+1} a_M) , \qquad (7.195)$$

and

$$Y_n^M = \frac{1}{\eta_{M+1}} P_n(k_{M+1} a_M) . \qquad (7.196)$$

At the interface at radius $a_l$, the impedance and admittance are

$$Z_n^{(l)} = \eta_{l+1} P_n(k_{l+1} a_l) \left[ 1 - V_n^{(l)} \frac{1 - Z_n^{(l+1)}/[\eta_{l+1} P_n(k_{l+1} a_{l+1})]}{1 - Z_n^{(l+1)}/[\eta_{l+1} Q_n(k_{l+1} a_{l+1})]} \right] \cdot$$
$$\left[ 1 - U_n^{(l)} \frac{1 - \eta_{l+1} P_n(k_{l+1} a_{l+1})/Z_n^{(l+1)}}{1 - \eta_{l+1} Q_n(k_{l+1} a_{l+1})/Z_n^{(l+1)}} \right]^{-1} \qquad (7.197)$$

and

$$Y_n^{(l)} = \frac{P_n(k_{l+1} a_l)}{\eta_{l+1}} \left[ 1 - V_n^{(l)} \frac{1 - \eta_{l+1} Y_n^{(l+1)}/P_n(k_{l+1} a_{l+1})}{1 - \eta_{l+1} Y_n^{(l+1)}/Q_n(k_{l+1} a_{l+1})} \right] \cdot$$
$$\left[ 1 - U_n^{(l)} \frac{1 - P_n(k_{l+1} a_{l+1})/[\eta_{l+1} Y_n^{(l+1)}]}{1 - Q_n(k_{l+1} a_{l+1})/[\eta_{l+1} Y_n^{(l+1)}]} \right]^{-1} , \qquad (7.198)$$

where $l$ begins at $M - 1$ and ends at $l = 0$. Note that if the core of the sphere is conducting, $Z_n^M \to 0$ and $Y_n^M \to \infty$, thus

$$Z_n^{(M-1)} = \eta_M P_n(k_M a_{M-1}) \frac{1 - V_n^{(M-1)}}{1 - U_n^{(M-1)} P_n(k_M a_M)/Q_n(k_M a_M)} \tag{7.199}$$

and

$$Y_n^{(M-1)} = \frac{P_n(k_M a_{M-1})}{\eta_M} \frac{1 - V_n^{(M-1)} Q_n(k_M a_M)/P_n(k_M a_M)}{1 - U_n^{(M-1)}}. \tag{7.200}$$

We use these expressions to compute the field scattered by dielectric and coated conducting spheres in the following sections.

### 7.9.1.3 Dielectric Sphere

A dielectric sphere comprises a single bounding curve on the interface $S_{01}$ between the free space region $R_0$ and the interior dielectric region $R_1$. Applying the PMCHWT formulation on $S_{01}$ yields a linear system of the form

$$\begin{bmatrix} Z^{EJ} & Z^{EM} \\ Z^{HJ} & Z^{HM} \end{bmatrix} \begin{bmatrix} I^J \\ I^M \end{bmatrix} \begin{bmatrix} V^E \\ V^H \end{bmatrix}, \tag{7.201}$$

where the sub-matrix blocks are

$$Z^{EJ} = L\left(\mathbf{f}_{S_{01}}, \mathbf{f}_{S_{01}}, R_0\right) + L\left(\mathbf{f}_{S_{10}}, \mathbf{f}_{S_{10}}, R_1\right), \tag{7.202}$$

$$Z^{EM} = K\left(\mathbf{f}_{S_{01}}, \mathbf{g}_{S_{01}}, R_0\right) + K\left(\mathbf{f}_{S_{10}}, \mathbf{g}_{S_{10}}, R_1\right), \tag{7.203}$$

$$Z^{HJ} = -K\left(\mathbf{g}_{S_{01}}, \mathbf{f}_{S_{01}}, R_0\right) - K\left(\mathbf{g}_{S_{10}}, \mathbf{f}_{S_{10}}, R_1\right), \tag{7.204}$$

$$Z^{HM} = \frac{\epsilon_0}{\mu_0} L\left(\mathbf{g}_{S_{01}}, \mathbf{g}_{S_{01}}, R_0\right) + \frac{\epsilon_1}{\mu_1} L\left(\mathbf{g}_{S_{10}}, \mathbf{g}_{S_{10}}, R_1\right). \tag{7.205}$$

Each block is understood to contain the $\hat{\mathbf{t}}$ and $\hat{\phi}$ components of all basis and testing functions. The right-hand side vectors are

$$V^E = V^E\left(\mathbf{f}_{S_{01}}, R_0\right) \tag{7.206}$$

and

$$V^H = V^H\left(\mathbf{g}_{S_{01}}, R_0\right), \tag{7.207}$$

and the solution vectors are

$$I^J = I_{S_{01}}^J \tag{7.208}$$

and

$$I^M = I_{S_{01}}^M. \tag{7.209}$$

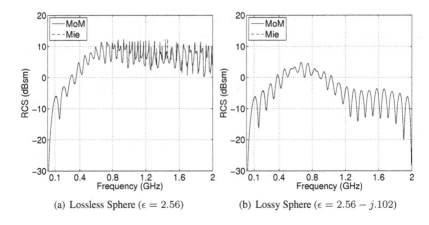

(a) Lossless Sphere ($\epsilon = 2.56$)          (b) Lossy Sphere ($\epsilon = 2.56 - j.102$)

**FIGURE 7.8:** Dielectric Sphere: Backscatter RCS versus Frequency

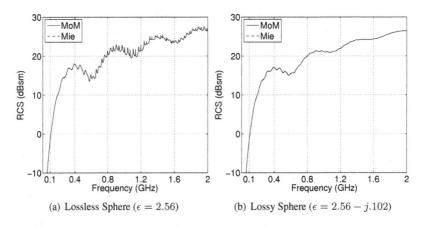

(a) Lossless Sphere ($\epsilon = 2.56$)          (b) Lossy Sphere ($\epsilon = 2.56 - j.102$)

**FIGURE 7.9:** Dielectric Sphere: Forward Scatter RCS versus Frequency

*RCS versus Frequency*

We first consider the RCS of a dielectric sphere of radius 0.5 meters for frequencies between 10 MHz and 2 GHz. For the Mie terms (7.170) and (7.171), we truncate the summation after 40 modes, and the BoR bounding curve comprises 180 segments of equal length. In Figure 7.8a we compare the Mie series to the MoM for a lossless sphere ($\epsilon_r = 2.56$) in backscattering ($\theta^i = \theta^s = 0$). The comparison is very good. A similar comparison is made in Figure 7.8b for a lossy sphere ($\epsilon_r = 2.56 - j.102$), where the comparison is again very good. The forward scattering ($\theta^i = 0, \theta^s = \pi$) for each case is summarized in Figures 7.9a and 7.9b, respectively, where the comparison is again quite good.

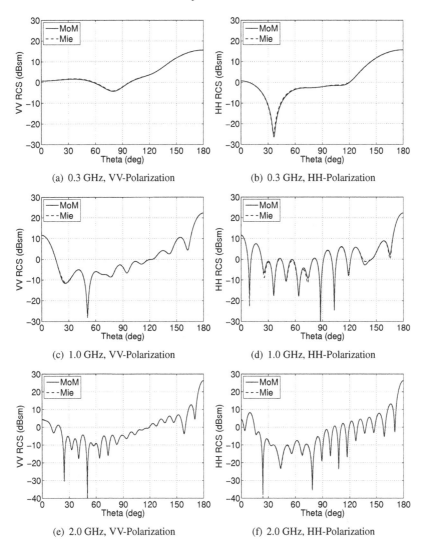

**FIGURE 7.10:** Lossless Dielectric Sphere ($\epsilon = 2.56$): Bistatic RCS

*Bistatic RCS*

Using the same parameters as before, we next consider the bistatic RCS at 0.3, 1.0 and 2.0 GHz, for bistatic angles from 0 to 180 degrees. The MoM results are compared to those of the Mie series for the lossless sphere in Figures 7.10a–7.10f. The comparison is very good. The results for the lossy sphere are compared in Figures 7.11a–7.11f, where the comparison is again very good.

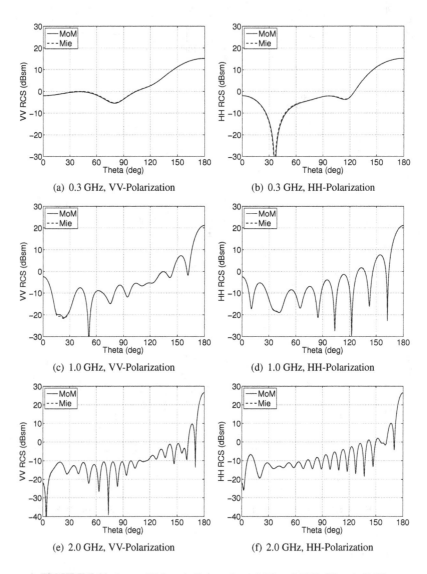

(a) 0.3 GHz, VV-Polarization

(b) 0.3 GHz, HH-Polarization

(c) 1.0 GHz, VV-Polarization

(d) 1.0 GHz, HH-Polarization

(e) 2.0 GHz, VV-Polarization

(f) 2.0 GHz, HH-Polarization

**FIGURE 7.11:** Lossy Dielectric Sphere ($\epsilon = 2.56 - j.102$): Bistatic RCS

### 7.9.1.4 Coated Sphere

A coated conducting sphere comprises two bounding curves: the interface $S_{01}$ between the free space region $R_0$ and dielectric region $R_1$, and the interface $S_{12}$ between the dielectric region $R_1$ and the core conducting region $R_2$. Applying the PMCHWT and CFIE formulations to this problem yields a linear system of the form

$$\begin{bmatrix} Z^{EJ} + Z^{nHJ} & Z^{EM} + Z^{nHM} \\ Z^{HJ} & Z^{HM} \end{bmatrix} \begin{bmatrix} I^J \\ I^M \end{bmatrix} \begin{bmatrix} V^E \\ V^H \end{bmatrix}, \tag{7.210}$$

where the sub-matrix blocks are

$$Z^{EJ} = \begin{bmatrix} \mathrm{L}\left(\mathbf{f}_{S_{01}}, \mathbf{f}_{S_{01}}, R_0\right) + \mathrm{L}\left(\mathbf{f}_{S_{10}}, \mathbf{f}_{S_{10}}, R_1\right) & \mathrm{L}\left(\mathbf{f}_{S_{10}}, \mathbf{f}_{S_{12}}, R_1\right) \\ \alpha_c \mathrm{L}\left(\mathbf{f}_{S_{12}}, \mathbf{f}_{S_{10}}, R_1\right) & \alpha_c \mathrm{L}\left(\mathbf{f}_{S_{12}}, \mathbf{f}_{S_{12}}, R_1\right) \end{bmatrix}, \tag{7.211}$$

$$Z^{nHJ} = \begin{bmatrix} 0 & 0 \\ (1 - \alpha_c)\eta_1 \mathrm{nK}\left(\mathbf{f}_{S_{12}}, \mathbf{f}_{S_{10}}, R_1\right) & (1 - \alpha_c)\eta_1 \mathrm{nK}\left(\mathbf{f}_{S_{12}}, \mathbf{f}_{S_{12}}, R_1\right) \end{bmatrix}, \tag{7.212}$$

$$Z^{EM} = \begin{bmatrix} \mathrm{K}\left(\mathbf{f}_{S_{01}}, \mathbf{g}_{S_{01}}, R_0\right) + \mathrm{K}\left(\mathbf{f}_{S_{10}}, \mathbf{g}_{S_{10}}, R_1\right) \\ \alpha_c \mathrm{K}\left(\mathbf{f}_{S_{12}}, \mathbf{g}_{S_{10}}, R_1\right) \end{bmatrix}, \tag{7.213}$$

$$Z^{nHM} = \begin{bmatrix} 0 \\ (1 - \alpha_c)\eta_1 \mathrm{nL}\left(\mathbf{f}_{S_{12}}, \mathbf{g}_{S_{10}}, R_1\right) \end{bmatrix}, \tag{7.214}$$

$$Z^{HJ} = \left[ -\mathrm{K}\left(\mathbf{g}_{S_{01}}, \mathbf{f}_{S_{01}}, R_0\right) - \mathrm{K}\left(\mathbf{g}_{S_{10}}, \mathbf{f}_{S_{10}}, R_1\right) \mid -\mathrm{K}\left(\mathbf{g}_{S_{10}}, \mathbf{f}_{S_{12}}, R_1\right) \right], \tag{7.215}$$

$$Z^{HM} = \left[ \tfrac{\epsilon_0}{\mu_0}\mathrm{L}\left(\mathbf{g}_{S_{01}}, \mathbf{g}_{S_{01}}, R_0\right) + \tfrac{\epsilon_1}{\mu_1}\mathrm{L}\left(\mathbf{g}_{S_{10}}, \mathbf{g}_{S_{10}}, R_1\right) \right], \tag{7.216}$$

where each operator is understood to contain the $\hat{\mathbf{t}}$ and $\hat{\boldsymbol{\phi}}$ components of all basis and testing functions. The right-hand side vectors are

$$V^E = \begin{bmatrix} V^E(\mathbf{f}_{S_{01}}, R_0) \\ 0 \end{bmatrix} \tag{7.217}$$

and

$$V^H = \begin{bmatrix} V^H(\mathbf{g}_{S_{01}}, R_0) \\ 0 \end{bmatrix}, \tag{7.218}$$

and the solution vectors are

$$I^J = \begin{bmatrix} I^J_{S_{01}} \\ I^J_{S_{12}} \end{bmatrix} \tag{7.219}$$

and

$$I^M = \begin{bmatrix} I^M_{S_{01}} \\ 0 \end{bmatrix}. \tag{7.220}$$

*RCS versus Frequency*

We now consider the RCS of a conducting sphere with dielectric coating for frequencies between 10 MHz and 2 GHz. The conducting core of the sphere has a radius of 0.5 meters, and the dielectric coating has a thickness of 5 centimeters, yielding an outermost radius of 0.55 meters. The bounding curve of the dielectric interface is divided into 180 segments, and the bounding curve of the PEC core is divided into 360 segments. The extra density on the innermost interface is chosen due to the shorter wavelength inside the dielectric coating. The remainder of the run parameters remain the same as before. In Figure 7.12a we compare the Mie series to the MoM in backscattering for a lossless coating ($\epsilon_r = 2.56$). The comparison is very good. In Figure 7.12b is a similar comparison for a lossy coating ($\epsilon_r = 2.56 - j.102$). The forward-scattering results for each case are compared in 7.13a and 7.13b, respectively, where the comparison is again quite good.

*Bistatic RCS*

Using the same parameters as before, we next consider the bistatic RCS at 0.3, 1.0, and 2.0 GHz, for bistatic angles from 0 to 180 degrees. The MoM results are compared to those of the Mie series for the lossless coating in Figures 7.14a–7.14f. The comparison is very good. The results for the lossy coating are compared in Figures 7.15a–7.15f, where the comparison is again very good.

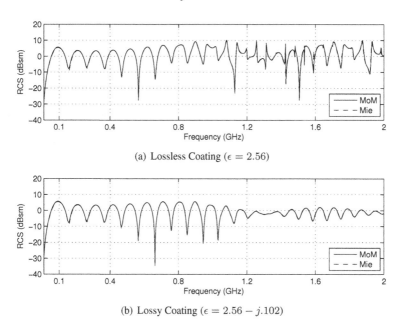

(a) Lossless Coating ($\epsilon = 2.56$)

(b) Lossy Coating ($\epsilon = 2.56 - j.102$)

**FIGURE 7.12:** Coated Sphere: Backscatter RCS versus Frequency

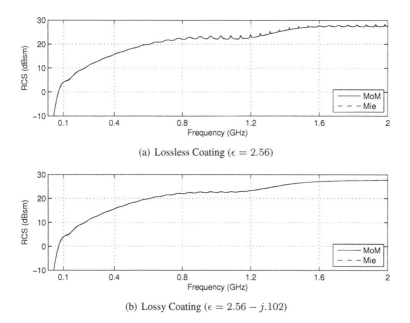

(a) Lossless Coating ($\epsilon = 2.56$)

(b) Lossy Coating ($\epsilon = 2.56 - j.102$)

**FIGURE 7.13:** Coated Sphere: Forward Scatter RCS versus Frequency

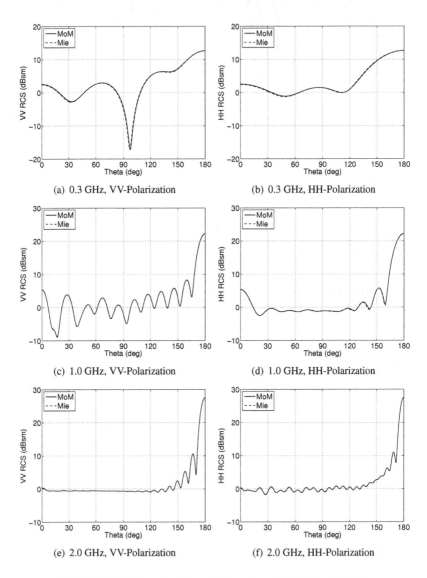

**FIGURE 7.14:** Lossless Coated Sphere ($\epsilon = 2.56$): Bistatic RCS

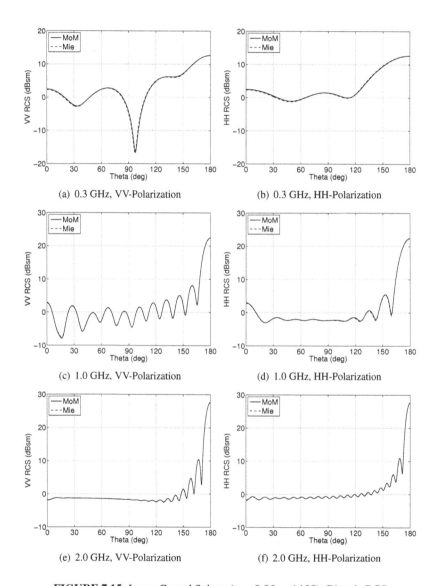

(a) 0.3 GHz, VV-Polarization

(b) 0.3 GHz, HH-Polarization

(c) 1.0 GHz, VV-Polarization

(d) 1.0 GHz, HH-Polarization

(e) 2.0 GHz, VV-Polarization

(f) 2.0 GHz, HH-Polarization

**FIGURE 7.15:** Lossy Coated Sphere ($\epsilon = 2.56 - j.102$): Bistatic RCS

### 7.9.2   EMCC Benchmark Targets

We will next use the MoM/BoR approach to compute the RCS of several benchmark radar targets described and measured by the EMCC in [9]. The objects considered are the ogive, double ogive, cone-sphere, and cone-sphere with gap. The first three are illustrated in Figures 7.16a–7.16c. The test articles used for the measurements were fabricated from aluminum using a numerically controlled mill. We note that in the range, the objects were positioned on a rotating platform in the $xy$ plane. Therefore, computed $\hat{\theta}$-polarization will be mapped to measured horizontal polarization, and computed $\hat{\phi}$-polarization mapped to measured vertical polarization. The comparisons are made using the measurements at 9.0 GHz, and in the plots that follow, measurement data have been shifted slightly in angle to better align them with the computed results.

#### 7.9.2.1   EMCC Ogive

The 10-inch EMCC ogive of Figure 7.16a can be expressed for $0 \leq \phi \leq 2\pi$ as

$$f(x) = \sqrt{1 - \left(\frac{x}{5}\right)^2 \sin^2(22.62°)} - \cos(22.62°)\,, \tag{7.221}$$

$$y(x) = \frac{f(x)\cos\phi}{1 - \cos(22.62°)}\,, \tag{7.222}$$

$$z(x) = \frac{f(x)\sin\phi}{1 - \cos(22.62°)}\,, \tag{7.223}$$

for $-5 \leq x \leq 5$ inches. The BoR bounding curve has 64 segments of equal length, and the MoM calculation is halted after 30 modes. The numerical results are compared to the EMCC measurements in Figures 7.17a and 7.17b for vertical and horizontal polarizations, respectively. The agreement is excellent, even at near-tip angles where the cross section is very low.

#### 7.9.2.2   EMCC Double Ogive

The 7.5-inch EMCC double ogive of Figure 7.16b can be expressed for $0 \leq \phi \leq 2\pi$ as

$$g(x) = \sqrt{1 - \left(\frac{x}{2.5}\right)^2 \sin^2(46.6°)} - \cos(46.6°)\,, \tag{7.224}$$

$$y(x) = \frac{g(x)\cos\phi}{1 - \cos(46.4°)}\,, \tag{7.225}$$

$$z(x) = \frac{g(x)\sin\phi}{1 - \cos(46.4°)}\,, \tag{7.226}$$

for $-2.5 \leq x \leq 0$ inches, and

$$f(x) = \sqrt{1 - \left(\frac{x}{5}\right)^2 \sin^2(22.62°) - \cos(22.62°)}, \qquad (7.227)$$

$$y(x) = \frac{f(x)\cos\phi}{1 - \cos(22.62°)}, \qquad (7.228)$$

$$z(x) = \frac{f(x)\sin\phi}{1 - \cos(22.62°)}, \qquad (7.229)$$

for $0 \leq x \leq 5$ inches. The BoR bounding curve has 64 segments of equal length, and the MoM calculation is halted after 30 modes. The numerical results are compared to the EMCC measurements in Figures 7.18a and 7.18b for vertical and horizontal polarizations, respectively. The agreement is again excellent.

### 7.9.2.3 EMCC Cone-Sphere

The 27-inch EMCC cone-sphere of Figure 7.16c can be expressed for $0 \leq \phi \leq 2\pi$ as

$$y(x) = 0.87145(x + 23.821)\cos\phi, \qquad (7.230)$$

$$z(x) = 0.87145(x + 23.821)\sin\phi, \qquad (7.231)$$

for $-23.821 \leq x \leq 0$ inches, and

$$y(x) = 2.947\sqrt{1 - \left(\frac{x - 0.359}{2.947}\right)^2}\cos\phi, \qquad (7.232)$$

$$z(x) = 2.947\sqrt{1 - \left(\frac{x - 0.359}{2.947}\right)^2}\sin\phi, \qquad (7.233)$$

for $0 \leq x \leq 3.306$ inches. The BoR bounding curve has 225 segments of equal length, and the MoM calculation is halted after 30 modes. The numerical results are compared to the EMCC measurements in Figures 7.19a and 7.19b for vertical and horizontal polarizations, respectively. The agreement is fairly good at all angles. The side specular in the measured data near 100 degrees is noticeably broader and of lesser amplitude than in the computed data. The reason for this differences is not known, and was not mentioned in [9].

### 7.9.2.4 EMCC Cone-Sphere with Gap

The EMCC cone-sphere with gap is identical to the cone-sphere except that a square groove of depth $\frac{1}{4}$ inch is cut into the target near the cone-sphere junction point. The gap can be expressed for $0 \leq \phi \leq 2\pi$ as

$$y(x) = 2.697\cos\phi, \qquad (7.234)$$

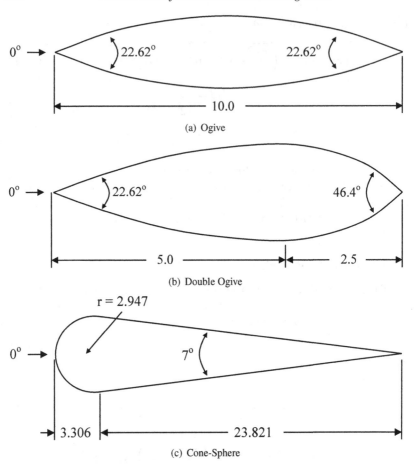

**FIGURE 7.16:** EMCC BoR Targets (Dimensions in Inches)

and

$$z(x) = 2.697 \sin \phi \,, \tag{7.235}$$

for $0 \le x \le 0.25$ inches. At 9.0 GHz, the dimensions of the gap are approximately $\lambda/5$. To generate the bounding curve, an additional 25 segments were added to the cone-sphere model in the vicinity of the gap, yielding a total of 250 boundary segments. The numerical results are compared to the EMCC measurements in Figures 7.20a and 7.20b for vertical and horizontal polarizations, respectively. The agreement is again excellent. The presence of the gap has a significant effect on the backscattered field, especially in horizontal polarization.

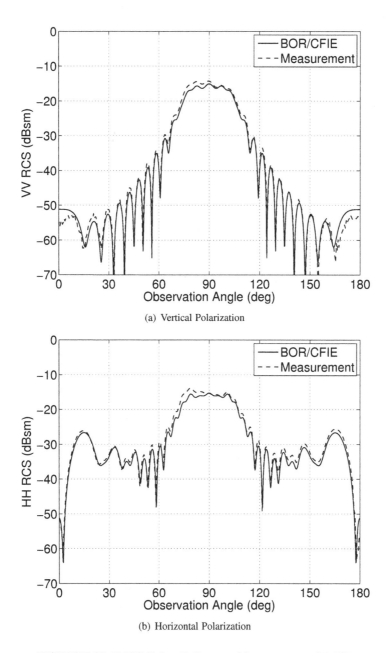

(a) Vertical Polarization

(b) Horizontal Polarization

**FIGURE 7.17:** EMCC Ogive: BoR versus Measurement at 9.0 GHz

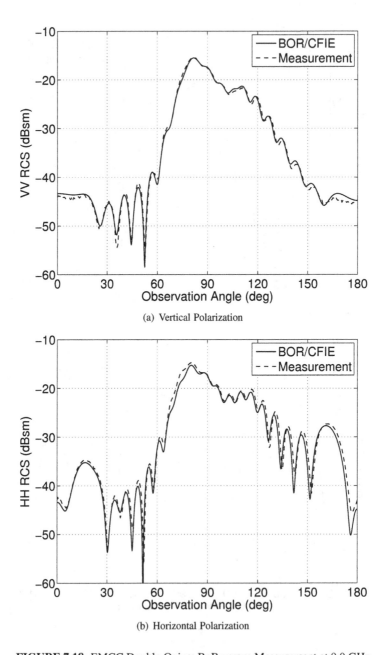

(a) Vertical Polarization

(b) Horizontal Polarization

**FIGURE 7.18:** EMCC Double Ogive: BoR versus Measurement at 9.0 GHz

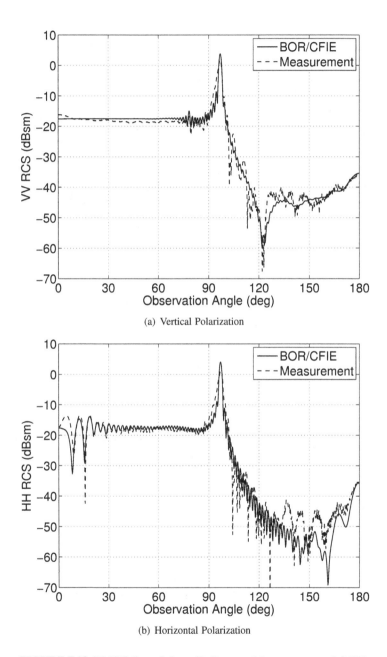

(a) Vertical Polarization

(b) Horizontal Polarization

**FIGURE 7.19:** EMCC Cone-Sphere: BoR versus Measurement at 9.0 GHz

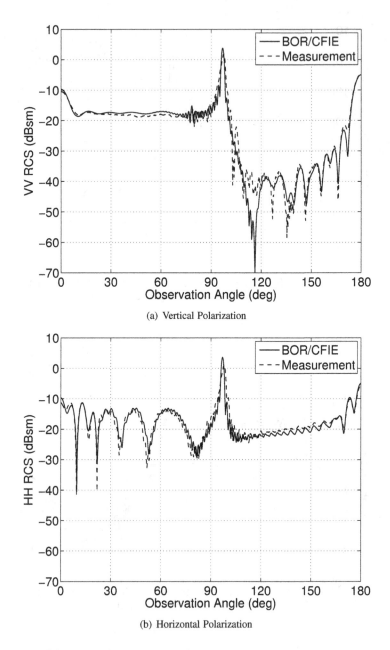

(a) Vertical Polarization

(b) Horizontal Polarization

**FIGURE 7.20:** EMCC Cone-Sphere with Gap: BoR versus Measurement at 9.0 GHz

### 7.9.3 Biconic Reentry Vehicle

We next consider a biconic reentry vehicle (RV) whose dimensions are outlined in Figure 7.21a. The RV has a large nose and deeply recessed rear cavity that should be significant sources of backscatter. The three-dimensional shape of the RV is illustrated at forward and rear aspects in Figures 7.21b and 7.21c, respectively. The total length of the BoR bounding curve for this object is approximately 3.2 meters.

When studying the backscattering behavior of a target, it is very often useful to consider its range profile over a wide range of monostatic angles. This

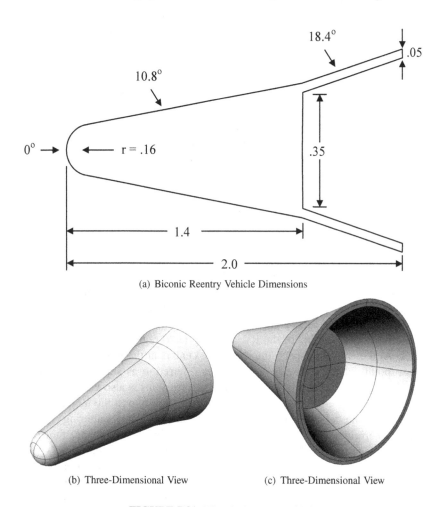

(a) Biconic Reentry Vehicle Dimensions

(b) Three-Dimensional View     (c) Three-Dimensional View

**FIGURE 7.21:** Biconic Reentry Vehicle

allows for isolation of the target's scattering features, often referred to as *scattering centers*. Therefore, we will compute the complex-valued scattered far field of the RV for frequencies ranging from 5 to 6 GHz (C-Band), for monostatic angles between 0 to 180 degrees. At 6 GHz, the length of the bounding curve of the RV is approximately 64 wavelengths. Our numerical model uses 577 boundary segments, which is approximately 10 segments per wavelength.

For visualization purposes, we compute the range profiles across all incident angles and stack them together to create a two-dimensional figure called a *range-aspect-intensity* (RAI) plot. RAI plots for vertical and horizontal polarizations are shown in Figures 7.22a and 7.22b, respectively. The choice of bandwidth results in a down-range resolution of approximately 15 centimeters, and a Hamming window is used to reduce range sidelobes. Clearly visible at forward-scattering angles are the large scattering responses from the nose and the rear base edge. At rear aspects, the multiple bounce returns from inside the cavity clearly dominate the response, giving rise to significant returns from down-range locations.

We can further isolate the scattering centers of the target by generating a two-dimensional *range-Doppler* image. Such an image is generated by computing the FFT of each range line over a small range of angles. If we assume that the Doppler frequency of each bin remains constant across the integration interval, then we can achieve a cross-range resolution given by [10]

$$\Delta_{cr} = \frac{\lambda_o}{2\Delta_\theta} \, , \tag{7.236}$$

where $\lambda_o$ is the average wavelength, and $\Delta_\theta$ is the angular extent of the integration. For an extent of about 6 degrees, the cross-range resolution is approximately 0.26 meters at 5.5 GHz. Range-Doppler images centered at observation angles of 3 and 138 degrees are shown in Figures 7.23a and 7.23b, respectively, for vertical polarization. A Hamming window is used in the cross-range dimension to reduce sidelobes. Superimposed onto each image is the two-dimensional outline of the RV, allowing us to register the scattering features. At the forward aspect, the scattering centers at the nose and stationary phase points of the base edge (sometimes referred to as slipping scatterers) are clearly visible. The joint between the two frustums is also visible, as well as the faint double-diffraction response located slightly past the target. At the rear aspect, the delayed returns dominate the image. These scattering centers appear to originate from off-body locations, obscuring the true shape of the target. Such features are often detrimental to automated target recognition (ATR) algorithms that use the object's shape as a discrimination feature. Real-world reentry vehicles often have hoses, propellant tanks, wiring, and assorted electronics in the rear cavity that give rise to additional multiple-bounce behavior. As a result, radar observations of such a target could result in an apparent length several times that of the true object.

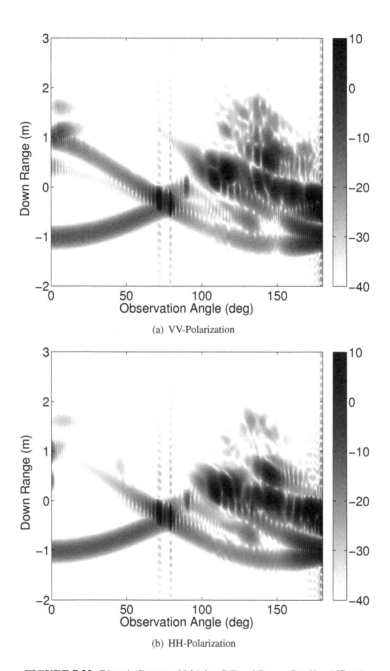

(a) VV-Polarization

(b) HH-Polarization

**FIGURE 7.22:** Biconic Reentry Vehicle: C-Band Range Profiles (dBsm)

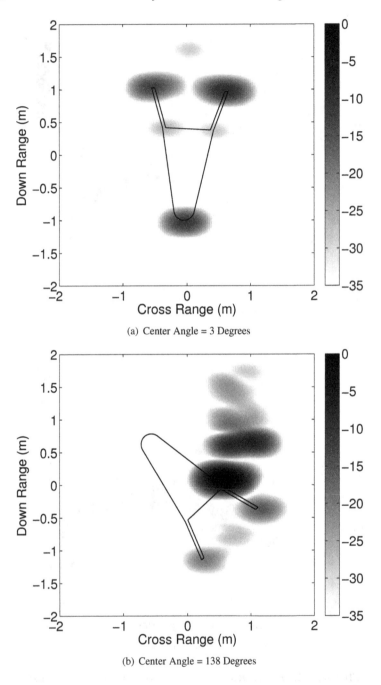

(a) Center Angle = 3 Degrees

(b) Center Angle = 138 Degrees

**FIGURE 7.23:** Biconic Reentry Vehicle: C-Band Range-Doppler Images (dBsm)

## 7.10    Treatment of Junctions

Thus far, our treatment has only considered interfaces between two regions. Points at which three or more regions meet are called *junctions*, and at these, the boundary conditions and orientation of vector basis functions must be considered more carefully. In this section, we will consider the treatment, placement, and orientation of the longitudinal and azimuthal vectors at several types of junctions in BoR problems. We also refer the reader to Section 8.6 for an in-depth discussion of the treatment of the boundary conditions and enforcement of integral equations at junctions of general configuration.

### 7.10.1    Orientation of Basis Functions

In Section 7.7, we examined the assignment and orientation of basis functions at an interface between two regions. We will now generalize this further by considering junctions between $N$ regions, where one or more of these regions may conducting. In a BoR geometry, junctions occur where an endpoint is shared by three or more boundary segments.

#### 7.10.1.1    Longitudinal Basis Vectors

According to Section 7.7, the longitudinally oriented basis functions assigned to a point on the interface between two dielectric regions must have opposing directions on each side, as shown in Figure 7.24a. Consider now the interface between three or more dielectric regions, as shown in Figure 7.24b. As Kirchoff's laws must be satisfied in each region, the basis vectors must flow past the common point in a clockwise (or counterclockwise) manner, as shown. Furthermore, the boundary conditions stipulate that these basis functions must

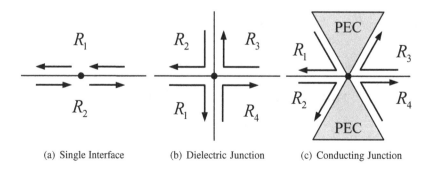

(a) Single Interface        (b) Dielectric Junction        (c) Conducting Junction

**FIGURE 7.24:** Orientation of Longitudinal Basis Vectors

have the same coefficient. If we now make one or more of the regions conducting, as in Figure 7.24c, this assignment will remain valid for dielectric interfaces that lie between conductors. Thus, the oriented basis functions in $R_1$ and $R_2$ will be combined into one unknown, as will the basis functions in $R_3$ and $R_4$.

### 7.10.1.2   Azimuthal Basis Vectors

At the intersection between two regions, as shown in Figure 7.25a, the azimuthally oriented basis functions assigned to the common point have a $+\hat{\phi}$ orientation in $R_1$, and a $-\hat{\phi}$ orientation in $R_2$. Under this arrangement, the boundary conditions at the interface are satisfied, and the basis functions on each side of the interface can be combined into a single unknown. If we now increase the number of dielectric regions meeting at this point to any odd number greater than or equal to three, this arrangement is no longer possible. This is illustrated in Figure 7.25b for a junction between three regions, where the boundary condition is not satisfied at the interface between $R_1$ and $R_3$. This implies that at the intersection of an odd number of dielectric regions, we must use half-triangles on each segment and assign separate coefficients to each. Increasing the number of regions to four, as in Figure 7.25c, we see that the boundary conditions are again satisfied, and we can again use full triangle functions and combine the unknowns into a single unknown. This approach is similar to that in [4], except that the approach taken there suggests using half-triangles for all azimuthal *and* longitudinal functions at a junction, regardless of the number of regions meeting there.

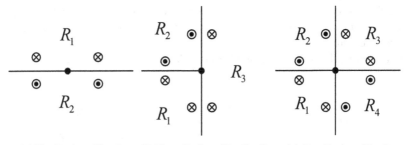

(a) Two Regions (Good)        (b) Three Regions (Not Good)        (c) Four Regions (Good)

**FIGURE 7.25:** Orientation of Azimuthal Basis Vectors

## 7.10.2 Examples with Junctions

We now consider several examples involving dielectric and dielectric/conducting objects with junctions.

### 7.10.2.1 Dielectric Sphere with Septum

We will first consider the dielectric sphere depicted in Figure 7.26a, where the sphere has been split into two hemispheres of equal size. There are now three dielectric regions: the free space region $R_0$ and the two dielectric halves $R_1$ and $R_2$. This object has a single junction at the intersection of the three regions.

*Results*

A good way to test the effectiveness of the junction treatment is to treat $R_1$ and $R_2$ as independent regions having identical dielectric properties. This effectively creates a homogeneous dielectric sphere, and the MoM/BoR results can be compared against the Mie series solution. The outermost radius of the sphere is $a_0 = 1\lambda_0$. The dielectric parameter of both regions is $\epsilon = 2.56$, and for the Mie solution, $N = 40$. The bounding curves are subdivided into 10 segments per wavelength in each region. The MoM results are compared to the Mie series in Figures 7.27a and 7.27b for vertical and horizontal polarization, respectively, where the comparison is very good.

### 7.10.2.2 Coated Sphere with Septum

We next consider a conducting sphere with a dielectric coating, depicted in Figure 7.26b. The coating is again separated into two partial hemispheres of equal size. There are three dielectric regions: the free space region $R_0$, and the two hemispherical dielectric regions $R_1$ and $R_2$. The conducting core comprises an additional region, $R_3$, which is not labeled. This object has two junc-

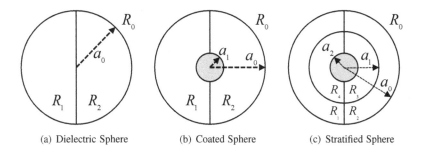

(a) Dielectric Sphere      (b) Coated Sphere      (c) Stratified Sphere

**FIGURE 7.26:** Spheres with a Hemispherical Division

tions: A junction at the intersection of $R_0$, $R_1$, and $R_2$, and another at the intersection of $R_1$, $R_2$, and $R_3$.

*Results*

$R_1$ and $R_2$ are again treated as independent regions having identical dielectric properties. The radius of the conducting core is $a_1 = 0.25\lambda_0$, and the radius of the coating is $a_0 = 1\lambda_0$. The dielectric parameter of both regions is $\epsilon = 5$, and for the Mie solution, $N = 40$. The bounding curves are subdivided into 10 segments per wavelength in each region. The MoM results are compared to the Mie series in Figures 7.27c and 7.27d for vertical and horizontal polarization, respectively, where the comparison is again very good.

### 7.10.2.3  Stratified Sphere with Septum

We next consider a conducting sphere with a stratified dielectric coating, depicted in Figure 7.26c. The stratified coatings are split into two partial hemispheres of equal size. There are five dielectric regions: the free space region $R_0$, the hemispherical regions of the outermost coating ($R_1$ and $R_2$) and the hemispherical regions of the innermost coating ($R_3$ and $R_4$). The conducting core comprises an additional region, $R_5$, which is not labeled. This object has three junctions: A junction at the intersection of $R_0$, $R_1$, and $R_2$, one at the intersection of $R_1$, $R_2$, $R_3$, and $R_4$, and another at the intersection of $R_3$, $R_4$, and $R_5$.

*Results*

For this example, regions $R_1$ and $R_2$ are assigned a dielectric parameter of $\epsilon = 5.0$, and regions $R_3$ and $R_4$ are assigned $\epsilon = 7.0$. The radius of the conducting core is $a_2 = 0.125\lambda_0$, the radius of the innermost coating is $a_1 = 0.6\lambda_0$, and the radius of the outermost coating is $a_0 = 1\lambda_0$. The bounding curves are subdivided into 10 segments per wavelength in each region, and for the Mie solution, $N = 40$. The MoM results are compared to the Mie series in Figures 7.27e and 7.27f for vertical and horizontal polarization, respectively, where the comparison is again very good.

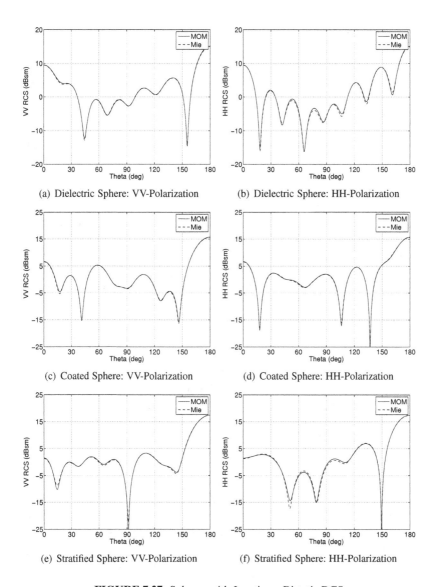

(a) Dielectric Sphere: VV-Polarization

(b) Dielectric Sphere: HH-Polarization

(c) Coated Sphere: VV-Polarization

(d) Coated Sphere: HH-Polarization

(e) Stratified Sphere: VV-Polarization

(f) Stratified Sphere: HH-Polarization

**FIGURE 7.27:** Spheres with Junctions: Bistatic RCS

### 7.10.2.4  Monoconic Reentry Vehicle with Dielectric Nose

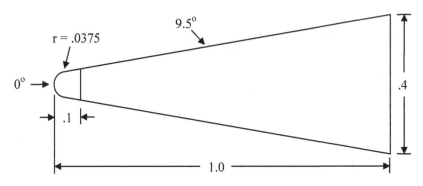

**FIGURE 7.28:** Monoconic Reentry Vehicle Dimensions (meters)

Finally, we will examine a monoconic reentry vehicle (RV) whose dimensions are outlined in Figure 7.28. The RV has a length of 1 meter and a base diameter of 0.4 meters, with a cone angle of 9.5 degrees. The dielectric section is 10 centimeters in length, and the nose tip has a radius of curvature of 3.75 centimeters. This problem has three regions: the free space region $R_0$, the dielectric (nose tip) region $R_1$, and the conducting (frustum) region $R_2$. As reentry vehicles typically have nose tips made of an ablative material, this is a somewhat realistic example.

*Results*

We compute the monostatic RCS of the RV at 5.0 GHz, and consider the nose tip with lossless ($\epsilon = 5$) and lossy ($\epsilon = 5.0 - j0.25$) dielectric parameters. The interfaces are constructed as follows: 223 segments for conducting frustum ($S_{02}$), 25 segments for the nose ($S_{01}$), and 19 segments for the interface (septum) $S_{12}$ between the frustum and nose. The RV with a lossless nose tip is compared to a perfectly conducting RV of the same dimensions in Figures 7.29a and 7.29b for vertical and horizontal polarizations, respectively. The RV with the dielectric nose has a significantly higher RCS at many aspect angles, particularly below 30 degrees. This is due to the many multiple-bounces inside the nose. A similar comparison is done for the RV with lossy nose in Figures 7.29a and 7.29b. The RCS has been significantly reduced from the lossless case, although it is still significantly higher at nose-forward aspect angles, particularly in horizontal polarization.

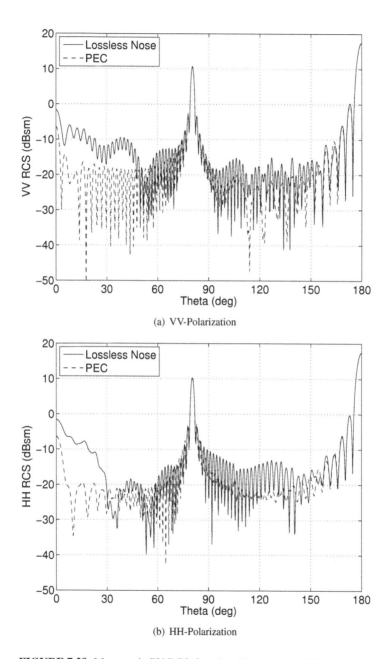

(a) VV-Polarization

(b) HH-Polarization

**FIGURE 7.29:** Monoconic RV RCS: Lossless Nose versus PEC at 5.0 GHz

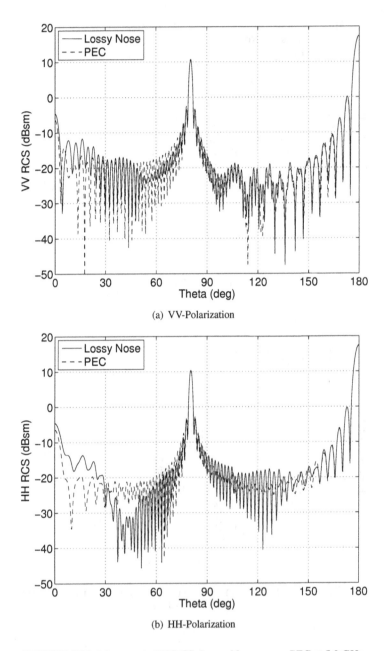

(a) VV-Polarization

(b) HH-Polarization

**FIGURE 7.30:** Monoconic RV RCS: Lossy Nose versus PEC at 5.0 GHz

# References

[1] R. Mautz and R. Harrington, "H-field, E-field and combined solutions for bodies of revolution," Tech. Rep. Report RADC-TR-77-109, Rome Air Development Center, Griffiss AFB, N.Y., March 1977.

[2] L. N. Medgyesi-Mitschang and J. M. Putnam, "Electromagnetic scattering from axially inhomogeneous bodies of revolution," *IEEE Trans. Antennas Propagat.*, vol. 32, pp. 796–806, March 1984.

[3] M. Abramowitz and I. Stegun, *Handbook of Mathematical Functions.* National Bureau of Standards, 1966.

[4] L. N. Medgyesi-Mitschang and J. M. Putnam, "Combined field integral equation formulation for inhomogeneous two- and three-dimensional bodies: The junction problem," *IEEE Trans. Antennas Propagat.*, vol. 39, pp. 667–672, May 1991.

[5] K. Faison, "The Electromagnetics Code Consortium," *IEEE Antennas Propagat. Magazine*, vol. 30, pp. 19–23, February 1990.

[6] G. T. Ruck, ed., *Radar Cross Section Handbook.* Plenum Press, 1970.

[7] B. R. Mahafza, *Introduction to Radar Analysis.* CRC Press, 1998.

[8] A. W. Doerry, "Catalog of window taper functions for sidelobe control," Tech. Rep. SAND2017-4042, Sandia National Laboratories, April 2017.

[9] A. C. Woo, H. T. G. Wang, M. J. Schuh, and M. L. Sanders, "Benchmark radar targets for the validation of computational electromagnetics programs," *IEEE Antennas Propagat. Magazine*, vol. 35, pp. 84–89, February 1993.

[10] A. Ausherman, A. Kozma, J. Waker, H. M. Jones, and E. C. Poggio, "Developments in radar imaging," *IEEE Trans. Aerospace Electron. Syst.*, vol. 20, pp. 363–400, July 1984.

# Chapter 8

## Three-Dimensional Problems

In this chapter we will use the Method of Moments to solve surface integral equation problems involving three-dimensional (3D) objects of arbitrary shape. This will allow us to solve many problems of practical interest such as those found in electromagnetic interference (EMI), electronics packaging, scattering, radar cross section, and antenna design. This area has received significant attention in the literature in the last thirty years, and many different methods of treating 3D surfaces of general shape have been considered. We will consider one of the most commonly used approaches, where the surfaces are described using flat, planar triangles, and the currents expanded using the well-known Rao-Wilton-Glisson (RWG) basis functions. In Chapter 7, we treated several dielectric and dielectric/conducting junctions of limited type. In this chapter, we will use the method outlined in [1] to treat junctions with a much more general configuration.

Historically, the 3D problems that could be solved were limited to those of small electrical size, due to limitations in CPU time and system memory. Solving larger problems required access to supercomputer-class hardware, which was not available to most people. As computers have continued to become more powerful and less expensive over the last twenty or so years, much larger problems can now be solved on most desktop computers. This has also opened up areas of research that had not been attempted or considered before. For example, the development of compressive techniques, such as the Adaptive Cross Approximation (ACA), the Multi-Level Adaptive Cross Approximation (MLACA), and the Fast Multipole Method (FMM), has further increased the tractability of larger problems. Those methods, discussed in the chapters that follow, will further build on the material presented in this chapter.

The approach in this chapter comprises three general steps: We will first consider the discretized EFIE, MFIE, and nMFIE of Section 3.6.2.2 in a region-wise context, and derive expressions for computing the required matrix elements. Next, we enforce the boundary conditions along each interface and at junctions, so that all linearly dependent unknowns are combined. The overdetermined matrix system will then be reduced by applying the integral equation formulations described in Section (3.6.3.1). We will then consider several numerical examples.

## 8.1   Modeling of Three-Dimensional Surfaces

Let us consider the digital construction and representation of three-dimensional objects. First, a description of the object to be modeled is required. If the object is purely conceptual, line drawings or engineering blue prints are drafted, or if the actual object is available, drawings can be made via physical measurement. Next, a model of the object is constructed using a computer-aided design (CAD) software. It is common for this step to often overlap with the first, as the conceptual design and surface modeling of many objects can be done completely inside the CAD program, with no pen or paper required. The result of this step is a digital, curved surface model, which is highly accurate. Most modern modeling programs such as Rhinoceros 3D [2] represent free-form objects using a mathematically exact description such as non-uniform rational B-splines (NURBS) [3]. This approach preserves necessary information such as the surface normal and radius of curvature at every point. NURBS is a part of many industry-wide standard CAD file formats and modeling engines such as IGES, STEP, ACIS, and Parasolid. Once the curved surface model is developed, it must then be reduced to a set of geometric primitives that can be processed. Among the most commonly used primitives are planar triangles (or *facets*), which are flexible and reasonably good at following the curvature of most realistic shapes [4]. Furthermore, numerical quadratures have been developed for the triangle (see Section 12.2), and analytic solutions to potential integrals exist for triangular subdomains [5, 6, 7, 8]. The model is then discretized or *tesselated* into a trianglular mesh that conforms to the surfaces of interest, and most CAD programs can generate meshes of very high quality. Due to their flexibility, complex shapes such as ground and air vehicles can be represented with great precision by facet models. Figures 8.1a, 8.1b, and 8.1c depict a sphere, a missile, and a main battle tank represented entirely by triangular facets.

### 8.1.1   Facet File

A common way to describe a triangular surface mesh is via the finite element connectivity file (or *facet file*) [9]. This file comprises a *node list* which contains the $(x, y, z)$ components of all nodes in the mesh, and a *facet list* that comprises the indexes of the three nodes that define each triangle. Additional information regarding each triangle, such as its dielectric interface assignment, is often added to the facet list. An example is provided in Figure 8.2, which comprises a square plate in the $xy$ plane constructed from 9 nodes and 8 triangles. In this example, a one-based indexing was used in the facet list.

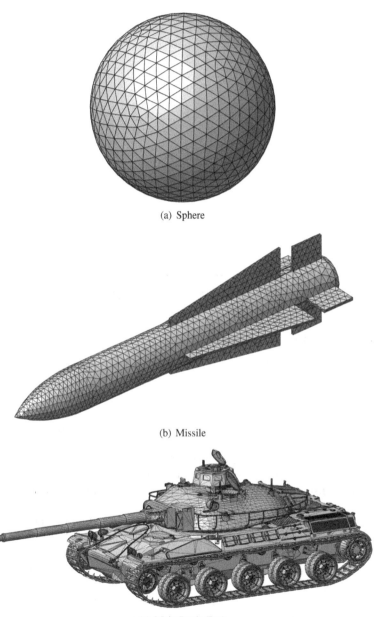

(a) Sphere

(b) Missile

(c) Main Battle Tank

**FIGURE 8.1:** Example 3D Facet Models

```
9
-1.0  1.0  0.0
 0.0  1.0  0.0
 1.0  1.0  0.0
-1.0  0.0  0.0
 0.0  0.0  0.0
 1.0  0.0  0.0
-1.0 -1.0  0.0
 0.0 -1.0  0.0
 1.0 -1.0  0.0
8
4 2 1
4 5 2
5 3 2
5 6 3
7 5 4
7 8 5
8 6 5
8 9 6
```

**FIGURE 8.2:** Example Facet File

## 8.1.2  Edge-Finding Algorithm

For a surface mesh, we need a way to identify and register all unique triangle edges. To illustrate an edge-finding algorithm, we will use the facet file described in Figure 8.2. The numbering of the nodes and facets in this model is illustrated in Figures 8.3a, and 8.3b, respectively. Our first task is to identify the connections between each node and each facet that references it. This will allow us to determine which facets share an edge. To do this, we create two lists for each node, a *node connectivity list* and a *facet connectivity list*. The first list contains all unique connections between a node and other nodes, and the second list contains the facets having that node as a vertex. We next loop over individual triangles and identify the nodes that make up the triangle, and sort them into ascending order. We then add the index of the current facet to the facet connectivity list of each node. For the node of lowest index, we add to its list the two higher node indexes if they are not already in the list. For the second node, the index of the third node is added to its list if not already present, and nothing is done for the third node. After this operation is completed, the node and facet connectivity lists are as shown in Table 8.1. Note that the node connectivity list contains no redundant links. This results in an empty list for node 9 since it is already referenced in the list for nodes 6 and 8.

At this point, the connectivity lists contain the information needed to find all the unique edges in the model. The remaining task is to go through the

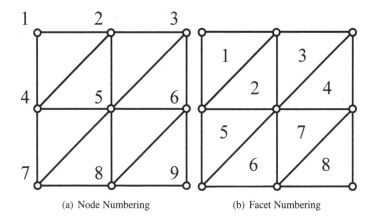

(a) Node Numbering    (b) Facet Numbering

**FIGURE 8.3:** Simple Flat Plate Geometry

**TABLE 8.1:** Node and Facet Connectivity Lists

| Node # | Connected Nodes | Connected Facets |
|--------|-----------------|------------------|
| 1 | 2 4 | 1 |
| 2 | 4 5 3 | 1 2 3 |
| 3 | 5 6 | 3 4 |
| 4 | 5 7 | 1 2 5 |
| 5 | 6 7 8 | 2 3 4 5 6 7 |
| 6 | 8 9 | 4 7 8 |
| 7 | 8 | 5 6 |
| 8 | 9 | 6 7 8 |
| 9 | - | 8 |

connectivity list for each node and create an edge for each entry. The facets common to the endpoints of each edge are recorded. If an edge belongs to only one triangle, it is called a *boundary edge*. Edges shared by two triangles are called *regular* edges (or just edges), and those shared by three or more facets are *junction* edges. Additional information regarding each edge can be stored at this step, such as its length, its endpoints, and which facets share it. Similarly, for each triangle we also store each of the unique edge indexes. This is useful for assigning basis function indexes later.

## 8.1.2.1 Shared Nodes

The edge-finding algorithm requires that facets that are connected to each other use common node indexes; otherwise, there is no logical connectivity and the algorithm will not find any edges. Some CAD programs generate redundant nodes for each triangle, and some triangle file formats such as Stereolithography (.stl) store the nodes of each triangle separately. The CAD modeler should ensure that these redundant nodes are collapsed into single nodes. Some CAD programs have a function that will *join* or *weld* meshes to remove coincident nodes before finalizing the geometry, and some advanced MoM solver codes will do this automatically when reading the facet file.

## 8.2   Expansion of Surface Currents

On each dielectric interface, the electric and magnetic currents are expanded as a sum of weighted basis functions as in (3.182) and (3.183). In the expansion, we use the Rao-Wilton-Glisson (RWG) basis function [10], which has become the most commonly used basis function in 3D problems since its introduction. The RWG function, as illustrated in Figure 8.4, is defined as

$$\mathbf{f}_n(\mathbf{r}) = \frac{L_n}{2A_n^+}\boldsymbol{\rho}_n^+(\mathbf{r}) \qquad \mathbf{r} \text{ in } T_n^+ \, , \tag{8.1}$$

$$\mathbf{f}_n(\mathbf{r}) = \frac{L_n}{2A_n^-}\boldsymbol{\rho}_n^-(\mathbf{r}) \qquad \mathbf{r} \text{ in } T_n^- \, , \tag{8.2}$$

$$\mathbf{f}_n(\mathbf{r}) = 0 \qquad \text{otherwise} \, , \tag{8.3}$$

where $T_n^+$ and $T_n^-$ are a pair of triangles that share edge $n$, and $L_n$ is the length of edge $n$. On $T_n^+$, the vector $\boldsymbol{\rho}_n^+(\mathbf{r})$ points *toward* the vertex $\mathbf{v}^+$ opposite the edge, and is

$$\boldsymbol{\rho}_n^+(\mathbf{r}) = \mathbf{v}^+ - \mathbf{r} \quad \mathbf{r} \text{ in } T_n^+ \, , \tag{8.4}$$

and on the $T_n^-$, $\boldsymbol{\rho}_n^-(\mathbf{r})$ points *away* from the opposite vertex $\mathbf{v}^-$, and is

$$\boldsymbol{\rho}_n^-(\mathbf{r}) = \mathbf{r} - \mathbf{v}^- \quad \mathbf{r} \text{ in } T_n^- \, . \tag{8.5}$$

These basis functions are assigned only to *interior* edges shared by two or more adjacent triangles (we will discuss how RWG functions are assigned at junctions shortly). The RWG function has no component normal to any edge other than the one to which it is assigned, and the component of the function normal to that edge is unity. Kirchoff's law is satisfied along each edge, and as boundary edges have no current flowing across them, they are not used. For the Galerkin testing procedure, we will use as testing functions the same RWG functions used to expand the current.

The RWG functions comprise *first-order* or *linear* basis functions, however basis functions of higher order on curvilinear (non-planar) triangles have also been studied, such as those considered by the authors in [11].

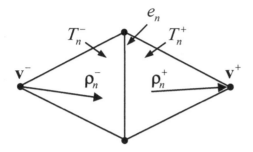

**FIGURE 8.4:** RWG Basis Function

## 8.2.1 Divergence of the RWG Function

To compute the divergence of the RWG function, let us consider $\mathbf{f}_n(\mathbf{r})$ on $T_n^-$. If we now consider $T_n^-$ in cylindrical coordinates, where $\mathbf{v}^-$ is the origin, $\mathbf{f}_n(\mathbf{r})$ can be written as a vector having only a radial component, yielding

$$\mathbf{f}_n(\mathbf{r}) = \frac{L_n}{2A_n} r \, \hat{\mathbf{r}} \,. \tag{8.6}$$

Computing the divergence in cylindrical coordinates then yields

$$\nabla \cdot \mathbf{f}_n(\mathbf{r}) = \frac{1}{r} \frac{\partial}{\partial r} \left[ \frac{L_n}{2A_n} r^2 \right] = \frac{L_n}{A} \quad \text{in } T_n^-, \tag{8.7}$$

and a similar calculation,

$$\nabla \cdot \mathbf{f}_n(\mathbf{r}) = -\frac{L_n}{A} \quad \text{in } T_n^+. \tag{8.8}$$

As the divergence of the current is proportional to the surface charge density through the equation of continuity, we see from (8.7) and (8.8) that the total charge density associated with adjacent triangle pairs is zero. As there is no accumulation of charges on an edge, the RWG function is said to be *divergence conforming* [11].

## 8.2.2 Assignment and Orientation of Basis Functions

Consider a problem of general configuration, comprising multiple dielectric and conducting regions, where there may be one or more junctions between regions. Let us now assign basis functions to the various interfaces on a region-wise basis. If $e_n$ is a regular edge on the interface $S_{lm}$ between the two dielectric regions $R_l$ and $R_m$, we assign one electric and magnetic basis function to the edge in each region, and the functions are oriented so that they flow in opposite directions on each side of the interface, as illustrated in Figure 7.24a. If $e_n$ is a junction edge, a similar assignment is made, except that the basis functions are oriented such that the current flows past the edge in a counterclockwise (or clockwise) manner, as shown in Figure 7.24b. These assignments ensure that Kirchoff's law is satisfied at $e_n$ in each region. If the edge lies on a conducting interface, or at the junction of multiple interfaces, where one or more of those interfaces are conducting, we omit the magnetic basis function, as magnetic currents vanish on conductors. Similarly, no electric or magnetic basis functions are assigned inside a closed conducting region. Doing this for all non-boundary edges leads to the discretized matrix system in (3.194), where the basis functions are oriented correctly but linear dependencies at the interfaces have not yet been taken into account. Enforcement of the boundary conditions is discussed in detail in Section 8.6.

## 8.3 EFIE

The EFIE (3.178) contains $\mathcal{L}$ and $\mathcal{K}$ operators, which were discretized and tested via the MoM in (3.185) and (3.186). In this section we formulate expressions for both operators, using RWG basis and testing functions.

### 8.3.1 $\mathcal{L}$ Operator

Substitution of the electric basis and testing functions into (3.185) yields in Region $R_l$ matrix elements given by

$$Z_{mn}^{EJ(l)} = \mathrm{L}(\mathbf{f}_m, \mathbf{f}_n) \,, \tag{8.9}$$

where

$$\mathrm{L}(\mathbf{f}_m, \mathbf{f}_n) = \int_{\mathbf{f}_m} \int_{\mathbf{f}_n} \left[ j\omega\mu_l \, \mathbf{f}_m(\mathbf{r}) \cdot \mathbf{f}_n(\mathbf{r}') - \frac{j}{\omega\epsilon_l} \nabla \cdot \mathbf{f}_m(\mathbf{r}) \, \nabla' \cdot \mathbf{f}_n(\mathbf{r}') \right] \frac{e^{-jk_l r}}{4\pi r} \, d\mathbf{r}' \, d\mathbf{r} \,. \tag{8.10}$$

The inner and outer integrals in (8.10) are each performed over two triangles. However, as each triangle supports a maximum of three basis and testing functions, the computations involving one source and testing triangle can contribute to a maximum of nine matrix elements. Thus, matrix filling will be most efficient by performing a loop over source and testing triangle pairs, where all possible combinations of basis and testing functions are considered for each pair.

### 8.3.1.1 Non-Near Terms

When the source and testing triangles are separated by a sufficient distance, (8.10) can be implemented as written. For a source and testing triangle, the contribution to (8.10) can be computed using an $M$-point numerical quadrature on each triangle, yielding

$$I = \frac{L_m L_n}{4\pi} \sum_{p=1}^{M} \sum_{q=1}^{M} w_p w_q \left[ \frac{j\omega\mu_l}{4} \boldsymbol{\rho}_m^{\pm}(\mathbf{r}_p) \cdot \boldsymbol{\rho}_n^{\pm}(\mathbf{r}_q') \pm \frac{j}{\omega\epsilon_l} \right] \frac{e^{-jk_l R_{pq}}}{R_{pq}} , \quad (8.11)$$

where $p$ and $q$ refer to the testing and source coordinates, respectively, and the quadrature weights $w_p$ and $w_q$ have been normalized with respect to triangle area. The distance $R_{pq}$ is

$$R_{pq} = \sqrt{(x_p - x_q)^2 + (y_p - y_q)^2 + (z_p - z_q)^2} . \quad (8.12)$$

The signs of the terms (8.11) depend on the orientation of the basis and testing functions on each triangle.

### 8.3.1.2 Near and Self Terms

When the source and testing triangles are close together or overlapping, the integrations must be treated carefully. In this case, we will use a similar approach taken for the near self terms in Section 6.1.1, where the integral was rewritten as a sum of bounded and singular terms. To do so, let us rewrite the 3D Green's function in region $R_l$ as

$$\frac{e^{-jk_l r}}{r} = \left[ \frac{e^{-jk_l r}}{r} - \frac{1}{r} \right] + \frac{1}{r} , \quad (8.13)$$

which is often referred to in the literature as a *singularity extraction*. The first term on the right-hand side is well behaved, having the limit

$$\lim_{r \to 0} \left[ \frac{e^{-jk_l r}}{r} - \frac{1}{r} \right] = -jk_l . \quad (8.14)$$

This portion can be used as-is in (8.11). Inserting the second term on the right-hand side of (8.13) into (8.10) yields singular potential integrals of the form

$$I_1 = \int_T \int_{T'} \boldsymbol{\rho}_m^{\pm}(\mathbf{r}) \cdot \boldsymbol{\rho}_n^{\pm}(\mathbf{r}') \frac{1}{r} \, d\mathbf{r}' \, d\mathbf{r} \quad (8.15)$$

and

$$I_2 = \int_T \int_{T'} \frac{1}{r} \, d\mathbf{r}' \, d\mathbf{r} . \quad (8.16)$$

We will consider (8.15) and (8.16) separately for the self and near terms.

*Self Terms*

When $T$ and $T'$ overlap, the inner and outer integrations in (8.15) and (8.16) can be computed analytically. For brevity, we will consider the integrals in terms of the basis vectors $\boldsymbol{\rho}_{m,n}(\mathbf{r}) = \boldsymbol{\rho}_{m,n}^{-}(\mathbf{r})$ on $T^{-}$, and the corresponding results for $\boldsymbol{\rho}_{m,n}^{+}(\mathbf{r})$ can be obtained by a change of sign. To begin, we convert $\boldsymbol{\rho}_{m,n}(\mathbf{r})$ to simplex coordinates (see Section 12.2.1), yielding

$$\boldsymbol{\rho}_{m,n}(\mathbf{r}) = (1 - \lambda_1 - \lambda_2)\mathbf{v}_1 + \lambda_1 \mathbf{v}_2 + \lambda_2 \mathbf{v}_3 - \mathbf{v}_{m,n} \,, \qquad (8.17)$$

where $\mathbf{v}_1, \mathbf{v}_2$ and $\mathbf{v}_3$ are the triangle nodes and $\mathbf{v}_{m,n}$ is the node opposite edge $m$ or $n$ on the triangle. Substitution of (8.17) into (8.15) yields

$$I_1 = \int_T \int_{T'} \left[ (1 - \lambda_1 - \lambda_2)\mathbf{v}_1 + \lambda_1 \mathbf{v}_2 + \lambda_2 \mathbf{v}_3 - \mathbf{v}_m \right]$$
$$\cdot \left[ (1 - \lambda_1' - \lambda_2')\mathbf{v}_1 + \lambda_1' \mathbf{v}_2 + \lambda_2' \mathbf{v}_3 - \mathbf{v}_n \right] \frac{1}{r} \, d\lambda_1' \, d\lambda_2' \, d\lambda_1 \, d\lambda_2 \,. \qquad (8.18)$$

Expansion of all the terms and performing some lengthy term collection yield the integrand

$$\Big[ \lambda_1 \lambda_1' (a_{11} - 2a_{12} + a_{22}) + \lambda_1 \lambda_2' (a_{11} - a_{13} - a_{12} + a_{23})$$
$$+ \lambda_2 \lambda_2' (a_{11} - 2a_{13} + a_{33}) + \lambda_1' \lambda_2 (a_{11} - a_{12} - a_{13} + a_{23})$$
$$+ \lambda_1 (-a_{11} + a_{1n} + a_{12} - a_{2n}) + \lambda_2 (-a_{11} + a_{1n} + a_{13} - a_{3n})$$
$$+ \lambda_1' (-a_{11} + a_{1m} + a_{12} - a_{2m}) + \lambda_2' (-a_{11} + a_{1m} + a_{13} - a_{3m})$$
$$+ a_{11} - a_{1n} - a_{1m} + a_{mn} \Big] \frac{1}{r} \,, \qquad (8.19)$$

where

$$a_{11} = \mathbf{v}_1 \cdot \mathbf{v}_1, \qquad a_{12} = \mathbf{v}_1 \cdot \mathbf{v}_2, \qquad a_{13} = \mathbf{v}_1 \cdot \mathbf{v}_3, \qquad (8.20)$$
$$a_{22} = \mathbf{v}_2 \cdot \mathbf{v}_2, \qquad a_{23} = \mathbf{v}_2 \cdot \mathbf{v}_3, \qquad a_{33} = \mathbf{v}_3 \cdot \mathbf{v}_3, \qquad (8.21)$$
$$a_{1n} = \mathbf{v}_1 \cdot \mathbf{v}_n, \qquad a_{1m} = \mathbf{v}_1 \cdot \mathbf{v}_m, \qquad a_{23} = \mathbf{v}_2 \cdot \mathbf{v}_3, \qquad (8.22)$$
$$a_{2n} = \mathbf{v}_2 \cdot \mathbf{v}_n, \qquad a_{2m} = \mathbf{v}_2 \cdot \mathbf{v}_m, \qquad a_{3n} = \mathbf{v}_3 \cdot \mathbf{v}_n, \qquad (8.23)$$
$$a_{3m} = \mathbf{v}_3 \cdot \mathbf{v}_m, \qquad a_{mn} = \mathbf{v}_n \cdot \mathbf{v}_n. \qquad (8.24)$$

This results in a set of integrals of the form

$$\int_T \int_{T'} \lambda_i \lambda_j' \frac{1}{|\mathbf{r} - \mathbf{r}'|} \, dT \, dT', \qquad (8.25)$$

where the various permutations were evaluated analytically by Eibert and

Hansen in [8, 12]. These are

$$\frac{1}{4A^2} \int_T \int_{T'} \lambda_1 \lambda_1' \frac{1}{|\mathbf{r} - \mathbf{r}'|} \, dT \, dT' = \frac{\ln_1}{20 l_1} + \frac{l_1^2 + 5 l_2^2 - l_3^2}{120 l_2^3} \ln_2$$

$$+ \frac{l_1^2 - l_2^2 + 5 l_3^2}{120 l_3^3} \ln_3 + \frac{l_3 - l_1}{60 l_2^2} + \frac{l_2 - l_1}{60 l_3^2} \, , \qquad (8.26)$$

$$\frac{1}{4A^2} \int_T \int_{T'} \lambda_1 \lambda_2' \frac{1}{|\mathbf{r} - \mathbf{r}'|} \, dT \, dT' = \frac{3 l_1^2 + l_2^2 - l_3^2}{80 l_1^3} \ln_1 + \frac{l_1^2 + 3 l_2^2 - l_3^2}{80 l_2^3} \ln_2$$

$$+ \frac{\ln_3}{40 l_3} + \frac{l_3 - l_2}{40 l_1^2} + \frac{l_3 - l_1}{40 l_2^2} \, , \qquad (8.27)$$

and

$$\frac{1}{4A^2} \int_T \int_{T'} \lambda_1' \frac{1}{|\mathbf{r} - \mathbf{r}'|} \, dT \, dT' = \frac{\ln_1}{8 l_1} + \frac{l_1^2 + 5 l_2^2 - l_3^2}{48 l_2^3} \ln_2 +$$

$$\frac{l_1^2 - l_2^2 + 5 l_3^2}{48 l_3^3} \ln_3 + \frac{l_3 - l_1}{24 l_2^2} + \frac{l_2 - l_1}{24 l_3^2} \, , \qquad (8.28)$$

where

$$l_1 = |\mathbf{v}_2 - \mathbf{v}_3| \, , \qquad (8.29)$$

$$l_2 = |\mathbf{v}_3 - \mathbf{v}_1| \, , \qquad (8.30)$$

$$l_3 = |\mathbf{v}_1 - \mathbf{v}_2| \, , \qquad (8.31)$$

and

$$\ln_1 = \log \frac{(l_1 + l_2)^2 - l_3^2}{l_2^2 - (l_3 - l_1)^2} \, , \qquad (8.32)$$

$$\ln_2 = \log \frac{(l_2 + l_3)^2 - l_1^2}{l_3^2 - (l_1 - l_2)^2} \, , \qquad (8.33)$$

$$\ln_3 = \log \frac{(l_3 + l_1)^2 - l_2^2}{l_1^2 - (l_2 - l_3)^2} \, . \qquad (8.34)$$

All integrals involving terms in the integrand of (8.19) can be obtained via (8.26-8.28) by permuting the vertex indices appropriately. The corresponding solution to (8.16) is

$$\frac{1}{4A^2} \int_T \int_{T'} \frac{1}{|\mathbf{r} - \mathbf{r}'|} \, dT \, dT' = \frac{\ln_1}{3 l_1} + \frac{\ln_2}{3 l_2} + \frac{\ln_3}{3 l_3}. \qquad (8.35)$$

Note that (8.35) does not depend on any basis function, so it only needs to be calculated once per triangle.

*Near Terms*

When $T$ and $T'$ are close together, (8.15) and (8.16) must be treated carefully to achieve maximum accuracy. Following our approach in computing the near terms in the thin wire EFIE, we will compute the outermost integral numerically and the innermost analytically. For the innermost integrals, we will use the method outlined in [5], which contains analytic expressions for potential integrals over $N$ sided planar polygons. As this method involves the triangle edges, consider the line segment $C$ with endpoints $\mathbf{r}^-$ and $\mathbf{r}^+$ located in $S$ (the plane of the polygon) as illustrated in Figure 8.5. The projections of source and observation vectors $\mathbf{r}'$ and $\mathbf{r}$ onto $S$ are $\rho'$ and $\rho$, respectively. The quantities in the figure are computed as follows:

$$\boldsymbol{\rho}^{\pm} = \mathbf{r}^{\pm} - \hat{\mathbf{n}}(\hat{\mathbf{n}} \cdot \mathbf{r}^{\pm}) \, , \tag{8.36}$$

$$\hat{\mathbf{l}} = \frac{\rho^+ - \rho^-}{|\rho^+ - \rho^-|} \, , \tag{8.37}$$

$$\hat{\mathbf{u}} = \hat{\mathbf{l}} \times \hat{\mathbf{n}} \, , \tag{8.38}$$

$$l^{\pm} = (\boldsymbol{\rho}^{\pm} - \boldsymbol{\rho}) \cdot \hat{\mathbf{l}} \, , \tag{8.39}$$

$$P^0 = |(\boldsymbol{\rho}^{\pm} - \boldsymbol{\rho}) \cdot \hat{\mathbf{u}}| \, , \tag{8.40}$$

$$P^{\pm} = |(\boldsymbol{\rho}^{\pm} - \boldsymbol{\rho})| = \sqrt{(P^0)^2 + (l^{\pm})^2} \, , \tag{8.41}$$

$$\hat{\mathbf{P}}^0 = \frac{(\boldsymbol{\rho}^{\pm} - \boldsymbol{\rho}) - l^{\pm}\hat{\mathbf{l}}}{P^0} \, , \tag{8.42}$$

$$R^0 = \sqrt{(P^0)^2 + d^2} \, , \tag{8.43}$$

and

$$R^{\pm} = \sqrt{(P^{\pm})^2 + d^2}. \tag{8.44}$$

The distance from the observation point to $S$ is

$$d = \hat{\mathbf{n}} \cdot (\mathbf{r} - \mathbf{r}^{\pm}). \tag{8.45}$$

Using the above, integrals of the form

$$I_1 = \int_{T'} \frac{\rho' - \rho}{r} \, d\mathbf{r}' \tag{8.46}$$

are evaluated via a sum over the $N$ polygon edges given by

$$I_1 = \frac{1}{2} \sum_{i=1}^{N} \hat{\mathbf{u}}_i \left[ (R_i^0)^2 \log \frac{R_i^+ + l_i^+}{R_i^- + l_i^-} + l_i^+ R_i^+ - l_i^- R_i^- \right]. \tag{8.47}$$

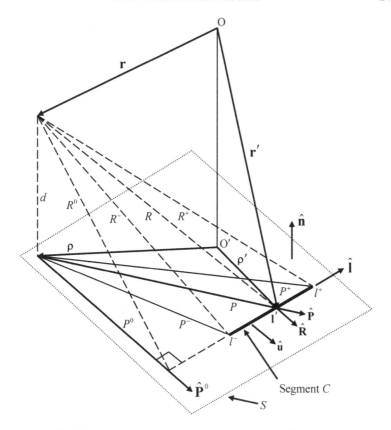

**FIGURE 8.5:** Geometric Quantities for Polygon Segment $C$

The integral

$$I_2 = \int_{T'} \frac{1}{r} \, d\mathbf{r}' \tag{8.48}$$

is evaluated similarly, yielding

$$I_2 = \sum_{i=1}^{N} \hat{\mathbf{P}}_i^0 \cdot \hat{\mathbf{u}}_i \left[ P_i^0 \log \frac{R_i^+ + l_i^+}{R_i^- + l_i^-} - |d| \left( \tan^{-1} \frac{P_i^0 l_i^+}{(R_i^0)^2 + |d| R_i^+} \right. \right.$$

$$\left. \left. - \tan^{-1} \frac{P_i^0 l_i^-}{(R_i^0)^2 + |d| R_i^-} \right) \right]. \tag{8.49}$$

Note that if the observation point $\rho$ lies anywhere along an edge, the contribution from that edge to (8.47) is zero. Additionally, (8.47) and (8.49) suffer from numerical issues when $\mathbf{r}$ lies in $S$ along the extension of an edge ($R_i^0 = 0$). In such cases, the observation should be moved a very small distance away from the edge in the plane, i.e.,

$$\mathbf{r} = \mathbf{r} + \epsilon \hat{\mathbf{u}} . \tag{8.50}$$

To evaluate the innermost integral of (8.15), we write the basis function $\rho_n(\mathbf{r}')$ as

$$\rho_n(\mathbf{r}') = \rho' - \rho_n, \tag{8.51}$$

where $\rho_n$ is the projection of $\mathbf{v}_n$ on $S$. We can then write

$$\rho' - \rho_n = (\rho' - \rho) + (\rho - \rho_n) , \tag{8.52}$$

which leads to

$$\int_{T'} \frac{\rho_n(\mathbf{r}')}{r} \, d\mathbf{r}' = \int_{T'} \frac{\rho' - \rho}{r} \, d\mathbf{r}' + (\rho - \rho_n) \int_{T'} \frac{1}{r} \, d\mathbf{r}' . \tag{8.53}$$

The two integrals on the right can now be evaluated via (8.47) and (8.49), and the innermost integral of (8.16) can be computed directly via (8.49). Integrals involving $\rho_n^+(\mathbf{r})$ are obtained similarly, with a change of sign.

Assuming the integration was exact, the source and observation triangles in (8.15) and (8.16) could be interchanged and the result would be the same. However, in this case the results will not be the same, as one integral is analytic and the other is numerical. To enforce this symmetry, we perform an additional integration with the triangles swapped and average the results.

*Duffy Transform*

The Duffy transform [13, 14, 15, 16] is a method of regularizing integrals over a triangle with a $1/r$ singularity at a vertex, allowing a numerical quadrature to be used. The region of integration is converted from a triangle to a rectangle following the transformation

$$\int_T \frac{f(\mathbf{r}')}{r(\mathbf{r}')} \, d\mathbf{r}' = \int_0^1 \int_0^1 \frac{f(u, \gamma)}{r(u, \gamma)} |J(u, \gamma)| \, du \, d\gamma , \tag{8.54}$$

where $J(u, \gamma)$ is the Jacobian. To illustrate, let us apply the transform to the integral

$$\int_T \frac{1}{r} \, d\mathbf{r}' = \int_0^1 \int_0^1 \frac{|J(u, \gamma)|}{r(u, \gamma)} \, du \, d\gamma , \tag{8.55}$$

where the triangle is defined by the points $\mathbf{v}_1 = (x_1, y_1)$, $\mathbf{v}_2 = (x_2, y_2)$, and $\mathbf{v}_3 = (x_3, y_3)$, and the vector $\mathbf{r}'$ on $T'$ is

$$\mathbf{r}' = (1 - u)(1 - \gamma)\mathbf{v}_1 + u\mathbf{v}_2 + \gamma(1 - u)\mathbf{v}_3 . \tag{8.56}$$

The Jacobian is

$$J(u,\gamma) = (1-u)\Big[ \big(x_2 - x_1 + \gamma[x_1 - x_3]\big)\big(y_3 - y_1\big)$$
$$-\big(y_2 - y_1 + \gamma[y_1 - y_3]\big)\big(x_3 - x_1\big)\Big], \tag{8.57}$$

and fixing the observation (singular) point $\mathbf{r}$ at $(x_1, y_1)$, $r$ can be written as

$$r(x,y) = \sqrt{(x_1 - x)^2 + (y_1 - y)^2} = \sqrt{a(u,\gamma) + b(u,\gamma)} = r(u,\gamma)\,, \tag{8.58}$$

where

$$a(u,\gamma) = x_1^2 + u x_1 x_2 + u^2 x_2^2 + 2(1-u)(u x_2 - x_1)\Big[(1-\gamma)x_1 + \gamma x_3\Big]$$
$$+(1-u)^2\Big[(1-\gamma)x_1 + \gamma x_3\Big]^2, \tag{8.59}$$

and

$$b(u,\gamma) = y_1^2 + u y_1 y_2 + u^2 y_2^2 + 2(1-u)(u y_2 - y_1)\Big[(1-\gamma)y_1 + \gamma y_3\Big]$$
$$+(1-u)^2\Big[(1-\gamma)y_1 + \gamma y_3\Big]^2. \tag{8.60}$$

The resulting $(1-u)$ term in the numerator acts to cancel the singularity in the denominator, allowing the integral to be performed using an numerical quadrature rule on a rectangular subdomain. When the observation point is located in the interior of the triangle, it can be divided into three sub-triangles with the new vertex placed at the singular point.

This method does suffer from some disadvantages. Since it is still a purely numerical method, compute time will be higher due to the number of numerical computations required at each quadrature point. The transformation is also only accurate for triangles with reasonably good aspect ratios, where thinner triangles require increasing numbers of quadrature points to retain the same level of accuracy [16]. In spite of these drawbacks, the method remains attractive in cases where the function $f(\mathbf{r})$ is not integrable analytically.

*Other Singularity Extractions*

It was suggested in [16] that because the first term in (8.13) has a discontinuous derivative at $r = 0$, using standard numerical quadrature will produce inaccurate results. They suggest an alternative singularity extraction, given by

$$\frac{e^{-jkr}}{r} = \left[\frac{e^{-jkr}}{r} - \frac{1}{r} + \frac{k^2 r}{2}\right] + \frac{1}{r} - \frac{k^2 r}{2}\,, \tag{8.61}$$

where the first term now has two continuous derivatives. They conclude that using (8.61) results in more accurate results than (8.13) for the same number of quadrature points. It is this author's experience that while this may be true, the singularity extraction in (8.13) is still accurate enough for most three-dimensional problems. It is also worth noting that the appendix of [16] also contains expressions for the evaluation of integrals of the form

$$\int_T r^n \, d\mathbf{r}' \,, \tag{8.62}$$

$$\int_T r^n \left[ \mathbf{r}' - \mathbf{v} \right] \, d\mathbf{r}' \,, \tag{8.63}$$

$$\int_T \nabla r^n \, d\mathbf{r}' \qquad \mathbf{r} \notin T \,, \tag{8.64}$$

and

$$\int_T \left[ \nabla r^n \right] \left[ \mathbf{r}' \times \mathbf{v} \right] \, d\mathbf{r}' \qquad \mathbf{r} \notin T \,, \tag{8.65}$$

where $\mathbf{v}$ is any vertex of $T$.

*Example Singular Integral Evaluation*

Let us compute (8.16) using (8.35) and (8.49). For the latter case, we apply a Gaussian quadrature rule of order $N$ (Section 12.2.4) to compute the outermost integral. The integration domain is an equilateral triangle with sides of length 1. The results are summarized in Table 8.2. The comparison matches to two significant digits at higher quadrature orders, however the results demonstrate that for overlapping triangles (8.35) remains the superior choice. We next calculate (8.16) over a pair of co-planar equilateral triangles with sides of length 1, which share a common edge. For comparison, we will use Gaussian quadrature to compute both the inner and outer integrals, and then (8.49) to compute the innermost integral analytically, leaving the outer integration unchanged. The results are summarized in Table 8.3. The comparison is fairly good, suggesting that either method may be sufficient in this case. We now make the internal angle between triangle faces much smaller. In this case we choose an angle of 10 degrees, and the results are summarized in Table 8.4. The results from quadrature alone compare poorly to those using the analytic inner integration. It is obvious from this example that analytic inner integrations should be used to calculate all near matrix elements. To determine when to apply the near-integration formulas, the programmer could compare the distances between triangle centers. If that distance is below some fraction of a wavelength, such as $\lambda/4 - \lambda/2$, the near integration is carried out.

**TABLE 8.2:** Overlapping Equilateral Triangle Double Integration

| Order 1 | Order 2 | Order 3 | Order 4 | Order 5 | Eqn. (8.35) |
|---------|---------|---------|---------|---------|-------------|
| 0.98767 | 0.85005 | 0.84709 | 0.83107 | 0.82830 | 0.82392 |

**TABLE 8.3:** Near Equilateral Triangle Double Integration (Co-planar)

| Integration Method | Order 1 | Order 2 | Order 3 | Order 4 | Order 5 |
|--------------------|---------|---------|---------|---------|---------|
| All Quadrature | 0.32475 | 0.34364 | 0.34724 | 0.35330 | 0.35794 |
| Quadrature + (8.49) | 0.32922 | 0.34331 | 0.34553 | 0.34903 | 0.35009 |

**TABLE 8.4:** Near Equilateral Triangle Double Integration (Non Co-planar)

| Integration Method | Order 1 | Order 2 | Order 3 | Order 4 | Order 5 |
|--------------------|---------|---------|---------|---------|---------|
| All Quadrature | 3.7357 | 2.1169 | 4.4860 | 1.5794 | 1.4878 |
| Quadrature + (8.49) | 0.86129 | 0.73681 | 0.73302 | 0.72283 | 0.72166 |

## 8.3.2 $\mathcal{K}$ Operator

Substitution of the magnetic basis and electric testing functions into (3.186) yields in Region $R_l$ matrix elements given by

$$Z_{mn}^{EM(l)} = \mathrm{K}(\mathbf{f}_m, \mathbf{g}_n) , \qquad (8.66)$$

where

$$
\mathrm{K}(\mathbf{f}_m, \mathbf{g}_n) = \frac{1}{2} \int_{\mathbf{f}_m, \mathbf{g}_n} \mathbf{f}_m(\mathbf{r}) \cdot \left[ \hat{\mathbf{n}}_l(\mathbf{r}) \times \mathbf{g}_n(\mathbf{r}) \right] d\mathbf{r}
$$
$$
- \frac{1}{4\pi} \int_{\mathbf{f}_m} \mathbf{f}_m(\mathbf{r}) \cdot \int_{\mathbf{g}_n} \left[ (\mathbf{r} - \mathbf{r}') \times \mathbf{g}_n(\mathbf{r}') \right] \frac{1 + jk_l r}{r^3} e^{-jk_l r} \, d\mathbf{r}' \, d\mathbf{r} .
$$
$$(8.67)$$

The first term on the right-hand side of (8.67) is used when the source and testing triangles overlap, and the second term when they do not.

### 8.3.2.1 Non-Near Terms

When the source and testing triangles are separated by a sufficient distance, (8.67) can be implemented as written. For a source and testing triangle, the contribution to (8.67) can be obtained using an $M$-point numerical quadrature, yielding

$$
\eta_0 \frac{L_m L_n}{8 A_m} \sum_{p=1}^{M} w_p \, \boldsymbol{\rho}_m^{\pm}(\mathbf{r}_p) \cdot \left[ \hat{\mathbf{n}}_l(\mathbf{r}_p) \times \boldsymbol{\rho}_n^{\pm}(\mathbf{r}_p) \right] \qquad (8.68)
$$

when the triangles overlap, and

$$-\eta_0 \frac{L_m L_n}{16\pi} \sum_{p=1}^{M} \sum_{q=1}^{M} w_p w_q \, \boldsymbol{\rho}_m^{\pm}(\mathbf{r}_p) \cdot \left[ (\mathbf{r}_p - \mathbf{r}_q') \times \boldsymbol{\rho}_n^{\pm}(\mathbf{r}_q') \right] \frac{1 + jk_l R_{pq}}{R_{pq}^3} e^{-jk_l R_{pq}}$$

(8.69)

when they do not. The quadrature weights are again normalized with respect to triangle area, and the variable signs depend on the orientation of the basis and testing functions on each triangle. The $\eta_0$ scale factor of the magnetic basis function has also been included in both terms.

### 8.3.2.2  Near Terms

When the source and testing triangles are closely separated, the second term on the right-hand side of (8.67) exhibits strongly singular behavior, and special care must be taken when computing the integration. In this case, we will use an approach similar to what was used for the $\mathcal{L}$ operator. Following [17], we will rewrite the innermost integral, perform a singularity extraction, and evaluate the singular terms analytically. In addition to the quantities specified previously in Figure 8.5, we define the additional quantities shown in Figure 8.6, where $\boldsymbol{\rho}_n(\mathbf{r}') = \boldsymbol{\rho}_n^-(\mathbf{r}')$ on $T_n^-$.

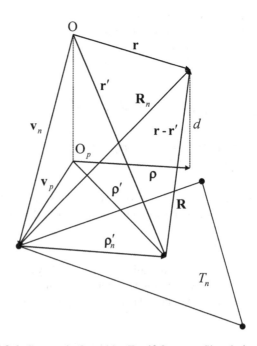

**FIGURE 8.6:** Geometric Quantities For $\mathcal{K}$ Operator Singularity Extraction

Making the definition

$$\mathbf{R}_n = \mathbf{r} - \mathbf{v}_n \,, \tag{8.70}$$

where $\mathbf{v}_n$ is the vertex opposite edge $n$ on $T_n$, and noting that

$$\mathbf{r} - \mathbf{r}' = \mathbf{R}_n - \boldsymbol{\rho}_n(\mathbf{r}') \,, \tag{8.71}$$

we can then write

$$(\mathbf{r} - \mathbf{r}') \times \boldsymbol{\rho}_n(\mathbf{r}') = \mathbf{R}_n \times \boldsymbol{\rho}_n(\mathbf{r}') \,. \tag{8.72}$$

Since $\mathbf{R}_n$ is a constant vector, the innermost integral in the second part of (8.67) can be written as

$$\int_{T_n} \left[ (\mathbf{r} - \mathbf{r}') \times \boldsymbol{\rho}_n(\mathbf{r}') \right] \frac{1 + jk_l r}{r^3} e^{-jk_l r} \, d\mathbf{r}' = \mathbf{R}_n \times \int_{T_n} \boldsymbol{\rho}_n(\mathbf{r}') \frac{1 + jk_l r}{r^3} e^{-jk_l r} \, d\mathbf{r}' \,. \tag{8.73}$$

We next perform a singularity extraction following [17], resulting in

$$\mathbf{R}_n \times \int_{T_n} \boldsymbol{\rho}_n(\mathbf{r}') \frac{1 + jk_l r}{r^3} e^{-jk_l r} \, d\mathbf{r}' = \mathbf{R}_n \times \left[ \int_{T_n} \boldsymbol{\rho}_n(\mathbf{r}') \right.$$

$$\left. \cdot \frac{(1 + jk_l r)e^{-jk_l r} - (1 + \frac{1}{2}k_l^2 r^2)}{r^3} + \mathbf{a}_n(\mathbf{r}) + \frac{k_l^2}{2} \mathbf{b}_n(\mathbf{r}) \right], \tag{8.74}$$

where

$$\mathbf{a}_n(\mathbf{r}) = \int_{T_n} \frac{\boldsymbol{\rho}_n(\mathbf{r}')}{r^3} \, d\mathbf{r}' \tag{8.75}$$

and

$$\mathbf{b}_n(\mathbf{r}) = \int_{T_n} \frac{\boldsymbol{\rho}_n(\mathbf{r}')}{r} \, d\mathbf{r}' \,. \tag{8.76}$$

The first term on the right-hand side in (8.74) is well behaved and can be evaluated using numerical quadrature. The remaining two integrals $\mathbf{a}_n(\mathbf{r})$ and $\mathbf{b}_n(\mathbf{r})$ are the singular contributions and will be evaluated analytically. The integral $\mathbf{b}_n(\mathbf{r})$ can be computed via (8.53). The integral $\mathbf{a}_n(\mathbf{r})$ can be computed by rewriting it following (8.52), which yields

$$\int_{T_n} \frac{\boldsymbol{\rho}_n(\mathbf{r}')}{r^3} \, d\mathbf{r}' = \int_{T_n} \frac{\boldsymbol{\rho}' - \boldsymbol{\rho}}{r^3} \, d\mathbf{r}' + (\boldsymbol{\rho} - \boldsymbol{\rho}_n) \int_{T_n} \frac{1}{r^3} \, d\mathbf{r}' \,. \tag{8.77}$$

The solutions to these integrals are [17]

$$\int_{T_n} \frac{\boldsymbol{\rho}' - \boldsymbol{\rho}}{r^3} \, d\mathbf{r}' = -\sum_{i=1}^{N} \hat{\mathbf{u}}_i \log \frac{R_i^+ + l_i^+}{R_i^- + l_i^-} \tag{8.78}$$

and

$$\int_{T_n} \frac{1}{r^3} \, d\mathbf{r}' = -\sum_{i=1}^{N} \hat{\mathbf{P}}_i^0 \cdot \hat{\mathbf{u}}_i \frac{1}{|d|} \left[ \tan^{-1} \frac{|d| l_i^+}{P_i^0 R_i^+} - \tan^{-1} \frac{|d| l_i^-}{P_i^0 R_i^-} \right.$$

$$\left. + \tan^{-1} \frac{l_i^-}{P_i^0} - \tan^{-1} \frac{l_i^+}{P_i^0} \right] \tag{8.79}$$

for $d \neq 0$, and

$$\int_{T_n} \frac{1}{r^3} \, d\mathbf{r}' = -\sum_{i=1}^{N} \hat{\mathbf{P}}_i^0 \cdot \hat{\mathbf{u}}_i \left[ \frac{l_i^+}{P_i^0 R_i^+} - \frac{l_i^-}{P_i^0 R_i^-} \right] \tag{8.80}$$

for $d = 0$. The sums in (8.78) and (8.80) may also have numerical problems when $\mathbf{r}$ lies in $S$ along the extension of an edge. In these cases, the observation point can again be modified following (8.50). Results for $\rho_n^+(\mathbf{r}')$ on $T_n^+$ are also obtained using (8.74) with a change of sign.

*Example Singular Integral Evaluation*

Let us compare Gaussian quadrature to the analytic solution (8.78–8.80) when computing the singular portion of (8.74). For this comparison, we compute the scalar quantity

$$\left| \mathbf{a}_n(\mathbf{r}) + \frac{k^2}{2} \mathbf{b}_n(\mathbf{r}) \right|. \tag{8.81}$$

The integrations are performed over an equilateral triangle with sides of length $1\lambda$, and quadratures of orders 3, 4, and 5 are used. We first place the observation point above the center of the triangle and vary the separation distance, with the results plotted in Figure 8.7a. The observation point is then placed above

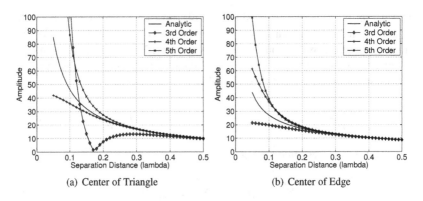

(a) Center of Triangle          (b) Center of Edge

**FIGURE 8.7**: K Operator Near Integration

the midpoint of an edge, and the distance varied again, with results shown in Figure 8.7b. In both cases, the comparison is good at larger distances but grows increasingly poor at closer ones, as expected. These results suggest that the singularity extraction method should be applied for triangle pairs separated by $\lambda/2$ or less.

### 8.3.3 Excitation

The right-hand side vector elements for the EFIE are given by (3.191). Using an $N$-point numerical quadrature, the contribution to (3.191) from a single testing triangle can be written as

$$\frac{L_m}{2} \sum_{p=1}^{M} w_p \boldsymbol{\rho}_m^{\pm}(\mathbf{r}_p) \cdot \mathbf{E}^i(\mathbf{r}_p) , \tag{8.82}$$

where the quadrature weight is normalized with respect to triangle area, and the variable sign depends on the orientation of the testing function on the triangle.

#### 8.3.3.1 Plane Wave Excitation

For incident plane waves of $\hat{\boldsymbol{\theta}}^i$- or $\hat{\boldsymbol{\phi}}^i$-polarization having an electric field of unit amplitude, (3.191) in the exterior region $R_0$ can be written as

$$V_m^{E(\theta,\phi)} = \frac{L_m}{2A_m} (\hat{\boldsymbol{\theta}}^i, \hat{\boldsymbol{\phi}}^i) \cdot \int_{f_m} \boldsymbol{\rho}_m^{\pm}(\mathbf{r}) e^{jk_0 \mathbf{r} \cdot \hat{\mathbf{r}}^i} \, d\mathbf{r} , \tag{8.83}$$

and following (8.82), the contribution from a single testing triangle is

$$\frac{L_m}{2} (\hat{\boldsymbol{\theta}}^i, \hat{\boldsymbol{\phi}}^i) \cdot \sum_{p=1}^{M} w_p \, \boldsymbol{\rho}_m^{\pm}(\mathbf{r}_p) \cdot e^{jk_0 \mathbf{r}_p \cdot \hat{\mathbf{r}}^i} . \tag{8.84}$$

#### 8.3.3.2 Planar Antenna Excitation

As planar antennas have many practical applications, there is often the need to simulate them during the design phase, making a feedpoint model of the antenna terminals necessary. In some applications, a high-fidelity feed model may be needed to obtain accurate results. Though such specialized models are beyond the scope of this text, the delta-gap model remains useful in computing impedance and radiation patterns in many cases. To develop a delta-gap model, consider the planar bowtie antenna illustrated in Figure 8.8a. We divide the antenna into two halves along edge $m$ and connect a voltage generator of amplitude $V_{in}$ to the triangles sharing the edge, as shown in Figure 8.8b. If we assume that a very small gap of width $d$ exists between the two triangles, the electric field in the gap is

$$\mathbf{E}^i = \frac{V_{in}}{d} \hat{\mathbf{t}}_m , \tag{8.85}$$

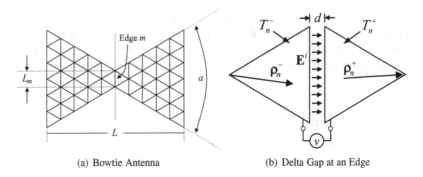

(a) Bowtie Antenna                    (b) Delta Gap at an Edge

**FIGURE 8.8:** Delta-Gap Feed Model

where $\hat{\mathbf{t}}_m$ is the vector normal to edge $m$ in the plane. Inserting (8.85) into (3.191), and using the fact that the normal component of the RWG function is unity along the edge, (3.191) evaluates to

$$V_m^{E(l)} = L_m V_{in} \ . \tag{8.86}$$

Once the MoM linear system has been solved, the input impedance may be desired. As the electric current vector element $I_m^{J(l)}$ comprises the linear current density flowing across edge $m$, the total electric current flowing across the edge is then

$$I_{in} = L_m I_m^{J(l)} \tag{8.87}$$

and the input impedance is

$$Z_{in} = \frac{V_{in}}{I_{in}} = \frac{V_{in}}{L_m I_m^{J(l)}} \ . \tag{8.88}$$

## 8.4 MFIE

The MFIE (3.179) also has $\mathcal{K}$ and $\mathcal{L}$ operators, which were discretized and tested via the MoM in (3.187) and (3.188). Substitution of the electric basis and magnetic testing functions into (3.187) yields in Region $R_l$ matrix elements given by

$$Z_{mn}^{HJ(l)} = -\mathrm{K}(\mathbf{g}_m, \mathbf{f}_n) \ , \tag{8.89}$$

and substitution of the magnetic basis and testing functions into (3.188) yields

$$Z_{mn}^{HM(l)} = \frac{\epsilon_l}{\mu_l} \mathrm{L}(\mathbf{g}_m, \mathbf{g}_n) \ , \tag{8.90}$$

where the expressions for L and K were developed previously in Sections 8.3.1 and 8.3.2, respectively. Note that the L elements must be scaled by $\eta_0^2$ to account for the scale factors of the magnetic basis and testing functions.

## 8.4.1 Excitation

The right-hand side vector elements for the MFIE are given by (3.192). Using an $N$-point numerical quadrature, the contribution to (3.192) from a single testing triangle can be written as

$$\frac{L_m}{2} \sum_{p=1}^{M} w_p \boldsymbol{\rho}_m^{\pm}(\mathbf{r}_p) \cdot \mathbf{H}^i(\mathbf{r}_p) \,, \tag{8.91}$$

where the quadrature weight is normalized with respect to triangle area, and the variable sign depends on the orientation of the testing function on the triangle.

### 8.4.1.1 Plane Wave Excitation

Given $\hat{\boldsymbol{\theta}}^i$ and $\hat{\boldsymbol{\phi}}^i$-polarized incident electric fields, the corresponding magnetic fields are $-\hat{\boldsymbol{\phi}}^i$ and $\hat{\boldsymbol{\theta}}^i$-polarized, respectively. Thus in the exterior region $R_0$ we can re-use the elements of the EFIE right-hand side, yielding MFIE right-hand side elements given by

$$V_m^{H,\theta} = -V_m^{E,\phi} \tag{8.92}$$

and

$$V_m^{H,\phi} = V_m^{E,\theta} \,. \tag{8.93}$$

## 8.5 nMFIE

The nMFIE (3.180) contains the $\hat{\mathbf{n}} \times \mathcal{L}$ and $\hat{\mathbf{n}} \times \mathcal{K}$ operators, which were discretized and tested via the MoM in (3.189) and (3.190). In this section we formulate expressions for both operators, using RWG basis and testing functions.

## 8.5.1 $\hat{\mathbf{n}} \times \mathcal{K}$ Operator

Substituting the electric basis and testing functions into (3.189) yields in Region $R_l$ matrix elements given by

$$Z_{mn}^{nHJ(l)} = \text{nK}(\mathbf{f}_m, \mathbf{f}_n) \,, \tag{8.94}$$

where

$$
nK(\mathbf{f}_m, \mathbf{f}_n) = \frac{1}{2} \int_{\mathbf{f}_m, \mathbf{f}_n} \mathbf{f}_m(\mathbf{r}) \cdot \mathbf{f}_n(\mathbf{r}) \, d\mathbf{r}
$$
$$
+ \frac{1}{4\pi} \int_{\mathbf{f}_m} \mathbf{f}_m(\mathbf{r}) \cdot \left[ \hat{\mathbf{n}}_l(\mathbf{r}) \times \int_{\mathbf{f}_n} \left[ (\mathbf{r} - \mathbf{r}') \times \mathbf{f}_n(\mathbf{r}') \right] \frac{1 + jk_l r}{r^3} e^{-jk_l r} \, d\mathbf{r}' \right] d\mathbf{r} .
$$

$$(8.95)$$

The first term on the right-hand side of (8.95) is used when the source and testing triangles overlap, and the second term when they do not. We note that (8.95) is almost identical to (8.67), except for the sign change on the second term and the re-location of $\hat{\mathbf{n}}_l(\mathbf{r})$.

### 8.5.1.1 Non-Near Terms

When the source and testing triangles are separated by a sufficient distance, (8.95) can be implemented as written. For a source and testing triangle, the contribution to (8.95) can be obtained using an $M$-point numerical quadrature, yielding

$$
\frac{L_m L_n}{8 A_m} \sum_{p=1}^{M} w_p \, \boldsymbol{\rho}_m^{\pm}(\mathbf{r}_p) \cdot \boldsymbol{\rho}_n^{\pm}(\mathbf{r}_p) \tag{8.96}
$$

when the triangles overlap, and

$$
\frac{L_m L_n}{16\pi} \sum_{p=1}^{M} \sum_{q=1}^{M} w_p w_q \, \boldsymbol{\rho}_m^{\pm}(\mathbf{r}_p) \cdot \left[ \hat{\mathbf{n}}_l(\mathbf{r}_p) \times (\mathbf{r}_p - \mathbf{r}_q') \times \boldsymbol{\rho}_n^{\pm}(\mathbf{r}_q') \right] \frac{1 + jk_l R_{pq}}{R_{pq}^3} e^{-jk_l R_{pq}}
$$

when they do not.

### 8.5.1.2 Near Terms

The innermost integral in the second part of (8.95) is identical to that of (8.67). To save time, the matrix fill process can use this fact to compute the innermost integral of the K and nK terms simultaneously.

### 8.5.2 $\hat{\mathbf{n}} \times \mathcal{L}$ Operator

Substituting the magnetic basis and electric testing functions into (3.190) yields in Region $R_l$ matrix elements given by

$$
Z_{mn}^{nHM(l)} = nL(\mathbf{f}_m, \mathbf{g}_n) , \tag{8.97}
$$

where

$$
\mathrm{nL}(\mathbf{f}_m, \mathbf{g}_n) = j\omega\epsilon_l \int_{\mathbf{f}_m} \mathbf{f}_m(\mathbf{r}) \cdot \left[ \hat{\mathbf{n}}_l(\mathbf{r}) \times \int_{\mathbf{g}_n} \mathbf{g}_n(\mathbf{r}') \frac{e^{-jk_l r}}{4\pi r} \, d\mathbf{r}' \right] d\mathbf{r}
$$

$$
- \frac{j}{\omega\mu_l} \oint_{\delta\mathbf{f}_m} \hat{\mathbf{c}}_m(\mathbf{r}) \cdot \left[ \mathbf{f}_m(\mathbf{r}) \times \hat{\mathbf{n}}_l(\mathbf{r}) \right] \int_{\mathbf{g}_n} \nabla' \cdot \mathbf{g}_n(\mathbf{r}') \frac{e^{-jk_l r}}{4\pi r} \, d\mathbf{r}' \, d\mathbf{r} \,, \quad (8.98)
$$

and (3.206) has been used to re-write the second part of (3.190). Note that since $\hat{\mathbf{c}}_m(\mathbf{r})$ is normal to the boundary in the plane of each triangle, it may have a different value along edge $m$ in $T_m^-$ than it does in $T_m^+$. Thus, the contour integral in the second part (8.5.2) must be performed over the boundaries of $T_m^-$ and $T_m^+$ separately (a total of six edges).

### 8.5.2.1 Non-Near Terms

When the source and testing triangles are separated by a sufficient distance, (8.5.2) can be implemented as written. For a source and testing triangle, the contribution to (8.5.2) can be broken into two parts. The first term comprises an $N$-point numerical quadrature over the source and testing triangles, given by

$$
I_1 = j\omega\epsilon_l\eta_0 \frac{L_m L_n}{16\pi} \sum_{p=1}^{M} \sum_{q=1}^{M} w_p w_q \boldsymbol{\rho}_m^{\pm}(\mathbf{r}_p) \cdot \left[ \hat{\mathbf{n}}_l(\mathbf{r}_p) \times \boldsymbol{\rho}_n^{\pm}(\mathbf{r}_q') \right] \frac{e^{-jk_l R_{pq}}}{R_{pq}} \,.
$$

(8.99)

The second term comprises a two-dimensional quadrature over the source triangle, and a one-dimensional quadrature over the three edges of the testing triangle. This term can be written as

$$
I_2 = \pm \frac{j\eta_0}{\omega\mu_l} \frac{L_m L_n}{8\pi} \sum_{e=1}^{3} \sum_{p=1}^{M} \sum_{q=1}^{M} w_{e,p} w_q \mathbf{c}_m(\mathbf{r}_{e,p}) \cdot \left[ \boldsymbol{\rho}_m^{\pm}(\mathbf{r}_{e,p}) \times \hat{\mathbf{n}}_l(\mathbf{r}_{e,p}) \right] \frac{e^{-jk_l R_{pq}}}{R_{pq}} \,,
$$

(8.100)

where the quadrature weights are again normalized, and the variable signs depend on the orientation of the basis and testing functions on each triangle. The $\eta_0$ scale factor of the magnetic basis function has also been included in both terms. Experience has shown that for a mesh having at least 10 edges per wavelength, 3-5 quadrature points per edge are sufficient for computing (8.100).

### 8.5.2.2 Near and Self Terms

For the near and self terms, we will compute the innermost integrals analytically, and the outermost numerically. As the innermost integrals of both terms in (8.5.2) have the same form as (8.46) and (8.48), these expressions will support computation of the L and nL near terms simultaneously.

### 8.5.3 Excitation

The right-hand side vector elements for the nMFIE are given by (3.193). Using an $N$-point numerical quadrature, the contribution to (3.193) from a single testing triangle can be written as

$$\frac{L_m}{2} \sum_{p=1}^{M} w_p \rho_m^{\pm}(\mathbf{r}_p) \cdot \left[ \hat{\mathbf{n}}_l(\mathbf{r}_p) \times \mathbf{H}^i(\mathbf{r}_p) \right], \qquad (8.101)$$

where the quadrature weight is normalized with respect to triangle area, and the variable sign depends on the orientation of the testing function on the triangle.

#### 8.5.3.1 Plane Wave Excitation

Given $\hat{\theta}^i$ and $\hat{\phi}^i$-polarized incident electric fields, the corresponding magnetic fields are $-\hat{\phi}^i$ and $\hat{\theta}^i$-polarized, respectively. Thus, for an electric field of unit amplitude, the nMFIE right-hand side vectors in the exterior region $R_0$ can be written as

$$V_m^{nH(\theta,\phi)} = \frac{L_m}{2\eta_0 A_m} \int_{\mathbf{f}_m} \rho_m^{\pm}(\mathbf{r} \cdot \left[ \hat{\mathbf{n}}_0(\mathbf{r}) \times (-\hat{\phi}^i, \hat{\theta}^i) \right] e^{jk_0 \mathbf{r} \cdot \hat{\mathbf{r}}^i} \, d\mathbf{r}, \qquad (8.102)$$

and following (8.101), the contribution to (8.102) from a single testing triangle is

$$\frac{L_m}{2\eta_0} \sum_{p=1}^{M} w_p \, \rho_m^{\pm}(\mathbf{r}_p) \cdot \left[ \hat{\mathbf{n}}_0(\mathbf{r}_p) \times (-\hat{\phi}^i, \hat{\theta}^i) \right] e^{jk_0 \mathbf{r}_p \cdot \hat{\mathbf{r}}^i}. \qquad (8.103)$$

## 8.6 Enforcement of Boundary Conditions

The matrix system (3.194) and basis functions have been defined region-wise, as discussed in Sections 3.6.3.1 and 8.2.2. However, the boundary conditions stipulate that there are linearly dependent unknowns that must be combined before the system can be solved. In this section, we will present a generalized procedure for treating the boundary conditions along interfaces and junctions of general type, following very closely the method outlined in [1].

### 8.6.1 Classification of Edges and Junctions

In treating the boundary conditions and imposing the desired integral equation formulation, we classify an edge $e_n$ into one of three types:

1. **Dielectric edge or junction**: $e_n$ lies on the intersection of two or more dielectric surfaces but does not touch any conducting surfaces.

2. **Conducting edge or junction**: $e_n$ lies on the intersection of open or closed conducting surfaces but does not touch any dielectric surfaces.

3. **Composite conducting-dielectric junction**: $e_n$ lies on the intersection of at least one open or closed conducting surface and at least one dielectric surface.

We will consider the rules for treating boundary conditions at each type of edge or junction in this section.

### 8.6.1.1 Dielectric Edges and Junctions

The boundary conditions along dielectric interfaces require that the oriented basis functions assigned to regular edges or junctions have the same coefficient. Therefore, following Section 7.10.1.1, all basis functions of the same type (electric and magnetic) assigned to $e_n$ are combined. Thus, the first rule is as follows:

- **Rule 1**: Because the coefficients of the oriented basis functions of the same type assigned to a dielectric edge or junction have the same value, these unknowns must be combined into a single unknown. This is accomplished by combining the corresponding columns of the system matrix.

### 8.6.1.2 Conducting Edges and Junctions

We next consider the boundary conditions at conducting edges or junctions, which may contain a mixture of open and closed surfaces. Since the magnetic field is not necessarily continuous across a conducting surface, the electric current **J** must have independent values on the opposite sides of the surface. Thus, the electric basis functions assigned to a conducting edge or junction cannot be combined. There is, however, one exception to this rule: when all conducting surfaces connected to $e_n$ are open, and $e_n$ is completely in the interior of a dielectric region. Consider an open conducting surface $S$ lying completely inside a homogeneous dielectric region, where $e_n$ comprises a regular edge on $S$ that has been assigned the oriented electric basis functions $\mathbf{f}_1$ and $\mathbf{f}_2$. In a homogeneous region, these basis functions produce the same fields with opposite signs due to their orientation. Therefore, the fields can be generated by only one basis function, and one of the basis functions can be removed. Generalizing this to a junction of $N$ conducting surfaces, we get our second rule:

- **Rule 2**: As the magnetic field is not necessarily continuous across conducting surfaces, the electric basis functions assigned to a conducting

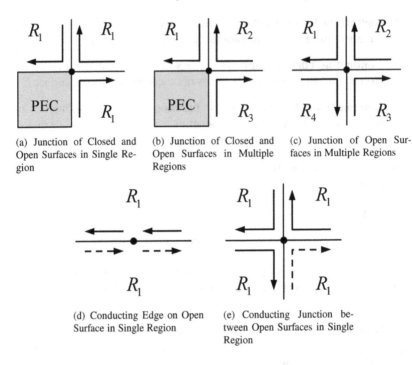

(a) Junction of Closed and Open Surfaces in Single Region

(b) Junction of Closed and Open Surfaces in Multiple Regions

(c) Junction of Open Surfaces in Multiple Regions

(d) Conducting Edge on Open Surface in Single Region

(e) Conducting Junction between Open Surfaces in Single Region

**FIGURE 8.9:** Basis Functions at Conducting Edges and Junctions

edge or junction are independent. In addition, if all the surfaces meeting at $e_n$ are open conducting surfaces, and $e_n$ lies completely in the interior of a homogeneous dielectric region, one of the electric basis functions assigned to $e_n$ is removed.

Several types of conducting edges and junctions that can be encountered are illustrated in Figures 8.9a–8.9e. In each figure, solid lines indicate open conducting surfaces, and the shaded areas are closed conducting regions. The basis functions drawn with dashed lines in Figures 8.9d and 8.9e indicate those that can be removed.

### 8.6.1.3   Composite Conducting-Dielectric Junctions

Now let $e_n$ comprise a composite conducting-dielectric junction. Again, because the magnetic field is not continuous across open metallic surfaces, the electric current must have independent values on the opposite sides of any open conducting surface meeting at $e_n$. Combining this with the previously stated rules, we obtain our third rule:

- **Rule 3**: Unknowns on opposite sides of open conducting surfaces meet-

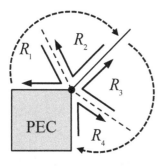

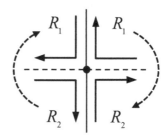

(a) Closed and Open Conductors, Multiple Dielectric Regions

(b) Open Conductors, Two Dielectric Regions

**FIGURE 8.10:** Basis Functions at Composite Junctions

ing at a conducting-dielectric junction cannot be combined. In addition, unknowns on opposite sides of dielectric surfaces that lie in between open or closed conductors must be combined into a single unknown.

Two examples of composite junctions are illustrated in Figures 8.10a and 8.10b. Solid lines indicate open conducting surfaces, dashed lines indicate dielectric interfaces, and shaded areas are closed conducting regions. The curved, dashed arrows indicate where basis functions are combined.

## 8.6.2 Reducing the Overdetermined System

Having combined all linearly dependent unknowns at edges and junctions by combining the columns of the system matrix, the linear system is now overdetermined. To obtain a well-defined system, the number of equations must be reduced to that of the remaining unknowns. The way this is done depends on the integral equation formulation. Our approach here is the EFIE-CFIE-PMCHWT approach of Section 3.6.3.1, where the EFIE and CFIE are applied at conducting edges and junctions, and PMCHWT is applied on dielectric edges and junctions. We will describe the approach in detail in this section.

### 8.6.2.1 PMCHWT at Dielectric Edges and Junctions

For a regular edge on the interface between two dielectric regions, the PMCHWT formulation is simply a summation of the EFIEs and MFIEs on each side of the interface. This can be generalized for dielectric junctions as follows. Let $e_n$ be a dielectric edge or junction. We then sum the adjacent

EFIEs and MFIEs, respectively, as

$$\sum_{l=1}^{M} \text{EFIE}_n^{(l)} , \qquad (8.104)$$

and

$$\sum_{l=1}^{M} \text{MFIE}_n^{(l)} , \qquad (8.105)$$

where $M$ is the number of dielectric regions meeting at $e_n$, and $\text{EFIE}_n^{(l)}$ and $\text{MFIE}_n^{(l)}$ comprise the EFIE and MFIE tested with the RWG function assigned to $e_n$ in Region $R_l$. Doing this combines the corresponding rows of the system matrix.

### 8.6.2.2 EFIE and CFIE at Conducting Edges and Junctions

Let $e_n$ be an edge on an open conducting surface, or a junction between several open conducting surfaces. If $e_n$ is completely inside one dielectric region, by Rule 2 we have also removed one of the unknowns associated with the electric basis functions assigned to $e_n$. In addition, we now remove the EFIE from the system equations, which was tested with the corresponding electric basis function.

Next, let $e_n$ be an edge on a closed conducting surface, or at the junction of a closed conduction surface and at least one open conducting surface. For each remaining electric basis function assigned to $e_n$, we combine the EFIE tested by the corresponding testing function on the open surfaces, with the CFIE (3.181) tested on the closed conducting surfaces.

To illustrate, consider the junction in Figure 8.11a, where an open conducting surface (solid line) connects to a closed conducting region (shaded area). There are two dielectric regions $R_1$ and $R_2$ in addition to the closed conducting region $R_3$. As $S_{12}$ (or $S_{21}$) is an open surface, the fields are not necessarily continuous across it, and following Rule 2, the electric basis functions $\mathbf{f}_1$ and $\mathbf{f}_2$ are considered to be independent. The choices of integral equations are as follows: The testing function associated with $\mathbf{f}_1$ will test in $R_1$ the EFIE on $S_{12}$, and the CFIE on $S_{13}$, which are added to a common matrix element. Likewise, the testing function associated with $\mathbf{f}_2$ will test in $R_2$ the EFIE on $S_{21}$ and the CFIE on $S_{23}$.

### 8.6.2.3 EFIE and CFIE at Composite Conducting-Dielectric Junctions

Finally, we consider general conducting-dielectric junctions. If no closed conducting surfaces meet at $e_n$, then by Rule 2, we combine the EFIEs between

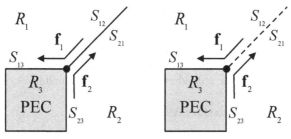

(a) Conducting Junction with Open and Closed Conducting Surfaces

(b) Composite Junction with Closed Conducting Surface

**FIGURE 8.11:** Integral Equation Treatment at Junctions

any two metallic surfaces via the sum

$$\sum_{l=1}^{M} \text{EFIE}_n^{(l)} \, , \tag{8.106}$$

where $M$ is the number of dielectric regions between the surfaces. If a closed conducting surface meets at $e_n$, then the CFIEs are combined between any two metallic surfaces in a similar manner. To illustrate, consider the composite junction of Figure 8.11b, which is similar to that of 8.11a, except that $S_{12}$ and $S_{21}$ now comprise a dielectric interface. The choice of integral equations are as follows: The testing function associated with $\mathbf{f}_1$ will test in $R_1$ the CFIE on $S_{12}$ and $S_{13}$. Likewise, the testing function associated with $\mathbf{f}_2$ will test in $R_2$ the CFIE on $S_{21}$ and $S_{23}$. As the boundary conditions on $S_{12}$ stipulate that $\mathbf{f}_1$ and $\mathbf{f}_2$ must have the same coefficient, the tested CFIEs that lie on and between $S_{13}$ and $S_{23}$ will contribute to the same matrix element.

## 8.7 Software Implementation Notes

In this section we discuss some elements of a practical software implementation of the techniques presented in this chapter, as well as important issues and considerations. Much of what is discussed herein will be further extended in Chapters 9 and 11.

## 8.7.1 Pre-Processing and Bookkeeping

In practice, the procedures outlined in Sections 8.2.2 and 8.6 condense into a set of pre-processing steps, resulting in a bookkeeping database used in the construction of the system matrix. As implementation details will differ between software codes, we will address these steps at a high level.

### 8.7.1.1 Region and Interface Assignments

The first step involves the construction of several lists. The first is a list of all unique dielectric regions and their dielectric parameters $\epsilon$ and $\mu$. Each region is assigned a unique ID, with the regions comprising free space and conductor assigned a special value (such as 0 or $-1$) used internally by the software. The second list describes all interfaces, where each interface is assigned a unique ID as well as the IDs of the regions that are interior and exterior to that interface (as determined by the surface normal). Each triangle is then assigned to an interface by appending the corresponding interface ID to each line of the connectivity list in the facet file. This is typically done when the facet file is constructed.

### 8.7.1.2 Geometry Processing

Applying the edge-finding algorithm (Section 8.1.2) to the facet file, we next identify all boundary, regular, and junction edges in the triangle mesh, as well as the triangles (and interfaces) that connect to each edge. With the connectivity between neighboring triangles established, we can perform basic error checking. Triangles connected at a regular edge that are assigned to the incorrect interfaces can be found, and the presence of boundary (naked) edges on supposed closed interfaces can be reported to the user.

### 8.7.1.3 Assignment and Orientation of Basis Functions

Each triangle is now assigned two data structures: one stores the indexes and orientations of the basis functions on the *interior* side of the triangle, and the other contains similar data for the *exterior* side. These structures are now populated by performing a loop over all non-boundary edges, where for each edge, we work in a local two-dimensional coordinate system. To illustrate, consider edge $e_n$ with endpoints $\mathbf{e}_1$ and $\mathbf{e}_2$, which lies at the junction of $M$ triangles. On triangle $T_m$, we denote the node opposite $e_n$ as $\mathbf{v}_m$. We then define the references axes in the local coordinate system as

$$\hat{\mathbf{z}} = \frac{\mathbf{e}_2 - \mathbf{e}_1}{|\mathbf{e}_2 - \mathbf{e}_1|} \,, \tag{8.107}$$

$$\hat{\mathbf{x}} = (\mathbf{v}_1 - \mathbf{e}_1) - \big[(\mathbf{v}_1 - \mathbf{e}_1) \cdot \hat{\mathbf{z}}\big]\hat{\mathbf{z}} \,, \tag{8.108}$$

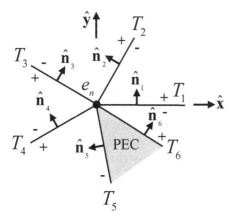

**FIGURE 8.12:** Local Reference Coordinates at Junction

and
$$\hat{\mathbf{y}} = \hat{\mathbf{z}} \times \hat{\mathbf{x}},$$ (8.109)

where the plane of $T_1$ lies along the local $\hat{\mathbf{x}}$ axis, and all other triangles at the junction are sorted in a counterclockwise manner as shown in Figure 8.12. A walk through all triangles in the counterclockwise direction is now all that is required. The normal vector on each triangle is used to determine whether we populate the data structure for its interior or exterior side. The vector orientations of the basis functions at $e_n$ are assigned in back-to-front manner on each triangle, as shown with the plus and minus signs. The rules outlined in Section 8.6.1 are enforced at each step. The indexes assigned to each basis function are taken from a running count of the number of electric or magnetic basis functions already assigned, where the counts are either incremented or left alone at each step during the walk on $e_n$, depending on how the rules are applied. Once all edges have been processed, we add to each region a pointer to each triangle on its boundary, and to the data structure assigned to the side of the triangle facing into the region.

## 8.7.2  Matrix and Right-Hand Side Fill

Matrix filling is performed on a per-region basis, as the integral operators depend on the wavenumber in each region, and only those triangles and basis functions assigned to each region are considered. Similarly, right-hand side elements are computed only in those regions where an impressed field exists. For scattering problems, this is the free space region $R_0$. The bookkeeping information generated during the pre-processing steps is utilized to determine the orientation and index of the basis and testing functions on the source and testing triangles, as well as the direction of the normal vector on each interface.

## 8.7.3  Parallelization

There are many parts of an MoM solver that can be parallelized effectively. The routines that compute the system matrix and right-hand side vectors are obvious candidates. Multiple right-hand sides can also be solved in parallel, as well as the matrix-vector product in an iterative solver. We now discuss some approaches to these problems for shared memory and distributed memory systems.

### 8.7.3.1  Shared Memory Systems

On a shared memory system, all geometry and book-keeping information as well as the system matrix are addressable by all threads. As system memory is almost always at a premium, the obvious choice is to allocate space for a single system matrix, and to then do the matrix fill in parallel. The matrix can be computed efficiently by looping over triangle pairs, where the integral operators are computed on separate threads. This requires thread synchronization every time a value is added to the system matrix, though as the time needed is small compared to that spent computing the operators, a simple mutex will suffice in resolving thread contention. For matrix factorization and right-hand side solving, LAPACK routines are used. Many of these are often single-threaded, though some implementations such as the Intel Math Kernel Library (MKL) have built-in parallelism. As the storage needed for the right-hand side vectors is small compared to the system matrix, multiple right-hand sides can be solved simultaneously via threads. This is ideal for monostatic scattering problems, where the number of incidence angles may exceed the number of available processors. If the number of right-hand sides is small, individual right-hand side vectors can be computed by performing the required quadratures and assignments in parallel.

### 8.7.3.2  Distributed Memory Systems

The design and implementation of a solver on distributed memory systems depends on several factors. The most important considerations are the maximum size of the problem to be solved, the number and type of processors and available RAM present on each node, and the speed and topology of the network connections. In most cases, each node will have enough memory to load and store the geometry and create the required bookkeeping data. It is the system matrix that is too large to be stored on a single node. In such a case, the matrix can be divided into blocks having roughly $N/N_n$ rows and columns, where $N$ is the total number of unknowns and $N_g$ is the number of compute nodes. Each node is then responsible for computing and storing a subset of these blocks. After matrix filling is complete, a block LU factorization following Section 4.1.3 can be applied. For this, one option is to use the ScaLAPACK library [18], which implements a parallel, distributed memory *LU* factorization. A distributed algorithm designed specifically for an ACA-compressed block matrix is outlined in detail later in Section 9.6.3.2.

## 8.7.4 Triangle Mesh Considerations

One of the most important elements in any MoM simulation is the quality of the triangle mesh that is used. We have already discussed at length the importance of obtaining accurate integrations for the matrix elements, however the model itself is also key to obtaining a good solution. Generating a good model of a realistic object is not a trivial task. Significant time and effort may be invested generating a triangular mesh that accurately describes the object and its various dielectric interfaces and junctions. Such a model must also have a sufficient number of unknowns at the frequency of interest, and must also be watertight and free of any T-junctions or other errors.

### 8.7.4.1 Aspect Ratio

Most computer-aided design (CAD) tools have the capability to generate a triangle mesh from a curved surface, however the quality of these meshes can vary. It is well known that meshes having triangles with poor aspect ratios yield a more poorly conditioned system matrix. Thus, when building a mesh, care should be taken to ensure that each triangle has a reasonable aspect ratio and that it does not have small internal angles. An example of such a mesh is shown in Figure 8.13a, where we have meshed a circular plate using a very simple algorithm. The triangles in this mesh grow progressively smaller and thinner the closer they are to the center. Though it is not generally possible to create a mesh comprising perfectly equilateral triangles, many meshing tools will allow the user to analyze and adjust the distribution of aspect ratios in a mesh. In areas where the triangles are too thin, these tools often allow for restructuring or rebuilding the triangles to improve their shape. A better mesh is illustrated in Figure 8.13b, where the aspect ratios are improved.

### 8.7.4.2 T-Junctions

In Section 8.1.2.1 we discussed the importance of having shared nodes between triangles. If nodes are not joined at an edge, the surface will not have logical connectivity at the edge, and the edge-finding algorithm will not locate it. On interfaces that bound a closed region, the mesh must be free of any boundary or naked edges. Such a mesh is referred to as "watertight." Some CAD programs may generate meshes where the nodes of one triangle are not co-

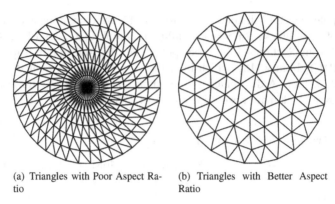

(a) Triangles with Poor Aspect Ratio

(b) Triangles with Better Aspect Ratio

**FIGURE 8.13:** Examples of Triangle Mesh Aspect Ratio

located with those of adjacent triangles. Consider the meshes shown in Figure 8.14. The mesh in Figure 8.14a is properly constructed, however the mesh in Figure 8.14b has several disjointed intersections, which are called *T-junctions*. These connections cannot be logically resolved through the connectivity list, and will result in boundary edges where there should be none. When the number of triangles in the model grows to be quite high, T-junctions can become difficult to find and eliminate. Though it is sometimes necessary to remove T-junctions manually, most CAD programs can automatically locate and highlight these edges, and in some cases repair the problem with little or no user intervention.

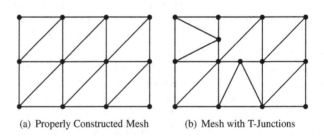

(a) Properly Constructed Mesh

(b) Mesh with T-Junctions

**FIGURE 8.14:** Triangle Mesh Connectivities

## 8.8 Numerical Examples

In this section we will use the methods outlined in this chapter to compute the radar cross section of several test articles. We will first consider a series of conducting, dielectric, and coated spheres, and then a set of benchmark targets that were constructed and measured by the Electromagnetic Code Consortium (EMCC) [19]. We will then consider the input impedance of several antennas including the well-known bowtie and Archimedean spiral. We will then consider the monoconic reentry vehicle discussed previously in Section 7.10.2.4. The Rhinoceros 3D CAD software was used to generate NURBS models of each object, which were then meshed using Altair Hypermesh$^{TM}$. In each case, meshes are generated with at least ten edges (unknowns) per wavelength in each dielectric region, PMCHWT is applied on dielectric interfaces, and the CFIE is applied on closed conductors with $\alpha = 0.5$.

### 8.8.1 Serenity

The examples in this chapter are computed using the Tripoint Industries code *Serenity*, which is part of our *lucernhammer* suite of radar cross section codes. Written in C++, *Serenity* implements the methods outlined in this chapter, and supports multiple dielectric regions and junctions of general configuration. It has an automatic edge-finding algorithm, adjustable Gaussian quadrature integration, and carefully written near and self term routines for all matrix elements.

*Serenity* implements a full matrix solver, as well as compressed matrix solvers which are discussed in the following chapters. The examples in this section are solved using the full matrix solver, which is implemented on shared memory systems using POSIX threads (pthreads), with parallelized matrix fill, LU factorization, and right-hand side solve.

### 8.8.2 Compute Platform

Computations in this Section were carried out on a Dell Precision T7900 workstation, with dual twelve-core Intel Xeon CPUs (E5-2690 v3) at 2.6 GHz with 256 GB of RAM, running Ubuntu Linux. *Serenity* was compiled using the GNU C++ Compiler (Version 7.5.0). The Intel Math Kernel Library (2019 version) was used for all BLAS and LAPACK functions. *Serenity* used 24 simultaneous computation threads in performing the matrix fill, LU factorization, and solution of right-hand sides.

## 8.8.3   Spheres

In this section we will compute the bistatic RCS of several conducting, fully dielectric, and coated spheres using the MoM, and compare the results to those obtained from the Mie series.

### 8.8.3.1   Conducting Sphere

In this section, we consider a conducting sphere with a radius of 0.5 meters. The analytic expressions for the scattered field of a conducting sphere were summarized previously in Section 7.9.1.1. For the MoM simulation, a facet model comprising 5120 near-equilateral triangles was constructed.

*Monostatic RCS versus Frequency*

We first compute the monostatic RCS for frequencies ranging from 10 to 750 MHz. In Figures 8.15a and 8.15b is compared the RCS obtained from the Mie series RCS to that of the MoM using the EFIE and nMFIE, respectively. The comparisons are good at lower frequencies, though as the frequency increases the nMFIE results become corrupted by internal resonances. In Figure 8.15c are plotted the results from the CFIE, where the resonance problems have been eliminated. In Figure 8.15d are plotted the condition numbers obtained from the LAPACK function cgecon for each case. We note the periodic internal resonances in the EFIE and nMFIE results, which correspond with the corruptions seen in Figures 8.15a and 8.15b. The condition number for the CFIE remains smooth, as expected. We also note that the EFIE solution becomes singular at low frequencies, which is due to low-frequency breakdown of the discretized matrix elements resulting from the $\mathcal{L}$ operator. As $\omega \to 0$, the surface currents decouple into solenoidal (divergence-free) and irrotational (curl-free) components, and the contribution from the vector part of (8.10) becomes lost due to the frequency scaling [20]. This problem can be mitigated by using basis functions that model these components separately, such as the Loop-Star basis functions discussed in [21].

*Bistatic RCS*

Using the same model and simulation parameters as before, we compute the bistatic RCS at 250, 500, and 750 MHz. The CFIE results are compared to those from the Mie series in Figures 8.16a–8.16f. The comparison is very good at all frequencies.

*Near Field*

Using (3.108), we next compute the near field of the conducting sphere illuminated by an $\hat{\mathbf{x}}$-polarized plane wave incident from the $+\hat{\mathbf{z}}$ direction at 750

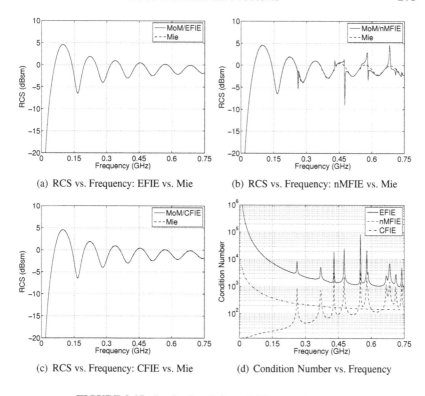

(a) RCS vs. Frequency: EFIE vs. Mie

(b) RCS vs. Frequency: nMFIE vs. Mie

(c) RCS vs. Frequency: CFIE vs. Mie

(d) Condition Number vs. Frequency

**FIGURE 8.15:** Conducting Sphere: RCS versus Frequency

MHz. In Figures 8.17a and 8.17b is shown the real part of the scattered and total electric field, respectively, in the $yz$ plane.

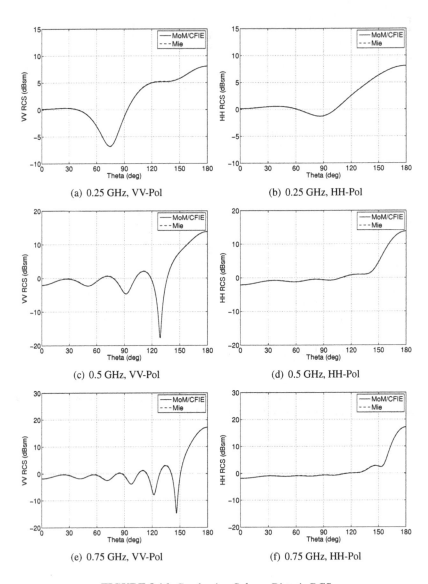

**FIGURE 8.16:** Conducting Sphere: Bistatic RCS

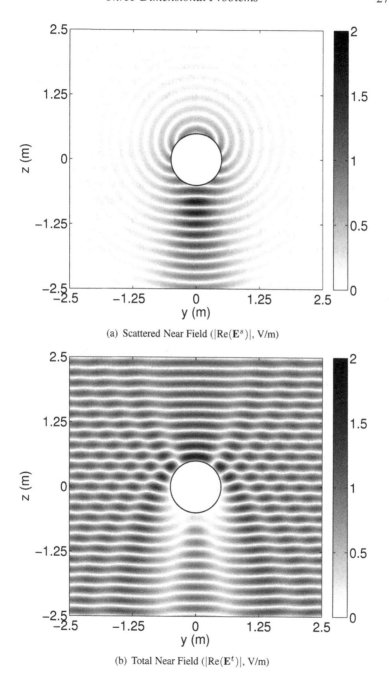

(a) Scattered Near Field ($|\mathrm{Re}(\mathbf{E}^s)|$, V/m)

(b) Total Near Field ($|\mathrm{Re}(\mathbf{E}^t)|$, V/m)

**FIGURE 8.17:** Conducting Sphere: Near Field at 750 MHz

### 8.8.3.2   Dielectric Sphere

In this section, we consider a dielectric sphere with a radius of 0.5 meters. The analytic expressions for the scattered field of a stratified sphere were summarized previously in Section 7.9.1.2. For the MoM simulation, we re-use the facet model from Section 8.8.3.1, and the PMCHWT formulation is applied.

*Monostatic RCS versus Frequency*

We first compute the RCS for frequencies ranging from 10 to 750 MHz. In Figure 8.18a we compare the RCS obtained by the Mie series to that of the MoM in backscattering ($\theta^i = \theta^s = 0$) for a lossless sphere ($\epsilon_r = 2.56$). The comparison is very good. In Figure 8.18b is a similar comparison for a lossy sphere ($\epsilon_r = 2.56 - j.102$), where the comparison is again very good. The forward-scattering RCS ($\theta^i = 0, \theta^s = \pi$) for each case is summarized in Figures 8.19a and 8.19b, respectively, where the comparison is again quite good.

*Bistatic RCS*

Using the same simulation parameters as before, we consider the bistatic RCS at 250, 500, and 750 MHz. The MoM results are compared to those of the Mie series for the lossless sphere in Figures 8.20a–8.20f. The comparison is very good. The results for the lossy sphere are compared in Figures 8.21a–8.21f, where the comparison is again very good.

*Near Field*

Using (3.108) and (3.112), we compute the near field of the lossless dielectric sphere illuminated by an $\hat{\mathbf{x}}$-polarized plane wave incident from the $+\hat{\mathbf{z}}$ direction at 750 MHz. Figures 8.22a and 8.22b show the real part of the scattered and total electric field, respectively, in the $yz$ plane. To obtain the combined image, the total near fields are computed separately in the external and internal dielectric regions, and are then overlaid.

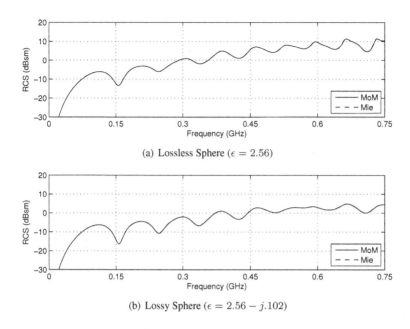

(a) Lossless Sphere ($\epsilon = 2.56$)

(b) Lossy Sphere ($\epsilon = 2.56 - j.102$)

**FIGURE 8.18:** Dielectric Sphere: Backscatter RCS vs. Frequency

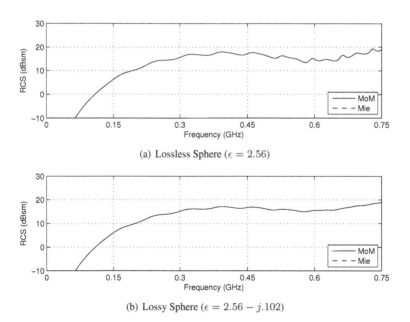

(a) Lossless Sphere ($\epsilon = 2.56$)

(b) Lossy Sphere ($\epsilon = 2.56 - j.102$)

**FIGURE 8.19:** Dielectric Sphere: Forward Scatter RCS vs. Frequency

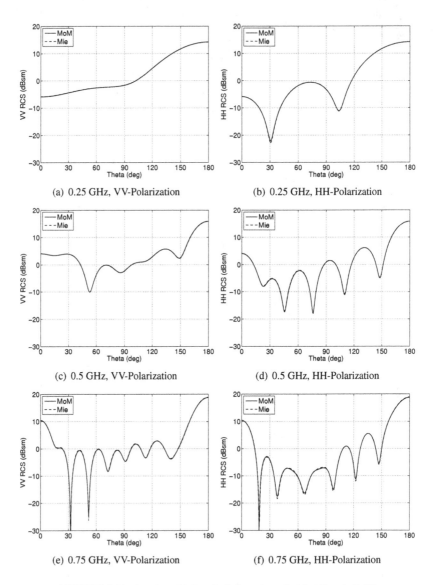

(a) 0.25 GHz, VV-Polarization

(b) 0.25 GHz, HH-Polarization

(c) 0.5 GHz, VV-Polarization

(d) 0.5 GHz, HH-Polarization

(e) 0.75 GHz, VV-Polarization

(f) 0.75 GHz, HH-Polarization

**FIGURE 8.20:** Lossless Dielectric Sphere ($\epsilon = 2.56$): Bistatic RCS

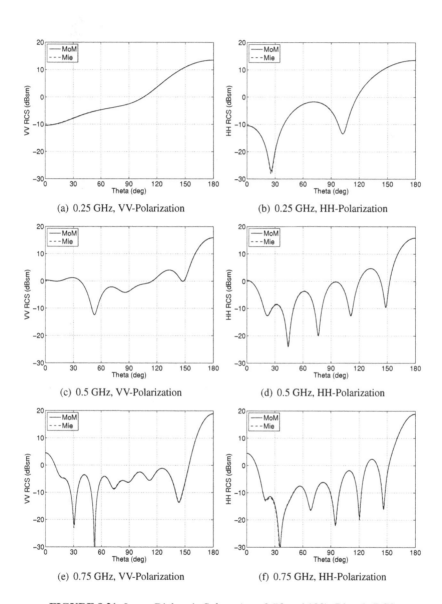

(a) 0.25 GHz, VV-Polarization

(b) 0.25 GHz, HH-Polarization

(c) 0.5 GHz, VV-Polarization

(d) 0.5 GHz, HH-Polarization

(e) 0.75 GHz, VV-Polarization

(f) 0.75 GHz, HH-Polarization

**FIGURE 8.21:** Lossy Dielectric Sphere ($\epsilon = 2.56 - j.102$): Bistatic RCS

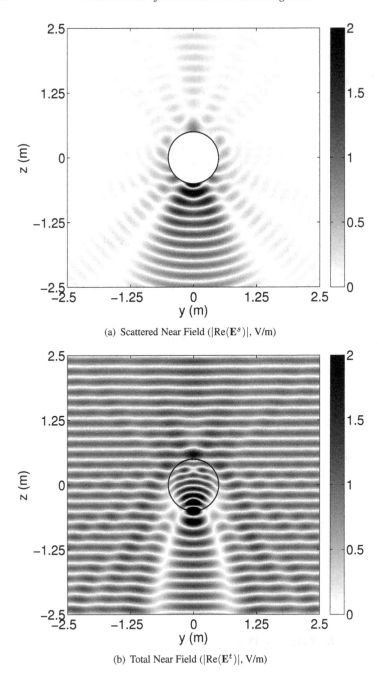

(a) Scattered Near Field ($|\mathrm{Re}(\mathbf{E}^s)|$, V/m)

(b) Total Near Field ($|\mathrm{Re}(\mathbf{E}^t)|$, V/m)

**FIGURE 8.22:** Dielectric Sphere: Near Field at 750 MHz

### 8.8.3.3 Coated Sphere

We next consider a conducting sphere with a thick dielectric coating. The radius of the conducting core is 0.25 meters, and is modeled using a mesh with 1280 near-equilateral triangles. The thickness of the dielectric coating is 0.25 meters, yielding an outermost radius of 0.5 meters. To describe the exterior surface, we again use the facet model from Section 8.8.3.1. This yields a total of 6400 triangles for both the inner and outer surfaces. The PMCHWT formulation is applied to the dielectric outer surface, and CFIE to the conducting core.

*Monostatic RCS versus Frequency*

We first compute the RCS of the coated sphere for frequencies ranging from 10 to 750 MHz. In Figure 8.23a we compare the RCS obtained by the Mie series to the MoM in backscattering ($\theta^i = \theta^s = 0$) for a lossless coating ($\epsilon_r = 2.56$). The comparison is very good. In Figure 8.23b is a similar comparison for a lossy coating ($\epsilon_r = 2.56 - j.102$). The forward-scattering RCS ($\theta^i = 0, \theta^s = \pi$) for each case is summarized in 8.24a and 8.24b, respectively, where the comparison is again quite good.

*Bistatic RCS*

Using the same simulation parameters as before, we next consider the bistatic RCS at 0.25, 0.5, and 0.75 GHz. The MoM results are compared to the Mie series for the lossless coating in Figures 8.25a–8.25f. The comparison is very good. The results for the lossy coating are compared in Figures 8.26a–8.26f, where the comparison is again very good.

*Near Field*

Using (3.108) and (3.112), we compute the near field of the lossless coated sphere illuminated by an $\hat{\mathbf{x}}$-polarized plane wave incident from the $+\hat{\mathbf{z}}$ direction at 750 MHz. Figures 8.27a and 8.27b show the real part of the scattered and total electric field, respectively, in the $yz$ plane. As with the dielectric sphere, the total near fields are computed separately in the external and internal dielectric regions, and are then overlaid.

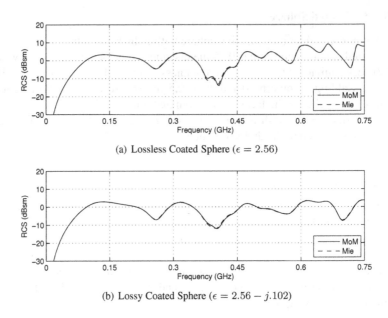

(a) Lossless Coated Sphere ($\epsilon = 2.56$)

(b) Lossy Coated Sphere ($\epsilon = 2.56 - j.102$)

**FIGURE 8.23:** Coated Sphere: Backscatter RCS vs. Frequency

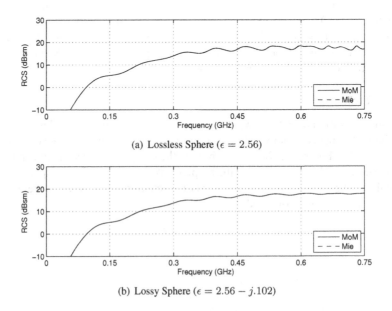

(a) Lossless Sphere ($\epsilon = 2.56$)

(b) Lossy Sphere ($\epsilon = 2.56 - j.102$)

**FIGURE 8.24:** Coated Sphere: Forward Scatter RCS vs. Frequency

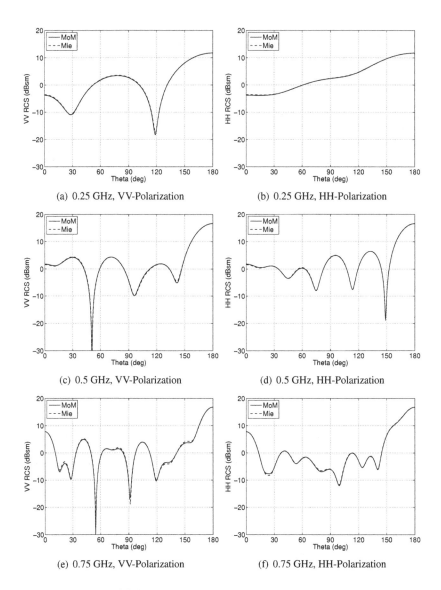

(a) 0.25 GHz, VV-Polarization

(b) 0.25 GHz, HH-Polarization

(c) 0.5 GHz, VV-Polarization

(d) 0.5 GHz, HH-Polarization

(e) 0.75 GHz, VV-Polarization

(f) 0.75 GHz, HH-Polarization

**FIGURE 8.25:** Lossless Coated Sphere ($\epsilon = 2.56$): Bistatic RCS

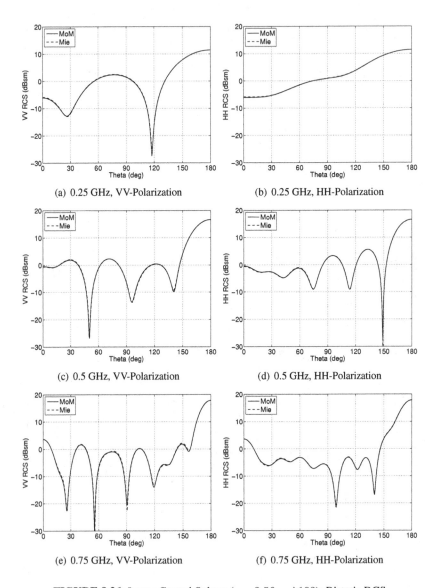

(a) 0.25 GHz, VV-Polarization

(b) 0.25 GHz, HH-Polarization

(c) 0.5 GHz, VV-Polarization

(d) 0.5 GHz, HH-Polarization

(e) 0.75 GHz, VV-Polarization

(f) 0.75 GHz, HH-Polarization

**FIGURE 8.26:** Lossy Coated Sphere ($\epsilon = 2.56 - j.102$): Bistatic RCS

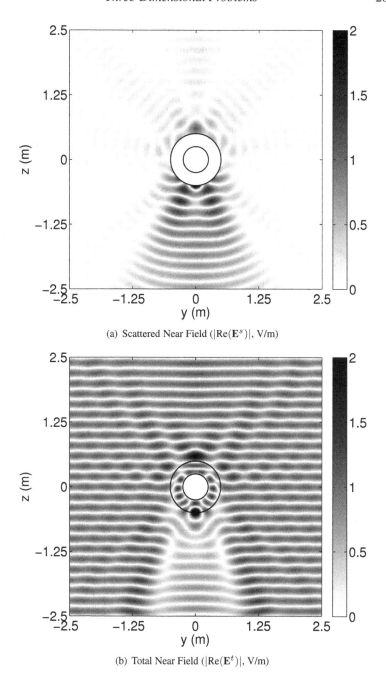

(a) Scattered Near Field ($|\mathrm{Re}(\mathbf{E}^s)|$, V/m)

(b) Total Near Field ($|\mathrm{Re}(\mathbf{E}^t)|$, V/m)

**FIGURE 8.27:** Coated Sphere: Near Field at 750 MHz

### 8.8.4 EMCC Plate Benchmark Targets

We next consider the EMCC benchmark plate radar targets that were described and measured in [22]. Because tip and edge diffractions were of primary interest in that paper, the measurements were performed using a conical cut 10 degrees from the plane of each target. The test articles were fabricated from aluminum foil or thin pieces of high-density foam coated with conducting paint. Thus, for each of our surface models we construct a thin sheet of planar facets, and for the MOM, the EFIE is used. In the plots that follow, the measurements were shifted in angle slightly to align them with the numerical results.

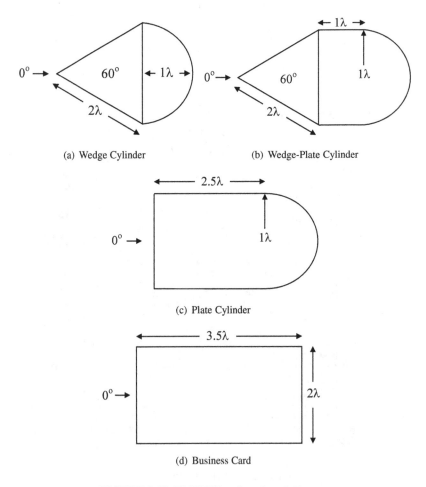

(a) Wedge Cylinder        (b) Wedge-Plate Cylinder

(c) Plate Cylinder

(d) Business Card

**FIGURE 8.28:** EMCC Plate Benchmark Targets

### 8.8.4.1 Wedge Cylinder

The wedge cylinder is shown in Figure 8.28a. The corresponding facet model comprises 3072 triangles, and is shown in Figure 8.29. The monostatic RCS from the MoM is compared to the EMCC measurements in Figures 8.33a and 8.33b for vertical and horizontal polarizations, respectively. The comparisons are very good for both polarizations, but are noticeably better in the horizontal case. In [22] the measurements were adjusted by 2dB before plotting; however that adjustment was not made here.

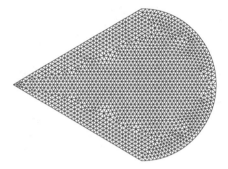

**FIGURE 8.29:** Facet Model of the Wedge Cylinder

### 8.8.4.2 Wedge-Plate Cylinder

The wedge-plate cylinder is shown in Figure 8.28b. The corresponding facet model comprises 4656 triangles, and is shown in Figure 8.30. The monostatic RCS from the MoM is compared to the EMCC measurements in Figures 8.33c and 8.33d for vertical and horizontal polarizations, respectively. The comparisons are very good for both polarizations, but again slightly better in the horizontal case.

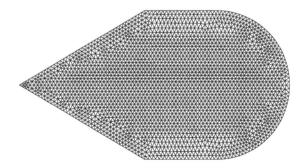

**FIGURE 8.30:** Facet Model of the Wedge-Plate Cylinder

### 8.8.4.3 Plate Cylinder

The plate cylinder is shown in Figure 8.28c. The corresponding facet model comprises 5058 triangles, and is shown in Figure 8.31. The monostatic RCS from the MoM is compared to the EMCC measurements in Figures 8.33e and 8.33f for vertical and horizontal polarizations, respectively. The comparison is quite good, however the measurement has a higher amplitude than expected between $-45° \leq \theta \leq 45°$ and $100° \leq \theta \leq 160°$ in Figure 8.33e, the cause of which is unknown.

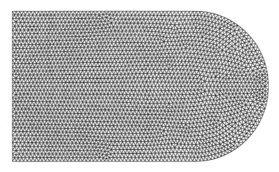

**FIGURE 8.31:** Facet Model of the Plate Cylinder

### 8.8.4.4 Business Card

The business card is shown in Figure 8.28d. The corresponding facet model comprises 5662 triangles, and is shown in Figure 8.32. The monostatic RCS from the MoM is compared to the EMCC measurements in Figures 8.33g and 8.33h for vertical and horizontal polarizations, respectively. The comparison is very good for horizontal polarization, though the computed results are lower than the measurement for vertical polarization, which was also noted in [22].

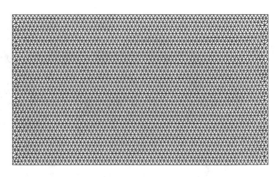

**FIGURE 8.32:** Facet Model of the Business Card

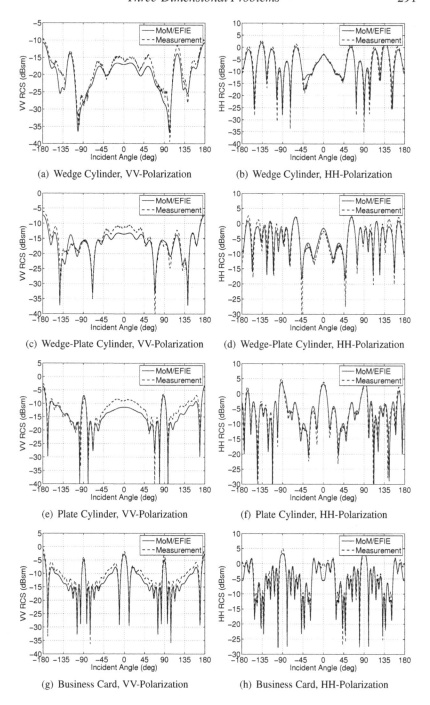

(a) Wedge Cylinder, VV-Polarization

(b) Wedge Cylinder, HH-Polarization

(c) Wedge-Plate Cylinder, VV-Polarization

(d) Wedge-Plate Cylinder, HH-Polarization

(e) Plate Cylinder, VV-Polarization

(f) Plate Cylinder, HH-Polarization

(g) Business Card, VV-Polarization

(h) Business Card, HH-Polarization

**FIGURE 8.33:** EMCC Plate Benchmark Targets: MoM vs. Measurement

## 8.8.5   Strip Dipole Antenna

We next consider the input impedance of a thin strip dipole antenna whose dimensions are outlined in Figure 8.34. This type of geometry is representative of thin foil chaff used as a radar countermeasure. The strip considered has a length $L = 1$ meter and width $w = 3$ mm, so Figure 8.34 is not to scale. The facet model comprises 400 triangles, and it is fed at the centermost edge ($e_m$) using the delta-gap model outlined in Section 8.3.3.2. As the strip is considered to be very thin, the EFIE is used. The computed input resistance and reactance are shown in Figure 8.35 for frequencies between 10 MHz and 1 GHz. The first resonance is at approximately 150 MHz, where the dipole is $\lambda/2$ in length.

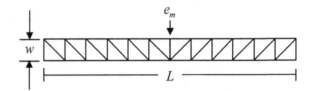

**FIGURE 8.34:** Strip Dipole Dimensions

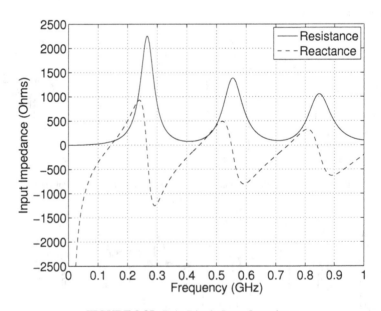

**FIGURE 8.35:** Strip Dipole Input Impedance

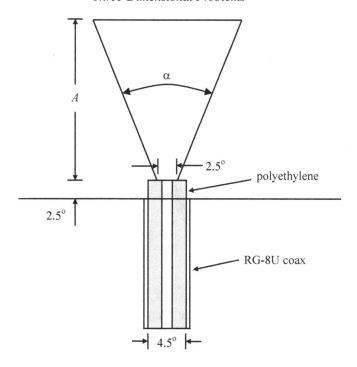

**FIGURE 8.36:** Bowtie Monopole Measurement Setup

## 8.8.6 Bowtie Antenna

The planar bowtie antenna depicted in Figure 8.8a is used extensively in applications where a wide operating bandwidth is necessary. A study of the input impedance and radiation characteristics of the bowtie antenna was presented previously in [23] for various flare angles and lengths. Their measurement setup is illustrated in Figure 8.36, where a half-bowtie was fed and measured against a ground plane comprising a conducting disk 8 feet in diameter. The feed utilized a RG-8U coaxial cable that was trimmed and inserted through a hole in the disk. The fed end of the bowtie was truncated to facilitate an electrical connection to the center conductor of the cable. Measurements were obtained at a fixed frequency of 500 MHz using a slotted line with sliding probe and transmission line charts. Each antenna was fabricated to the maximum electrical length and physically cut down as the measurements were made. Therefore, the electrical dimensions of the feedpoint and ground plane remained constant.

For the MoM simulation, we will model a full bowtie in free space, and take into account that the input impedance obtained will be approximately twice that of a half-bowtie over a finite conducting ground plane. Where the

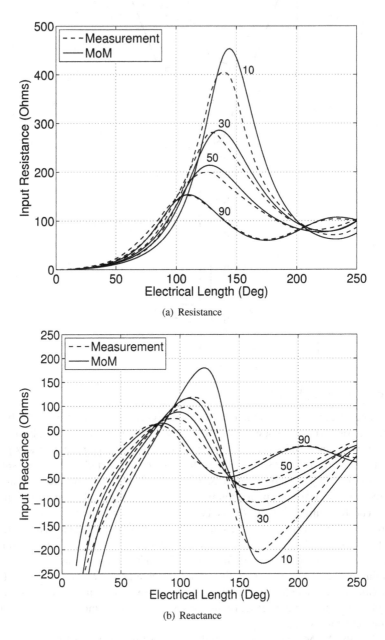

(a) Resistance

(b) Reactance

**FIGURE 8.37:** Bowtie Input Impedance: EFIE vs. Measurement

physical item was cut down to size at each step, in our case it makes more sense to fix the dimensions of the antenna and vary the frequency. We consider flare angles of 10, 30, 50, and 90 degrees, and the corresponding facet models comprise 826, 2010, 3462, and 7472 triangles, respectively. Each facet model has a half-length of 1 meter and an edge length of 2 cm at the feedpoint (see Figure 8.8a). In Figures 8.37a and 8.37b we compare the input resistance and reactance obtained from MoM with a delta-gap feed to the measurements in [23]. The comparison is fair, considering the differences between the physical and computer models. The disparity between the delta-gap feed model and the physical setup is an obvious difference, and likely contributes to the observed offset between the resonant frequencies as well as the impedance amplitudes. A similar observation was made in [24]. If we had instead used the magnetic frill in Section 5.2.2 and reduced the physical length of the model rather than changing the frequency, the comparison might be better, though it would be more time intensive in terms of setup and execution.

### 8.8.7   Archimedean Spiral Antenna

We next consider the characteristics of the Archimedean spiral antenna [25, 26], which has desirable broadband performance in terms of its input impedance as well as its circularly polarized radiation pattern. The geometry of this antenna is described in Figure 8.38a. The radius of each arm is linearly proportional to the angle and can be written as

$$r = b\phi + r_1 \qquad (8.110)$$

and

$$r = b(\phi + \pi) + r_1 , \qquad (8.111)$$

where $r_1$ is a starting radius and $b$ is a constant that depends on the width $w$ and spacing $s$ of the arms of the antenna. If we choose a self-complimentary spiral with $s = w$, then

$$b = \frac{2w}{\pi} . \qquad (8.112)$$

If the spiral has an infinite number of turns, then by Babinet's principle, the input impedance of the antenna should be approximately half that of the surrounding medium. If that medium is free space, then

$$Z_{in} = \frac{\eta_0}{2} \approx 188.4 \, \Omega . \qquad (8.113)$$

When a voltage is applied to the center terminals of this antenna, it behaves in a manner similar to a two-wire transmission line and gradually transitions into a radiating structure where the circumference of the spiral is one wavelength [26]. In theory, an antenna with the largest number of turns should have the

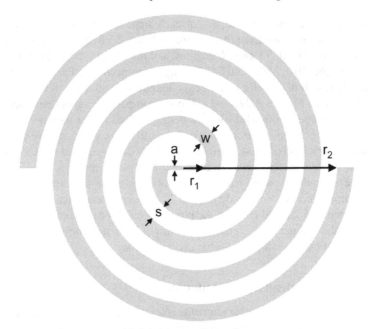

(a) Spiral Antenna Dimensions

(b) Example Spiral Antenna Mesh

**FIGURE 8.38:** Archimedean Spiral Antenna

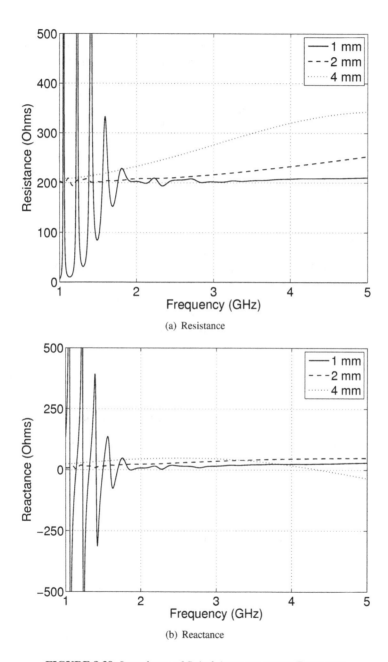

(a) Resistance

(b) Reactance

**FIGURE 8.39:** Impedance of Spiral Antennas versus Frequency

broadest bandwidth, however the available space will limit its maximum size. A method of connecting the antenna to the feedline must also be devised. All of these considerations will influence the final input impedance, which we do not expect will agree perfectly with the theoretical value of $Z_{in}$. For the numerical model, we choose a starting radius $r_1 = w$ and bridge the arms of the spiral at the center using a flat strip of width $a = w/2$. A heavier triangulation is used near the center, and a lower density is applied to the arms. This tessellation scheme is illustrated in Figure 8.38b, where the arms each comprise two and a half turns ($0 \leq \phi \leq \frac{5}{2}\pi$).

We will now attempt to find the dimensions of a spiral antenna having the flattest impedance characteristics between 1 and 5 GHz. To do so, we construct three spiral antennas, each with 8 turns per arm ($0 \leq \phi \leq 16\pi$), with widths $w$ of 1, 2, and 4 mm. A delta-gap source is used to excite the edge at the center of the strip bridging the arms. The input resistance and reactance of each antenna are shown in Figures 8.39a and 8.39b, respectively. The antenna of 1 mm width has a nearly constant input resistance of approximately 210 Ohms between 2 and 5 GHz, and an input reactance of around 20 Ohms. The other two antennas have a wider variation throughout this range, though their impedances are somewhat smoother at lower frequencies, which is expected given their larger overall dimensions. We expect that by adding more turns to the 1-mm antenna, its usable range could be extended down to 1 GHz with little impact on its higher-frequency performance. Using the 1-mm design as a starting point, a laboratory model could now be fabricated and a more accurate estimate of the impedance measured using a vector network analyzer (VNA).

### 8.8.8   Monoconic Reentry Vehicle with Dielectric Nose

Finally, we examine the monoconic reentry vehicle with dielectric nose, described previously in Section 7.10.2.4. The facet model of this object comprises three parts: the nose, septum, and frustum, which are modeled using 1944, 472, and 5824 triangles, respectively. These parts are shown in an exploded view in Figure 8.40. Because multiple bounces occur inside the dielectric nose, giving rise to more complex fields in that region, the level of facetization on the nose and septum is higher than on the frustum. We consider two cases: a lossless nose ($\epsilon = 5$), and a lossy nose ($\epsilon = 5 - j.25$).

We first compare the monostatic RCS obtained by *Serenity* and *Galaxy* for frequencies ranging from 100 MHz to 1.1 GHz. The results for the lossless nose are compared at an incident angle of $\theta = \pi/4$ for vertical and horizontal polarizations in Figures 8.41a and 8.41b, respectively, where the comparison is fairly good. A similar comparison for the lossy nose is made at $\theta = 3\pi/4$ in Figures 8.42a and 8.42b, respectively, which is again fairly good.

We next compare the monostatic RCS for all angles at 1.1 GHz. The results for lossless nose are compared for vertical and horizontal polarizations in

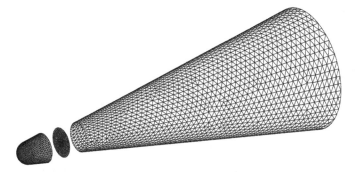

**FIGURE 8.40:** Monoconic RV with Dielectric Nose (Exploded View)

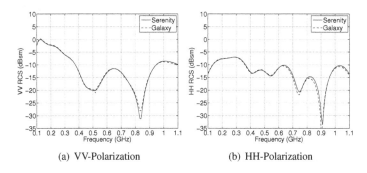

(a) VV-Polarization  (b) HH-Polarization

**FIGURE 8.41:** Monoconic RV with Lossless Nose: RCS vs. Frequency at $\theta = \pi/4$

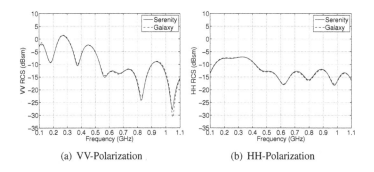

(a) VV-Polarization  (b) HH-Polarization

**FIGURE 8.42:** Monoconic RV with Lossy Nose: RCS vs. Frequency at $\theta = 3\pi/4$

Figures 8.43a and 8.43b, respectively. The comparison is fairly good. A similar comparison for the lossy nose is made in Figures 8.43c and 8.43d, respectively, where the comparison is fairly good. We observe that with the lossy nose tip, the RCS has been reduced slightly at forward aspect angles, as expected.

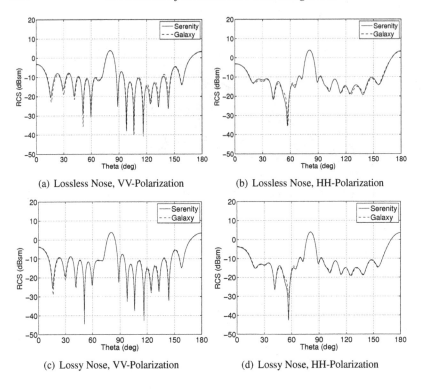

(a) Lossless Nose, VV-Polarization

(b) Lossless Nose, HH-Polarization

(c) Lossy Nose, VV-Polarization

(d) Lossy Nose, HH-Polarization

**FIGURE 8.43:** Monoconic RV RCS at 1.1 GHz

### 8.8.9   Summary of Examples

Run metrics for the examples in this section are summarized in Table 8.5. Listed are the total number of electric and magnetic basis functions ($N_J$ and $N_M$), memory ($M_Z$, in MB) required to store the system matrix, matrix fill time ($T_Z$, in seconds), and LU factorization time ($T_{LU}$, in seconds) using the LAPACK function `cgetrf`. As some examples involved solutions across a number of frequencies, the timing data shown reflects the average across all frequencies.

The Dell workstation used for these examples has 256 GB of RAM. Using 8 bytes for each single-precision complex value, a matrix with approximately 185,000 unknowns might be handled before all memory was exhausted. The actual limit is probably much less, as it is unlikely that a single, contiguous array of that size would be available in RAM. In the following chapters, we will demonstrate the solution of problems much larger than this limit on the same workstation, using compressive techniques.

**TABLE 8.5:** Run Metrics

| Object | $N_J$ | $N_M$ | $M_{\mathbf{Z}}$ | $T_{\mathbf{Z}}$ | $T_{\mathrm{LU}}$ |
|---|---|---|---|---|---|
| PEC Sphere (EFIE) | 7680 | 0 | 450 | 13 | 1 |
| PEC Sphere (nMFIE) | 7680 | 0 | 450 | 13 | 1 |
| PEC Sphere (CFIE) | 7680 | 0 | 450 | 14 | 1 |
| Dielectric Sphere | 7680 | 7680 | 1800 | 43 | 7 |
| Coated Sphere | 9600 | 7680 | 2278 | 52 | 10 |
| Wedge Cylinder | 4537 | 0 | 157 | 5 | 0.5 |
| Wedge-Plate Cylinder | 6893 | 0 | 362 | 11 | 1 |
| Plate Cylinder | 7486 | 0 | 427 | 14 | 1 |
| Business Card | 8383 | 0 | 536 | 18 | 2 |
| Strip Dipole | 399 | 0 | 1.2 | 0.1 | 0.1 |
| Bowtie (10°) | 1151 | 0 | 10 | 0.3 | 0.1 |
| Bowtie (30°) | 2911 | 0 | 65 | 1.9 | 0.2 |
| Bowtie (50°) | 5067 | 0 | 196.9 | 5.7 | 0.5 |
| Bowtie (90°) | 11013 | 0 | 925 | 27 | 3 |
| Spiral | 4333 | 0 | 143 | 7 | 0.5 |
| RV | 12340 | 2896 | 1171 | 40 | 8 |

# References

[1] P. Ylä-Oijala, M. Taskinen, and J. Sarvas, "Surface integral equation method for general composite metallic and dielectric structures with junctions," *Progress in Electromagnetics Research*, vol. 52, pp. 81–108, 2005.

[2] Rhinoceros 3D, Robert McNeel and Associates, rhino3d.com.

[3] L. Piegl and W. Tiller, *The NURBS Book*. Springer, 1995.

[4] J. Foley, A. van Dam, S. Feiner, and J. Hughes, *Computer Graphics: Principles and Practice*. Addison-Wesley, 1996.

[5] D. Wilton, S. Rao, A. Glisson, D. Schaubert, M. Al-Bundak, and C. M. Butler, "Potential integrals for uniform and linear source distributions on polygonal and polyhedral domains," *IEEE Trans. Antennas Propagat.*, vol. 32, pp. 276–281, March 1984.

[6] S. Caorsi, D. Moreno, and F. Sidoti, "Theoretical and numerical treatment of surface integrals involving the free-space Green's function," *IEEE Trans. Antennas Propagat.*, vol. 41, pp. 1296–1301, September 1993.

[7] R. D. Graglia, "On the numerical integration of the linear shape functions times the 3-D Green's function or its gradient on a plane triangle," *IEEE Trans. Antennas Propagat.*, vol. 41, pp. 1448–1454, October 1993.

[8] T. F. Eibert and V. Hansen, "On the calculation of potential integrals for linear source distributions on triangular domains," *IEEE Trans. Antennas Propagat.*, vol. 43, pp. 1499–1502, December 1995.

[9] J. Jin, *The Finite Element Method in Electromagnetics*. John Wiley and Sons, 1993.

[10] S. Rao, D. Wilton, and A. Glisson, "Electromagnetic scattering by surfaces of arbitrary shape," *IEEE Trans. Antennas Propagat.*, vol. 30, pp. 409–418, May 1982.

[11] R. D. Graglia, D. R. Wilton, and A. F. Peterson, "Higher order interpolatory vector bases for computational electromagnetics," *IEEE Trans. Antennas Propagat.*, vol. 45, pp. 329–342, March 1997.

[12] D. Sievers, T. F. Eibert, and V. Hansen, "Correction to 'on the calculation of potential integrals for linear source distributions on triangular domains'," *IEEE Trans. Antennas Propagat.*, vol. 53, p. 3113, September 2005.

[13] M. G. Duffy, "Quadrature over a pyramid or cube of integrands with a singularity at a vertex," *SIAM J. Numer. Anal*, vol. 19, pp. 1260–1262, December 1982.

[14] A. Tzoulis and T. Eibert, "Review of singular potential integrals for method of moments solutions of surface integral equations," *Advances in Radio Science*, vol. 2, pp. 97–99, 2004.

[15] W. C. Chew, J. M. Jin, E. Michielssen, and J. Song, *Fast and Efficient Algorithms in Computational Electromagnetics*. Artech House, 2001.

[16] P. Ylä-Oijala and M. Taskinen, "Calculation of CFIE impedance matrix elements with RWG and n x RWG functions," *IEEE Trans. Antennas Propagat.*, vol. 51, pp. 1837–1846, August 2003.

[17] R. E. Hodges and Y. Rahmat-Samii, "The evaluation of MFIE integrals with the use of vector triangle basis functions," *Microw. Opt. Tech. Lett.*, vol. 14, pp. 9–14, January 1997.

[18] L. S. Blackford, J. Choi, A. Cleary, E. D'Azevedo, J. Demmel, I. Dhillon, J. Dongarra, S. Hammarling, G. Henry, A. Petitet, K. Stanley, D. Walker, and R. C. Whaley, *ScaLAPACK Users' Guide*. Philadelphia, PA: Society for Industrial and Applied Mathematics, 1997.

[19] K. Faison, "The Electromagnetics Code Consortium," *IEEE Antennas Propagat. Magazine*, vol. 30, pp. 19–23, February 1990.

[20] J. S. Zhao and W. C. Chew, "Integral equation solution of Maxwell's equations from zero frequency to microwave frequencies," *IEEE Trans. Antennas Propagat.*, vol. 48, pp. 1635–1645, October 2000.

[21] J. F. Lee, R. Lee, and R. J. Burkholder, "Loop star basis functions and a robust preconditioner for EFIE scattering problems," *IEEE Trans. Antennas Propagat.*, vol. 51, pp. 1855–1863, August 2003.

[22] A. C. Woo, H. T. G. Wang, M. J. Schuh, and M. L. Sanders, "Benchmark plate radar targets for the validation of computational electromagnetics programs," *IEEE Antennas Propagat. Magazine*, vol. 34, pp. 52–56, December 1992.

[23] G. H. Brown and J. O. M. Woodward, "Experimentally determined radiation characteristics of conical and triangular antennas," *RCA Review*, pp. 425–452, 1952.

[24] C. Leat, N. Shuley, and G. Stickley, "Triangular-patch model of bowtie antennas: Validation against Brown and Woodward," *IEE Proc.-Microw. Antennas Propag.*, vol. 145, pp. 465–470, December 1998.

[25] W. L. Curtis, "Spiral antennas," *IRE Trans. Antennas Propagat.*, vol. 8, pp. 298–306, May 1960.

[26] J. A. Kaiser, "The Archimedean two-wire spiral antenna," *IRE Trans. Antennas Propagat.*, vol. 8, pp. 312–323, May 1960.

# Chapter 9

## Adaptive Cross Approximation

In previous chapters we discussed and observed the issues and complexities encountered in applying the Method of Moments to larger problems. Foremost among these are limited system RAM for storing the MoM system matrix, and the LU factorization whose computational complexity scales as $O(N^3)$. If the number of right-hand sides is small, the use of iterative techniques (Section 4.2) instead of LU factorization may yield a shorter run time, however radar cross section problems often have thousands of right-hand sides. Additionally, the MoM system matrix is often poorly conditioned, and so preconditioning is typically needed. However, preconditioners carry with them their own memory and CPU overhead, and even then the number of iterations per right-hand side may be very high or the solution may not converge at all. On the other hand, if the system matrix does not fit in RAM, historically there have been few options besides changing the geometry to reduce the number of unknowns, or using virtual memory (disk) and an out-of-core solver. The latter option, however, may make the problem intractable due to the far slower read/write speed of hard disks. With the increasing availability of multi-core CPUs and parallel processing, the overall computational burden has been lessened, but the difficulties with memory requirements remain. These issues motivate the search for new techniques and algorithms that solve the system directly, avoiding the issues that plague iterative solvers, while lessening the memory requirements.

In this chapter we investigate the Adaptive Cross Approximation (ACA), a method by which a significant portion of the MoM system matrix is compressed and stored in memory directly. A direct LU factorization is then performed, optimized using a CPU or GPU-accelerated Level 3 BLAS. The ACA is particularly attractive for radar cross section problems, as it permits solution of many right-hand sides simultaneously. As the ACA is a purely matrix-based method, requiring no knowledge about the underlying problem, our discussion in this chapter focuses primarily on the requisite linear algebra and changes required to add ACA capability to an existing full matrix 3D MoM solver. In Chapter 10, we will then extend the ACA further to a Multi-Level Adaptive Cross Approximation (MLACA), which improves compressibility further at the expense of additional run time. In Chapter 11, we will then examine the Fast Multipole Method (FMM), which shares some fundamental concepts with the ACA and MLACA but compresses and solves the MoM matrix system in a different way.

---

## 9.1   Rank Deficiency

The matrix system that results from MoM discretization and testing has full rank. However, it is well known that if the basis functions are partitioned into spatially localized groups, the matrix becomes a block structure where the off-diagonal blocks are *rank-deficient* [1]. To begin, let us consider a matrix block $\mathbf{Z}_{m \times n}$ having $m$ rows and $n$ columns, which can be decomposed using the singular-value decomposition (SVD) as

$$\mathbf{Z}_{m \times n} = \mathbf{U}_{m \times m} \mathbf{\Sigma}_{m \times n} \mathbf{V}^*_{n \times n} \, , \qquad (9.1)$$

where the diagonals of $\mathbf{\Sigma}$ comprise the singular values, and $\mathbf{U}$ and $\mathbf{V}^*$ contain the left and right-singular vectors, respectively. As $\mathbf{Z}$ is rank-deficient, the singular values in $\mathbf{\Sigma}$ decrease quickly below machine precision. To illustrate, consider the two roughly spherical groups of basis functions in Figure 9.1, having diameter $a$ and centers separated by distance $D$.

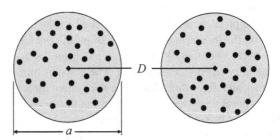

**FIGURE 9.1:** Groups of Basis Functions

We now define the matrix elements $Z_{mn}$ as the 3D Green's function

$$Z_{mn} = G(\mathbf{r}, \mathbf{r}') = \frac{e^{-jkR_{mn}}}{R_{mn}} \, , \qquad (9.2)$$

where $R_{mn} = |\mathbf{r}_m - \mathbf{r}_n|$. We now compute $\mathbf{Z}_{m \times n}$ for spherical regions of diameter $a = 1\lambda$ having 256 randomly distributed points, where the separation distance $D$ is assigned values of 0, $2\lambda$ and $4\lambda$. This corresponds to overlapping groups, groups whose bounding spheres just touch, and well-separated groups. In Figure 9.2 are plotted the normalized singular values of $\mathbf{Z}$ for each case, with

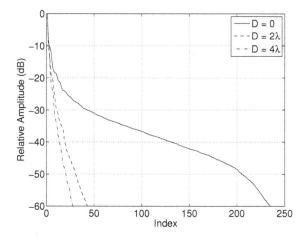

**FIGURE 9.2:** Singular Values versus Group Separation $D$

a minimum threshold of $10^{-6}$. This is very close to machine precision when using 32-bit floating point variables. We see that when the groups overlap, the singular values fall off very slowly. However, as the separation increases, they fall off much faster, and when $D = 4\lambda$, only the first 28 singular values are above the threshold. This is an intuitive result, given that as the separation between groups becomes large, there is less variation in $R_{mn}$ and $Z$ begins to take on redundant information. This analysis implies that for well separated groups, $Z_{m \times n}$ could be approximated as

$$Z_{m \times n} \approx U_{m \times k} \Sigma_{k \times k} V_{k \times n}^* = Z_u^{m \times k} Z_v^{k \times n}, \qquad (9.3)$$

where the *effective* rank $k \ll m, n$ is chosen based on a threshold $\tau$ of largest to smallest singular values. Only $Z_u$ and $Z_v$ would be stored explicitly.

### 9.1.1 Limitations of Using SVD For Compression

Assuming that a block structure for the MoM matrix exists, a naive approach for compressing the off-diagonal blocks would be to simply fill and compress them using the SVD. However, as matrix filling is a time consuming operation and the SVD itself scales as $O(N^3)$, this would be computationally infeasible for larger problems. The ACA, introduced in Section 9.2, alleviates these issues for large problem sizes. In the following chapters, we will revisit the SVD in our discussion of MLACA and FMM, as for smaller group sizes it is often easier and more efficient to apply it directly.

## 9.2   Adaptive Cross Approximation

The Adaptive Cross Approximation, first presented by Bebendorf in [2], efficiently computes a compressed representation of a rank-deficient matrix block $\mathbf{Z}_{m \times n}$ that satisfies the condition

$$\|\mathbf{R}_{m \times n}\| = \|\mathbf{Z}_{m \times n} - \tilde{\mathbf{Z}}_{m \times n}\| \leq \epsilon \|\mathbf{Z}_{m \times n}\|, \qquad (9.4)$$

where $\mathbf{R}$ is the error matrix and $\|\mathbf{Z}\|$ is the matrix Frobenius norm. It is a simple, method-agnostic, fast rank-revealing algorithm which iteratively constructs the product matrices $\mathbf{Z}_u$ and $\mathbf{Z}_v$ in (9.3) by selecting only $k$ rows and columns from $\mathbf{Z}$, until the residual norm falls below a chosen threshold $\tau_{ACA}$. Thus for small $k$, only a small fraction of $\mathbf{Z}$ needs to be computed, with final storage requirements for $\mathbf{Z}_u$ and $\mathbf{Z}_v$ comprising only $k(m + n)$, versus $mn$ for $\mathbf{Z}$.

The ACA was first applied to electromagnetic problems in [3], and the procedure from that paper is summarized in Algorithm 6 with some modifications. In this algorithm, $\mathbf{I} = [I_1, \ldots, I_m]$ and $\mathbf{J} = [J_1, \ldots, J_n]$ are arrays containing the indexes of selected rows and columns of $\mathbf{Z}_{m \times n}$. The vector $\mathbf{u}_k$ is the $k^{\text{th}}$ column of the matrix $\mathbf{Z}_u$ and $\mathbf{v}_k$ is the $k^{\text{th}}$ row of the matrix $\mathbf{Z}_v$. $\tilde{\mathbf{R}}(I_1, :)$ and $\tilde{\mathbf{R}}(:, J_1)$ indicate the $I_1^{\text{th}}$ row and $J_1^{\text{th}}$ column of $\tilde{\mathbf{R}}$, respectively. $\tilde{\mathbf{Z}}^{(k)}$ comprises the matrix $\tilde{\mathbf{Z}}$ at the $k^{\text{th}}$ iteration. Step 12 as originally presented in [3] had an error, and has been corrected following [4].

We note that the convergence criteria in step 13 is based on an *estimate* of the norm and is not completely accurate. Though this author has found these criteria to work fine in practice, these criteria were studied further in [5] and [6], where some alternatives are proposed.

There are two immediately obvious advantages to the ACA:

1. The ACA iteratively assembles $\mathbf{Z}_u$ and $\mathbf{Z}_v$ by adding more rows and columns until an acceptable error threshold is reached. The ACA itself has relatively low computational overhead.

2. Complete knowledge of $\mathbf{Z}$ is not necessary. Only $k$ rows and columns must be computed.

### 9.2.1   Modifications

Based on our experience with the ACA in practice, we have made several modifications to the algorithm as presented originally.

**Algorithm 6** Adaptive Cross Approximation

1: **procedure** INITIALIZATION
2:　　Set the $k = 1, \tilde{\mathbf{Z}} = 0$
3:　　Find a nonzero row $\mathbf{Z}(I_1, :)$ of $\mathbf{Z}$. If none are found, ACA is done.
4: **end procedure**
5: **procedure** ITERATION
6:　　Update the $I_k{}^{\text{th}}$ row of the error matrix: $\tilde{\mathbf{R}}(I_k, :) = \mathbf{Z}(I_k, :) - \sum_{l=1}^{k-1} (\mathbf{u}_l)_{I_k} \mathbf{v}_l$
7:　　If all $\|\tilde{\mathbf{R}}(I_k, j)\| = 0, j \neq J_1, \ldots, J_{k-1}$, ACA is done.
8:　　Find the $k^{\text{th}}$ column index $J_k : |\tilde{\mathbf{R}}(I_k, J_k)| = \max_j(|\tilde{\mathbf{R}}(I_k, j)|), j \neq J_1, \ldots, J_{k-1}$
9:　　$\mathbf{v}_k = \tilde{\mathbf{R}}(I_k, :)/\tilde{\mathbf{R}}(I_k, J_k)$
10:　　Initialize the $J_k{}^{\text{th}}$ column of the error matrix: $\tilde{\mathbf{R}}(:, J_k) = \mathbf{Z}(:, J_k) - \sum_{l=1}^{k-1} (\mathbf{v}_l)_{J_k} \mathbf{u}_l$
11:　　$\mathbf{u}_k = \tilde{\mathbf{R}}(:, J_k)$
12:　　$\|\tilde{\mathbf{Z}}^{(k)}\|^2 = \|\tilde{\mathbf{Z}}^{(k-1)}\|^2 + 2\mathrm{Re}\big\{ \sum_{j=1}^{k-1} (\mathbf{u}_j^H \mathbf{u}_k)(\mathbf{v}_j^H \mathbf{v}_k) \big\} + \|\mathbf{u}_k\|^2 \|\mathbf{v}_k\|^2$
13:　　Convergence check: If $\|\mathbf{u}_k\| \|\mathbf{v}_k\| \leq \epsilon \|\tilde{\mathbf{Z}}^{(k)}\|$, ACA is done.
14:　　Find the next row index $I_{k+1} : |\tilde{\mathbf{R}}(I_{k+1}, J_k)| = \max_i(|\tilde{\mathbf{R}}(i, J_k)|), i \neq I_1, \ldots, I_k$
15:　　$k = k + 1$
16: **end procedure**

#### 9.2.1.1 Initialization

When originally initializing the ACA in [3], the first row of $\mathbf{Z}$ was used. However, this approach fails if that row is zero or the entire block $\mathbf{Z}$ is zero, which is possible for problems having multiple dielectric regions or certain geometric configurations. During matrix filling, we will have *a priori* knowledge of some of these zero blocks, and these can be zeroed explicitly without doing any work (see Section 9.6.1.1). However, during the LU factorization some blocks will no longer be zero due to fill-in, and each row in those blocks must be tested explicitly.

#### 9.2.1.2 Early Termination

A second modification was added in step 7. During the ACA iteration, the values in $\tilde{\mathbf{R}}(I_k, :)$ computed in step 6 can become very small. As the search in step 8 is done from an ever decreasing pool of unused columns, it is possible that the largest available $\tilde{\mathbf{R}}(I_k, J_k) \approx 0$ and the operation in step 9 will fail. Thus, if this condition is reached, the ACA algorithm is terminated.

### 9.2.1.3   Pathological Failure Case

A case where the ACA fails was identified when computing an off-diagonal matrix block for a dielectric cube. This occurred when the source and testing groups comprised co-planar groups of facets on a cube face. For this block, (3.184) takes on the form

$$\mathbf{Z} = \begin{bmatrix} Z^{EJ} & Z^{EM} \\ Z^{HJ} & Z^{HM} \end{bmatrix} = \begin{bmatrix} Z^{EJ} & 0 \\ 0 & Z^{HM} \end{bmatrix}, \tag{9.5}$$

and (3.186) and (3.187) evaluate to zero due to the co-planar geometry. The failure occurs because the ACA does not know which rows are zero. During initialization, the first-row selection in step 3 will simply choose the first row of $\mathbf{Z}$, whose nonzero portion comprises that taken from sub-block $Z^{EJ}$. As a consequence, the column search in step 8 will only select columns from this nonzero portion, whose nonzero content is again from $Z^{EJ}$ only. Thus, the ACA will *never* select rows and columns with elements from sub-block $Z^{HM}$. A similar condition would occur if the ACA was initialized with a row from the bottom part of $\mathbf{Z}$, where elements from sub-block $Z^{EJ}$ would now be excluded.

Our solution is to *force* the ACA to select content from both $Z^{EJ}$ and $Z^{HM}$. We first rearrange $\mathbf{Z}$ so it has the block structure in (9.5). This is done during the clustering and basis mapping step (Section 9.3) by ordering all electric basis functions before magnetic basis functions in the mapping for each block. We then pass a hint to the ACA driver routine, containing the number of magnetic basis functions in the source and testing groups. If these are nonzero, this driver will alternate between two instances of the ACA, which collectively add to $\mathbf{Z}_u$ and $\mathbf{Z}_v$ but are restricted to choosing only rows from the upper or lower parts of $\mathbf{Z}$. As the number of basis functions of each type may differ, as in the case of composite geometry, it is possible that one instance will run out of rows before the other. In this case, the remaining instance continues normally until convergence is reached.

### 9.2.2   QR/SVD Recompression

The rows and columns of $\mathbf{U}$ and $\mathbf{V}$ as determined by ACA are not strictly orthonormal. Thus, significant additional compression can be obtained by applying the QR/SVD recompression method prescribed in [7], which requires QR decompositions of $\mathbf{U}$ and $\mathbf{V}$ and an SVD of size $k \times k$. This method begins by computing

$$\mathbf{Z}_u^{m \times k} = \mathbf{Q}_u^{m \times k} \mathbf{R}_u^{k \times k} \tag{9.6}$$

and

$$\mathbf{Z}_v^{H, n \times k} = \mathbf{Q}_v^{n \times k} \mathbf{R}_v^{k \times k}. \tag{9.7}$$

We then compute the SVD of the product

$$\mathbf{R}_u^{k \times k} \times \mathbf{R}_v^{H,k \times k} = \mathbf{U}^{k \times k} \mathbf{\Sigma}^{k \times k} \mathbf{V}^{*,k \times k} , \qquad (9.8)$$

and we then determine a new effective rank $r \leq k$ by finding all $diag(\mathbf{\Sigma}) \geq \tau \mathbf{\Sigma}(1,1)$. Thus, 9.8 becomes

$$\mathbf{R}_u^{k \times k} \times \mathbf{R}_v^{H,k \times k} = \mathbf{U}^{k \times r} \mathbf{\Sigma}^{r \times r} \mathbf{V}^{*,r \times k} . \qquad (9.9)$$

We then compute the re-compressed versions of $\mathbf{Z}_u$ and $\mathbf{Z}_v$ as

$$\mathbf{Z}_u^{m \times r} = \mathbf{Q}_u^{m \times k} \mathbf{U}^{k \times r} , \qquad (9.10)$$

and

$$\mathbf{Z}_v^{r \times n} = \mathbf{\Sigma}^{r \times r} \left[ \mathbf{Q}_v^{n \times k} \mathbf{R}_v^{H,k \times r} \right]^H . \qquad (9.11)$$

Note that the matrix $\mathbf{\Sigma}$ may be multiplied with either $\mathbf{U}$ or $\mathbf{V}$ to yield $\mathbf{Z}_u$ or $\mathbf{Z}_v$. For the regular ACA, either form may be used, however both will be required for the MLACA described later in Section 10.1.1.

After the ACA algorithm has generated an initial estimate of $\mathbf{Z}_u$ and $\mathbf{Z}_v$, this recompression step is invoked immediately to compress them further. Through extensive testing it has been found that good results can be achieved with a tolerance $\tau_{SVD} \approx 10 \, \tau_{ACA}$, and for practical problems, the QR/SVD recompression typically yields additional compression of 20 to 30 percent over the ACA alone. Though this step does add more computational overhead, this is amortized later during the block LU factorization as the off-diagonal block matrices are smaller and their products require fewer total operations.

# 9.3 Clustering Techniques

In the present treatment we limit the discussion to RWG basis functions, which are assigned to triangle edges. Following the full matrix approach from Chapter 8, where after all triangles have been analyzed, junctions found and basis functions assigned, we are left with a single, large group of basis functions. We now wish to partition these basis functions into $M$ spatially localized groups, and there are several potential clustering algorithms to choose from. The balanced binary tree and octree use axis-aligned bounding boxes to recursively divide the target's bounding volume, and a cobblestone algorithm was used in [8] to cluster groups for the ACA. However, this author's experience has shown that a $K$-means clustering algorithm yields groups with the most consistent size, and better overall compression and performance.

Given $N$ points $(\mathbf{r}_1, \mathbf{r}_2, \ldots, \mathbf{r}_N)$, $K$-means attempts to partition these points into $M$ groups $(S_1, S_2, \ldots, S_M)$ by minimizing the inter-cluster sum of squares, given by

$$\sum_i^M \sum_{r_j \in S_i} \| \mathbf{r}_k - \boldsymbol{\mu}_i \|^2, \qquad (9.12)$$

where $\boldsymbol{\mu}_i$ is the mean of group $S_i$. Theoretically this problem is NP-hard, however it can be implemented practically as an iterative algorithm. Given a total number of basis functions $N$ and a number of groups $M$, the algorithm is initialized by choosing at random the mid-points of $M$ surface edges as the initial mean for each group. Alternatively, a target group size $N_g$ could be specified, and the number of groups computed as $M = N/N_g$. At each iteration, edges are assigned to groups based on the minimum squared distance of their midpoint from each group's mean in (9.12). The means of each group are then re-computed using the newly assigned points, and the process repeated. For most practical problems, it was observed that the groups do not change much past 30 or so iterations. Though the amount of work grows significantly per iteration with the number of basis functions, the assignment step is easily parallelized.

Illustrated in Figure 9.3 are the results of the octree, cobblestone, and $K$-means algorithms applied to the EMCC cone-sphere target described in Section 7.9.2.3. The surface mesh has 79496 facets and 119244 edges, and each algorithm has a target group size of 2000 edges. The octree and cobblestone clustering in Figures 9.3a and 9.3b clearly yield groups having a greater overall variation in size than the $K$-means clustering in Figure 9.3c.

Though $K$-means clustering was found to perform the best of the alternatives for this use case, in some cases it still yields groups having a number of basis functions much larger (or smaller) than the initial target and final average. However, it is the size of the largest group $N_{max}$ that is of most concern, as the largest diagonal block will be of size $N_{max} \times N_{max}$. This quantity also affects the size of many work temporary spaces, which must not grow too large. This is of particular concern when operating on GPUs, which often have a much more limited memory space than the host system. Thus, for a group whose size exceeds the original target by more than 50 percent, it is split into two new groups via $K$-means. Finally, groups are sorted into ascending order by size, which allows for a more efficient use of workspace memory as the LU factorization moves along the diagonal. We will discuss CPU and GPU workspace management in more detail in Sections 9.6.2.1 and 9.6.2.2.

Once the groups have been assembled, a global basis function re-mapping is performed. This is a simple loop over all groups, where the electric basis functions in a group are added to the map first, followed by any magnetic basis functions (see Section 9.2.1.3). Once finished, the basis functions in each group will have global linear indexing, and the new map is used to re-order the original indexing assignments made elsewhere.

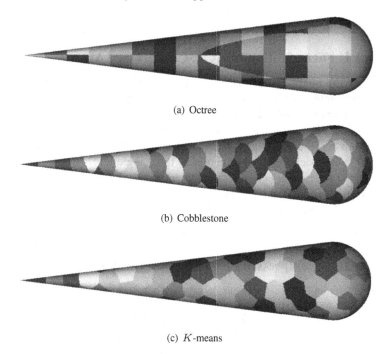

(a) Octree

(b) Cobblestone

(c) $K$-means

**FIGURE 9.3:** Different Clustering Techniques

## 9.3.1 Target Group Size For ACA

Proper selection of the target group size $N_g$ for the ACA is based on several factors. As the diagonal blocks are not compressed in ACA and are stored in full form, $N_g$ cannot be made too large or the memory requirements for the diagonals will be too large. Similarly, the performance of the Level 3 BLAS matrix-matrix product cgemm in optimized BLAS libraries such as Intel MKL and NVIDIA cuBLAS grow with increasing block size, but will eventually plateau.

Conversely, we cannot make $N_g$ too small, or the compressibility of the off-diagonal blocks will suffer, and we will not exploit the full potential of the optimized BLAS. Thus, our experience has shown the most favorable $N_g$ for the ACA on current hardware ranges between 2500 to 3500 unknowns. This balances the compression of off-diagonal blocks, the storage requirement of diagonal blocks, and the efficiency of block operations. We will consider the benefits of larger groups later in Section 10.4.2.1, where we compare the ACA to the MLACA.

## 9.4    LU Factorization of ACA-Compressed Matrix

After the block MoM matrix has been filled, with off-diagonal blocks stored in compressed form, the matrix system must be solved. In [3] the system was iteratively solved using a preconditioned GMRES algorithm. It was then shown in [8] that if a block LU factorization is applied to the compressed system matrix, the off-diagonal blocks of the **LU** matrix are also compressible. This approach, referred to herein as the "single-level ACA", is particularly suited to efficient computer implementation, as block updates in the LU factorization (and RHS solutions) are performed in rank-reduced form using a compressed matrix representation. This approach is described in detail in this section.

### 9.4.1    ACA-Compressed Block LU Factorization

Consider an ACA-compressed system matrix, written as

$$\mathbf{Z} = \begin{bmatrix} \mathbf{Z}_{11} & [\mathbf{Z}_u\mathbf{Z}_v]_{12} & [\mathbf{Z}_u\mathbf{Z}_v]_{13} \\ [\mathbf{Z}_u\mathbf{Z}_v]_{21} & \mathbf{Z}_{22} & [\mathbf{Z}_u\mathbf{Z}_v]_{23} \\ [\mathbf{Z}_u\mathbf{Z}_v]_{31} & [\mathbf{Z}_u\mathbf{Z}_v]_{32} & \mathbf{Z}_{33} \end{bmatrix}, \tag{9.13}$$

where a $3 \times 3$ block matrix is used for illustrative purposes. In practice, the matrix may have hundreds of block rows and columns. Assuming the off-diagonal blocks of the **LU** matrix are also compressible, it can be written similarly as

$$[\mathbf{LU}] = \begin{bmatrix} [\mathbf{LU}]_{11} & [\mathbf{U}_u\mathbf{U}_v]_{12} & [\mathbf{U}_u\mathbf{U}_v]_{13} \\ [\mathbf{L}_u\mathbf{L}_v]_{21} & [\mathbf{LU}]_{22} & [\mathbf{U}_u\mathbf{U}_v]_{23} \\ [\mathbf{L}_u\mathbf{L}_v]_{31} & [\mathbf{L}_u\mathbf{L}_v]_{32} & [\mathbf{LU}]_{33} \end{bmatrix}. \tag{9.14}$$

In Section 4.1.3, a right-looking block LU factorization for a general block matrix was discussed. One of the core elements in this algorithm was that as it progresses down the diagonal to the right, blocks in the trailing sub-matrix are continuously updated. However, any update operation made to a *compressed* block first requires that block to be expanded (uncompressed) back to full size, the update performed, and the updated block recompressed. Computationally, this would be inefficient, and approximation errors would be compounded by repeated recompression of blocks. However, a simple reordering of operations will make this much more efficient for a compressed matrix. Instead of iteratively updating the trailing sub-matrix, at each step we perform a total update of the diagonal and its trailing row and column blocks, using completed blocks from previous rows and columns. In this way, the off-diagonal blocks are recompressed only once during the factorization.

---

**Algorithm 7** ACA-Compressed Block LU Factorization

1: **for** $b = 1$ to $M$ **do**
2: $\quad \mathbf{A} = \mathbf{Z}_{bb} - \sum_{s=1}^{b-1} [\mathbf{L}_u \mathbf{L}_v]_{bs} [\mathbf{U}_u \mathbf{U}_v]_{sb}$
3: $\quad [\mathbf{LU}]_{bb} = \mathrm{LU}(\mathbf{A})$ (scalar LU factorization)
4: $\quad$ **for** $s = b + 1$ to $M$ **do**
5: $\qquad \mathbf{L}_{sb} = \Big[ [\mathbf{Z}_u \mathbf{Z}_v]_{sb} - \sum_{p=1}^{b-1} [\mathbf{L}_u \mathbf{L}_v]_{sp} [\mathbf{U}_u \mathbf{U}_v]_{pb} \Big] \mathbf{U}_{bb}^{-1}$
6: $\qquad$ Compress $\mathbf{L}_{sb}$ via ACA and store $[\mathbf{L}_u \mathbf{L}_v]_{sb}$
7: $\qquad \mathbf{U}_{bs} = \mathbf{L}_{bb}^{-1} \Big[ [\mathbf{Z}_u \mathbf{Z}_v]_{bs} - \sum_{p=1}^{b-1} [\mathbf{L}_u \mathbf{L}_v]_{bp} [\mathbf{U}_u \mathbf{U}_v]_{ps} \Big]$
8: $\qquad$ Compress $\mathbf{U}_{bs}$ via ACA and store $[\mathbf{U}_u \mathbf{U}_v]_{bs}$
9: $\quad$ **end for**
10: **end for**

---

The modified approach is summarized in Algorithm 7, where the matrix products with $\mathbf{U}_{bb}^{-1}$ and $\mathbf{L}_{bb}^{-1}$ comprise a triangular back-substitution. Note that in steps 2, 5 and 7, the updates are made to full-size matrix blocks. In step 2, the diagonal $\mathbf{Z}_{bb}$ is already full, and in steps 5 and 7, $\mathbf{Z}_{sb}$ and $\mathbf{Z}_{bs}$ are first expanded to full size. However, the block updates in steps 2, 5 and 7 are very fast, as the matrix products are performed in rank-reduced form using the Level 3 BLAS. To illustrate, consider the product of ACA compressed matrices

$$\mathbf{A}^{m_1 \times n_1} \mathbf{B}^{m_2 \times n_2} = \mathbf{A}_u^{m_1 \times k_1} \mathbf{A}_v^{k_1 \times n_1} \mathbf{B}_u^{m_2 \times k_2} \mathbf{B}_v^{k_2 \times n_2} , \qquad (9.15)$$

where $n_1 = m_2$. The innermost product is performed first, which is

$$\mathbf{T}_1^{k_1 \times k_2} = \mathbf{A}_v^{k_1 \times n_1} \mathbf{B}_u^{m_2 \times k_2} , \qquad (9.16)$$

and so

$$\mathbf{A}^{m_1 \times n_1} \mathbf{B}^{m_2 \times n_2} = \mathbf{A}_u^{m_1 \times k_1} \mathbf{T}_1^{k_1 \times k_2} \mathbf{B}_v^{k_2 \times n}. \qquad (9.17)$$

Which product in (9.17) is performed next depends on $\min(k_1, k_2)$. If $k_1 < k_2$, we compute

$$\mathbf{T}_2^{k_1 \times n_2} = \mathbf{T}_1^{k_1 \times k_2} \mathbf{B}_v^{k_2 \times n_2} \qquad (9.18)$$

and then

$$\mathbf{A}^{m_1 \times n_1} \mathbf{B}^{m_2 \times n_2} = \mathbf{A}_u^{m_1 \times k_1} \mathbf{T}_2^{k_1 \times n_2}. \qquad (9.19)$$

Otherwise, we compute

$$\mathbf{T}_3^{m_1 \times k_2} = \mathbf{A}_u^{m_1 \times k_1} \mathbf{T}_1^{k_1 \times k_2} \qquad (9.20)$$

and then

$$\mathbf{A}^{m_1 \times n_1} \mathbf{B}^{m_2 \times n_2} = \mathbf{T}_3^{m_1 \times k_2} \mathbf{B}_v^{k_2 \times n_2}. \qquad (9.21)$$

Note that in practice, $\mathbf{L}_{sb}$ and $\mathbf{U}_{bs}$ comprise temporary workspaces whose contents are discarded after being compressed by the ACA driver in steps 6 and 8.

**9.4.1.1   Compressibility of the LU Matrix**

Though the **LU** matrix blocks are compressible, in general they are less compressible than the blocks of the MoM matrix **Z** for the same $\tau_{ACA}$. Thus, during factorization the size of the **LU** matrix will typically increase after each row/column update, requiring extra attention to memory management. It is not uncommon for the final size of the **LU** matrix to be more than double that of the MoM matrix. Experience has also shown that this size increase is much greater for problems involving dielectrics, versus those with only conductors.

---

## 9.5   Solution of the ACA-Compressed Matrix System

Monostatic scattering problems often have thousands of incident angles, and solving for each RHS individually can be very time consuming. Let us instead group all the RHS and solution vectors into matrices $\mathbf{B}_{N \times N_{RHS}}$ and $\mathbf{X}_{N \times N_{RHS}}$ where $N$ is the number of unknowns and $N_{RHS}$ the number of incident angles. Given the $3 \times 3$ matrix in (9.14), the linear matrix system has the form

$$\begin{bmatrix} [\mathbf{LU}]_{11} & [\mathbf{U}_u\mathbf{U}_v]_{12} & [\mathbf{U}_u\mathbf{U}_v]_{13} \\ [\mathbf{L}_u\mathbf{L}_v]_{21} & [\mathbf{LU}]_{22} & [\mathbf{U}_u\mathbf{U}_v]_{23} \\ [\mathbf{L}_u\mathbf{L}_v]_{31} & [\mathbf{L}_u\mathbf{L}_v]_{32} & [\mathbf{LU}]_{33} \end{bmatrix} \begin{bmatrix} [\mathbf{X}_u\mathbf{X}_v]_1 \\ [\mathbf{X}_u\mathbf{X}_v]_2 \\ [\mathbf{X}_u\mathbf{X}_v]_3 \end{bmatrix} = \begin{bmatrix} [\mathbf{B}_u\mathbf{B}_v]_1 \\ [\mathbf{B}_u\mathbf{B}_v]_2 \\ [\mathbf{B}_u\mathbf{B}_v]_3 \end{bmatrix}. \quad (9.22)$$

Note that the block structure of the system matrix, resulting from the ACA clustering, also defines the block structures of **X** and **B**. It was shown in [8] that when there are many closely-spaced incident angles, **X** and **B** are also highly compressible. Thus, we can solve all right-hand sides simultaneously using the block back-substitution in Algorithm 8. As in the LU factorization, all matrix products are in rank-reduced form, and $\mathbf{B}_b$ and $\mathbf{X}_b$ each comprise a temporary workspace that is discarded after compression.

As with the **LU** matrix, the blocks of the solution matrix **X** are generally less compressible than those of the RHS matrix **B**. However, as the overall size of **X** is typically far less than the **LU** matrix, this does not cause issues in practice.

If there is only a single right-hand side, the RHS blocks in (9.22) will have a single column. Thus, the matrix system reduces to

$$\begin{bmatrix} [\mathbf{LU}]_{11} & [\mathbf{U}_u\mathbf{U}_v]_{12} & [\mathbf{U}_u\mathbf{U}_v]_{13} \\ [\mathbf{L}_u\mathbf{L}_v]_{21} & [\mathbf{LU}]_{22} & [\mathbf{U}_u\mathbf{U}_v]_{23} \\ [\mathbf{L}_u\mathbf{L}_v]_{31} & [\mathbf{L}_u\mathbf{L}_v]_{32} & [\mathbf{LU}]_{33} \end{bmatrix} \begin{bmatrix} \mathbf{x}_1 \\ \mathbf{x}_2 \\ \mathbf{x}_3 \end{bmatrix} = \begin{bmatrix} \mathbf{b}_1 \\ \mathbf{b}_2 \\ \mathbf{b}_3 \end{bmatrix}, \quad (9.23)$$

and the back-substitution is now performed as in Algorithm 9. Note that when

---

**Algorithm 8** ACA-Compressed Block-RHS Solution

1: Given $\mathbf{LU\,X} = \mathbf{B}$, first solve $\mathbf{UX} = \mathbf{L}^{-1}\mathbf{B}$:
2: **for** $b = 1$ to $M$ **do**
3:     $\mathbf{B}_b = \mathbf{L}_{bb}^{-1}\big[[\mathbf{B}_u\mathbf{B}_v]_b - \sum_{s=1}^{b-1}[\mathbf{L}_u\mathbf{L}_v]_{bs}[\mathbf{B}_u\mathbf{B}_v]_s\big]$
4:     Compress $\mathbf{B}_b$ via ACA and store updated $[\mathbf{B}_u\mathbf{B}_v]_b$
5: **end for**
6: Then solve $\mathbf{X} = \mathbf{U}^{-1}\big[\mathbf{L}^{-1}\mathbf{B}\big]$
7: **for** $b = M$ to 1 **do**
8:     $\mathbf{X}_b = \mathbf{U}_{bb}^{-1}\big[[\mathbf{B}_u\mathbf{B}_v]_b - \sum_{s=b+1}^{M}[\mathbf{U}_u\mathbf{U}_v]_{bs}[\mathbf{X}_u\mathbf{X}_v]_s\big]$
9:     Compress $\mathbf{X}_b$ via ACA and store $[\mathbf{X}_u\mathbf{X}_v]_b$
10: **end for**

---

**Algorithm 9** Single RHS Solution

1: Given $\mathbf{LU\,x} = \mathbf{b}$, first solve $\mathbf{Ux} = \mathbf{L}^{-1}\mathbf{b}$:
2: **for** $b = 1$ to $M$ **do**
3:     $\mathbf{b}_b = \mathbf{L}_{bb}^{-1}\big[\mathbf{b}_b - \sum_{s=1}^{b-1}[\mathbf{L}_u\mathbf{L}_v]_{bs}\,\mathbf{b}_s\big]$
4: **end for**
5: Then solve $\mathbf{x} = \mathbf{U}^{-1}\big[\mathbf{L}^{-1}\mathbf{b}\big]$
6: **for** $b = M$ to 1 **do**
7:     $\mathbf{x}_b = \mathbf{U}_{bb}^{-1}\big[\mathbf{b}_b - \sum_{s=b+1}^{M}[\mathbf{U}_u\mathbf{U}_v]_{bs}\,\mathbf{b}_s\big]$
8: **end for**

---

performing the matrix-vector products in steps 3 and 4, it is more efficient to form the product with the $\mathbf{V}$ matrix first. That is, for the product $\mathbf{A}_u\mathbf{A}_v\mathbf{b}$, first compute $\mathbf{c} = \mathbf{A}_v\mathbf{b}$ and then $\mathbf{A}_u\mathbf{c}$.

## 9.6   Software Implementation Notes

In this section we will discuss practical implementation of the ACA in software. In particular, we will describe several elements of how this was done in our *Serenity* MoM solver. Note that the material presented here comprises only programming suggestions, based on the author's own implementation.

### 9.6.1   Software Class Support

In this section we will discuss some of the software classes implemented in our *Serenity* solver in support of ACA operations. As *Serenity* is written in the C++ programming language, these classes make use of that language's object-oriented features, such as inheritance and polymorphism. In particular, the use of virtual functions has enabled functionality that would have been difficult or impossible to implement in other languages, while reducing the total codebase size.

#### 9.6.1.1   Element Engine Class

The ElementEngine class is responsible for computing matrix elements in the MoM and RHS sub-blocks. At a minimum, it must be capable of computing a full diagonal block, and selected rows and columns of any off-diagonal matrix block or RHS block. For testing and debugging purposes, our implementation supports these operations for all matrix blocks. This was critical in identifying and correcting the issue identified in Section 9.2.1.3.

During initialization, the element engine performs several key bookkeeping operations that facilitate these capabilities. For each basis function group, it determines the dielectric regions in which those basis functions have support, and maps each basis function to its corresponding triangle pair. A list of those triangles is then constructed for each region, where each entry references a triangle and edges having basis functions that exist *only in that region* and *only in that group*.

Filling full matrix blocks in their entirety is carried out in a straightforward manner, following very closely the treatment in Chapter 8 and in particular Section 8.7. A loop over source and testing triangle pairs is still performed, using only those triangles and corresponding edges that were added to the lists for the source and testing group.

Filling of individual rows and columns comprises a slightly different operation. When filling row $r$, basis function $r$ is a testing function and the two testing triangles are obtained from that function's triangle pair mapping. An innermost loop is then performed over all triangles in the source group. When filling column $c$, basis function $c$ is a source function and the source triangles are similarly obtained from that function's triangle mapping. An outermost loop is then performed over all triangles in the testing group.

Because the element engine maps a group's basis functions to triangles and triangles with edges having basis functions on a *per region* basis, this information is used to determine whether a matrix block is zero before any computation is performed. In this case, the block would be explicitly set to zero without needing to invoke the ACA driver.

A significant amount of code from our original full matrix *Serenity* solver was re-used in the element engine. Its routine for computing diagonal matrix blocks, for example, is almost identical to the original full matrix fill routine.

### 9.6.1.2 Matrix Classes

In prior chapters, little attention was paid to a software class structure for working with the MoM matrix system. This is because one is not really needed for smaller problems that don't have a block structure. In this case, the programmer can simply allocate a single, linear memory buffer for the matrix and address that array directly to perform all matrix operations. However, with a compressed block matrix system this is no longer possible, and a more elaborate matrix class system is needed.

*Abstract Matrix Class*

`AbstractMatrix` is an abstract class that defines the virtual functions (interface) used by the ACA driver when compressing a matrix block $Z_{m \times n}$. These functions are re-implemented by all classes that derive from `AbstractMatrix`. These functions are:

1. `getNumRows()`: Returns the number of rows $m$ in $Z$.

2. `getNumColumns()`: Returns the number of columns $n$ in $Z$.

3. `getRow(r)`: Returns row $r$ of $Z$.

4. `getColumn(c)`: Returns column $c$ of $Z$.

5. `getNumStored()`: Returns the total number of data elements stored.

This is the most critical class in our software implementation. The ACA driver requires only a reference to an `AbstractMatrix` object, which provides the selected rows and columns of $Z$ via the abstract interface.

*Full Matrix Class*

The `FullMatrix` class is used to store a regular, uncompressed matrix block of size $m \times n$, and `getNumStored()` returns $mn$. It returns rows and columns by simply copying them from its internal storage.

*ACA Matrix Class*

The `ACAMatrix` class stores the **U** and **V** components of a compressed matrix block, which are stacked together into a single array in memory. `getNumStored()` returns $(m + n)k$. The ACA driver routine, discussed in Section 9.2, constructs matrix blocks of this type. Note that this class does not necessarily need to implement `getRow` or `getColumn`, however doing so is very useful for debugging purposes.

*BlockMatrix Class*

The `BlockMatrix` class is the master, high-level class that stores and manages $M$ block sub-rows and $N$ block sub-columns. It provides a function `getBlock(sr, sc)` which returns a pointer to the matrix block on sub-row $sr$ and sub-column $sc$, which are of type `FullMatrix` or `ACAMatrix`. These pointers are simply stored in a flat array of size $M \times N$, stored in row-major format. For the `BlockMatrix` class, `getNumStored()` returns the sum of that function for all stored matrix blocks. It also provides a `getPercentCompressed()` function which compares the result of `getNumStored()` against $mn$.

*ACA Translator Class*

As mentioned previously, the ACA driver and underlying ACA routines are only provided as reference to an `AbstractMatrix` class, which is queried to obtain the matrix dimensions and selected rows and columns. However, the source of these rows and columns differs depending on which step of the solution is being performed. During LU factorization, the source comprises a `FullMatrix` class which simply copies the data from internal storage. However, when filling the system matrix or a right-hand side matrix, the rows and columns are computed on the fly by the element engine, which needs additional information regarding the source and testing groups, and the incident angles to be considered. The `ACAMatrixTranslator` and `ACARHSTranslator` classes interface directly with the element engine, providing all necessary extra information, while presenting the same `AbstractMatrix` interface to the ACA driver.

## 9.6.2   Shared Memory Processing

On shared memory systems, *Serenity* performs all operations in parallel using POSIX Threads. *Serenity* implements two thread classes: one for operation on the CPU, and one for operation on the GPU. These classes have many similarities, and are discussed in more detail in this section.

### 9.6.2.1 ACA CPU Thread Class

The ACACPUThread class is used on systems where no GPU is available, and performs all calculations on the CPU. For matrix operations, *Serenity* uses the CPU-optimized BLAS from the Intel MKL.

*Workspace Management*

As modern computer systems have multi-core CPUs supporting a large number of threads, the scratch space used by each thread must be managed carefully. Specifically, an attempt should be made to keep these spaces as small as possible, and avoid reallocations which lead to memory fragmentation over time. Our strategy is to assign to each thread resizable work arrays, which are initially empty. Prior to using an array, it is reallocated to a larger size if necessary, based on the space currently needed. As a result, resizing of a workspace occurs only a few times until it reaches a certain size, beyond which resizing is infrequent and memory fragmentation minimized.

*Matrix Filling*

When filling the system matrix, each thread computes matrix blocks asynchronously from other threads, choosing from blocks that are not yet filled. This step requires only a mutex-protected counter; no other coordination between threads is needed. For approximating off-diagonal blocks, the ACA driver is provided workspaces to store the rows and columns computed by the element engine, accumulate $\mathbf{U}$ and $\mathbf{V}$, and perform the QR/SVD recompression once the ACA is completed. The size of these arrays is increased on demand as needed.

*LU Factorization*

The LU factorization requires more coordination between threads as the factorization moves along the diagonal in Algorithm 7. A single thread performs the diagonal update and the scalar LU factorization in step 3. As this updated diagonal is required for the row and column updates, a barrier is used to restrict progress to step 4 until this thread is finished. Threads are now free to perform updates of the trailing row and column blocks (steps 4 to 9), doing so asynchronously by choosing from blocks that are not yet updated. Another barrier is used at the end of this loop so that all rows and columns are fully updated before moving to the next diagonal. During the update, each thread will need workspaces for storing $\mathbf{L}_{sb}$ and $\mathbf{U}_{bs}$, as well as the intermediate products $\mathbf{T}_1$, $\mathbf{T}_2$, and $\mathbf{T}_3$ from Section 9.4.1. As before, these arrays are resized as needed.

*Block RHS Solution*

Filling of the RHS blocks is also done in parallel, with each thread choosing from blocks that are not yet filled. The solution of each RHS in Algorithm 8 is performed by a single thread. For radar-cross section problems this comprises two threads, one for each of the vertical and horizontal polarizations. Each thread will need workspaces for storing $\mathbf{B}_b$ and $\mathbf{X}_b$, as well as the intermediate products $\mathbf{T}_1$, $\mathbf{T}_2$, and $\mathbf{T}_3$ previously mentioned in Section 9.4.1.

### 9.6.2.2  ACA GPU Thread Class

The `ACAGPUThread` class derives from the `ACACPUThread` class, and shares much of the same code, but performs a majority of the matrix filling, factorization and RHS solution on the GPU. Our implementation is designed to operate with NVIDIA GPUs using the NVIDIA CUDA programming language and the GPU-optimized cuBLAS library. Operating on the GPU requires additional programming considerations, as it has its own on-board memory space. Data to be processed must first be uploaded to the GPU (device) memory, any processing done, and the results downloaded back to the CPU (host) memory. Most GPUs are connected to the host system through an expansion interface, most often a PCI-E expansion slot. As these have a much lower bandwidth than the memory bus, the programmer must carefully manage data transfer to and from the GPU so that it remains busy and not stalled, waiting for data. Fortunately, NVIDIA GPUs support simultaneous upload and download of data, and overlap of different computations on the device via independent compute "streams". *Serenity* leverages this capability, with GPU threads operating independently from the others, and a minimal amount of coordination between steps. On systems with multiple GPUs, *Serenity* schedules an equal number of GPU threads on the host for each device. For systems supporting a total of $N_t$ host threads and $N_{GPU}$ GPUs, this is $N_t/N_{GPU}$ threads per device.

*GPU Workspace Management*

Each `ACAGPUThread` needs a workspace in device memory. This memory differs from host memory in that it is typically of limited size, so it must be managed carefully to avoid depletion. Our strategy is to allocate one large, contiguous block of device memory during initialization, and then assign portions of of this block to each thread as needed. As there is no reallocation, workspace assignments must remain fixed during certain operations. Additionally, as each thread operates asynchronously, there is no *a priori* knowledge regarding which portions of an operation will be completed by a thread. Thus, the workspaces assigned to each thread must be sized based on the maximum possible space that *may* be needed. As a consequence, available workspace on the device may not be sufficient to support all threads, and some may go idle. Our approach makes all possible effort to minimize idle threads.

*Matrix Filling*

NVIDIA GPUs have Streaming Multiprocessor (SM) units, each of which contains many CUDA processing cores having floating point and integer processors. These CUDA cores are optimized to perform simple, parallel mathematical operations and do not perform instruction prefetching or speculative execution like a general-purpose CPU. Therefore, they are most efficient when executing the same operations in parallel, and complex code with excessive branching and divergence should be avoided. Our matrix fill implementation computes all far matrix elements on the GPU, as that code comprises loops over source and testing triangles and has very few conditional logic statements. As the routines for computing near and self terms have much more branching logic, they are computed by the host thread on the CPU. When computing a row, column or diagonal block, both operations are overlapped, with the GPU and CPU operating simultaneously. When the CPU is finished, it synchronizes with the GPU and downloads the results, which are then combined with the host's results. Each thread requires a device workspace of size $N_{max} \times N_{max}$ for matrix filling, sufficient for storing of all possible rows and columns and diagonal blocks.

Our CUDA code is a simplified version of the original CPU far-matrix element code. Information about triangle and edge geometry, as well as triangle-to-basis and basis-to-triangle mapping, are assembled by the element engine and uploaded to the device at initialization. On the device, each CUDA thread assembles a single matrix element via edge pairs instead of face pairs. Though this results in some redundant calculation, there is no inter-thread communication which simplifies the GPU code. For a more in-depth explanation at the code-level, the reader is referred to [9] and [10], where similar approaches to matrix filling were taken. For more detailed information on CUDA programming, the reader should consult the NVIDIA CUDA Programmer's Guide [11].

*LU Factorization*

The LU factorization on the GPU follows the same steps as on the CPU, however for GPU processing, blocks must be sent to the GPU prior to any operations. Workspaces for these blocks, as well as the intermediate products (see Section 9.4.1) are all located in device memory and must be sized properly.

Diagonal blocks are full, and a single shared workspace of size $N_b \times N_{max}$ is sufficient for storing the diagonal on each device. Each diagonal is updated and LU factored on both the host and one device by a single thread, following the approach in [12]. On multi-GPU systems, the factored diagonal is then copied to all other devices and remains until the trailing row and column update is complete. We note also that during the update of the diagonal, and trailing rows and columns, threads will repeatedly access the **U** blocks directly above the diagonal and the **L** blocks to its left. Staging these blocks on each device prior to the update avoids having to upload them every time they are needed. Fortunately, the size of these blocks is known beforehand, so the shared storage needed on each device can be sized exactly. However, staging blocks in this way reduces the device workspace left for GPU threads, and the space needed increases as the factorization moves to the right. Thus, if device workspace starts being depleted, staging is disabled.

As the host threads operate asynchronously, it is not known ahead of time which trailing block each thread will update. Therefore, each thread is assigned a block of workspace sufficient for handling any blocks it may encounter. If workspace is not available for all threads, those without will remain idle. However, as the groups were sorted into ascending order by size, workspaces are smallest at the beginning, and grow larger as the factorization progresses down the diagonal. Thus, if any threads are idled, it will occur later in the factorization, rather than at the start.

For each host thread, workspace is needed to store the full-size $\mathbf{L}_{sb}$ and $\mathbf{U}_{bs}$ blocks. As groups are sorted by size, the maximum possible size of these blocks is $N_b \times N_{max}$, increasing to $N_{max} \times N_{max}$ on the last diagonal. Workspace is also needed for both of the product matrices to be multiplied (or one, if the other was staged), and the intermediate products $\mathbf{T}_1$, $\mathbf{T}_2$ and $\mathbf{T}_3$. These can be sized by searching all matrix blocks above and to the right of the diagonal, and all blocks to the left and below the diagonal, and finding the maximum compressed block size $C_{max}$ and the maximum rank $k_{max}$. Two spaces of size $C_{max}$ are reserved (or one, if staging), as well as a space of size $k_{max} \times k_{max}$ for $\mathbf{T}_1$ and a space of size $N_{max} \times k_{max}$ for $\mathbf{T}_2$ and $\mathbf{T}_3$.

All matrix operations in Algorithm 7 are carried out asynchronously on the GPU using CUDA streams and the cuBLAS library. The host threads are responsible for uploading any matrix blocks not already staged to the device and making the appropriate calls to cuBLAS. They then wait for the block update to complete before downloading the finished block to host memory. ACA compression of the updated blocks (steps 6 and 8) is then performed on the CPU.

*Block RHS Solution*

The RHS solution on the GPU follows the same steps as on the CPU, with one thread per RHS, and all matrix operations performed on the device. Workspace for each thread must be sufficient to handle products involving the largest **LU** matrix block as well as RHS matrix block. A workspace of size $N_{max} \times N_{max}$ is needed to store the diagonal, and workspace of size $N_{max} \times N_{RHS}$ for storing the full size $\mathbf{B}_b$ and $\mathbf{X}_b$. Let us define $C_{RHS}^{max}$ as the maximum compressed block size and $k_{RHS}^{max}$ the maximum rank of all RHS matrix blocks. Spaces of size $C_{max}$ and $C_{RHS}^{max}$ are needed for storing product matrices, a space of size $k_{max} \times k_{RHS}^{max}$ for $\mathbf{T}_1$ and a space of size $\max(N_{max}, N_{RHS}) \times \max(k_{max}, k_{RHS}^{max})$ for $\mathbf{T}_2$ and $\mathbf{T}_3$.

### 9.6.3 Distributed Memory Processing

Efficient operation on a distributed memory system requires knowledge about the CPU and network configuration of that system. The implementation discussed in this section was designed for a cluster-type system having multiple compute nodes, each with multi-core CPUs and a large memory pool, connected via a fast network fabric. Communication between compute nodes is done using the OpenMPI message passing library [13], a high-performance implementation of the Message Passing Interface (MPI) standard [14].

Compared to a shared memory system, a distributed memory platform has the benefit of a much larger potential memory pool as well as many more processors. For us to effectively utilize the memory space, we must break up the problem into smaller chunks and distribute them relatively equally across all nodes. However, this means that during processing, each node will require data stored on another node, which must be transmitted across the network. As the network is many times slower than direct memory access on each node, the programmer must analyze and modify core algorithms to minimize data transfer when possible. To effectively utilize the multiple cores on each node, these algorithms must also have the appropriate parallelization strategy, so that threads remain busy and not idle waiting for data. We will next consider modifications to the block LU factorization, block RHS solution, and shared memory implementation that allow for efficient operation on distributed memory systems.

#### 9.6.3.1 Parallelization Strategy

When using MPI, the application is launched simultaneously on all compute nodes, all of which have multiple processing cores. In this context, each process is analogous to a thread in the shared memory implementation. Thus, to make effective use of the cores on each node, there are two valid paths that can be taken in designing the application:

1. A single-threaded program launched multiple times on each node. Parallelization occurs only at the MPI level.

2. A multi-threaded program launched once on each node. Parallelization occurs at the MPI level *and* the process level. This is called a *hybrid* approach.

During development, we experimented with both approaches. and the hybrid approach was found to have the best overall performance. At the process level, all high-level ACA operations are performed by a single thread. Matrix filling is done by a separate pool of work threads, and BLAS functions done in parallel by the Intel MKL. This approach eliminates costly inter-process MPI communication and the same data being stored into memory multiple times on each node. We have also adapted Algorithms 7 and 8 to better suit distributed memory operation. These modifications are discussed in Sections 9.6.3.2 and 9.6.3.3.

### 9.6.3.2 Block LU Factorization Using MPI

We note that in Algorithm 7, updates of a $\mathbf{U}$ block to the right of the diagonal require access to all previous $\mathbf{U}$ blocks in that column. Similarly, updates of an $\mathbf{L}$ block below the diagonal require access to all previous $\mathbf{L}$ blocks in that row. Thus, our strategy is to assign each diagonal and all $\mathbf{U}$ blocks above that diagonal to individual nodes in a round-robin fashion. In a similar fashion, we assign all $\mathbf{L}$ blocks to the left of each diagonal to individual nodes in a reverse round-robin scheme. As a result, the only non-local blocks needed during a row/column update are the diagonal and those located above and to the left of the diagonal. During matrix filling, each node fills only local blocks, and the local block matrix structure is sparse. As an example, consider a $7 \times 7$ block matrix and 5 MPI processes. The block-to-process map in this case comprises the matrix $\mathbf{M}$ given by

$$
\mathbf{M} = \begin{bmatrix}
0 & 1 & 2 & 3 & 4 & 0 & 1 \\
3 & 1 & 2 & 3 & 4 & 0 & 1 \\
2 & 2 & 2 & 3 & 4 & 0 & 1 \\
1 & 1 & 1 & 3 & 4 & 0 & 1 \\
0 & 0 & 0 & 0 & 4 & 0 & 1 \\
4 & 4 & 4 & 4 & 4 & 0 & 1 \\
3 & 3 & 3 & 3 & 3 & 3 & 1 \\
2 & 2 & 2 & 2 & 2 & 2 & 2
\end{bmatrix}.
\tag{9.24}
$$

A modified version of the compressed block LU factorization is presented in Algorithm 10 for MPI processing. For each iteration of the outer loop, all off-diagonal $\mathbf{U}$ blocks directly above the diagonal and $\mathbf{L}$ blocks to its left are disseminated to all nodes via broadcasts. The diagonal $\mathbf{Z}_{bb}$ is then updated on the node where it is local, and the updated diagonal is then broadcast to all nodes. The remainder of the update operations follow Algorithm 7, where only local blocks are updated by each node.

### 9.6.3.3 Block-RHS Solution Using MPI

For the block-RHS solution, RHS blocks local to each node are reconstructed up front and then iteratively updated. To minimize communication

**Algorithm 10** ACA-Compressed Block LU Factorization Using MPI

1: **for** $b = 1$ to $M$ **do**
2:     **for** $k = 1$ to $b - 1$ **do**
3:         Broadcast compressed $\mathbf{U}_{kb}$
4:         Broadcast compressed $\mathbf{L}_{bk}$
5:     **end for**
6:     **if** $\mathbf{Z}_{bb}$ is local **then**
7:         $\mathbf{A} = \mathbf{Z}_{bb} - \sum_{s=1}^{b-1} [\mathbf{L}_u \mathbf{L}_v]_{bs} [\mathbf{U}_u \mathbf{U}_v]_{sb}$
8:         $[\mathbf{LU}]_{bb} = \text{LU}(\mathbf{A})$    (scalar LU factorization)
9:     **end if**
10:    Broadcast $[\mathbf{LU}]_{bb}$
11:    **for** $s = b + 1$ to $M$ **do**
12:        **if** $\mathbf{L}_{sb}$ is local **then**
13:            $\mathbf{L}_{sb} = \left[ [\mathbf{Z}_u \mathbf{Z}_v]_{sb} - \sum_{p=1}^{b-1} [\mathbf{L}_u \mathbf{L}_v]_{sp} [\mathbf{U}_u \mathbf{U}_v]_{pb} \right] \mathbf{U}_{bb}^{-1}$
14:            Compress $\mathbf{L}_{sb}$ via ACA and store $[\mathbf{L}_u \mathbf{L}_v]_{sb}$
15:        **end if**
16:        **if** $\mathbf{U}_{bs}$ is local **then**
17:            $\mathbf{U}_{bs} = \mathbf{L}_{bb}^{-1} \left[ [\mathbf{Z}_u \mathbf{Z}_v]_{bs} - \sum_{p=1}^{b-1} [\mathbf{L}_u \mathbf{L}_v]_{bp} [\mathbf{U}_u \mathbf{U}_v]_{ps} \right]$
18:            Compress $\mathbf{U}_{bs}$ via ACA and store $[\mathbf{U}_u \mathbf{U}_v]_{bs}$
19:        **end if**
20:    **end for**
21: **end for**

and simplify programming, a new mapping is generated to rearrange the **LU** matrix post-factorization. These assign each row to a single node, in a round-robin fashion. Following the example from Section 9.6.3.2, these new mappings would comprise matrices $\mathbf{M}_{LU}$ and $\mathbf{M}_{RHS}$ given by

$$\mathbf{M}_{LU} = \begin{bmatrix} 0 & 0 & 0 & 0 & 0 & 0 & 0 \\ 1 & 1 & 1 & 1 & 1 & 1 & 1 \\ 2 & 2 & 2 & 2 & 2 & 2 & 2 \\ 3 & 3 & 3 & 3 & 3 & 3 & 3 \\ 4 & 4 & 4 & 4 & 4 & 4 & 4 \\ 0 & 0 & 0 & 0 & 0 & 0 & 0 \\ 1 & 1 & 1 & 1 & 1 & 1 & 1 \end{bmatrix}, \quad \mathbf{M}_{RHS} = \begin{bmatrix} 0 \\ 1 \\ 2 \\ 3 \\ 4 \\ 0 \\ 1 \end{bmatrix}. \quad (9.25)$$

Application of $\mathbf{M}_{LU}$ to the completed **LU** matrix comprises a shuffling step where nodes exchange blocks as necessary. An updated version of the ACA-compressed block-RHS solution is now presented in Algorithm 11 for MPI. We note that the only communications required here are the broadcasts in steps 8 and 23, comprising the most recently completed RHS block.

---

**Algorithm 11** ACA-Compressed Block-RHS Solution Using MPI

---

1:  Given $\mathbf{LU\,X} = \mathbf{B}$, first solve $\mathbf{UX} = \mathbf{L}^{-1}\mathbf{B}$:
2:  **for** $b = 1$ to $M$ **do**
3:      **if** $\mathbf{B}_b$ is local **then**
4:          $\mathbf{B}_b = \mathbf{L}_{bb}^{-1}\mathbf{B}_b$
5:          Compress $\mathbf{B}_b$ via ACA and store updated $[\mathbf{B}_u\mathbf{B}_v]_b$
6:      **end if**
7:      **if** $b < M$ **then**
8:          Broadcast compressed $\mathbf{B}_b$
9:          **for** $s = b + 1$ to $M$ **do**
10:             **if** $\mathbf{B}_s$ is local **then**
11:                 $\mathbf{B}_s = \mathbf{B}_s - [\mathbf{L}_u\mathbf{L}_v]_{sb}[\mathbf{B}_u\mathbf{B}_v]_b$
12:             **end if**
13:         **end for**
14:     **end if**
15: **end for**
16: Next solve $\mathbf{X} = \mathbf{U}^{-1}\left[\mathbf{L}^{-1}\mathbf{B}\right]$
17: **for** $b = M$ to $1$ **do**
18:     **if** $\mathbf{B}_b$ is local **then**
19:         $\mathbf{X}_b = \mathbf{U}_{bb}^{-1}\mathbf{B}_b$
20:         Compress $\mathbf{X}_b$ via ACA and store updated $[\mathbf{X}_u\mathbf{X}_v]_b$
21:     **end if**
22:     **if** $b > 1$ **then**
23:         Broadcast compressed $\mathbf{X}_b$
24:         **for** $s = b - 1$ to $1$ **do**
25:             **if** $\mathbf{B}_s$ is local **then**
26:                 $\mathbf{B}_s = \mathbf{B}_s - [\mathbf{U}_u\mathbf{U}_v]_{sb}[\mathbf{X}_u\mathbf{X}_v]_b$
27:             **end if**
28:         **end for**
29:     **end if**
30: **end for**

---

## 9.7 Numerical Examples

In this section we compute the radar cross section of several test articles using the *Serenity* solver code, which implements the ACA algorithm outlined in this chapter. We will consider conducting, dielectric, and coated spheres, several benchmark targets with measured data, and the monoconic RV with dielectric nose. The electrical size will be much larger than those in prior chapters. Triangle meshes are constructed with at least $\lambda/10$ unknowns per wavelength in each dielectric region. The CFIE is applied on closed conductors with $\alpha = 0.5$, and PMCHWT applied to dielectric interfaces. For the ACA, the target group size is $N_g = 2500$ basis functions, and $\tau_{ACA} = 10^{-5}$.

### 9.7.1 Compute Platform

Computations in this section were carried out on the Dell workstation described previously in Section 8.8.2. This workstation is equipped with dual NVIDIA GTX 1080 Ti GPUs, each having 11 GB of device memory. For matrix filling, near and self matrix and self elements are computed on the CPU, and far matrix elements on the GPUs. Block updates in the LU factorization, as well as the RHS solution, are all performed on the GPUs.

### 9.7.2 Adaptive ACA Tolerance

When compressing off-diagonal blocks, the most straightforward approach is to use a single $\tau_{ACA}$ for all blocks. However, we know that the amplitudes of interactions between groups decreases with separation, and so it is reasonable to assume that the importance (and the accuracy requirement) of the resulting blocks also decreases with distance.

Consider group $A$ with bounding sphere of diameter $d_A$ centered at $\mathbf{r}_A$, and group $B$ with bounding sphere of diameter $d_B$ centered at $\mathbf{r}_B$. We define the separation $D$ between these groups as

$$D = |\mathbf{r}_A - \mathbf{r}_B| - d_A - d_B \,. \tag{9.26}$$

When compressing the off-diagonal blocks resulting from the group pair $A$ and $B$, *Serenity* will compute a new tolerance $\tau'_{ACA}$ based on the following heuristics:

$$\tau'_{ACA} = \tau_{ACA} \qquad\qquad D < 1\lambda \,, \tag{9.27}$$

$$\tau'_{ACA} = 2\,\tau_{ACA} \qquad\quad 2\lambda > D \geq 1\lambda \,, \tag{9.28}$$

$$\tau'_{ACA} = 5\,\tau_{ACA} \qquad\quad 5\lambda > D \geq 2\lambda \,, \tag{9.29}$$

$$\tau'_{ACA} = 10\,\tau_{ACA} \qquad 10\lambda > D \geq 5\lambda \,, \tag{9.30}$$

and

$$\tau'_{ACA} = 20\,\tau_{ACA} \qquad\qquad D \geq 10\lambda . \qquad\qquad (9.31)$$

Based on numerical experimentation, these choices reduce memory overhead significantly without noticeably affecting the solution accuracy.

### 9.7.3   Spheres

In this section we will consider the bistatic RCS of conducting, fully dielectric, and coated spheres. These spheres have the same dimensions as the spheres in Section 8.8.3, but a much larger electrical size.

#### 9.7.3.1   Conducting Sphere

We first consider a conducting sphere with a radius of 0.5 meters at frequencies of 1.5, 3.0, and 6.0 GHz. For the MoM, we use models comprising 20480, 81920 and 327680 triangles, resulting in 30720, 122880 and 491520 unknowns, respectively. The bistatic RCS computed by the ACA is compared to the Mie series in Figures 9.4a–9.4f. The agreement is excellent.

#### 9.7.3.2   Dielectric Sphere

We next consider a dielectric sphere with a radius of 0.5 meters at frequencies of 1.5, and 3.0 GHz. For the MoM simulation, we re-use the models from Section 9.7.3.1, however there are now an equal number of magnetic basis functions. This results in 61440 at 1.5 GHz and 245760 unknowns at 3.0 GHz. In Figures 9.5a–9.5d is compared the bistatic RCS computed by the ACA to the Mie series for a lossless sphere ($\epsilon_r = 2.56$) at 1.5 and 3.0 GHz. The agreement is excellent. A similar comparison is made for a lossy sphere ($\epsilon_r = 2.56 - j.102$) at 3.0 GHz in Figures 9.6a–9.6b, where the agreement is again very good.

#### 9.7.3.3   Coated Sphere

We next consider the coated sphere from Section 8.8.3.3 at 1.5 and 3.0 GHz. At 1.5 GHz, the surface meshes for the conducting core and the coating comprise 5120 and 20480 triangles, resulting in 38400 electric and 30720 magnetic basis functions, respectively, for a total of 69120 unknowns. At 3.0 GHz, they comprise 20480 and 81920 triangles, resulting in 153600 electric and 122880 magnetic basis functions, respectively, for a total of 276480 unknowns. In Figures 9.7a–9.7d is compared the bistatic RCS computed by the ACA to the Mie series for a lossless coating ($\epsilon_r = 2.56$) at 1.5 and 3.0 GHz. The agreement is excellent. A similar comparison is shown in Figures 9.8a–9.8b for a lossy coating ($\epsilon_r = 2.56 - j.102$) at 3.0 GHz, where the agreement is again very good.

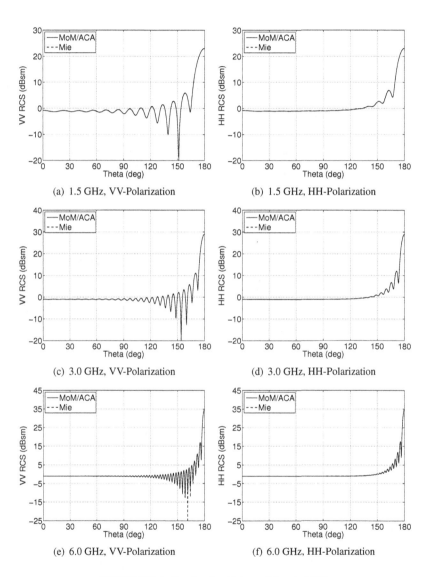

(a) 1.5 GHz, VV-Polarization

(b) 1.5 GHz, HH-Polarization

(c) 3.0 GHz, VV-Polarization

(d) 3.0 GHz, HH-Polarization

(e) 6.0 GHz, VV-Polarization

(f) 6.0 GHz, HH-Polarization

**FIGURE 9.4:** Conducting Sphere: Bistatic RCS

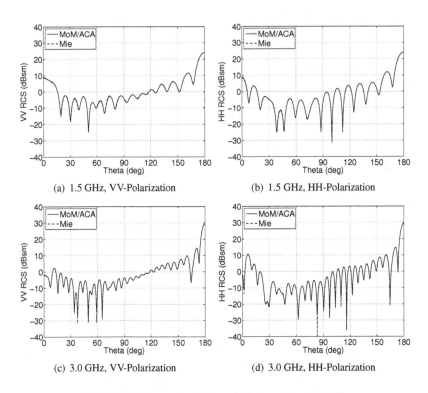

(a) 1.5 GHz, VV-Polarization      (b) 1.5 GHz, HH-Polarization

(c) 3.0 GHz, VV-Polarization      (d) 3.0 GHz, HH-Polarization

**FIGURE 9.5:** Lossless Dielectric Sphere: Bistatic RCS

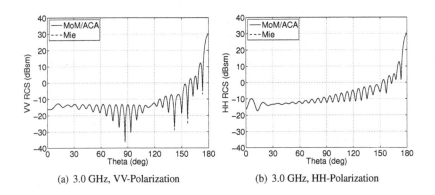

(a) 3.0 GHz, VV-Polarization      (b) 3.0 GHz, HH-Polarization

**FIGURE 9.6:** Lossy Dielectric Sphere: Bistatic RCS

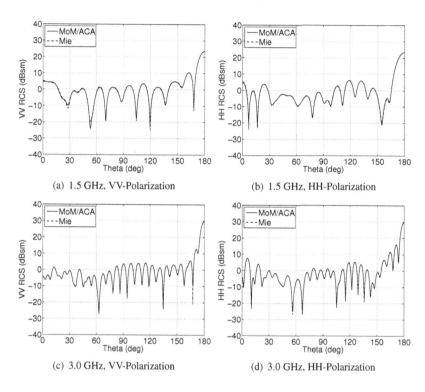

**FIGURE 9.7:** Lossless Coated Sphere: Bistatic RCS

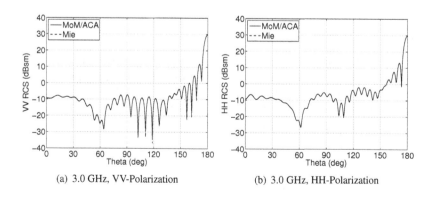

**FIGURE 9.8:** Lossy Coated Sphere: Bistatic RCS

## 9.7.4 EMCC Benchmark Targets

In this section, we again consider the EMCC benchmark radar targets previously discussed in Section 7.9.2, as well as the NASA almond from [15] and the cube and prism from [16].

### 9.7.4.1 EMCC Ogive

The EMCC ogive was previously described in Section 7.9.2.1, and the facet model used in this example comprises 21760 triangles, resulting in 32640 unknowns. The results from the ACA are compared to the EMCC measurements at 9.0 GHz in Figures 9.9a and 9.9b for vertical and horizontal polarizations, respectively. The agreement is very good. Surface current maps for $\theta^i = \pi/4$ are shown in Figures 9.16a and 9.16b for vertical and horizontal polarizations, respectively.

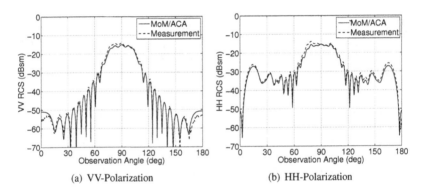

(a) VV-Polarization  (b) HH-Polarization

**FIGURE 9.9:** EMCC Ogive: MoM/ACA vs. Measurement at 9.0 GHz

### 9.7.4.2 EMCC Double Ogive

The EMCC double ogive was previously described in Section 7.9.2.2, and the facet model used in this example comprises 36240 triangles, resulting in 54360 unknowns. The results from the ACA are compared to the EMCC measurements at 9.0 GHz in Figures 9.10a and 9.10b for vertical and horizontal polarizations, respectively. The agreement is again very good. Surface current maps for $\theta^i = \pi/4$ are shown in Figures 9.16c and 9.16d for vertical and horizontal polarizations, respectively.

### 9.7.4.3 EMCC Cone-Sphere

The EMCC cone-sphere was previously described in Section 7.9.2.3, and the facet model used in this example comprises 79496 triangles, resulting in

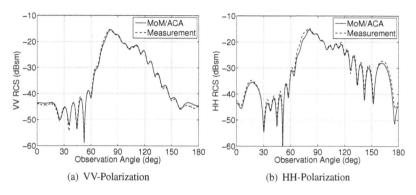

(a) VV-Polarization        (b) HH-Polarization

**FIGURE 9.10:** EMCC Double Ogive: MoM/ACA vs. Measurement at 9.0 GHz

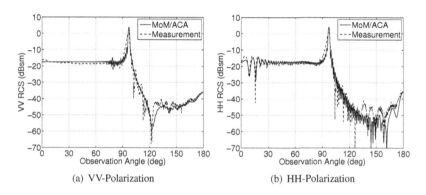

(a) VV-Polarization        (b) HH-Polarization

**FIGURE 9.11:** EMCC Cone-Sphere: MoM/ACA vs. Measurement at 9.0 GHz

119244 unknowns. The results from the ACA are compared to the EMCC measurements at 9.0 GHz in Figures 9.11a and 9.11b for vertical and horizontal polarizations, respectively. The agreement is fairly good at all angles, although there is some disagreement at angles above 120 degrees. Surface current maps for $\theta^i = \pi/4$ are shown in Figures 9.16e and 9.16f for vertical and horizontal polarizations, respectively.

### 9.7.4.4 EMCC Cone-Sphere with Gap

The EMCC cone-sphere with gap was previously described in Section 7.9.2.4, and the facet model used in this example comprises 82528 triangles, resulting in 123792 unknowns. The results from the ACA are compared to the EMCC measurements at 9.0 GHz in Figures 9.12a and 9.12b for vertical and horizontal polarizations, respectively. The agreement is fairly good at all an-

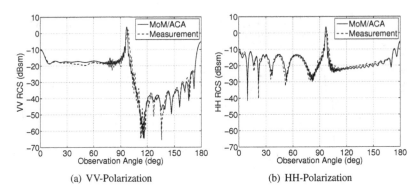

**(a) VV-Polarization**          **(b) HH-Polarization**

**FIGURE 9.12:** EMCC Cone-Sphere with Gap: MoM/ACA vs. Measurement at 9.0 GHz

gles. Surface current maps on for $\theta^i = \pi/4$ are shown in Figures 9.16g and 9.16h for vertical and horizontal polarizations, respectively.

### 9.7.4.5 NASA Almond

The 9.936-inch NASA almond of Figure 9.13 is a target comprising a mostly smooth surface that supports specular and creeping wave scattering, and a pointed end that gives rise to tip diffraction effects. It can be expressed for $0 \leq \phi \leq 2\pi$ as

$$x = dt \,, \tag{9.32}$$

$$y(t) = 0.193333d\sqrt{1 - \left(\frac{t}{0.416667}\right)^2}\cos\phi \,, \tag{9.33}$$

$$z(t) = 0.64444d\sqrt{1 - \left(\frac{t}{0.416667}\right)^2}\sin\phi \,, \tag{9.34}$$

for $-0.41667 < t < 0$ inches, and

$$x = dt \,, \tag{9.35}$$

$$y(t) = 4.83345d\left[\sqrt{1 - \left(\frac{t}{2.08335}\right)^2} - 0.96\right]\cos\phi \,, \tag{9.36}$$

$$z(t) = 1.61115d\left[\sqrt{1 - \left(\frac{t}{2.08335}\right)^2} - 0.96\right]\sin\phi \,, \tag{9.37}$$

for $0 < t < 0.58333$, where $d = 9.936$ inches. The facet model for this example comprises 46984 triangles, resulting in 70476 unknowns. Details of the

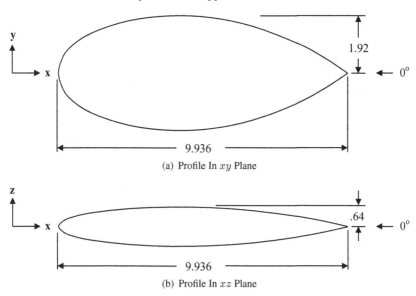

(a) Profile In $xy$ Plane

(b) Profile In $xz$ Plane

**FIGURE 9.13:** NASA Almond

surface mesh for the pointed and rounded ends are shown in Figures 9.14a and 9.14b, respectively. Monostatic observations are made in the $xy$ plane of Figure 9.13a. The results from the ACA are compared to the EMCC measurements at 7 GHz in Figures 9.15a and 9.15b for vertical and horizontal polarizations, respectively. The computed results were lower than the measurements by approximately 1.5 dB across a large range of angles, and were adjusted by that amount in each figure. After this adjustment, the comparison is very good. Results at 9.92 GHz are shown in Figures 9.15c and 9.15d for vertical and horizontal polarizations, respectively. A similar bias was also observed in this case, and after adjustment the comparison is again fairly good. However, the difference is much greater in this case, particularly in VV polarization. The reason for the difference is not known, though a similar difference was also noted in [15]. The authors of that paper suggested the cause may have been the density of the unknowns $(6/\lambda)$ in the CAD model that was used. As we use more than double this density in our model (about $13/\lambda$), we suspect the cause to be something else.

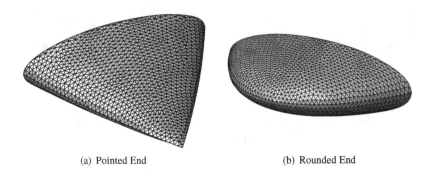

(a) Pointed End                              (b) Rounded End

**FIGURE 9.14:** Almond Facet Model Detail

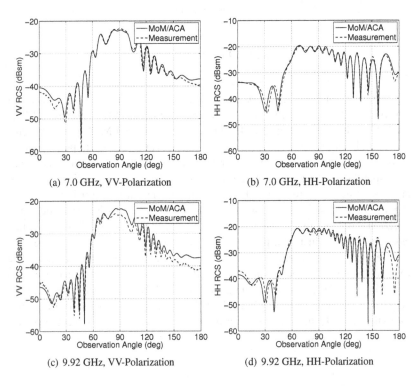

(a) 7.0 GHz, VV-Polarization                 (b) 7.0 GHz, HH-Polarization

(c) 9.92 GHz, VV-Polarization                (d) 9.92 GHz, HH-Polarization

**FIGURE 9.15:** NASA Almond: MoM/ACA vs. Measurement

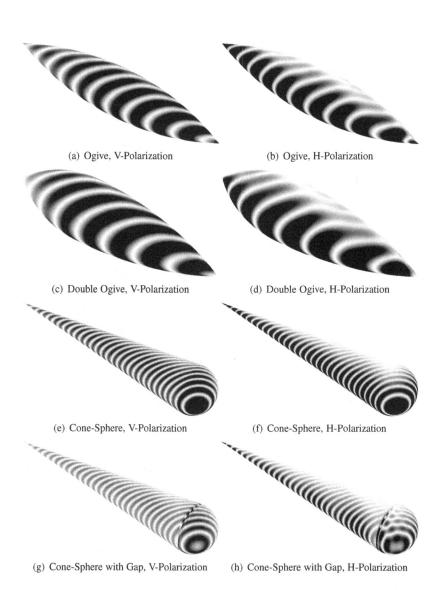

(a) Ogive, V-Polarization

(b) Ogive, H-Polarization

(c) Double Ogive, V-Polarization

(d) Double Ogive, H-Polarization

(e) Cone-Sphere, V-Polarization

(f) Cone-Sphere, H-Polarization

(g) Cone-Sphere with Gap, V-Polarization

(h) Cone-Sphere with Gap, H-Polarization

**FIGURE 9.16:** Surface Currents on EMCC Test Targets at 9.0 GHz

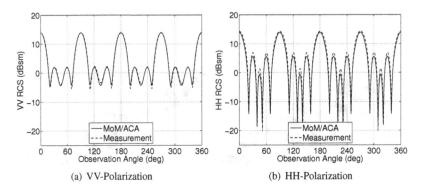

(a) VV-Polarization                                    (b) HH-Polarization

**FIGURE 9.17:** EMCC Cube: MoM/ACA vs. Measurement at 0.43 GHz

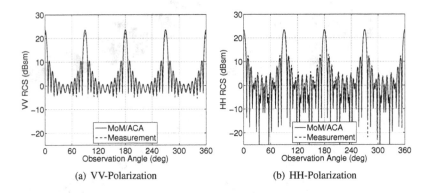

(a) VV-Polarization                                    (b) HH-Polarization

**FIGURE 9.18:** EMCC Cube: MoM/ACA vs. Measurement at 1.3 GHz

### 9.7.4.6   EMCC Cube

The EMCC cube is a high-precision, conducting cube with sides of length 1 meter and axis-aligned faces [16]. Monostatic observations are made in the $xy$ plane at 0.43 and 1.3 GHz, for azimuth angles from 0 to 360 degrees. For the cube, facet models with 4542 and 41818 triangles were constructed, yielding a total of 6813 and 62727 unknowns, respectively. The results from the ACA are compared to the EMCC measurements at 0.43 GHz in Figures 9.17a and 9.17b, and at 1.3 GHz in Figures 9.18a and and 9.18b, respectively. The comparison is very good at all angles and frequencies.

### 9.7.4.7   EMCC Prism

The EMCC prism [16] has a base with dimensions shown in Figure 9.19, and an overall thickness of 30.48 cm. Monostatic observations are made at an

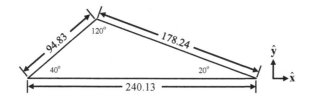

**FIGURE 9.19:** EMCC Prism (Dimensions in Centimeters)

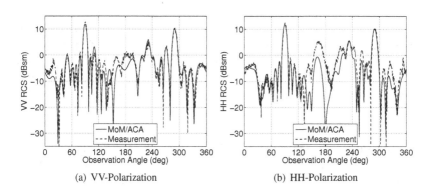

(a) VV-Polarization     (b) HH-Polarization

**FIGURE 9.20:** EMCC Prism: MoM/ACA vs. Measurement at 0.43 GHz

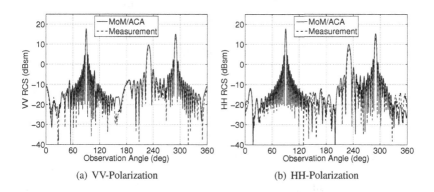

(a) VV-Polarization     (b) HH-Polarization

**FIGURE 9.21:** EMCC Prism: MoM/ACA vs. Measurement at 1.3 GHz

elevation of 10 degrees, at 0.43 and 1.3 GHz, for azimuth angles from 0 to 360 degrees. The purpose of this object and measurement setup is to capture traveling wave effects for three different wedge angles [16].

For the prism, facet models with 3610 and 31830 triangles were constructed, yielding a total of 5415 and 47745 unknowns, respectively. The results from the ACA are compared to the EMCC measurements at 0.43 GHz in Figures 9.20a and 9.20b, and at 1.3 GHz in Figures 9.21a and and 9.21b, respectively. The comparison is fairly good at and near specular angles, but there is a fair amount of disagreement elsewhere, particularly near 180 degrees azimuth. Comparisons in [16] against results obtained from the Fast Illinois Solver Code (FISC) [17] showed similar disagreements.

### 9.7.5    Dielectric Cube and Ogive

Next, we will consider a small cube and ogive, two fully dielectric benchmark targets that were fabricated and measured in [18]. These targets were milled from a solid block of polyethylene, which was measured to have an average permittivity $\epsilon \approx 2.35$. As the measured data was taken from a scanned version of the document, only the HH-polarized RCS was visible enough to allow extraction.

#### 9.7.5.1    Small Polyethylene Cube

The small polyethylene cube has a side length of approximately 1.181 inches, with an error of $\pm 0.002$ inches [18]. We consider the RCS at 6.0, 8.0, 10.0, 12.0, 13.0, 14.0, 16.0 and 18.0 GHz. For the cube, a facet model with 10800 facets was constructed, comprising 10 edges per wavelength at 18.0 GHz, and a total of 32400 unknowns. This model was used at all frequencies.

The computed and measured results are compared for horizontal polarization in Figures 9.22a–9.22h. The comparison is fairly good at all frequencies, though the simulated results consistently underpredict the RCS. This was also observed by the author in [18] using results from a full-matrix MoM solver.

#### 9.7.5.2    Polyethylene Ogive

The polyethylene ogive has the same exterior dimensions as the EMCC ogive described previously in Section 7.9.2.1, with an error of $\pm 0.005$ inches. We consider the RCS at 2.0, 4.0, 6.0, 8.0, 10.0, 12.0, 14.0, 16.0 and 18.0 GHz. For convenience, a single facet model was constructed having 10 edges per wavelength at 18.0 GHz, and was used at all frequencies. This model had 77624 facets, resulting in 116436 edges and 232872 unknowns. The computed and measured results are compared for horizontal polarization in Figures 9.23a–9.23h. The comparison is quite good at all frequencies, and across all incident angles except for those near tip-on incidence where the RCS is much lower than at broadside. The memory requirements versus frequency are summarized in Table 9.1. Interestingly, the amount of compression is observed to decrease versus frequency.

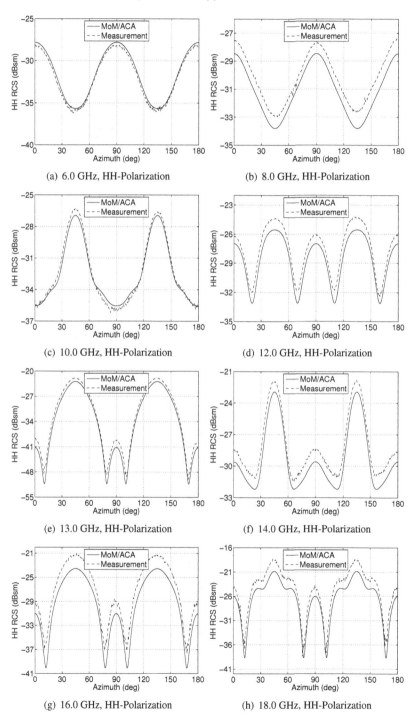

(a) 6.0 GHz, HH-Polarization

(b) 8.0 GHz, HH-Polarization

(c) 10.0 GHz, HH-Polarization

(d) 12.0 GHz, HH-Polarization

(e) 13.0 GHz, HH-Polarization

(f) 14.0 GHz, HH-Polarization

(g) 16.0 GHz, HH-Polarization

(h) 18.0 GHz, HH-Polarization

**FIGURE 9.22:** Small Polyethylene Cube: RCS

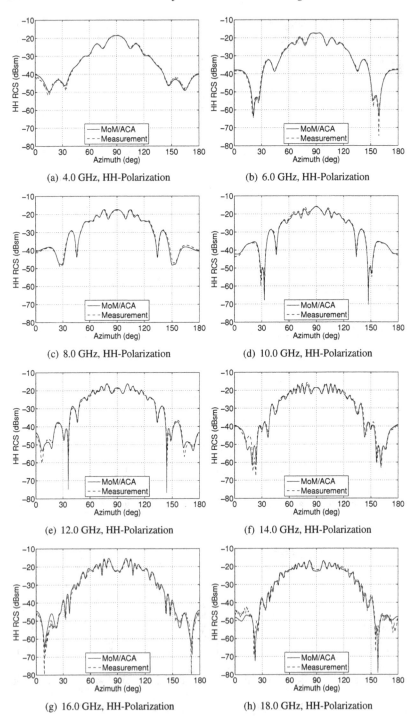

(a) 4.0 GHz, HH-Polarization

(b) 6.0 GHz, HH-Polarization

(c) 8.0 GHz, HH-Polarization

(d) 10.0 GHz, HH-Polarization

(e) 12.0 GHz, HH-Polarization

(f) 14.0 GHz, HH-Polarization

(g) 16.0 GHz, HH-Polarization

(h) 18.0 GHz, HH-Polarization

**FIGURE 9.23:** Polyethylene Ogive: RCS

## 9.7.6 UT Austin Benchmark Targets

We will next consider several conducting, dielectric and composite targets from the Austin Benchmark Suites for Computational Electromagnetics, a publicly available benchmark suite for RCS simulations [19]. The targets fabricated for this suite were 3D-printed in resin, measured to have an average permittivity of $\epsilon = 3.0 - j0.1$. The printed parts were then smoothed by hand. Bulk dielectric parts comprise the bare resin, and a thin layer of highly conductive silver paint was applied to create the conducting surfaces [20, 21].

### 9.7.6.1 PEC Almond

We first consider the PEC almond, which was fabricated to have the same external dimensions as in Figure 9.13. We consider the monostatic RCS at 3.5, 5.125, 7.0 and 10.25 GHz, and re-use the facet model from Section 9.7.4.5. The computed and measured results are compared for vertical and horizontal polarizations in Figures 9.25a–9.25h, where the comparison is fairly good.

### 9.7.6.2 Solid Resin Almond

We next consider a solid resin almond of the same dimensions, again using the same facet model. We consider the monostatic RCS RCS at 2.58, 5.125, 7.0 and 10.25 GHz. The computed and measured results are compared for vertical and horizontal polarizations in Figures 9.25a–9.25h, where the comparison is again fairly good.

### 9.7.6.3 Closed-Tail Almond

We next consider the closed-tail almond of the same exterior dimensions. In this case, a solid resin almond was fabricated and then split as shown in Figure 9.24. The 4.14-inch ellipsoidal portion coated with conducting paint, and the two halves joined together with a thin layer of epoxy. For the MoM simulation, new facet models were constructed, having 31892, 28246, and 3876 facets, comprising the exterior of the solid resin tip, conducting rear, and the septum between the halves, respectively. The computed and measured results are compared for vertical and horizontal polarizations in Figures 9.27a–9.27h, where the comparison is again fairly good.

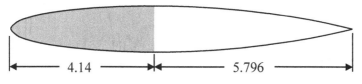

**FIGURE 9.24:** Closed-Tail Almond

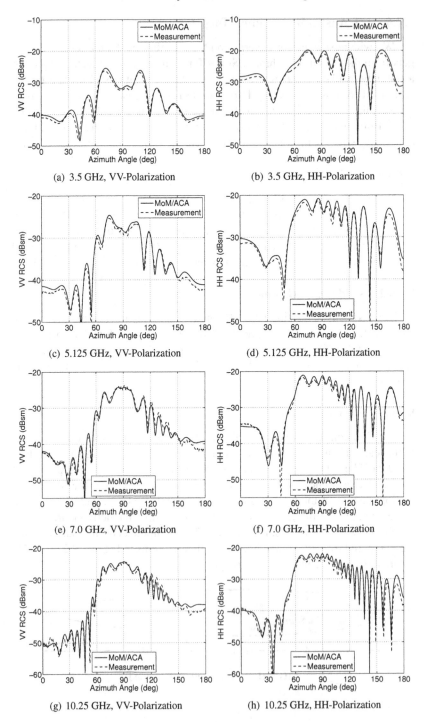

(a) 3.5 GHz, VV-Polarization

(b) 3.5 GHz, HH-Polarization

(c) 5.125 GHz, VV-Polarization

(d) 5.125 GHz, HH-Polarization

(e) 7.0 GHz, VV-Polarization

(f) 7.0 GHz, HH-Polarization

(g) 10.25 GHz, VV-Polarization

(h) 10.25 GHz, HH-Polarization

**FIGURE 9.25:** Austin PEC Almond: RCS

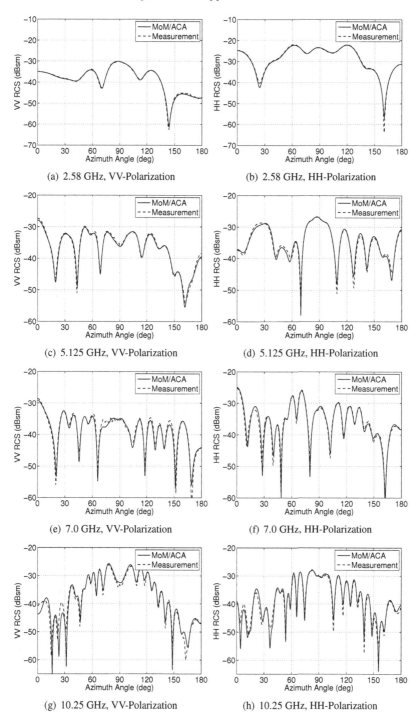

(a) 2.58 GHz, VV-Polarization

(b) 2.58 GHz, HH-Polarization

(c) 5.125 GHz, VV-Polarization

(d) 5.125 GHz, HH-Polarization

(e) 7.0 GHz, VV-Polarization

(f) 7.0 GHz, HH-Polarization

(g) 10.25 GHz, VV-Polarization

(h) 10.25 GHz, HH-Polarization

**FIGURE 9.26:** Austin Solid Resin Almond: RCS

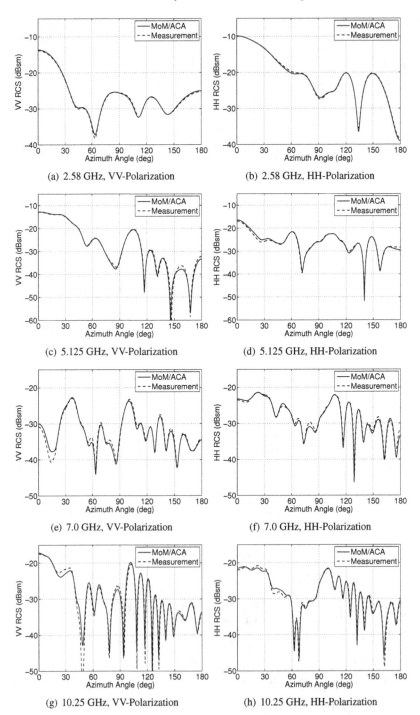

(a) 2.58 GHz, VV-Polarization

(b) 2.58 GHz, HH-Polarization

(c) 5.125 GHz, VV-Polarization

(d) 5.125 GHz, HH-Polarization

(e) 7.0 GHz, VV-Polarization

(f) 7.0 GHz, HH-Polarization

(g) 10.25 GHz, VV-Polarization

(h) 10.25 GHz, HH-Polarization

**FIGURE 9.27:** Austin Closed-Tail Almond: RCS

### 9.7.6.4   Open-Tail Almond

We next consider the open-tail almond, similar to the closed-tail almond in Figure 9.24. In this case, a solid resin almond was fabricated, the 5.796-inch tip portion covered in masking tape, and the 4.14-inch ellipsoidal portion sprayed with conducting paint [21]. This yields an open rear cavity with a thin, conducting wall. For the MoM simulation, the facet models comprising the exterior surfaces of the closed-tail almond were reused. On the tip portion, comprising the air-dielectric interface, PMCHWT is applied as usual. The rear portion, however, comprises a thin conducting interface that borders two conducting regions. Thus, separate electric basis functions are assigned to each side of the interface and only the EFIE is applied. The computed and measured results are compared for vertical and horizontal polarizations in Figures 9.27a–9.27h, where the comparison is again fairly good.

### 9.7.6.5   EXPEDITE-RCS Aircraft

We next consider the EXPEDITE-RCS aircraft model described in [22], and depicted in Figures 9.28a and 9.28b from front and rear perspectives, respectively. This geometry was developed to create a complex, realistic aircraft model that could be used and disseminated in a collaborative manner. The 9.1875-inch long test article was fabricated and measured in the same manner as the almond in [21]. This particular model is watertight, with engine intake and exhaust cavities completely closed off.

A number of meshes for the aircraft are available in the UT Austin Benchmark suite. From these, a model having 155432 triangles and 233148 edges was chosen for the MoM simulation. This results in 233148 and 466296 unknowns for the PEC and dielectric cases, respectively. The computed and measured results for the PEC model are compared at 2.58, 5.125, 7.0 and 10.25 GHz in Figures 9.30a–9.30h, in vertical and horizontal polarizations, where the comparison is fairly good. A similar comparison is made for the solid resin model in Figures 9.31a–9.31h, where the comparison is again fairly good.

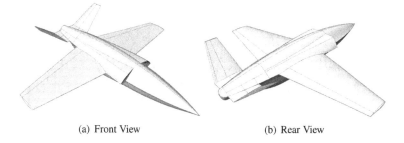

(a) Front View          (b) Rear View

**FIGURE 9.28:** Austin EXPEDITE-RCS Aircraft Model

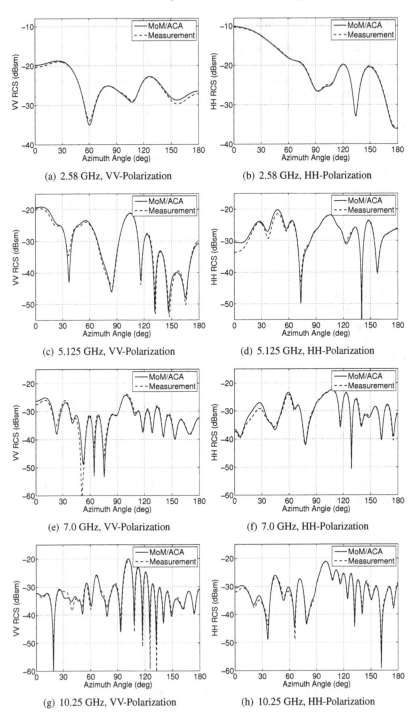

(a) 2.58 GHz, VV-Polarization

(b) 2.58 GHz, HH-Polarization

(c) 5.125 GHz, VV-Polarization

(d) 5.125 GHz, HH-Polarization

(e) 7.0 GHz, VV-Polarization

(f) 7.0 GHz, HH-Polarization

(g) 10.25 GHz, VV-Polarization

(h) 10.25 GHz, HH-Polarization

**FIGURE 9.29:** Austin Open-Tail Almond: RCS

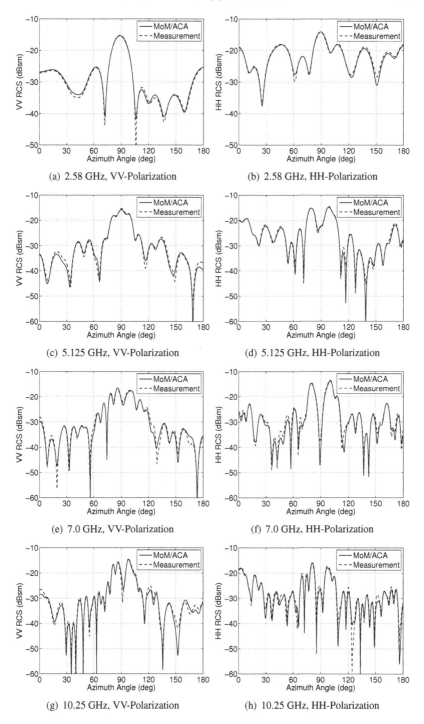

(a) 2.58 GHz, VV-Polarization

(b) 2.58 GHz, HH-Polarization

(c) 5.125 GHz, VV-Polarization

(d) 5.125 GHz, HH-Polarization

(e) 7.0 GHz, VV-Polarization

(f) 7.0 GHz, HH-Polarization

(g) 10.25 GHz, VV-Polarization

(h) 10.25 GHz, HH-Polarization

**FIGURE 9.30:** PEC EXPEDITE-RCS Aircraft: RCS

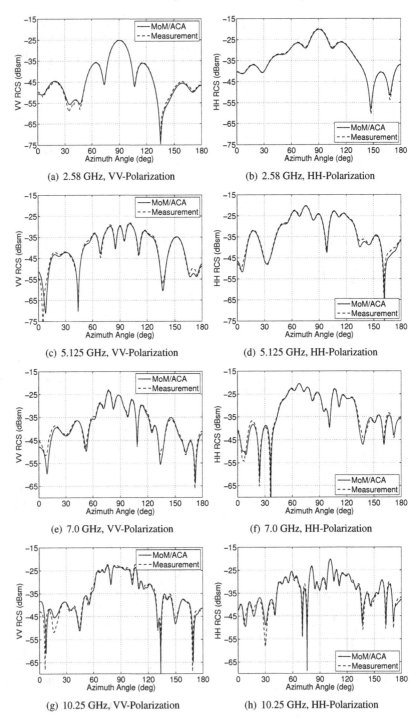

(a) 2.58 GHz, VV-Polarization

(b) 2.58 GHz, HH-Polarization

(c) 5.125 GHz, VV-Polarization

(d) 5.125 GHz, HH-Polarization

(e) 7.0 GHz, VV-Polarization

(f) 7.0 GHz, HH-Polarization

(g) 10.25 GHz, VV-Polarization

(h) 10.25 GHz, HH-Polarization

**FIGURE 9.31:** Solid Resin EXPEDITE-RCS Aircraft: RCS

## 9.7.7   Monoconic Reentry Vehicle

Finally, we will examine the monoconic reentry vehicle considered previously in Sections 7.10.2.4 and 8.8.8. We consider two configurations: the totally conducting RV, and the RV with a dielectric nose. As before, we will compare the results from Serenity to those from our *Galaxy* MoM/BoR solver, as there are no measurements available. For these examples, we use a value of $\tau_{ACA} = 10^{-4}$.

### 9.7.7.1   Conducting RV

We first consider the conducting RV, which has exterior dimensions identical to those in Figure 7.28. We compute the monostatic RCS at 5.0, 9.5 and 12.0 GHz, using facet models having 144592, 480592, and 776080 triangles. This results in 216888, 720888, and 1171872 unknowns, respectively. The results are compared for vertical and horizontal polarizations in Figures 9.32a–9.32f, where the comparison is very good.

### 9.7.7.2   RV with Dielectric Nose

We next consider the RV with a lossless dielectric nose ($\epsilon = 5$), with dimensions as in Figure 7.28. We compute the monostatic RCS at 5.0, 9.5 and 12.0 GHz, using facet models having 145760, 484384, and 781740 triangles, respectively. These comprise the total number of facets used for the nose, base and septum interfaces, resulting in a total of 225536, 751232, and 1213896 unknowns, respectively. The results are compared for vertical and horizontal polarizations in Figures 9.33a–9.33f, where the comparison is fairly good, although there is some disagreement at nose-forward viewing angles. We suspect that this is due to the facetization density on the nose and septum, though we did not investigate the issue further. Compared to the fully conducting RV, there is a significant increase in RCS at nose-forward angles due to multiple bounces inside the dielectric nose tip.

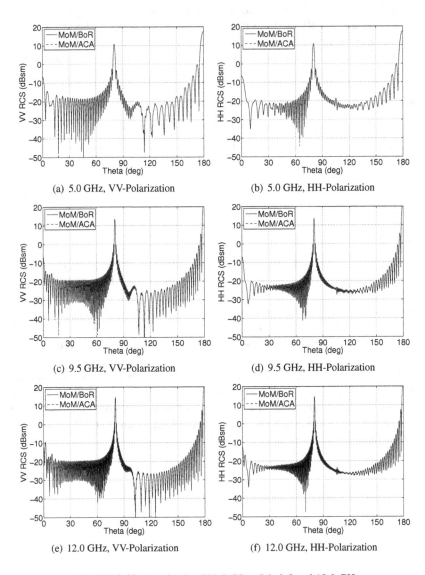

(a)  5.0 GHz, VV-Polarization

(b)  5.0 GHz, HH-Polarization

(c)  9.5 GHz, VV-Polarization

(d)  9.5 GHz, HH-Polarization

(e)  12.0 GHz, VV-Polarization

(f)  12.0 GHz, HH-Polarization

**FIGURE 9.32:** Conducting RV: RCS at 5.0, 9.5 and 12.0 GHz

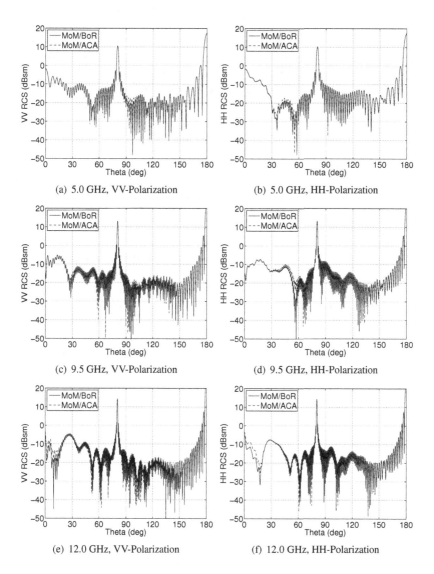

(a) 5.0 GHz, VV-Polarization

(b) 5.0 GHz, HH-Polarization

(c) 9.5 GHz, VV-Polarization

(d) 9.5 GHz, HH-Polarization

(e) 12.0 GHz, VV-Polarization

(f) 12.0 GHz, HH-Polarization

**FIGURE 9.33:** RV with Lossless Nose: RCS at 5.0, 9.5 and 12.0 GHz

### 9.7.8    Summary of Examples

Run metrics for the examples in this section are listed in Table 9.1. For problems that used the same facet model at several frequencies, only the results at the highest frequency are shown. Summarized are the operating frequency $f_0$, number of electric and magnetic basis functions ($N_J$ and $N_M$), storage for the ACA-compressed matrix ($M_\mathbf{Z}$), matrix fill time ($T_\mathbf{Z}$), storage for the ACA-compressed **LU** matrix ($M_\mathbf{LU}$), LU factorization time ($T_{LU}$), and amount of compression of the **LU** matrix ($C$). Frequency is in GHz, storage in GB, time in seconds, and compression in percent. Compression is computed as

$$C = (1 - M_\mathbf{LU}/M_0) \cdot 100\% \qquad (9.38)$$

where $M_0$ comprises the memory required to store the matrix in uncompressed form. We note that the level of compression is excellent overall, and increases with electrical size, exceeding 98 percent in some cases. The PEC RV, for example, having 1171872 basis functions at 12.0 GHz would require approximately 10 TB of memory for the **LU** matrix if stored in uncompressed form. Solving this directly would require supercomputing resources out of the reach of all but the most privileged users. Using the ACA this was reduced to 140 GB, and solved easily on our desktop workstation in under two hours.

In Figures 9.34a and 9.34b are plotted $M_\mathbf{Z}$ and $M_\mathbf{LU}$ versus the number of unknowns $N$, for all the test cases in Table 9.1. We see that the storage requirement of $M_\mathbf{Z}$ scales approximately as $O(N^{1.29} \log N)$, slightly less than the $O(N^{4/3} \log N)$ reported for the ACA previously in [3]. The storage requirement of $M_\mathbf{LU}$ scales approximately as $O(N^{1.39} \log N)$, slightly higher than $O(N^{4/3} \log N)$.

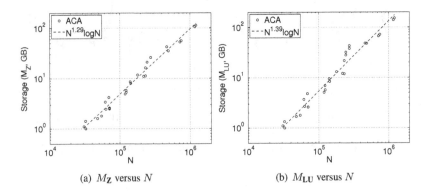

(a) $M_\mathbf{Z}$ versus $N$        (b) $M_\mathbf{LU}$ versus $N$

**FIGURE 9.34:** ACA Storage versus Number of Unknowns

TABLE 9.1: ACA Run Metrics

| Object | $f_0$ | $N_J$ | $N_M$ | $M_Z$ | $T_Z$ | $M_{LU}$ | $T_{LU}$ | $C(\%)$ |
|---|---|---|---|---|---|---|---|---|
| PEC Sphere | 1.5 | 30720 | 0 | 1.1 | 4.8 | 1.1 | 5.7 | 84.3 |
| PEC Sphere | 3.0 | 122880 | 0 | 5.8 | 20 | 6.9 | 35 | 93.9 |
| PEC Sphere | 6.0 | 491520 | 0 | 35 | 114 | 48 | 650 | 97.3 |
| Lossless Dielectric Sphere | 1.5 | 30720 | 30720 | 3.4 | 19 | 3.7 | 23 | 86.9 |
| Lossless Dielectric Sphere | 3.0 | 122880 | 122880 | 21 | 125 | 32 | 176 | 92.9 |
| Lossy Dielectric Sphere | 3.0 | 122880 | 122880 | 21 | 123 | 28 | 162 | 93.7 |
| Lossless Coated Sphere | 1.5 | 38400 | 30720 | 4.2 | 23 | 4.8 | 30 | 86.5 |
| Lossless Coated Sphere | 3.0 | 153600 | 122880 | 26 | 157 | 43 | 243 | 92.5 |
| Lossy Coated Sphere | 3.0 | 153600 | 122880 | 26 | 156 | 38 | 225 | 93.4 |
| Almond | 7.0 | 70476 | 0 | 2.5 | 10 | 2.6 | 15 | 93.2 |
| Almond | 9.92 | 70476 | 0 | 2.6 | 11 | 2.6 | 15 | 93.0 |
| Ogive | 9.0 | 32640 | 0 | 1.0 | 7.2 | 1.0 | 7.5 | 87.7 |
| Double Ogive | 9.0 | 54360 | 0 | 1.8 | 8.0 | 1.8 | 11 | 91.7 |
| Cone-Sphere | 9.0 | 119244 | 0 | 4.9 | 17 | 5.1 | 31 | 95.2 |
| Cone-Sphere with Gap | 9.0 | 123792 | 0 | 5.3 | 19 | 5.7 | 39 | 95.2 |
| PEC Cube | 0.43 | 6813 | 0 | 0.1 | 1 | 0.1 | 1 | 72.0 |
| PEC Cube | 1.3 | 62717 | 0 | 2.5 | 15 | 2.7 | 25 | 90.0 |
| Prism | 0.43 | 5415 | 0 | 0.063 | 1 | 0.062 | 1 | 72.0 |
| Prism | 1.3 | 47745 | 0 | 1.6 | 14 | 1.7 | 18 | 90.2 |
| Small Polyethylene Cube | 18.0 | 16200 | 16200 | 1.41 | 21 | 1.34 | 28 | 82.9 |
| Polyethylene Ogive | 18.0 | 116436 | 116436 | 16.2 | 110 | 21.6 | 234 | 94.7 |
| Austin PEC Almond | 10.25 | 70476 | 0 | 2.54 | 14.4 | 2.62 | 23 | 92.9 |
| Austin Solid Resin Almond | 9.25 | 70476 | 70476 | 8.53 | 65.1 | 9.44 | 121 | 93.6 |
| Austin Closed-Tail Almond | 10.25 | 95920 | 47752 | 7.99 | 62 | 8.37 | 142 | 94.6 |
| Austin Open-Tail Almond | 10.25 | 132460 | 47752 | 11.7 | 110 | 13.0 | 200 | 94.6 |
| PEC Aircraft | 10.25 | 233148 | 0 | 11.2 | 74 | 11.8 | 202 | 97.0 |
| Solid Resin Aircraft | 10.25 | 233148 | 233148 | 42.5 | 529 | 48.3 | 1238 | 97.0 |
| PEC RV | 5.0 | 219508 | 0 | 11 | 34 | 12 | 94 | 96.6 |
| PEC RV | 9.5 | 720888 | 0 | 53 | 171 | 67 | 1722 | 98.3 |
| PEC RV | 12.0 | 1171872 | 0 | 106 | 359 | 140 | 6500 | 98.6 |
| Dielectric Nose RV | 5.0 | 219508 | 6028 | 11 | 39 | 12 | 103 | 96.7 |
| Dielectric Nose RV | 9.5 | 730576 | 20656 | 56 | 204 | 72 | 1927 | 98.3 |
| Dielectric Nose RV | 12.0 | 1180548 | 33348 | 113 | 423 | 151 | 7365 | 98.6 |

# References

[1] E. Michielssen and A. Boag, "A multilevel matrix decomposition algorithm for analyzing scattering from large structures," *IEEE Trans. Antennas Propagat.*, vol. 44, pp. 1086–1093, August 1996.

[2] M. Bebendorf, "Approximation of boundary element matrices," *Numer. Math.*, vol. 86, pp. 565–589, June 2000.

[3] K. Zhao, M. N. Vouvakis, and J.-F. Lee, "The adaptive cross approximation algorithm for accelerated method of moments computations of EMC problems," *IEEE Trans. Electromagn. Compat.*, vol. 47, pp. 763–773, Nov 2005.

[4] X. Chen, C. Gu, J. Ding, Z. Li, and Z. Niu, "Multilevel fast adaptive cross-approximation algorithm with characteristic basis functions," *IEEE Trans. Antennas Propagat.*, vol. 63, pp. 3994–4002, September 2015.

[5] A. Heldring, E. Ubeda, and J. M. Rius, "On the convergence of the ACA algorithm for radiation and scattering problems," *IEEE Trans. Antennas Propagat.*, vol. 62, pp. 3806–3809, July 2014.

[6] A. Heldring, E. Ubeda, and J. M. Rius, "Stochastic estimation of the frobenius norm in the ACA convergence criterion," *IEEE Trans. Antennas Propagat.*, vol. 63, pp. 1155–1158, March 2015.

[7] M. Bebendorf and S. Kunis, "Recompression techniques for adaptive cross approximation," *J. Integral Equations Applications*, vol. 21, pp. 331–357, July 2009.

[8] J. Shaeffer, "Direct solve of electrically large integral equations for problem sizes to 1 M unknowns," *IEEE Trans. Antennas Propagat.*, vol. 56, pp. 2306–2313, August 2008.

[9] E. Lezar and D. Davidson, "GPU-accelerated method of moments by example: Monostatic scattering," *IEEE Antennas Propagat. Mag.*, vol. 52, pp. 120–135, December 2010.

[10] B. Virk, "Implementing Method of Moments on a GPGPU using NVIDIA CUDA," Master's thesis, Georgia Institute of Technology, May 2010.

[11] CUDA C Programming Guide, docs.nvidia.com/cuda/cuda-c-programming-guide.

[12] J. Dongarra, M. Gates, A. Haidar, J. Kurzak, P. Luszczek, S. Tomov, and I. Yamazaki, "Accelerating numerical dense linear algebra calculations with gpus," *Numerical Computations with GPUs*, pp. 1–26, 2014.

[13] OpenMPI, www.open-mpi.org.

[14] MPI Forum, www.mpi-forum.org.

[15] A. C. Woo, H. T. G. Wang, M. J. Schuh, and M. L. Sanders, "Benchmark radar targets for the validation of computational electromagnetics programs," *IEEE Antennas Propagat. Magazine*, vol. 35, pp. 84–89, February 1993.

[16] A. Greenwood, "Electromagnetic Code Consortium Benchmarks," Tech. Rep. AFRL-DE-TR-2001-1086, Air Force Research Laboratory, December 2001.

[17] J. Song, C. C. Lu, W. C. Chew, and S. W. Lee, "Fast Illinois solver code (FISC)," *IEEE Antennas Propagat. Mag.*, vol. 40, no. 3, pp. 27–34, 1998.

[18] J. M. Parks, "Scattering from Dielectric Bodies," Master's thesis, Air Force Institute of Technology, December 1997.

[19] github.com/UTAustinCEMGroup/AustinCEMBenchmarks.

[20] J. T. Kelley, D. A. Chamulak, C. C. Courtney, and A. E. Yilmaz, "Rye canyon radar cross-section measurements of benchmark almond targets," *IEEE Antennas Propagat. Mag.*, vol. 52, pp. 120–135, February 2020.

[21] J. T. Kelley, D. A. Chamulak, C. C. Courtney, and A. E. Yilmaz, "Measurements of non-metallic targets for the Austin RCS benchmark suite," in *Proc. Ant. Meas. Tech. Assoc. (AMTA) Symp*, 2019.

[22] J. T. Kelley, A. Maicke, D. A. Chamulak, C. C. Courtney, and A. E. Yilmaz, "Adding a reproducible airplane model to the Austin RCS benchmark suite," in *Proc. Applied Comp. Electromagnetics Society (ACES) Symp.*, July 2020.

# Chapter 10

## Multi-Level Adaptive Cross Approximation

In Chapter 9, basis functions were clustered into spatially localized, roughly same-sized groups via the $K$-means algorithm, imposing a block structure on the MoM matrix. It was then shown that the off-diagonal blocks in the system matrix, as well as the corresponding off-diagonal **LU** matrix blocks, can be compressed via the ACA+QR/SVD technique (Section 9.2.2). As $K$-means is not a hierarchical algorithm, a single set of basis function groups is created. Thus, we refer to this approach as the single-level ACA, or simply the ACA.

It was pointed out in [1] that the degrees of freedom (DoF) remain asymptotically constant when size of the source and testing groups are inversely varied. This concept was used to further extend the ACA in [2, 3] through a recursive subdivision of the basis functions in each group. This approach splits blocks into column sub-blocks, which are compressed separately via ACA. Each resulting **U** matrix, comprising the singular values and left-singular vectors of each compressed sub-block, are stacked together and compressed recursively via ACA. This multi-level approach, called the Multi-Level ACA (MLACA), was shown to greatly improve the overall rate of compression at the expense of increased calculation time. In that paper, however, only the system matrix was compressed, and an iterative solver was used for the solution. This author then showed in [4] that the MLACA is also capable of compressing the **LU** matrix blocks very efficiently, making a direct solution feasible.

In this Chapter, we present a further refinement of our algorithm in [4], which fuses elements from the ACA in Chapter 9 with the MLACA as presented in [2]. The geometry is partitioned using a semi-hierarchical $K$-means clustering, and the MoM system matrix **Z**, as well as the **LU** matrix, are compressed. Unlike in the ACA, the off-diagonal *and* diagonal blocks are compressed in the MLACA, yielding a much greater level of compression.

Many concepts and algorithms from the single-level ACA in Chapter 9, such as the algorithms for block-LU factorization and block-RHS solution, apply to the MLACA with almost no modifications needed. Other implementation details, such as software changes and additional classes needed for MLACA support, will be discussed in detail.

## 10.1    MLACA Compression of Matrix Blocks

In this section we will discuss hierarchical clustering of sub-groups and the MLACA compression algorithm. We will then apply it to the direct solution of the MoM matrix system via LU factorization in Section 10.2.

### 10.1.1    MLACA Fundamentals

MLACA is based on the idea that the number of degrees of freedom (DoF) remains asymptotically constant when the sizes of well-separated source and testing groups are inversely varied. Thus, an $L$ level MLACA further divides the basis functions in each single-level ACA group into $2^L$ spatially localized sub-groups. Each off-diagonal block $\mathbf{Z}^{m \times n}$, with $m$ rows and $n$ columns, now comprises $2^L$ row and column sub-blocks each having approximately $\frac{m}{2^L}$ and $\frac{n}{2^L}$ rows and columns. To begin the MLACA compression, the ACA is first applied to each block column of $\mathbf{Z}$ independently. The $\mathbf{U}$ matrices of the compressed block columns, comprising the singular values and left-singular vectors of each, are then joined in pairs column-wise, and divided in half row-wise. The ACA compression process is then repeated for the upper and lower halves, recursively, until only a single block row and column remains.

To illustrate, consider an $L = 2$ level MLACA, where operations on the first level $l = 0$ are depicted in Figure 10.1a. Each block column of $\mathbf{Z}$ is compressed separately using ACA+QR/SVD, yielding four $\mathbf{U}$ and $\mathbf{V}$ block matrix pairs. The singular values from the recompression step are retained in each $\mathbf{U}$ block. The $\mathbf{U}$ blocks, each comprising approximately $m$ rows and $k$ columns, where $k << n$, are stacked together to create the notional block matrix $\mathbf{U}_0^{(1)}$. The smaller $\mathbf{V}$ blocks, forming the notional block-diagonal matrix $\mathbf{V}_0^{(1)}$, each comprise approximately $k$ rows and $\frac{n}{2^L}$ columns. These $\mathbf{V}$ blocks are fully compressed and are stored. We then proceed as in Figure 10.1(b), where we now combine pair-wise the $\mathbf{U}$ column blocks and divide the result in half row-wise, yielding four smaller blocks, each having approximately $\frac{m}{2}$ rows and $2k$ columns. We then recurse to level $l = 1$, where the block columns of each half, each having approximately $\frac{m}{2^l}$ rows and $2k$ columns, are compressed in the same way as on the previous level. This recursion is repeated until the finest level is reached, where there is only a single block row and column. After applying the ACA on this level, the $\mathbf{U}$ and $\mathbf{V}$ matrix pair are stored and the MLACA compression of $\mathbf{Z}$ is complete.

As this algorithm stacks together and recursively compresses the $\mathbf{U}$ blocks, we refer to it as a $U$-type MLACA. An error analysis of MLACA was presented by other authors in [4]. Finally, we note that for $L = 0$, MLACA is identical to the single-level ACA.

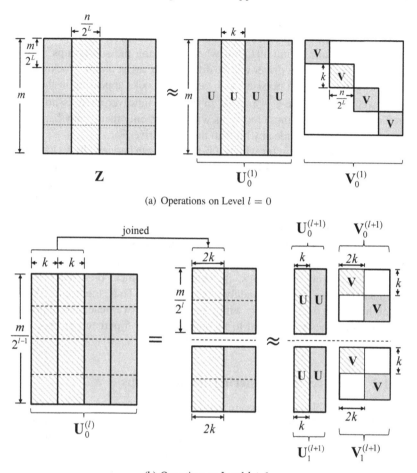

(a) Operations on Level $l = 0$

(b) Operations on Level $l + 1$

**FIGURE 10.1:** 2-Level $U$-Type MLACA

### 10.1.1.1 SVD-Based Compression on Higher Levels

In Section 9.2.1.3 was presented a failure case for the ACA, where the source and testing groups comprised co-planar groups of facets on a cube face. This problem was solved by arranging all electric basis functions in a group first, followed by any magnetic basis functions. Hints were then passed to the ACA driver, forcing the ACA to search the upper and lower halves of the matrix.

This potential failure case remains when using the MLACA, and must be addressed. When performing compressions on level $l = 0$, it can be avoided in the same manner as in the single-level ACA, by ordering the electric basis functions in each group block sub-column first, followed by magnetic functions. This is done at initialization when clustering the sub-groups, and the hints provided the ACA driver will be different for each column block. On levels $l > 0$, this problem persists, with zero sub-blocks appearing in the upper and lower halves of the $\mathbf{U}_0^{(l)}$ matrix. In this case however, there is no way to predict *a priori* where these locations will appear, only that it may happen, so it is impossible to provide any hints to the ACA driver.

Fortunately, knowing for which sub-blocks it *can* happen allows us to handle all possible cases. If either (or both) of the top-level source and testing sub-groups have only electric basis functions, we know it will not happen and we use the ACA+QR/SVD for compression at all levels. However, when both sub-groups have magnetic basis functions, we know the failure is possible. Thus, we use ACA+QR/SVD on level $l = 0$, as hints are available for the ACA driver. On levels $l > 0$, we instead apply the SVD directly. This approach would be inefficient on level $l = 0$ if applied to the column blocks of $\mathbf{Z}^{m \times n}$, each approximately of size $m \times \frac{n}{2^L}$. However, the blocks on levels $l > 0$ are much smaller, being approximately of size $\frac{m}{2^l} \times 2k$. Thus, the use of SVD on the higher levels does not carry with it unreasonable overhead, and in practice compression times were found to be higher by only 5 percent, or less, when applying the SVD directly versus ACA+QR/SVD.

## 10.1.2   Hierarchical Clustering of Sub-Groups

In MLACA, block columns are paired together before recursing to a higher level. Therefore, the basis functions in each pair must be clustered so that their union is itself spatially local. Similarly, when the joined blocks are split in half row-wise, the upper and lower halves should each be spatially local. Thus, we see a clear parent-child relationship between the block rows and columns as the algorithm moves between levels. This suggests that a binary tree-based subdivision of groups, where sub-groups on each level are divided in half spatially, is needed for the MLACA.

We must consider carefully how this subdivision is performed. In [2] a tree-based subdivision of the basis functions in angle space, based on the solid angle of a source group seen by a testing group, and vice versa, was used. It was said that this solid-angle approach is critical for the MLACA to obtain its gains in computational complexity and memory savings. However, this approach results in a local re-ordering of the rows and columns in each block that are necessarily different for other blocks. When using an iterative solver, this does not present any issues, as only elements from the input and output vectors are shuffled during the matrix-vector product with a block. However, this will cause difficulties in performing steps 2, 5, and 7 in Algorithm 7, as $\mathbf{Z}$, $\mathbf{L}$, and $\mathbf{U}$ will need to be unshuffled prior to each matrix product so that the global ordering is followed, and the results shuffled again prior to recompression. This would be very inefficient.

Our approach is to apply the same $K$-means algorithm used to create the single-level ACA groups, but in a recursive way. We first generate the single level groups, as in the ACA. For a problem with $N$ basis functions, this results in $M = N/N_g$ single-level groups, where $N_g$ is the target number of basis functions per group. For each group, then we apply $K$-means to create two spatially local sub-groups, and then recursively divide these in half via $K$-means, and so on, to a maximum level $L$. This comprises a per-group reordering, and so the global ordering can be adjusted to match and no shuffling is required later. This strategy was compared to the tree-based solid angle method when compressing off-diagonal blocks for a variety of test geometries. It was observed that the size of the compressed blocks differed by less than 2 percent in each case, validating the approach.

### 10.1.3 Compression of Diagonal Blocks

Another advantage of the hierarchical clustering of sub-groups is that it permits compression of the diagonal blocks. Consider the example diagonal block in Figure 10.2, which has $m$ rows and columns, and an $L = 3$ level MLACA is used. A MLACA compression of the diagonal block now comprises a recursive walk of the binary tree starting at the root ($l = 0$). On levels $l > 0$, the off-diagonal sub-blocks are compressed using an $L - l$ level MLACA, and recursion to a higher level occurs only for the diagonal sub-blocks. On the finest level, the diagonal sub-blocks, now each having roughly $\frac{m}{2^L}$ rows and columns, are stored in full form and the compression is complete.

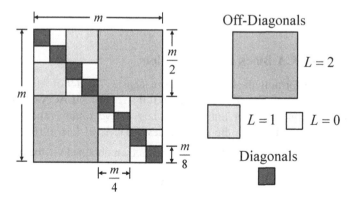

**FIGURE 10.2:** MLACA Compression of Diagonal Block For $L = 3$

## 10.2  Direct Solution of MLACA-Compressed Matrix System

The LU factorization of the MLACA-compressed MoM system matrix follows Algorithm 7, with a few modifications, as summarized in Algorithm 12. Each diagonal is first expanded to full form prior to its update in step 2 and later re-compressed in step 5, and the full version of $[\mathbf{LU}]_{bb}$ from step 4 is retained for the triangular back-substitution in steps 7 and 9. We now discuss the reconstruction of MLACA-compressed matrix blocks, and their products, as these must be optimized to yield a fast and efficient algorithm.

---

**Algorithm 12** MLACA-Compressed Block LU Factorization

---

1: **for** $b = 1$ to $M$ **do**
2:     Reconstruct $\mathbf{Z}_{bb}$
3:     $\mathbf{A} = \mathbf{Z}_{bb} - \sum_{s=1}^{b-1}[\mathbf{L}_u \mathbf{L}_v]_{bs}[\mathbf{U}_u \mathbf{U}_v]_{sb}$
4:     $[\mathbf{LU}]_{bb} = \mathrm{LU}(\mathbf{A})$    (scalar LU factorization)
5:     Compress and store $[\mathbf{LU}]_{bb}$ via MLACA, retain full $[\mathbf{LU}]_{bb}$
6:     **for** $s = b + 1$ to $M$ **do**
7:         $\mathbf{L}_{sb} = \left[ [\mathbf{Z}_u \mathbf{Z}_v]_{sb} - \sum_{p=1}^{b-1}[\mathbf{L}_u \mathbf{L}_v]_{sp}[\mathbf{U}_u \mathbf{U}_v]_{pb} \right] \mathbf{U}_{bb}^{-1}$
8:         Compress and store $\mathbf{L}_{sb}$ via MLACA
9:         $\mathbf{U}_{bs} = \mathbf{L}_{bb}^{-1} \left[ [\mathbf{Z}_u \mathbf{Z}_v]_{bs} - \sum_{p=1}^{b-1}[\mathbf{L}_u \mathbf{L}_v]_{bp}[\mathbf{U}_u \mathbf{U}_v]_{ps} \right]$
10:         Compress and store $\mathbf{U}_{bs}$ via MLACA
11:     **end for**
12: **end for**

---

### 10.2.1  MLACA Block Reconstruction

Prior to performing the block updates in steps 3, 7 and 9 of Algorithm 12, we must first expand $\mathbf{Z}_{bb}$, $\mathbf{Z}_{sb}$, and $\mathbf{Z}_{sb}$ to full size. For MLACA-compressed blocks this is done via recursion, where nodes on level $l$ reconstruct their portion of the matrix $\mathbf{U}_{0,1}^{l-1}$ on the previous level. On level $l = 0$ (the coarsest level), once $\mathbf{U}_0^{(1)}$ is available, the full matrix is then obtained via the product

$$\mathbf{Z} = \mathbf{U}_0^{(1)} \mathbf{V}_0^{(1)}. \tag{10.1}$$

Operations on each level are very fast, as each reconstruction step comprises matrix products in rank-reduced form.

## 10.2.2    Matrix Product and $V$-Type MLACA

Matrix products in the ACA block updates, discussed in Section 9.4.1 are very fast as they operate in rank-reduced form, requiring only a few calls to the Level 3 BLAS. However, now consider the product **AB** of two $U$-type MLACA matrices on level $l = 0$, as shown in Figure 10.3. Prior to performing the product, the compressed **U**-column blocks of each matrix have been reconstructed. It is clear that the innermost product $\mathbf{A}_v\mathbf{B}_u$ is not efficient, as each small diagonal block of $\mathbf{A}_v$ multiplies only a small portion of the block columns of $\mathbf{B}_u$. For the $L = 2$ level MLACA depicted, this product requires $2^L = 4$ cgemm calls.

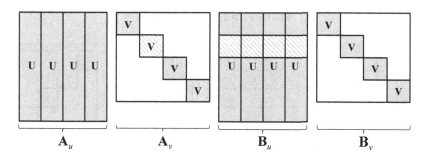

**FIGURE 10.3:** Product of $U$-Type MLACA Matrices

This innermost product can be reduced to a single operation if we modify the MLACA algorithm used to compress **A**. Consider again the level $l = 0$ operations depicted in Figure 10.1(a). If we instead compress the block rows of **Z**, we now have the configuration in Figure 10.4. The smaller diagonal **U** blocks of $\mathbf{U}_0^{(1)}$ are stored, and the larger **V** blocks comprising $\mathbf{V}_0^{(1)}$ are paired row-wise and split in half column-wise. These left and right halves are then recursively compressed in a manner analogous to the $U$-type MLACA, resulting in a variant we call the $V$-type MLACA.

The resulting product of $V$ and $U$-type MLACA matrices is depicted in Figure 10.5, where the innermost product has been reduced to a single operation. We note that we can force all MLACA matrix products to have this form by compressing all blocks below the diagonal using $V$-type MLACA, and all blocks above the diagonal using $U$-type MLACA.

In [4], the innermost product was performed by first fully reconstructing $\mathbf{A}_v$ and $\mathbf{B}_v$, prior to computing any intermediate products. This approach is summarized in Section 10.2.2.1. Since the publication of [4], a more efficient method was identified, wherein the first intermediate product is computed directly without needing to compute $\mathbf{A}_v$ or $\mathbf{B}_v$. This approach is presented in Section 10.2.2.2.

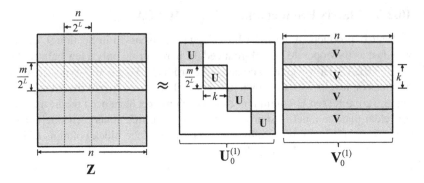

**FIGURE 10.4:** $V$-Type MLACA

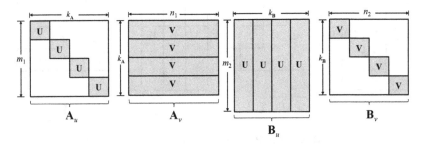

**FIGURE 10.5:** Product of $V$ and $U$-Type Matrices

### 10.2.2.1    Top-Level Matrix Product

The "top-level" matrix product begins by reconstructing the compressed row blocks of $\mathbf{A}_v$ and the compressed column blocks of $\mathbf{B}_v$, using the recursive approach in Section 10.2.1. The innermost product in Figure 10.5 then comprises a single operation, yielding

$$\mathbf{T}_1^{k_\mathrm{A} \times k_\mathrm{B}} = \mathbf{A}_v^{k_\mathrm{A} \times n_1} \mathbf{B}_u^{m_2 \times k_\mathrm{B}}, \tag{10.2}$$

where $\mathbf{T}_1^{k_\mathrm{A} \times k_\mathrm{B}}$ is a $2^L \times 2^L$ block matrix, and $k_\mathrm{A}$ and $k_\mathrm{B}$ are the sum of the ranks of all the compressed sub-blocks in $\mathbf{A}$ and $\mathbf{B}$, respectively. As in Section 9.4.1, whether the left or right product is performed first depends on $\min(k_\mathrm{A}, k_\mathrm{B})$. If $k_\mathrm{A} < k_\mathrm{B}$, we compute $\mathbf{T}_2 = \mathbf{T}_1 \mathbf{B}_v$ yielding

$$\mathbf{T}_2^{k_\mathrm{A} \times n_2} = \mathbf{T}_1^{k_\mathrm{A} \times k_\mathrm{B}} \mathbf{B}_v^{k_\mathrm{B} \times n_2}, \tag{10.3}$$

and then

$$[\mathbf{LU}]^{m_1 \times n_2} = \mathbf{A}_u^{m_1 \times k_\mathrm{A}} \mathbf{T}_2^{k_\mathrm{A} \times n_2}. \tag{10.4}$$

Otherwise, we compute $\mathbf{T}_3 = \mathbf{A}_u \mathbf{T}_1$, resulting in

$$\mathbf{T}_3^{m_1 \times k_\mathrm{B}} = \mathbf{A}_u^{m_1 \times k_\mathrm{A}} \mathbf{T}_1^{k_\mathrm{A} \times k_\mathrm{B}}, \tag{10.5}$$

and then

$$[\mathbf{LU}]^{m_1 \times n_2} = \mathbf{T}_3^{m_1 \times k_B} \mathbf{B}_v^{k_B \times n_2}. \qquad (10.6)$$

As $\mathbf{A}_u$ and $\mathbf{B}_v$ each comprise block diagonal matrices of $2^L$ blocks, the product $\mathbf{AB}$ is thus reduced to $2^{L+1} + 1$ matrix products.

### 10.2.2.2 Bottom-Level Matrix Product

The "top-level" approach outlined in Section 10.2.2.1 is reasonably efficient, however it becomes less attractive as the size of the groups is increased (see Section 10.4.2.1). This is because as $n_1$ and $m_2$ grow, the workspace needed to store $\mathbf{A}_v^{k_A \times n_1}$ and $\mathbf{B}_u^{m_2 \times k_B}$ becomes excessive. This is especially problematic when working on the GPU, as each host thread must have its own workspace for $\mathbf{A}_v$ and $\mathbf{B}_u$ as well as the other intermediate products. In particular, $\mathbf{T}_2^{k_A \times n_2}$ and $\mathbf{T}_3^{m_1 \times k_B}$ are of similar size, further increasing the GPU memory burden. Finally, the approach from Section 9.6.2.2, where all $\mathbf{L}_v$ left of the diagonal and $\mathbf{U}_u$ above the diagonal are pre-staged on the GPU, quickly depletes GPU memory and limits the number of threads that can work on the device.

However, we note that $\mathbf{A}_v$ and $\mathbf{B}_u$ comprise temporary matrices which are discarded after computing the intermediate product $\mathbf{T}_1$ in (10.2). Similarly, when reconstructing $\mathbf{A}_v$ and $\mathbf{B}_u$, on higher levels of recursion the matrices $\mathbf{U}_{0,1}^l$ ($U$-type) and $\mathbf{V}_{0,1}^l$ ($V$-type) are also temporary and discarded after reconstructing their portion of $\mathbf{U}_{0,1}^{l-1}$ or $\mathbf{V}_{0,1}^{l-1}$ on the next higher level. It is only on the finest level $L$ where the matrices $\mathbf{U}_{0,1}^L$ and $\mathbf{V}_{0,1}^L$ are actually stored.

Therefore, we can instead construct the product $\mathbf{T}_1$ directly by walking the tree structure of $\mathbf{A}$ and $\mathbf{B}$ simultaneously, bottom to top, starting on the finest level. To illustrate the order of operations, we will first work from the top level to the finest level. Let us define $\mathbf{V}^{(0)} = \mathbf{A}_v$ and $\mathbf{U}^{(0)} = \mathbf{B}_u$, and as depicted in Figure 10.6, the product $\mathbf{T}_1^{(0)} = \mathbf{V}^{(0)}\mathbf{U}^{(0)}$ on level $l = 0$. Let us now consider the compression of the left half of $\mathbf{V}^{(0)}$ and upper half of $\mathbf{U}^{(0)}$, as depicted. This corresponds to following only one side of the binary tree. The result on level $l = 1$ is shown in Figure 10.7, where $\mathbf{U}_V^1$ and $\mathbf{V}_U^1$ are stored, $\mathbf{V}_V^{(1)}$ and $\mathbf{U}_U^{(1)}$ are temporary matrices to be compressed, and $\mathbf{T}_1^{(1)} = \mathbf{V}_V^{(1)}\mathbf{U}_U^{(1)}$. Recursing in the same way again yields the result on level $l = 2$, shown in Figure 10.8. In this case, $\mathbf{U}_V^{(2)}$, $\mathbf{V}_V^{(2)}$, $\mathbf{U}_U^{(2)}$, and $\mathbf{V}_U^{(2)}$ are all stored and the recursion is finished. Note that at each step in the recursion, there is a second path which compresses the right half of $\mathbf{V}_V^l$ and the bottom half of $\mathbf{U}_U^l$, which would be depicted similarly.

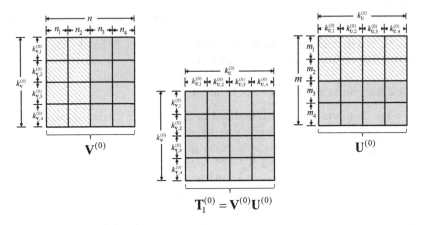

$$\mathbf{T}_1^{(0)} = \mathbf{V}^{(0)}\mathbf{U}^{(0)}$$

**FIGURE 10.6:** Reconstruction of $\mathbf{T}_1^{(0)}$

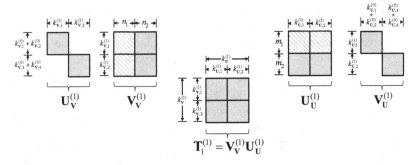

$$\mathbf{T}_1^{(1)} = \mathbf{V}_\mathbf{V}^{(1)}\mathbf{U}_\mathbf{U}^{(1)}$$

**FIGURE 10.7:** Reconstruction of $\mathbf{T}_1^{(1)}$

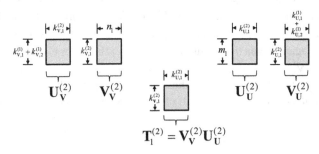

$$\mathbf{T}_1^{(2)} = \mathbf{V}_\mathbf{V}^{(2)}\mathbf{U}_\mathbf{U}^{(2)}$$

**FIGURE 10.8:** Reconstruction of $\mathbf{T}_1^{(2)}$

Having reached the bottom of the tree, we now work our way back to the top. Starting on level $l = 2$, we first compute

$$\mathbf{T}_1^{(2),k_{V,1}^{(2)} \times k_{U,1}^{(2)}} = \mathbf{V}_V^{(2),k_{V,1}^{(2)} \times n_1} \mathbf{U}_U^{(2),m_1 \times k_{U,1}^{(2)}}, \tag{10.7}$$

and then

$$\mathbf{T}_2^{(2),k_V^{(1)} \times k_{U,1}^{(2)}} = \mathbf{U}_V^{(2),k_V^{(1)} \times k_{V,1}^{(2)}} \mathbf{T}_1^{(2),k_{V,1}^{(2)} \times k_{U,1}^{(2)}}, \tag{10.8}$$

where $k_V^{(1)} = k_{V,1}^{(1)} + k_{V,2}^{(1)}$, and finally

$$\mathbf{T}_1^{(1),k_V^{(1)} \times k_U^{(1)}} = \mathbf{T}_2^{(2),k_V^{(1)} \times k_{U,1}^{(2)}} \mathbf{V}_U^{(2),k_{U,1}^{(2)} \times k_U^{(1)}}, \tag{10.9}$$

where $k_U^{(1)} = k_{U,1}^{(1)} + k_{U,2}^{(1)}$. We note that there is an equally sized contribution to (10.9) obtained by following the other recursive path. Moving now to level $l = 1$, and having already computed the intermediate product $\mathbf{T}_1^{(1)}$, we can immediately compute $\mathbf{T}_2^{(1)}$, which is

$$\mathbf{T}_2^{(1),k_V^{(0)} \times k_U^{(1)}} = \mathbf{U}_V^{(1),k_V^{(0)} \times k_V^{(1)}} \mathbf{T}_1^{(1),k_V^{(1)} \times k_U^{(1)}}, \tag{10.10}$$

and finally

$$\mathbf{T}_1^{(0),k_V^{(0)} \times k_U^{(0)}} = \mathbf{T}_1^{k_A \times k_B} = \mathbf{T}_2^{(1),k_V^{(0)} \times k_U^{(1)}} \mathbf{V}_U^{(1),k_U^{(1)} \times k_U^{(0)}}, \tag{10.11}$$

where

$$k_V^{(0)} = k_A = k_{V,1}^{(0)} + k_{V,2}^{(0)} + k_{V,3}^{(0)} + k_{V,4}^{(0)}, \tag{10.12}$$

and

$$k_U^{(0)} = k_B = k_{U,1}^{(0)} + k_{U,2}^{(0)} + k_{U,3}^{(0)} + k_{U,4}^{(0)}. \tag{10.13}$$

We note that at each step in the reconstruction, the dimensions of the intermediate products comprise the rank sum of matrices on that level, or on the parent level. As each product is relatively small, they require much less workspace and can be computed very quickly.

Having now computed $\mathbf{T}_1^{k_A \times k_B}$ in (10.2), the choice of whether to perform the left or right product first is made by minimizing the storage of $\mathbf{T}_2^{k_A \times n_2}$ and $\mathbf{T}_3^{m_1 \times k_B}$. As the top-level groups were sorted by size, each diagonal is of size $N_b \times N_b$, increasing to $N_{max} \times N_{max}$, and so we are always guaranteed maximum sizes of $m_1 = N_b$ and $n_2 = N_{max}$, or vice versa, depending on whether we are updating trailing $\mathbf{U}$ blocks or $\mathbf{L}$ blocks, respectively. Thus, if $m_1 > n_2$, we take the first path and compute $\mathbf{T}_2^{k_A \times n_2}$, otherwise we take the second path and compute $\mathbf{T}_3^{m_1 \times k_B}$.

### 10.2.3   MLACA Block-RHS Solution

For problems having many right-hand sides, we can again apply Algorithm 8 with a few changes as summarized in Algorithm 13. Prior to the back-substitutions in steps 4 and 10, we first reconstruct $[\mathbf{LU}]_{bb}$ to full size, which is fast and adds little overhead. Though the blocks of the $\mathbf{LU}$ matrix are MLACA-compressed, in this case we will use the single-level ACA to compress the RHS blocks. Thus, there will be matrix products between a $U$ or $V$-type MLACA compressed $\mathbf{LU}$ matrix block and single-level ACA compressed RHS matrix block, as illustrated in Figures 10.9a and 10.9b, respectively. In both cases, the middle product is carried out first, yielding

$$\mathbf{C}^{k_{\mathrm{A}} \times k_{\mathrm{B}}} = \mathbf{A}_v^{k_{\mathrm{A}} \times n_1} \mathbf{B}_u^{m_2 \times k_{\mathrm{B}}}. \tag{10.14}$$

If $\mathbf{A}$ is $U$-type, whether $\mathbf{A}_u\mathbf{C}$ or $\mathbf{CB}_v$ is performed next depends on which path results in the fewest total operations. If $\mathbf{A}$ is instead $V$-type, we perform the left-most product $\mathbf{A}_u\mathbf{C}$ first as $\mathbf{A}_u$ is a sparse block matrix. This yields

$$\mathbf{D}^{m_1 \times k_{\mathrm{B}}} = \mathbf{A}_u^{m_1 \times k_{\mathrm{A}}} \mathbf{C}^{k_{\mathrm{A}} \times k_{\mathrm{B}}} \tag{10.15}$$

and then finally

$$[\mathbf{AB}]^{m_1 \times n_2} = \mathbf{D}^{m_1 \times k_{\mathrm{B}}} \mathbf{B}_v^{k_{\mathrm{B}} \times n_2}. \tag{10.16}$$

---

**Algorithm 13** MLACA-Compressed Block-RHS Solution

---

1:  Given $\mathbf{LU}\,\mathbf{X} = \mathbf{B}$, first solve $\mathbf{UX} = \mathbf{L}^{-1}\mathbf{B}$:
2:  **for** $b = 1$ to $M$ **do**
3:      Reconstruct $[\mathbf{LU}]_{bb}$
4:      $\mathbf{B}_b = \mathbf{L}_{bb}^{-1}\big[[\mathbf{B}_u\mathbf{B}_v]_b - \sum_{s=1}^{b-1}[\mathbf{L}_u\mathbf{L}_v]_{bs}[\mathbf{B}_u\mathbf{B}_v]_s\big]$
5:      Compress $\mathbf{B}_b$ via ACA and store updated $[\mathbf{B}_u\mathbf{B}_v]_b$
6:  **end for**
7:  Then solve $\mathbf{X} = \mathbf{U}^{-1}\big[\mathbf{L}^{-1}\mathbf{B}\big]$
8:  **for** $b = M$ to $1$ **do**
9:      Reconstruct $[\mathbf{LU}]_{bb}$
10:      $\mathbf{X}_b = \mathbf{U}_{bb}^{-1}\big[[\mathbf{B}_u\mathbf{B}_v]_b - \sum_{s=b+1}^{M}[\mathbf{U}_u\mathbf{U}_v]_{bs}[\mathbf{X}_u\mathbf{X}_v]_s\big]$
11:      Compress $\mathbf{X}_b$ via ACA and store $[\mathbf{X}_u\mathbf{X}_v]_b$
12:  **end for**

---

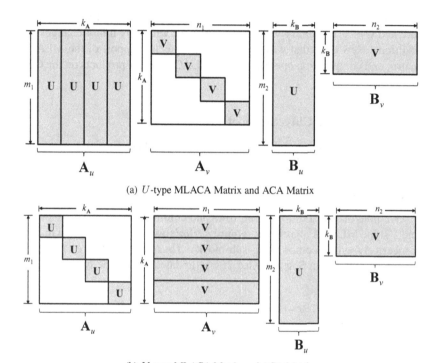

(a) $U$-type MLACA Matrix and ACA Matrix

(b) $V$-type MLACA Matrix and ACA Matrix

**FIGURE 10.9:** Product of MLACA Matrix and ACA Matrix

## 10.3   Software Implementation Notes

In this section we will discuss how MLACA support was implemented in our *Serenity* MoM solver. Much of what is discussed here builds upon what was done previously with the ACA in Section 9.6, and as before, what is presented here are only programming suggestions based on the author's experience.

MLACA support was added to *Serenity* several years after the ACA capability was originally developed. Almost all of the existing software architecture and C++ class library were reused as-is, or with only minor changes and additions. New code that was added comprises class support of the MLACA matrix and compression operations, and MLACA matrix products on the CPU and GPU.

### 10.3.1   Software Class Support

In this section we will discuss software classes that were implemented or updated in support of MLACA operations.

#### 10.3.1.1   Abstract Matrix Class

We note that in addition to full rows and columns, on levels $l < L$ the $U$- and $V$-type MLACA algorithms require *partial* rows and columns of a matrix, respectively, when compressing sub-blocks. Thus, the `AbstractMatrix`, previously defined in Section 9.6.1.2, was updated to provide two new functions:

1. `getPartialRow(`$r, c_1, c_2$`)`: Returns columns $c_1$ to $c_2$ from row $r$ of a matrix block.

2. `getPartialColumn(`$c, r_1, r_2$`)`: Returns rows $r_1$ to $r_2$ from column $c$ of a matrix block.

These functions are implemented by the `ElementEngine` class or the `MLACATranslator` class, as discussed in following the sections.

#### 10.3.1.2   Element Engine Class

The `ElementEngine` class, previously discussed in Section 9.6.1.1, now implements `getPartialRow()` and `getPartialColumn()`. During matrix filling, we note that only a subset of the triangles in the source group are needed to carry out these calculations, and these are known *a priori* due to the hierarchical clustering of basis functions in each group (Section 10.1.2). Thus, at initialization, the `ElementEngine` assembles for each group a list of triangles needed to populate a sub-row or sub-column when that group is a source group, minimizing compute time during the matrix fill.

### 10.3.1.3  MLACA Translator Class

The ACAMatrixTranslator class, described previously in Section 9.6.1.2, provides an AbstractMatrix class interface to the ACA driver while interfacing with the element engine. For MLACA-based matrix filling, we implement a MLACAMatrixTranslator class, which derives from ACAMatrixTranslator and implements the getPartialRow and getPartialColumn methods. Top-level ACA functions, such as the getRow and getColumn methods, simply use existing base class methods.

### 10.3.1.4  MLACAMatrix and MLACANode Classes

The ACAMatrix class, discussed previously in Section 9.6.1.2, is a container class that stores the **U** and **V** components of a single-level ACA compressed matrix block. Objects of this class are generated as a product of the ACA+QR/SVD compression (Section 9.2). Compression of a block using MLACA is a recursive operation, and as a result there is information at each node in the binary tree required to store and reconstruct the original block. Thus, in support of MLACA operation, we implement two new classes. The MLACANode class comprises a node in the MLACA binary tree. This class has functions that permit it to construct and store its compressed, notional $\mathbf{A}^{(l)}$ and $\mathbf{B}^{(l)}$ blocks from the matrix on the previous level. The MLACAMatrix class is a top-level container class, implementing all of the virtual functions of the AbstractMatrix class (see Section 9.6.1.2), as well as storing the root-level MLACANode.

Filling of off-diagonal blocks using the MLACA is similar to the single-level ACA. During the MLACA recursion, upper level MLACANode of the MLACA tree store only **U** or **V** matrices, if $V$- or $U$-type, respectively, whereas the leaf nodes store both **U** and **V**. These are stored in a single array in memory, with each MLACANode storing integer offsets into this array.

### 10.3.1.5  HMatrix Class

The BlockMatrix class, discussed previously in Section 9.6.1.2, is a high-level class responsible for storing pointers to all matrix blocks in row-major format. In that class, the diagonal blocks comprised FullMatrix blocks, and off-diagonals ACAMatrix blocks. Though this behavior was hard-coded, it was sufficient to support ACA operations.

In the MLACA, we now compress the diagonal blocks. Each diagonal block is subdivided recursively, with MLACAMatrix off-diagonal sub-blocks on each level, and FullMatrix diagonal sub-blocks on the finest level.

The diagonal sub-blocks on coarser levels, however, remained undefined. For these, a generalized block matrix container class similar to `BlockMatrix` is needed. As a result, the `BlockMatrix` class was refactored and replaced by the `HMatrix` class, where restrictions on the types of blocks that can be stored at any position have been removed.

## 10.3.2 Shared Memory Processing

For computing and solving the MLACA-compressed matrix system on shared memory systems, we derive from the same CPU and GPU thread classes implemented previously for the ACA, reusing most existing functionality. A few changes are needed, as well as the addition of new code, as discussed in Sections 10.3.2.2 and 10.3.2.3.

### 10.3.2.1 Reconstruction of Blocks and Intermediate Products

Prior to updating a MLACA block, we must first reconstruct the top-level $\mathbf{U}_0^{(1)}$ ($U$-type) or $\mathbf{V}_0^{(1)}$ ($V$-type) matrix. This is done on the CPU, following Section 10.2.1. Similarly, prior to any matrix product, the top-level intermediate product $\mathbf{T}_1$ in (10.2) is also computed on the CPU following Section 10.2.2.2.

Note that these reconstructions comprise a large number of small matrix products, however on the GPU, matrix products are most efficient on larger matrix products executed in single-kernel. Consequently, reconstruction on the GPU was found to be slower than on the CPU, and so these steps are always done on the CPU.

### 10.3.2.2 MLACA CPU Thread Class

The MLACA CPU thread class inherits from the ACA CPU thread class, re-using all of the existing functions for matrix filling, LU factorization and block-RHS solving. New functions are added for the matrix products outlined in (10.3-10.6) and (10.14-10.16).

### 10.3.2.3 MLACA GPU Thread Class

Similarly, the MLACA GPU thread class inherits from the ACA GPU thread class, reusing many of the existing functions, with new functions added to support the computation of MLACA matrix products on the GPU. The key differences involve workspace management, as discussed next.

*Matrix Filling*

In MLACA, the diagonal blocks are compressed. Thus, each thread needs a device workspace of size $S_{max}^L = N_{max}^L \times N_{max}^L$ for storing full diagonal blocks on level $l = L$, where $N_{max}^L$ is the largest sub-block size on that level. Additionally, each thread also needs a workspace of size $N_{max}$ for the rows and columns computed for all off-diagonal blocks on the top level $l = 0$. Thus, a single workspace space of size $\max(S_{max}^L, N_{max})$ is allocated for matrix-filling operations on the device.

*LU Factorization*

Staging of prior **L** and **U** blocks on the GPU is still done, however as $\mathbf{T}_1$ is computed on the CPU, only the top level $\mathbf{V}_0^{(1)}$ ($U$-type) and $\mathbf{U}_0^{(1)}$ ($V$-type) block-matrix portions are staged. As before, this is disabled if GPU workspace becomes limited.

For each thread, device workspaces of size $N_b \times N_{max}$ are again needed to store the full-size $\mathbf{L}_{sb}$ and $\mathbf{U}_{bs}$ blocks. Workspaces are also needed for the top level portions of the product matrices to be multiplied (or one, if the other was staged), and the intermediate products $\mathbf{T}_1$, $\mathbf{T}_2$ and $\mathbf{T}_3$. These are again sized by searching all matrix blocks above and to the right of the diagonal, and all blocks to the left and below the diagonal. In this case, we find the maximum top level block-matrix size $S_{max} = \max\left(\max(\mathbf{U}_0^{(1)}), \max(\mathbf{V}_0^{(1)})\right)$ and the maximum rank sum $K_{max}$. Two spaces of size $S_{max}$ are reserved for the top-level portions (or one, if staging), as well as a space of size $K_{max} \times K_{max}$ for $\mathbf{T}_1$ and a space of size $N_b \times K_{max}$ for $\mathbf{T}_2$ and $\mathbf{T}_3$.

### 10.3.3 Distributed Memory Processing

After implementing the code changes needed for MLACA-based shared memory processing, it was found that our existing distributed memory implementation for the single-level ACA, as discussed previously in Section 9.6.3, required very few code changes to support MLACA. The only new code implemented specifically for MPI operation was for serialization and deserialization of the MLACA matrix, required for message passing of that data type.

## 10.4 Numerical Examples

In this section we will compare the performance of the MLACA algorithm to the single-level ACA in computing the radar cross section of several test articles. We will again look at the conducting, dielectric, and coated sphere, benchmark targets with measured data, and the monoconic RV with dielectric nose considered previously in Section 9.7. The electrical size of these problems is again larger than in prior chapters, taking advantage of the extra compression afforded by MLACA. For each test case, our *Serenity* solver code is used, and triangle meshes are constructed with at least $\lambda/10$ unknowns per wavelength in each dielectric region. The CFIE is applied on closed conductors with $\alpha = 0.5$, and PMCHWT applied to dielectric interfaces.

### 10.4.1 Compute Platform

Calculations in this section were carried out using two versions of the *Serenity* solver. The first comprises the shared memory, GPU-accelerated version, executed on the Dell workstation described previously in Section 9.7.1. The second comprises a distributed memory, CPU-only *Serenity* solver that uses MPI for inter-process communication. The MPI data distribution and parallelization strategy was discussed previously in Section 9.6.3. The distributed memory system comprised a 50 node compute cluster, each node having dual twelve-core Intel Xeon CPUs (E5-2680 v3) at 2.50GHz (1200 total CPUs) and 128 GB RAM per node (6.4 TB total RAM), connected via a 10 gigabit Ethernet network fabric, running Red Hat Enterprise Linux.

### 10.4.2 Conducting Spheres

We first compare the memory and CPU requirements of the single-level ACA and MLACA, in both matrix filling and LU factorization, for perfectly electrically conducting (PEC) spheres of radius $a$ ranging from 1 to 32 $\lambda$. The GPU-accelerated *Serenity* solver is used. The number of triangles $N_T$ and unknowns $N$ for each sphere are summarized in Table 10.1. The target size of top-level basis function groups was $N_g = 4000$ unknowns, and $\tau_{ACA} = 10^{-4}$ for both ACA and MLACA. This $\tau_{ACA}$ was chosen so that the examples (except $a = 32\lambda$) would fit in available RAM. A much smaller value ($\tau_{ACA} \leq 10^{-5}$) would be required for an accurate solution. In this study, we only present results from MLACA levels up to $L = 4$, as the compression achieved for levels $L \geq 5$ was not significantly greater than for $L = 4$ and $N_g = 4000$. We will examine the effect of increasing the group size further in a later section.

Table 10.2 summarizes the matrix fill time ($T_Z$, in seconds) and the memory

**TABLE 10.1:** PEC Spheres: Dimensions and Number of Unknowns

| $a(\lambda)$ | 1 | 2 | 4 | 8 | 16 | 32 |
|---|---|---|---|---|---|---|
| $N_T$ | 5120 | 20480 | 81920 | 327680 | 1310720 | 5242880 |
| $N$ | 7680 | 30720 | 122880 | 491520 | 1966080 | 7864320 |

**TABLE 10.2:** PEC Spheres: Memory ($M_Z$, GB), Fill Time ($T_Z$, seconds)

| | ACA | | $L = 1$ | | $L = 2$ | | $L = 3$ | | $L = 4$ | |
|---|---|---|---|---|---|---|---|---|---|---|
| $a(\lambda)$ | $M_Z$ | $T_Z$ | $M_Z$ | $T_Z$ | $M_Z$ | $T_Z$ | $M_Z$ | $T_Z$ | $M_Z$ | $T_Z$ |
| 1 | .093 | 1.2 | .067 | 1.6 | .052 | 2.1 | .047 | 3.0 | .051 | 4.4 |
| 2 | 1.0 | 5.6 | .68 | 5.4 | .44 | 6.4 | 0.32 | 8.1 | .26 | 10.9 |
| 4 | 5.9 | 19 | 3.7 | 19 | 2.5 | 21 | 1.8 | 28 | 1.4 | 38 |
| 8 | 30 | 90 | 20 | 90 | 14 | 110 | 11 | 158 | 10 | 246 |
| 16 | 174 | 634 | 135 | 715 | 112 | 954 | 97 | 1472 | 81 | 2345 |
| 32* | 1560 | 4680 | 1373 | 6120 | 1125 | 9000 | 896 | 12960 | 748 | 23400 |

\* Results Estimated.

**TABLE 10.3:** PEC Spheres: Memory ($M_{LU}$, GB), LU Factorization Time ($T_{LU}$, seconds)

| | ACA | | $L = 1$ | | $L = 2$ | | $L = 3$ | | $L = 4$ | |
|---|---|---|---|---|---|---|---|---|---|---|
| $a(\lambda)$ | $M_{LU}$ | $T_{LU}$ | $M_{LU}$ | $T_{LU}$ | $M_{LU}$ | $T_{LU}$ | $M_{LU}$ | $T_{LU}$ | $M_{LU}$ | $T_{LU}$ |
| 1 | .093 | .5 | .067 | .9 | .053 | .8 | .048 | 1.0 | .052 | 1.2 |
| 2 | 1.0 | 7.6 | .69 | 7.5 | .46 | 7.7 | .34 | 7.7 | .27 | 9.4 |
| 4 | 6.5 | 41 | 4.1 | 51 | 2.8 | 46 | 2.1 | 48 | 1.7 | 53 |
| 8 | 37 | 562 | 25 | 566 | 18 | 563 | 14 | 568 | 12 | 579 |
| 16 | 248 | 21485 | 191 | 19594 | 151 | 21028 | 126 | 21333 | 110 | 21840 |

requirements of the MoM matrix $\mathbf{Z}$ ($M_Z$, in GB). The matrix for $a = 32\lambda$ would not fit in available RAM, however the final matrix size and fill time can be accurately estimated by *Serenity* during the fill operation. Thus, the results for this case were obtained by allowing *Serenity* to run until memory was exhausted.

Similarly, Table 10.3 summarizes the LU factorization time and the memory requirements of the **LU** matrix ($M_{LU}$, in GB). To see better the differences, in Table 10.4 are summarized the percent difference (decrease) in the size of the $\mathbf{Z}$ and **LU** matrix when using MLACA versus the single-level ACA. The additional compression obtained using MLACA is substantial, with its effectiveness increasing with level. From these results, we note two details: For

**TABLE 10.4:** Percent Difference (decrease) in $M_Z$ and $M_{LU}$: MLACA versus Single-Level ACA

| | $M_Z$ | | | | | | $M_{LU}$ | | | | |
|---|---|---|---|---|---|---|---|---|---|---|---|
| $a(\lambda)$ | 1 | 2 | 4 | 8 | 16 | 32 | 1 | 2 | 4 | 8 | 16 |
| $L=1$ | 32 | 38 | 47 | 40 | 25 | 13 | 32 | 37 | 45 | 59 | 26 |
| $L=2$ | 57 | 78 | 82 | 73 | 43 | 32 | 55 | 74 | 80 | 88 | 49 |
| $L=3$ | 66 | 103 | 108 | 93 | 57 | 54 | 64 | 99 | 102 | 107 | 65 |
| $L=4$ | 138 | 117 | 124 | 100 | 73 | 70 | 57 | 115 | 117 | 117 | 77 |

$a \geq 4$, MLACA compresses the LU blocks to a greater degree than it does the Z blocks. We also see that the percent increase in compression peaks at $4 \geq a \geq 2$, and then decreases somewhat. This is due to the diagonal blocks, where as the problem size (and number of groups) increase, comprise an increasingly smaller percentage of the total number of matrix blocks.

Table 10.5 summarizes the percent difference (increase) in matrix fill time $T_Z$ and LU factorization time $T_{LU}$. We have omitted results for $a < 4$ as these runs were so short that system overhead and other effects caused large variances in total run time. The increases in $T_Z$ is as expected, where the additional compression steps add significantly to the run time. However the increases in $T_{LU}$ are far more moderate, particularly for $a = 8\lambda$ and $a = 16\lambda$ where the increase in run time is only a few percent for all $L$. This suggests that the block updates (prior to recompression) consume the most time in the LU factorization, and that our MLACA block LU factorization is almost as efficient as our single-level ACA factorization.

In Figures 10.10a and 10.10b are plotted the memory requirements $M_Z$ and $M_{LU}$ versus the number of unknowns. In both figures, we observe that for larger problem sizes ($a \geq 8$), the ACA and MLACA both have a complexity of roughly $O(N^{1.4} \log N)$, somewhat higher than the $O(N^{1.28} \log N)$ complexity observed for the ACA with smaller problems in Section 9.7.8.

#### 10.4.2.1 Variation of Target Group Size

We next examine the performance of MLACA versus the target group size $N_g$ for the $16\lambda$ sphere having $N = 1966080$ unknowns. Summarized in Table 10.6 are the group sizes considered, and the resulting minimum ($N_{min}$), maximum ($N_{max}$), and average ($N_{avg}$) number of basis functions per group. We note that while $N_g \approx N_{avg}$ in each case, there is some variation about the average. This occurs because the algorithm we used to generate the sphere does not create facets of exactly the same size, and because the $K$-means algorithm

**TABLE 10.5:** Percent Difference (increase) in Fill Time $T_Z$ and LU Factorization Time $T_{LU}$: MLACA versus Single-Level ACA

|  | $T_Z$ | | | | $T_{LU}$ | | |
|---|---|---|---|---|---|---|---|
| $a(\lambda)$ | 4 | 8 | 16 | 32 | 4 | 8 | 16 |
| $L = 1$ | 0 | 0 | 12 | 27 | 22 | 1 | 9 |
| $L = 2$ | 10 | 20 | 40 | 63 | 11 | 0 | 2 |
| $L = 3$ | 38 | 55 | 80 | 94 | 16 | 1 | 1 |
| $L = 4$ | 67 | 93 | 115 | 133 | 26 | 3 | 2 |

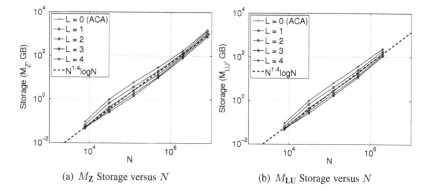

(a) $M_Z$ Storage versus $N$

(b) $M_{LU}$ Storage versus $N$

**FIGURE 10.10:** Storage versus Number of Unknowns $N$ for PEC Spheres

**TABLE 10.6:** $16\lambda$ PEC Sphere: Target Group Size ($N_g$), Minimum ($N_{min}$), Maximum ($N_{max}$), Average ($N_{avg}$)

| $N_g$ | $N_{min}$ | $N_{max}$ | $N_{avg}$ |
|---|---|---|---|
| 2500 | 1796 | 3516 | 2498 |
| 5000 | 3746 | 6853 | 4990 |
| 7500 | 6175 | 10165 | 7475 |
| 10000 | 8119 | 13732 | 9980 |
| 12500 | 10215 | 16083 | 12443 |

is iterative and does not guarantee the same number of basis functions in each group.

In Table 10.7 are summarized the memory requirements $M_Z$ and $M_{LU}$ for the single-level ACA and MLACA for levels $L \leq 5$. Notable are several trends in this data. For small $N_g$, the compression slows (or gets worse) with increasing $L$, and similar behavior occurs for small $L$ and increasing $N_g$. However,

**TABLE 10.7:** 16λ PEC Sphere: Memory ($M_Z$, GB), Memory ($M_{LU}$, GB) versus Target Group Size

| | ACA | | $L = 1$ | | $L = 2$ | | $L = 3$ | | $L = 4$ | | $L = 5$ | |
|---|---|---|---|---|---|---|---|---|---|---|---|---|
| $N_g$ | $M_Z$ | $M_{LU}$ | $M_Z$ | $M_{LU}$ | $M_Z$ | $M_{LU}$ | $M_Z$ | $M_{LU}$ | $M_Z$ | $M_{LU}$ | $M_Z$ | $M_{LU}$ |
| 2500 | 284 | 582 | 214 | 480 | 180 | 400 | 174 | 355 | 181 | 340 | 209 | 368 |
| 5000 | 247 | 525 | 190 | 408 | 152 | 317 | 117 | 262 | 103 | 233 | 105 | 229 |
| 7500 | 269 | 487 | 185 | 362 | 140 | 271 | 112 | 215 | 89 | 182 | 79 | 171 |
| 10000 | 295 | 487 | 197 | 347 | 136 | 251 | 106 | 193 | 87 | 158 | 71 | 142 |
| 12500 | 327 | 499 | 210 | 346 | 139 | 243 | 104 | 182 | 82 | 145 | 69 | 127 |

compression increases dramatically as both $N_g$ and $L$ are increased. In particular, we note that for $N_g = 12500$, $M_Z = 69$ GB and $M_{LU} = 127$ GB for the $L = 5$ MLACA, respectively, versus $M_Z = 327$ GB and $M_{LU} = 499$ GB for the single-level ACA.

### 10.4.3  Polyethylene Cone-Sphere

Next, we consider the RCS of a polyethylene conesphere, another benchmark target fabricated and measured in [5]. This object has the same dimensions as the EMCC cone-sphere described in Section 7.9.2.3, with an error of ±0.005 inches. The average permittivity $\epsilon \approx 2.35$, and only the HH-polarized RCS could be extracted from the figures in [5]. As with the polyethylene ogive in Section 9.7.5.2, a single facet model was constructed and used at all frequencies. This model had 371936 triangles, resulting in 557904 edges and 1115808 unknowns. For the MoM solution, an $L = 2$ level MLACA was used, with a target group size of $N_g = 5000$ unknowns. The memory requirements $M_Z$ and $M_{LU}$ versus frequency are summarized in Table 10.8. As was observed with the polyethylene ogive, the compression here also decreases versus frequency. The computed and measured results are compared for horizontal polarization in Figures 10.11a–10.11h, where tip-on incidence corresponds to an azimuth angle of 180 degrees. The comparison is fairly good at all frequencies and incident angles.

**TABLE 10.8:** Polyethylene Cone-Sphere: Memory ($M_Z$, GB), Memory ($M_{LU}$, GB)

| $f_0$ | 4.0 | 6.0 | 8.0 | 10.0 | 12.0 | 14.0 | 16.0 | 18.0 |
|---|---|---|---|---|---|---|---|---|
| $M_Z$ | 88 | 91 | 93 | 96 | 98 | 101 | 102 | 105 |
| $M_{LU}$ | 105 | 119 | 136 | 153 | 170 | 188 | 206 | 224 |

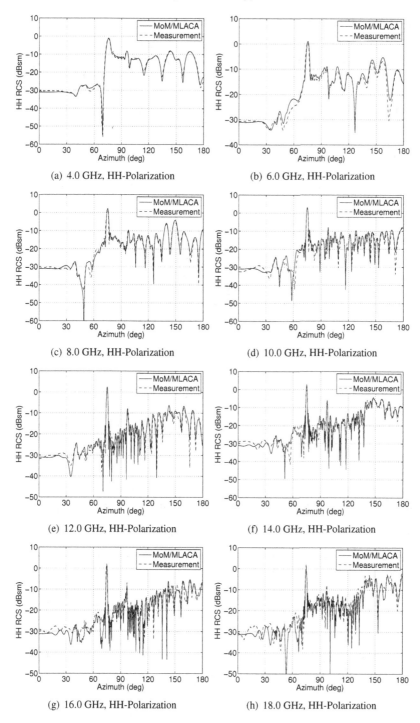

(a) 4.0 GHz, HH-Polarization

(b) 6.0 GHz, HH-Polarization

(c) 8.0 GHz, HH-Polarization

(d) 10.0 GHz, HH-Polarization

(e) 12.0 GHz, HH-Polarization

(f) 14.0 GHz, HH-Polarization

(g) 16.0 GHz, HH-Polarization

(h) 18.0 GHz, HH-Polarization

**FIGURE 10.11:** Polyethylene Cone-Sphere: RCS

### 10.4.4 Monoconic Reentry Vehicle

Next, we will examine the monoconic reentry vehicle considered previously in Sections 7.10.2.4, 8.8.8, and 9.7.7. In this case, we consider only the RV with dielectric nose, and we will compare the results from the single-level ACA to MLACA. We compute the monostatic RCS at 12.0, 14.0, 16.0, 20.0, and 24.0 GHz, where the number of triangles $N_T$, electric basis functions $N_J$, magnetic basis functions $N_M$, and total number of unknowns $N$ in each model are summarized in Table 10.9. For each example, $\tau_{ACA} = 10^{-5}$, and the target group size is $N_g = 10000$ unknowns. The results from the single-level ACA and $L = 5$ level MLACA are compared for vertical and horizontal polarizations across all frequencies in Figures 10.12a–10.12j, The comparison is very good.

**TABLE 10.9:** Monoconic Reentry Vehicle: Number of Triangles and Unknowns

| $f_0$(GHz) | 12.0 | 14.0 | 16.0 | 20.0 | 24.0 |
|---|---|---|---|---|---|
| $N_T$ | 787100 | 1065932 | 1167268 | 1842152 | 3252068 |
| $N_J$ | 1180548 | 1598780 | 1750780 | 2763076 | 4877898 |
| $N_M$ | 33348 | 46988 | 48220 | 80212 | 144420 |
| $N$ | 1213896 | 1645768 | 1799000 | 2843288 | 5022318 |

In Table 10.10 are summarized the memory requirements $M_Z$ and $M_{LU}$ for the single-level ACA and MLACA for levels $L \leq 5$. Using MLACA again results in a dramatic improvement in total memory usage versus the single-level ACA, particularly at higher frequencies when using more MLACA levels. To visualize these trends, the data are plotted in Figures 10.13a and 10.13b for $M_Z$ and $M_{LU}$, respectively. We note that for this particular problem, the $L = 5$ level MLACA yields the best compression, however the added compression decreased significantly between $L = 4$ and $L = 5$. It is unlikely that an $L = 6$ level MLACA would yield any additional benefit for this value of $N_g$.

**TABLE 10.10:** Monoconic Reentry Vehicle: Memory ($M_Z$, GB), Memory ($M_{LU}$, GB)

| $f_0$ | ACA | | $L = 1$ | | $L = 2$ | | $L = 3$ | | $L = 4$ | | $L = 5$ | |
|---|---|---|---|---|---|---|---|---|---|---|---|---|
| | $M_Z$ | $M_{LU}$ | $M_Z$ | $M_{LU}$ | $M_Z$ | $M_{LU}$ | $M_Z$ | $M_{LU}$ | $M_Z$ | $M_{LU}$ | $M_Z$ | $M_{LU}$ |
| 12.0 | 151 | 184 | 97 | 124 | 66 | 88 | 49 | 67 | 39 | 55 | 34 | 49 |
| 14.0 | 222 | 284 | 144 | 196 | 98 | 141 | 74 | 108 | 59 | 89 | 52 | 80 |
| 16.0 | 245 | 328 | 161 | 231 | 112 | 170 | 85 | 133 | 69 | 110 | 61 | 101 |
| 20.0 | 429 | 666 | 293 | 491 | 211 | 372 | 165 | 299 | 137 | 253 | 125 | 234 |
| 24.0 | 884 | 1841 | 634 | 1156 | 473 | 888 | 377 | 720 | 316 | 608 | 291 | 554 |

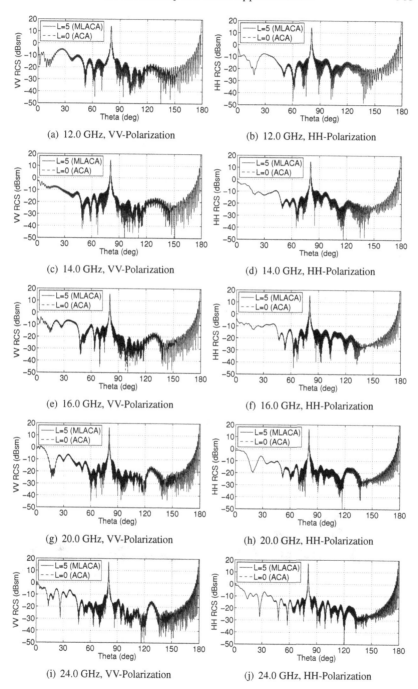

(a) 12.0 GHz, VV-Polarization

(b) 12.0 GHz, HH-Polarization

(c) 14.0 GHz, VV-Polarization

(d) 14.0 GHz, HH-Polarization

(e) 16.0 GHz, VV-Polarization

(f) 16.0 GHz, HH-Polarization

(g) 20.0 GHz, VV-Polarization

(h) 20.0 GHz, HH-Polarization

(i) 24.0 GHz, VV-Polarization

(j) 24.0 GHz, HH-Polarization

**FIGURE 10.12:** RV with Lossless Nose: RCS at 12.0, 14.0, 16.0, 20.0 and 24.0 GHz

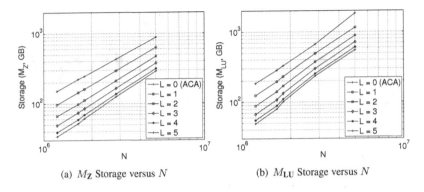

(a) $M_Z$ Storage versus $N$                    (b) $M_{LU}$ Storage versus $N$

**FIGURE 10.13:** Storage versus $N$ For the Monoconic Reentry Vehicle

### 10.4.4.1    EMCC Prism

Finally, we will again consider the EMCC Prism from Section 9.7.4.7, this time at a frequency of 9.2 GHz. For the prism, a facet model with 1481992 triangles was constructed, yielding a total of 2222988 unknowns. For the MoM solution, an $L = 3$ level MLACA was used, with a target group size of $N_g = 5000$ unknowns. The results from the MLACA are compared to the EMCC measurements in Figures 10.14a and 10.14b, for vertical and horizontal polarizations, respectively. The comparison is fairly good in vertical polarization, however the MoM underpredicts the RCS in horizontal polarization at many angles. The cause for this disagreement is unknown, and there was not a similar comparison available in [6] for cross-referencing.

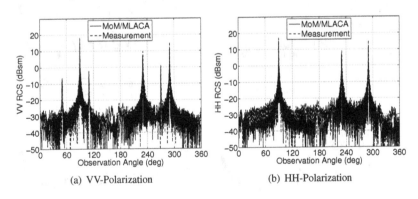

(a) VV-Polarization                         (b) HH-Polarization

**FIGURE 10.14:** EMCC Prism: MoM/MLACA vs. Measurement at 9.2 GHz

# References

[1] E. Michielssen and A. Boag, "A multilevel matrix decomposition algorithm for analyzing scattering from large structures," *IEEE Trans. Antennas Propagat.*, vol. 44, pp. 1086–1093, August 1996.

[2] J. Tamayo, A. Heldring, and J. Rius, "Multilevel adaptive cross approximation (MLACA)," *IEEE Trans. Antennas Propagat.*, vol. 59, pp. 4600–4608, December 2011.

[3] J. Tamayo, *Multilevel Adaptive Cross Approximation and Direct Evaluation Method for Fast and Accurate Discretization of Electromagnetic Integral Equations*. PhD thesis, Polytechnic University of Catalonia, 2011.

[4] W. C. Gibson, "Efficient solution of electromagnetic scattering problems using multilevel adaptive cross approximation (MLACA) and LU factorization," *IEEE Trans. Antennas Propagat.*, vol. 68, pp. 3815–3823, May 2020.

[5] J. M. Parks, "Scattering from Dielectric Bodies," Master's thesis, Air Force Institute of Technology, December 1997.

[6] A. Greenwood, "Electromagnetic Code Consortium Benchmarks," Tech. Rep. AFRL-DE-TR-2001-1086, Air Force Research Laboratory, December 2001.

# Chapter 11

## The Fast Multipole Method

In the previous two chapters, we applied the ACA and the MLACA to a direct solution of the MoM system matrix in compressed form using LU factorization. Overall, the compression levels were very good, with both methods having a storage complexity that scaled approximately as $O(N^{4/3} \log N)$. Compared to the ACA, MLACA was found to yield a much greater amount of compression, particularly for larger problems, at the expense of additional compute time in compressing the matrix blocks.

In this chapter we introduce the Fast Multiple Method (FMM), which also exploits the rank-deficient nature of the MoM system matrix but approaches the solution in a different way. Whereas the ACA and MLACA compute and store all matrix blocks, with off-diagonal blocks in compressed form, in the FMM only the diagonal and a fraction of the off-diagonal blocks are computed and stored directly. The remaining matrix elements are instead computed on the fly in aggregated form as part of the matrix-vector product in an iterative solver. In this way, the FMM achieves a $O(N^{3/2})$ complexity [1] at the expense of an iterative solution and extended run time. If the geometry is then divided in a recursive, hierarchical way, this yields a Multi-Level Fast Multipole Algorithm (MLFMA), which has an improved $O(N \log N)$ complexity [2] at the expense of a more complex software implementation. It is the MLFMA to which we devote much of our attention in this chapter.

The treatment that follows is novel in that it fuses several elements of the ACA with MLFMA, in particular its use of a block matrix approach and SVD-based compression of the off-diagonal matrix blocks and the FMM radiation/receive functions. This approach, which we call the MLFMA/SVD, retains the $O(N \log N)$ storage complexity, but yields a much greater memory savings versus the regular MLFMA.

## 11.1 The N-Body Problem

Consider the $N$-body problem, encountered often in many areas of applied physics. The most classic example is found in the motion of celestial objects interacting via gravitational attraction. For example, if we wish to know the position and velocity of a star at some point in time, the differential equations of motion must be numerically integrated in time. This requires a calculation of the gravitational potential due to all other nearby stars at each time step in the integration. Similarly, as the other stars are also moving, the gravitational forces on them, and their position and velocity, must also be computed. This results in an $N \times N$ problem at each point in time, where $N$ is the total number of stars. When $N$ is small, a brute-force pairwise calculation of the forces can be carried out at each time step. However, when $N$ becomes very large, the number of calculations in the brute-force approach may grow so large as to be impractical or impossible.

The Fast Multipole Method (FMM), originally introduced by Greengard and Rokhlin in [3], was successful in reducing the complexity of this $N$-body problem. In this method, the gravitational potentials of a group of closely spaced particles are re-written in terms of their multipole expansions. These are combined into a single function representing the potential of the entire group. The interaction between that group and any particles far away from that group can then be computed using the multipole expansion. Thus, the FMM allows for grouping together many nearby particles and treating them as if they are a single particle. The MoM system matrix can be viewed as another type of $N$-body problem, where each matrix element comprises the field radiated by one basis function and received by another.

As we will see, the FMM can be applied to vector electromagnetic problems, however the multipole expansions are specific to the Green's function and must be worked out specifically for the electromagnetic case. This differs from the ACA approach, which is method agnostic and requires no knowledge of the underlying problem.

## 11.2   Matrix-Vector Product

Let us first consider a matrix-vector product in the context of the Method of Moments, written as

$$\mathbf{x} = \mathbf{Z}\mathbf{y}, \tag{11.1}$$

where the system matrix $\mathbf{Z}$ represents the interactions (fields) between source and testing functions, and the right-hand side vector $\mathbf{y}$ comprises the excitation of each basis function. The rows of the result vector $\mathbf{x}$ comprised the coherent sum of those fields "received" by each testing function due to the fields "radiated" by all basis functions. Noting that the amplitude of the interaction between a source and testing pair diminishes quickly versus distance, let us di-

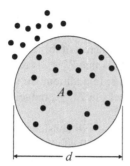

**FIGURE 11.1:** Near and Far Points

vide the space around each basis function into two regions: a *near*-field region having stronger interactions, and a *far*-field region with weaker interactions. This is illustrated in Figure 11.1, where point $A$ is at the center of a shaded circular region of diameter $d$. All points inside the circular region are said to lie in the *near* region, and those outside in the *far* region. Note that though they are similarly named, these should not be confused with the near and far field regions described in Section 3.5.

Consider next row $m$ of $\mathbf{Z}$, which we denote as $\mathbf{z}_{m*}$. The product of this row with the column vector $\mathbf{y}$ yields the column vector element $x_m$ given by

$$x_m = \mathbf{z}_{m*} \, \mathbf{y} \, , \tag{11.2}$$

which comprises the fields received by testing function $m$ due to all basis functions. As some basis functions are in the near region of testing function $m$ and the others are in the far region, the two vectors can be re-ordered and divided so that (11.2) now reads as

$$x_m = \mathbf{z}_{m*} \, \mathbf{y} = \mathbf{z}_{m*}^{near} \, \mathbf{y}^{near} + \mathbf{z}_{m*}^{far} \, \mathbf{y}^{far}, \tag{11.3}$$

where the elements in the sub-vectors $\mathbf{z}_{m*}^{near}$ and $\mathbf{y}^{near}$ are due to the source functions in the region near to testing function $m$, and those in $\mathbf{z}_{m*}^{far}$ and $\mathbf{y}^{far}$ are due to those in the far region. Extending this concept, if we now write the system matrix as

$$\mathbf{Z} = \mathbf{Z}^{near} + \mathbf{Z}^{far}, \tag{11.4}$$

where $\mathbf{Z}^{near}$ and $\mathbf{Z}^{far}$ are sparse, the matrix-vector product can be written as the sum of a *near product* and a *far product*, yielding

$$\mathbf{x} = \mathbf{x}^{near} + \mathbf{x}^{far} = \mathbf{Z}^{near}\mathbf{y} + \mathbf{Z}^{far}\mathbf{y}, \tag{11.5}$$

where $\mathbf{x}^{near}$ and $\mathbf{x}^{far}$ are sparse vectors. In the FMM, the matrix-vector product comprises two steps: a *near* product with $\mathbf{Z}^{near}$, where the elements are computed and stored explicitly, and a *far* product with $\mathbf{Z}^{far}$, which is computed in a different way on the fly. We will discuss the practical implementation of these steps throughout the rest of this chapter. As we will see, one of the primary benefits of the FMM is that $\mathbf{Z}^{near}$ is extremely sparse, greatly reducing the memory requirements. In addition, we will use the SVD to compress some of the off-diagonal blocks in $\mathbf{Z}^{near}$, reducing the memory footprint even further.

In the rest of this section, we will consider the far part of the matrix-vector product in more detail. First we will address the concepts of the addition theorem and wave translation, which allow us to treat groups of basis functions as a single group. Following this, we will then develop the FMM-specific expressions used to compute the far matrix elements.

### 11.2.1 Addition Theorem

Consider again the three dimensional Green's function (3.39), written as

$$G(\mathbf{r}, \mathbf{r}') = \frac{e^{-jk|\mathbf{r}-\mathbf{r}'|}}{4\pi|\mathbf{r}-\mathbf{r}'|} , \qquad (11.6)$$

where $\mathbf{r}'$ and $\mathbf{r}$ are the source and field points, respectively. If we now add to $\mathbf{r}$ a offset $\mathbf{x}$, (11.6) can be written as

$$\frac{e^{-jk|\mathbf{r}-\mathbf{r}'+\mathbf{x}|}}{4\pi|\mathbf{r}-\mathbf{r}'+\mathbf{x}|} = \frac{e^{-jk|\mathbf{R}+\mathbf{x}|}}{4\pi|\mathbf{R}+\mathbf{x}|} , \qquad (11.7)$$

where $\mathbf{R} = \mathbf{r} - \mathbf{r}'$. Provided that $|\mathbf{x}| < |\mathbf{R}|$, we now define the *addition theorem* for spherical waves, given by [1, 2]

$$\frac{e^{-jk|\mathbf{R}+\mathbf{x}|}}{4\pi|\mathbf{R}+\mathbf{x}|} = -\frac{jk}{4\pi}\sum_{l=0}^{\infty}(-1)^l(2l+1)j_l\big(k|\mathbf{x}|\big)h_l^{(2)}\big(k|\mathbf{R}|\big)P_l\big(\hat{\mathbf{x}}\cdot\hat{\mathbf{R}}\big) , \quad (11.8)$$

where $j_l(x)$ a spherical Bessel function of the first kind, $P_l(x)$ a Legendre polynomial of order $l$, and $h_l^{(2)}(x)$ is a spherical Hankel function of the second kind, i.e.,

$$h_l^{(2)}(x) = \sqrt{\frac{\pi}{2x}}H_{l+1/2}^{(2)}(x) = j_l(x) - jn_l(x) , \qquad (11.9)$$

where $j_l(x)$ and $n_l(x)$ are spherical Bessel functions of the first and second kinds, respectively. Examining (11.8) closely, we see that it is a way of writing a wave radiating from a point as if it were radiating from a different point nearby. Following Stratton [4], we can convert (11.8) to a surface integral on the unit sphere via

$$4\pi(-j^l)j_l\big(k|\mathbf{x}|\big)P_l\big(\hat{\mathbf{x}}\cdot\hat{\mathbf{R}}\big) = \oint_1 e^{-jk\hat{\mathbf{k}}\cdot\mathbf{x}}P_l\big(\hat{\mathbf{k}}\cdot\hat{\mathbf{R}}\big)\,dS , \qquad (11.10)$$

where $\hat{\mathbf{k}}$ is the radially oriented unit vector on the sphere. Using this expression allows us to then write

$$\frac{e^{-jk|\mathbf{R}+\mathbf{x}|}}{4\pi|\mathbf{R}+\mathbf{x}|} = \frac{k}{(4\pi)^2} \oint_1 e^{-jk\hat{\mathbf{k}}\cdot\mathbf{x}} \sum_{l=0}^{\infty} (-j)^{l+1}(2l+1)h_l^{(2)}\left(k|\mathbf{R}|\right)P_l(\hat{\mathbf{k}}\cdot\hat{\mathbf{R}})\,dS\,,$$

(11.11)

where the integration and summation have been exchanged. This exchange will be legitimate provided that we truncate the summation at some finite order $L$, which limits the maximum value of $|\mathbf{R}|$. Making this truncation, we can now write (11.11) as

$$\frac{e^{-jk|\mathbf{R}+\mathbf{x}|}}{4\pi|\mathbf{R}+\mathbf{x}|} = \oint_1 e^{-jk\hat{\mathbf{k}}\cdot\mathbf{x}}\, T_L(k,\hat{\mathbf{k}},\mathbf{R})\,dS\,,$$

(11.12)

where

$$T_L(k,\hat{\mathbf{k}},\mathbf{R}) = \frac{k}{(4\pi)^2} \sum_{l=0}^{L} (-j)^{l+1}(2l+1)h_l^{(2)}(k|\mathbf{R}|)P_l(\hat{\mathbf{k}}\cdot\hat{\mathbf{R}})\,.$$

(11.13)

We refer to $T_L(k,\hat{\mathbf{k}},\mathbf{R})$ as the *transfer function*, as it allows us to convert outgoing spherical waves at a source point to a set of incoming spherical waves at a field point. In practice, the integration on the unit sphere in (11.12) is carried out numerically with quadratures in the $\theta$ and $\phi$ dimensions. We will discuss the types of quadratures and required sampling rates in more detail in a later section.

## 11.2.2 Wave Translation

Let us illustrate the addition theorem by calculating the Green's function between a source point $\mathbf{r}'$ and observation point $\mathbf{r}$. We first define the vector

$$\mathbf{v} = \mathbf{r} - \mathbf{r}',$$

(11.14)

and then introduce two points $\mathbf{a}$ and $\mathbf{b}$ that are very close to $\mathbf{r}$ and $\mathbf{r}'$, respectively, as shown in Figure 11.2. We can now write (11.14) as

$$\mathbf{v} = (\mathbf{r} - \mathbf{a}) + (\mathbf{a} - \mathbf{b}) - (\mathbf{r}' - \mathbf{b})\,,$$

(11.15)

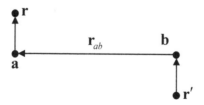

**FIGURE 11.2:** Wave Translation

or

$$\mathbf{v} = \mathbf{r}_{ra} + \mathbf{r}_{ab} - \mathbf{r}_{r'b} \; . \tag{11.16}$$

Using (11.16), we can now write (11.12) as

$$\frac{e^{-jk|\mathbf{r}-\mathbf{r}'|}}{4\pi|\mathbf{r}-\mathbf{r}'|} = \int_1 e^{-jk\hat{\mathbf{k}}\cdot(\mathbf{r}_{ra}-\mathbf{r}_{r'b})} T_L(k,\hat{\mathbf{k}},\mathbf{r}_{ab})\, dS \; , \tag{11.17}$$

where

$$T_L(k,\hat{\mathbf{k}},\mathbf{r}_{ab}) = \frac{k}{(4\pi)^2} \sum_{l=0}^{L} (-j)^{l+1} (2l+1) h_l^{(2)}\big(k|\mathbf{r}_{ab}|\big) P_l\big(\hat{\mathbf{k}}\cdot\hat{\mathbf{r}}_{ab}\big) \; . \tag{11.18}$$

This is an important result, as the transfer function depends only on $\mathbf{r}_{ab}$. If we were to move $\mathbf{r}$ and $\mathbf{r}'$ a small distance from their previous location, the transfer function remains unchanged. Therefore, we can compute the interaction between *any* two points near **a** and **b** using the *same* transfer function. Let us now write (11.17) as

$$\frac{e^{-jk|\mathbf{r}-\mathbf{r}'|}}{4\pi|\mathbf{r}-\mathbf{r}'|} = \int_1 R(\mathbf{r},\hat{\mathbf{k}})\, T_L(k,\hat{\mathbf{k}},\mathbf{r}_{ab}) T(\mathbf{r}',\hat{\mathbf{k}})\, dS \; , \tag{11.19}$$

where at the source point $\mathbf{r}'$, we define the *radiation function*

$$T(\mathbf{r}',\hat{\mathbf{k}}) = e^{jk\hat{\mathbf{k}}\cdot\mathbf{r}_{r'b}}, \tag{11.20}$$

and at the field point $\mathbf{r}$, we define the *receive function*

$$R(\mathbf{r},\hat{\mathbf{k}}) = e^{-jk\hat{\mathbf{k}}\cdot\mathbf{r}_{ra}}. \tag{11.21}$$

If we move $\mathbf{r}'$ or $\mathbf{r}$ to another location, only the radiation or receive function changes. As a result, we can now compute a sum of the Green's functions evaluated at the field point $\mathbf{r}$ due to *many* source points $\mathbf{r}'_n$ located close to **b**. Using (11.19), this can be written as

$$\sum_{n=1}^{N} \frac{e^{-jk|\mathbf{r}-\mathbf{r}'_n|}}{4\pi|\mathbf{r}-\mathbf{r}'_n|} = \int_1 R(\mathbf{r},\hat{\mathbf{k}})\, T_L(k,\hat{\mathbf{k}},\mathbf{r}_{ab}) \sum_{n=1}^{N} T_n(\mathbf{r}'_n,\hat{\mathbf{k}})\, dS \; , \tag{11.22}$$

where the individual radiation functions are

$$T_n(\mathbf{r}'_n,\hat{\mathbf{k}}) = e^{jk\hat{\mathbf{k}}\cdot\mathbf{r}_{r'_n b}} \; . \tag{11.23}$$

It is (11.22) that allows us to quickly compute the far matrix-vector product. The radiation functions for all source points are coherently added (*aggregated*) to create a local field at **b**. This field is then *transmitted* via the transfer function yielding a local field at **a**. The local field is then multiplied with the receive function at the field point and integrated on the unit sphere (*disaggregated*) yielding the desired sum. In a practical FMM implementation, the basis and testing functions are clustered into groups. The radiation and receive functions for all basis and testing functions, as well as the transfer functions linking together all groups, are pre-computed. The far part of the matrix-vector product then comprises aggregations, transfers, and disaggregations between well-separated groups.

### 11.2.2.1 Complex Wavenumbers

In lossy dielectric regions having complex valued wavenumbers, the transfer function (11.18) remains valid, however special numerical routines are needed to compute the spherical Hankel function for non-integer orders and complex arguments. The subroutines by Amos [5] are ideal for this purpose, in particular the routines `cbesh` and `zbesh` for single- and double-precision complex variables, respectively. These routines are available via the Netlib repository[1], and are often integrated in software libraries for computing special functions.

## 11.2.3  Far Matrix Elements

Having discussed the addition theorem and the concepts of wave translation, we will now derive expressions for the radiation and receive functions used to compute the MoM matrix elements (3.185–3.190). Though the examples in this chapter use a three-dimensional approach with the RWG basis and testing functions, the expressions presented in this section are independent of basis function. We remind the reader that though the radiation and receive functions are derived in the context of a single source-field pair, the FMM matrix-vector product operates on aggregated (summed) radiation functions and individual matrix elements are not formed explicitly.

Note that we will allow the points **a** and **b** to remain undefined for now. They will be specified later when we discuss clustering the basis functions into groups.

### 11.2.3.1  EFIE

The EFIE (3.178) contains $\mathcal{L}$ and $\mathcal{K}$ operators, which were discretized and tested via the MoM in (3.185) and (3.186). We will derive the corresponding radiation and receive functions in this section.

---

[1]http://www.netlib.org/amos

$\mathcal{L}$ *Operator*

Let us focus our attention on the second term on the right in (3.185). As the FMM interactions are not in the near zone, we do not need to re-distribute the differential operators as in Section 3.6.2.3. Instead, we follow Section 3.4.4 and move them so that they operate only on the Green's function. This yields in region $R_l$ matrix elements given by

$$Z_{mn}^{EJ(l)} = \mathrm{L}(\mathbf{f}_m, \mathbf{f}_n) , \qquad (11.24)$$

where

$$\mathrm{L}(\mathbf{f}_m, \mathbf{f}_n) = j\omega\mu_l \int_{\mathbf{f}_m} \mathbf{f}_m(\mathbf{r}) \cdot \int_{\mathbf{f}_n} \mathbf{f}_n(\mathbf{r}') \left[1 - \frac{1}{k_l^2}\nabla\nabla'\right] G(\mathbf{r}, \mathbf{r}')\, d\mathbf{r}'\, d\mathbf{r} . \quad (11.25)$$

We next apply the differential operations to the Green's function in (11.17). This is accomplished by noting that the $T_L(k, \hat{\mathbf{k}}, \mathbf{r}_{ab})$ is not a function of $r$ or $r'$, and the vectors $\mathbf{r}_{ra}$ and $\mathbf{r}_{r'b}$ are radially directed vectors in field and source coordinates, respectively, with $\mathbf{a}$ and $\mathbf{b}$ as the origin. Inserting the result into (11.25) yields an expression of the form

$$\mathrm{L}(\mathbf{f}_m, \mathbf{f}_n) = \int_1 \mathbf{R}^{\mathcal{L}}(\mathbf{f}_m, \hat{\mathbf{k}}) \cdot T_L(k_l, \hat{\mathbf{k}}, \mathbf{r}_{ab})\mathbf{T}^{\mathcal{L}}(\mathbf{f}_n, \hat{\mathbf{k}})\, dS , \qquad (11.26)$$

where the radiation function for basis function $\mathbf{f}_n(\mathbf{r}')$ is

$$\mathbf{T}^{\mathcal{L}}(\mathbf{f}_n, \hat{\mathbf{k}}) = \left[1 - \hat{\mathbf{k}}\hat{\mathbf{k}}\right] \cdot \int_{\mathbf{f}_n} \mathbf{f}_n(\mathbf{r}')e^{jk_l\hat{\mathbf{k}}\cdot\mathbf{r}_{r'b}}\, d\mathbf{r}' \qquad (11.27)$$

and the receive function for testing function $\mathbf{f}_m(\mathbf{r})$ is

$$\mathbf{R}^{\mathcal{L}}(\mathbf{f}_m, \hat{\mathbf{k}}) = j\omega\mu_l \int_{\mathbf{f}_m} \mathbf{f}_m(\mathbf{r})e^{-jk_l\hat{\mathbf{k}}\cdot\mathbf{r}_{ra}}\, d\mathbf{r} . \qquad (11.28)$$

We note that the $\left[1 - \hat{\mathbf{k}}\hat{\mathbf{k}}\right]$ term in (11.27) removes the components of the integral along the radial vector $\hat{\mathbf{k}}$, leaving only $\hat{\theta}$ and $\hat{\phi}$ components.

$\mathcal{K}$ *Operator*

As the source and testing functions are well separated, the supports of $\mathbf{f}_m$ and $\mathbf{g}_n$ do not overlap and we retain only the first term on the right-hand side of (3.186). This yields in Region $R_l$ matrix elements given by

$$Z_{mn}^{EM(l)} = \mathrm{K}(\mathbf{f}_m, \mathbf{g}_n) , \qquad (11.29)$$

where

$$\mathrm{K}(\mathbf{f}_m, \mathbf{g}_n) = \int_{\mathbf{f}_m} \mathbf{f}_m(\mathbf{r}) \cdot \int_{\mathbf{g}_n} \mathbf{g}_n(\mathbf{r}) \times \nabla' G(\mathbf{r}, \mathbf{r}')\, d\mathbf{r}'\, d\mathbf{r} . \qquad (11.30)$$

Taking the gradient of (11.17) and re-arranging the cross product yields

$$K(\mathbf{f}_m, \mathbf{g}_n) = \int_1 \mathbf{R}^{\mathcal{K}}(\mathbf{f}_m, \hat{\mathbf{k}}) \cdot T_L(k_l, \hat{\mathbf{k}}, \mathbf{r}_{ab}) \mathbf{T}^{\mathcal{K}}(\mathbf{g}_n, \hat{\mathbf{k}}) \, dS , \qquad (11.31)$$

where the radiation function for basis function $\mathbf{g}_n(\mathbf{r}')$ is

$$\mathbf{T}^{\mathcal{K}}(\mathbf{g}_n, \hat{\mathbf{k}}) = \int_{\mathbf{g}_n} \mathbf{g}_n(\mathbf{r}') e^{jk_l \hat{\mathbf{k}} \cdot \mathbf{r}_{r'b}} \, d\mathbf{r}' \qquad (11.32)$$

and the receive function for testing function $\mathbf{f}_m(\mathbf{r})$ is

$$\mathbf{R}^{\mathcal{K}}(\mathbf{f}_m, \hat{\mathbf{k}}) = jk_l \, \hat{\mathbf{k}} \times \int_{\mathbf{f}_m} \mathbf{f}_m(\mathbf{r}) e^{-jk_l \hat{\mathbf{k}} \cdot \mathbf{r}_{ra}} \, d\mathbf{r} . \qquad (11.33)$$

The cross product with $\hat{\mathbf{k}}$ in (11.33) results in the receive function having only $\hat{\theta}$ and $\hat{\phi}$ components. An $\eta_0$ scale factor must also be applied to the radiation function to account for the magnetic basis function.

### 11.2.3.2  MFIE

The MFIE (3.179) also has $\mathcal{K}$ and $\mathcal{L}$ operators, which were discretized and tested via the MoM in (3.187) and (3.188). Substitution of the electric basis and magnetic testing functions into (3.187) yields in Region $R_l$ matrix elements given by

$$Z_{mn}^{HJ(l)} = -K(\mathbf{g}_m, \mathbf{f}_n) , \qquad (11.34)$$

and substitution of the magnetic basis and testing functions into (3.188) yields

$$Z_{mn}^{HM(l)} = \frac{\epsilon_l}{\mu_l} L(\mathbf{g}_m, \mathbf{g}_n) , \qquad (11.35)$$

where the L and K elements were defined in Section 11.2.3.1. An $\eta_0$ scale factor must also be applied to the radiation and receive functions to account for the presence of the magnetic basis functions, where appropriate.

### 11.2.3.3  nMFIE

The nMFIE (3.180) contains the $\hat{\mathbf{n}} \times \mathcal{L}$ and $\hat{\mathbf{n}} \times \mathcal{K}$ operators, which were discretized and tested via the MoM in (3.189) and (3.190). We derive their radiation and receive functions in this section.

#### $\hat{\mathbf{n}} \times \mathcal{K}$ Operator

As the basis and testing functions are well separated, the supports of $\mathbf{f}_m$ and $\mathbf{f}_n$ do not overlap and we retain only the first term on the right-hand side in (3.188). This yields in Region $R_l$ matrix elements given by

$$Z_{mn}^{nHJ(l)} = nK(\mathbf{f}_m, \mathbf{f}_n) , \qquad (11.36)$$

where

$$nK(\mathbf{f}_m, \mathbf{f}_n) = -\int_{\mathbf{f}_m} \mathbf{f}_m(\mathbf{r}) \cdot \left[ \hat{\mathbf{n}}_l(\mathbf{r}) \times \int_{\mathbf{f}_n} \mathbf{f}_n(\mathbf{r}) \times \nabla' G(\mathbf{r}, \mathbf{r}') \, d\mathbf{r}' \right] d\mathbf{r} \,. \quad (11.37)$$

Taking the gradient of (11.17) and re-arranging the cross products yields

$$nK(\mathbf{f}_m, \mathbf{f}_n) = \int_1 \mathbf{R}^{\hat{\mathbf{n}} \times \mathcal{K}}(\mathbf{f}_m, \hat{\mathbf{k}}) \cdot T_L(k_l, \hat{\mathbf{k}}, \mathbf{r}_{ab}) \mathbf{T}^{\hat{\mathbf{n}} \times \mathcal{K}}(\mathbf{f}_n, \hat{\mathbf{k}}) \, dS \,, \quad (11.38)$$

where the radiation function for basis function $\mathbf{f}_n(\mathbf{r}')$ is

$$\mathbf{T}^{\hat{\mathbf{n}} \times \mathcal{K}}(\mathbf{f}_n, \hat{\mathbf{k}}) = \int_{\mathbf{f}_n} \mathbf{f}_n(\mathbf{r}') e^{jk_l \hat{\mathbf{k}} \cdot \mathbf{r}_{r'b}} \, d\mathbf{r}' \quad (11.39)$$

and the receive function for testing function $\mathbf{f}_m(\mathbf{r})$ is

$$\mathbf{R}^{\hat{\mathbf{n}} \times \mathcal{K}}(\mathbf{f}_m, \hat{\mathbf{k}}) = -jk_l \, \hat{\mathbf{k}} \times \int_{\mathbf{f}_m} \left[ \mathbf{f}_m(\mathbf{r}) \times \hat{\mathbf{n}}_l(\mathbf{r}) \right] e^{-jk_l \hat{\mathbf{k}} \cdot \mathbf{r}_{ra}} \, d\mathbf{r} \,. \quad (11.40)$$

We note that due to the cross product with $\hat{\mathbf{k}}$, the receive function has only $\hat{\theta}$ and $\hat{\phi}$ components.

### $\hat{n} \times \mathcal{L}$ *Operator*

Substituting the magnetic basis and electric testing functions into (3.190) yields in Region $R_l$ matrix elements given by

$$Z_{mn}^{nHM(l)} = nL(\mathbf{f}_m, \mathbf{g}_n) \,, \quad (11.41)$$

where

$$nL(\mathbf{f}_m, \mathbf{g}_n) = j\omega\epsilon_l \int_{\mathbf{f}_m} \mathbf{f}_m(\mathbf{r}) \cdot \left[ \hat{\mathbf{n}}_l(\mathbf{r}) \times \int_{\mathbf{g}_n} \mathbf{g}_n(\mathbf{r}') \left[ 1 - \frac{1}{k_l^2} \nabla\nabla' \right] G(\mathbf{r}, \mathbf{r}') \, d\mathbf{r}' \right] d\mathbf{r} \,. \quad (11.42)$$

Applying the differential operators yields a result similar to the $\mathcal{L}$ operator in Section 11.2.3.1, which is

$$nL(\mathbf{f}_m, \mathbf{g}_n) = \int_1 \mathbf{R}^{\hat{\mathbf{n}} \times \mathcal{L}}(\mathbf{f}_m, \hat{\mathbf{k}}) \cdot T_L(k_l, \hat{\mathbf{k}}, \mathbf{r}_{ab}) \mathbf{T}^{\hat{\mathbf{n}} \times \mathcal{L}}(\mathbf{g}_n, \hat{\mathbf{k}}) \, dS \,, \quad (11.43)$$

where the radiation function for basis function $\mathbf{g}_n(\mathbf{r}')$ is

$$\mathbf{T}^{\hat{\mathbf{n}} \times \mathcal{L}}(\mathbf{g}_n, \hat{\mathbf{k}}) = \left[ 1 - \hat{\mathbf{k}}\hat{\mathbf{k}} \right] \cdot \int_{\mathbf{g}_n} \mathbf{g}_n(\mathbf{r}') e^{jk_l \hat{\mathbf{k}} \cdot \mathbf{r}_{r'b}} \, d\mathbf{r}' \quad (11.44)$$

and the receive function for testing function $\mathbf{f}_m(\mathbf{r})$ is

$$\mathbf{R}^{\hat{\mathbf{n}} \times \mathcal{L}}(\mathbf{f}_m, \hat{\mathbf{k}}) = j\omega\epsilon_l \int_{\mathbf{f}_m} \left[ \mathbf{f}_m(\mathbf{r}) \times \hat{\mathbf{n}}(\mathbf{r}) \right] e^{-jk_l \hat{\mathbf{k}} \cdot \mathbf{r}_{ra}} \, d\mathbf{r} \,. \quad (11.45)$$

We note that the radiation function has only $\hat{\theta}$ and $\hat{\phi}$ components. A $\eta_0$ scale factor must also be applied to account for the presence of the magnetic basis function.

### 11.2.4 Unit Sphere Decomposition

In the previous sections we noted that for each matrix element, the radiation or receive function had only $\hat{\theta}$ and $\hat{\phi}$ components. As a result, the inner product of the radiation and receive functions operates on only those components, and any $\hat{k}$-oriented components are disregarded. Therefore, when pre-computing these functions, only the scalar component in each dimension is stored. For example, when pre-computing the value of the receive function $\mathbf{R}^{\mathcal{L}}$ at each quadrature point on the unit sphere, one would calculate and store the scalar values $R^{\mathcal{L}}_{\theta}$ and $R^{\mathcal{L}}_{\phi}$, which are

$$R^{\mathcal{L}}_{\theta,\phi} = (\hat{\theta}, \hat{\phi}) \cdot \mathbf{R}^{\mathcal{L}} \qquad (11.46)$$

and similarly, one would store the scalar values $T^{\mathcal{L}}_{\theta}$ and $T^{\mathcal{L}}_{\phi}$ for the radiation function $\mathbf{T}^{\mathcal{L}}$.

Though the receive functions for the various operators have different forms, a brief review of the radiation functions in Sections 11.2.3.1–11.2.3.3 shows them to be identical when decomposed this way. Thus, only one radiation function is needed for each basis function. These have the form

$$\mathbf{T}^{J}(\mathbf{f}_n, \hat{\mathbf{k}}) = \left[ 1 - \hat{\mathbf{k}}\hat{\mathbf{k}} \right] \cdot \int_{\mathbf{f}_n} \mathbf{f}_n(\mathbf{r}')e^{jk_l\hat{\mathbf{k}}\cdot\mathbf{r}_{r'b}} \, d\mathbf{r}' \qquad (11.47)$$

for the electric basis functions, and

$$\mathbf{T}^{M}(\mathbf{g}_n, \hat{\mathbf{k}}) = \left[ 1 - \hat{\mathbf{k}}\hat{\mathbf{k}} \right] \cdot \int_{\mathbf{g}_n} \mathbf{g}_n(\mathbf{r}')e^{jk_l\hat{\mathbf{k}}\cdot\mathbf{r}_{r'b}} \, d\mathbf{r}' \qquad (11.48)$$

for the magnetic basis functions. In general, $\mathbf{T}^{J}$ and $\mathbf{T}^{M}$ are not necessarily the same since $\mathbf{f}_n$ and $\mathbf{g}_n$ may be different, however for RWG basis functions they are identical except for the $\eta_0$ scale factor in the magnetic basis function. Thus, we can compute and store only one radiation function for each RWG function.

---

## 11.3 One-Level Fast Multipole Algorithm

We will now consider the practical implementation details of the matrix-vector product in the context of a one-level Fast Multipole Algorithm (FMA). We will discuss grouping of basis functions into near and far groups, truncation of the transfer function, sampling rates on the unit sphere, and calculation and storage of radiation and receive functions and the near matrix. In Section 11.4, we will then utilize elements of the one-level algorithm in developing the Multi-Level Fast Multipole Algorithm (MLFMA), which is the preferred approach used in virtually all MoM solvers.

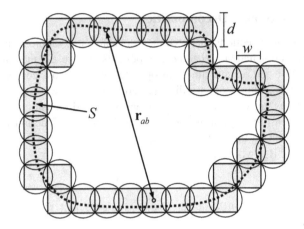

**FIGURE 11.3:** Spatial Subdivision and Grouping of Basis Functions

## 11.3.1   Clustering of Basis Functions

For most efficient programming and execution, a single sample rate and number of multipoles should be used to compute and store all radiation, receive, and transfer functions. Consequently, those values will depend on the diameter of the largest group. We also note that since the transfer function depends on the vector $\mathbf{r}_{ab}$ between group pairs, a total of $M^2$ transfer functions would be needed, assuming $M$ randomly centered groups. Thus, the $K$-means algorithm that was used to cluster basis functions for the ACA (Section 9.3) is not optimal in this case. Instead, let us instead enclose the entire surface $S$ by a bounding cube, which then is divided into a number of small, equally sized cubes with sides of length $w$ and a bounding sphere of diameter $d = \sqrt{3w^2}$. We set $w$ to some fraction of the free space wavelength, such as $\lambda_0/2$. Basis functions are sorted into cubes by testing the bounding box against the center of the edge a function is assigned to. When complete, non-empty cubes and their basis functions comprise the groups, and all empty cubes are discarded, resulting in an arrangement similar to the one depicted in Figure 11.3. The points $\mathbf{a}$ and $\mathbf{b}$, left undefined in Section 11.2.3, now comprise the centers of the cubes. As a result of this regularly spaced grid, there are now a limited number of unique transfer vectors $\mathbf{r}_{ab}$, greatly reducing the number of transfer functions that must be stored.

### 11.3.1.1   Classification of Near and Far Groups

To classify pairs of groups as *near* groups or *far* groups, we must define what it means to be *well separated* in the context of the FMM. Consider two groups with bounding boxes centered at $\mathbf{a}$ and $\mathbf{b}$. For the addition theorem

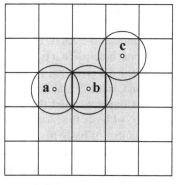

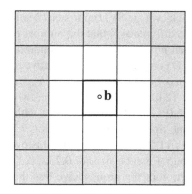

(a) Groups Near To Group **b**          (b) Groups Far From Group **b**

**FIGURE 11.4:** Near and Far Groups

of (11.8) to remain valid, the distance between $|\mathbf{r}_{ab}|$ between groups must be greater than or equal to at least one diameter $d$ [6]. Whether two groups are well separated is now a measure of the distance between their bounding spheres. Consider the groups as shown in Figure 11.4a, where the group at **b** comprises the box in the middle. Boxes sharing an edge with this one, such as the one at **a**, are separated by less than $d$. Those that share a corner, such as the one at **c**, are separated by exactly $d$. Thus, we label as *near* all boxes that share an edge or a corner. For the box at **b**, this includes itself and all the boxes it touches, which comprise the shaded boxes in Figure 11.4a. The *far* groups are those that are separated by at least one box, and comprise the shaded boxes in Figure 11.4b. The near-zone boxes that lie between the group at **b** and the far groups are often referred to as *buffer boxes*.

Following this clustering and classification strategy, the MoM matrix now comprises the sum of two sparse block matrices, $\mathbf{Z}^{near}$ and $\mathbf{Z}^{near}$. Basis functions can now be re-indexed following the same approach in Section 9.3.

## 11.3.2   Near Matrix

As $\mathbf{Z}^{near}$ comprises a block matrix, it can computed using the same software routines used to fill the ACA and MLACA block matrix, with a few modifications as discussed in Section 11.6.1. Compression of the off-diagonal blocks is discussed next.

### 11.3.2.1   Compression of Near Matrix

The diagonal blocks of $\mathbf{Z}^{near}$ are not compressible, and are computed and stored in full form. However, it is possible to compress some of the off-diagonal blocks via the SVD approach (Section 9.1) and store them in compressed form. The difference is that the number of basis functions per group in the FMM is often no more than a few hundred, versus thousands in the ACA, and so the SVD compression is not effective for all off-diagonal blocks. That is, for some matrix blocks of size $m \times n$, $k_{SVD}(m + n) \geq mn$ for a given $\tau_{SVD}$.

To illustrate, let us consider two examples from Chapter 9: the PEC Sphere at 6.0 GHz (Section 9.7.3.1) with 491520 basis functions, and the EMCC Cone-Sphere at 9 GHz (Section 9.7.4.3) with 119244 basis functions. In Figures 11.5a–11.5f are plotted histograms of the number of basis functions per group, for cube sizes of $0.25\lambda$, $0.375\lambda$ and $0.5\lambda$. We see that at $0.25\lambda$, a large portion of the groups have less than 20 basis functions. Off-diagonal blocks involving these groups are typically not compressible. As the cube size increases, we see that the average number increases, and at $0.5\lambda$ there are many groups with over 100 basis functions. For each case, we now compute all near matrix off-diagonal blocks, and perform an SVD-based compression to determine their compressibility. In Figures 11.6a–11.6f are plotted corresponding histograms of the compression of the blocks, in percent, for $\tau_{SVD} = 10^{-4}$. We note that in all cases, while there are many non-compressible blocks, this number decreases with increasing cube size, and the compressibility of individual blocks improves significantly. In Table 11.1 is summarized the compression of the the near matrix, taking into account the non-compressible diagonal blocks. It is small at $0.25\lambda$, but significant gains are obtained at $0.375\lambda$ and $0.5\lambda$. Though not as substantial as in the ACA and MLACA, which often yield over 98% compression, the near matrix size is still greatly reduced. This will be even more significant in Section 11.5, where we use the SVD to compress off-diagonal preconditioner matrix blocks, but with a less restrictive $\tau_{SVD}$.

In practice, our approach for computing off-diagonal blocks is to fill them completely, and then perform compression via direct application of the SVD. As the blocks are typically very small, this approach is more efficient than applying the ACA+QR/SVD technique. If $k_{SVD}(m + n) < mn$ for the given $\tau_{SVD}$, we retain the compressed version, otherwise the full version is stored.

**TABLE 11.1:** Near Matrix Percent Compression

|  | PEC Sphere | | | Cone-Sphere | | |
|---|---|---|---|---|---|---|
| Cube Size ($\lambda$) | 0.25 | 0.375 | 0.5 | 0.25 | 0.375 | 0.5 |
| Compression (%) | 0.3 | 15.6 | 35.6 | 9.4 | 38.3 | 53.5 |

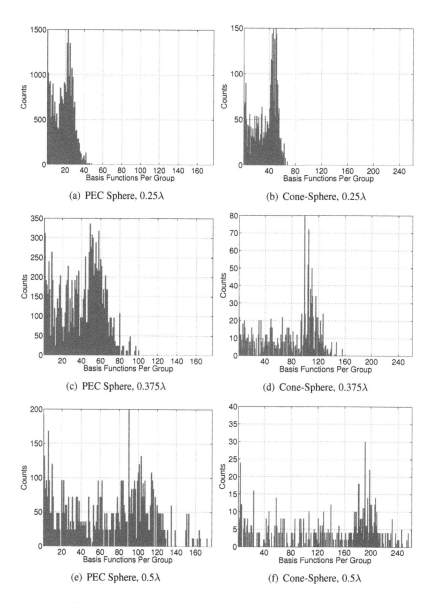

(a) PEC Sphere, 0.25λ

(b) Cone-Sphere, 0.25λ

(c) PEC Sphere, 0.375λ

(d) Cone-Sphere, 0.375λ

(e) PEC Sphere, 0.5λ

(f) Cone-Sphere, 0.5λ

**FIGURE 11.5:** Basis Functions Per Group For Different Cube Sizes

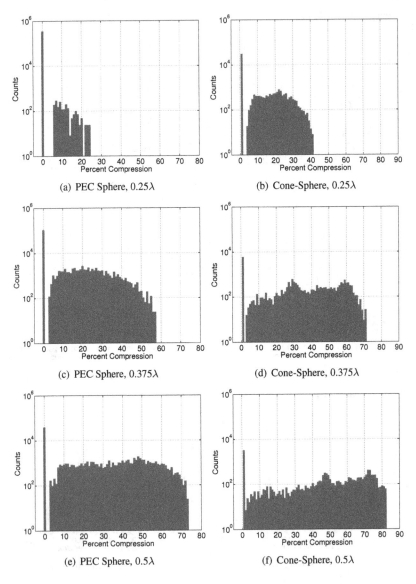

(a) PEC Sphere, 0.25λ

(b) Cone-Sphere, 0.25λ

(c) PEC Sphere, 0.375λ

(d) Cone-Sphere, 0.375λ

(e) PEC Sphere, 0.5λ

(f) Cone-Sphere, 0.5λ

**FIGURE 11.6:** Compression of Off-Diagonal Blocks For Different Cube Sizes

### 11.3.3  Number of Multipoles

An accurate evaluation of the integral (11.12) requires truncation of the transfer function (11.13) at some limit $L$, which we refer to as the number of *multipoles*. This limit depends on the size of the groups and the wavenumber, and expressions for it have been determined empirically. For real-valued $k$, Rokhlin prescribes for single precision calculations the formula [1]

$$L = kD + 5\log(kD + \pi) \,, \tag{11.49}$$

and for double precision calculations,

$$L = kD + 10\log(kD + \pi) \,, \tag{11.50}$$

where $D = |\mathbf{x}|$ in (11.12). The maximum possible value of $D$ to be accounted for occurs when $\mathbf{r} - \mathbf{r}'$ and $\mathbf{a} - \mathbf{b}$ are co-linear. With basis functions grouped by cubes, this value is equal to the bounding sphere diameter $d$, so we take $D = d$. Another expression was presented later by Chew and Song in [7] and is

$$L = kd + \beta(kd)^{1/3} \,, \tag{11.51}$$

where $\beta$ is the number of digits of accuracy required. They suggest that $\beta = 6$ is sufficient for most purposes. The same formula is also used in [8] for the optimal sampling of scattered fields on the sphere.

#### 11.3.3.1  Limiting $L$ for Transfer Functions

Though there are many different separations $|\mathbf{r}_{ab}|$ between groups, using a single value of $L$ from (11.51) for all $|\mathbf{r}_{ab}|$ will present numerical issues. This is because the spherical Hankel function $h_l^{(2)}(x)$ becomes highly oscillatory for fixed $x$ and increasing $l$. Therefore, we cannot take $L$ to be much larger than the argument of the Hankel function, which is $|k_l\mathbf{r}_{ab}|$ in region $R_l$. Our approach is to simply limit the value of $L$ to the minimum of either (11.51) or $|k_l\mathbf{r}_{ab}|$ in each region. Though this has the effect of reducing the accuracy of the expansion for groups having the smallest separations, in practice this does not appear to significantly affect the results.

#### 11.3.3.2  $L$ for Complex Wavenumbers

The expressions for $L$ in Section 11.3.3 were derived empirically for real-valued wavenumbers. However, in lossy dielectric regions having complex wavenumbers, the number of terms required in the addition theorem increases, and (11.51) is no longer valid. When the losses are small, $L$ increases only slightly, and the approximation

$$L = |kd| + \beta(|kd|)^{1/3} \tag{11.52}$$

is adequate in most cases. For large losses the increase in $L$ may be much larger, and it must be determined adaptively [9].

## 11.3.4    Integration on the Sphere

Numerical integration of (11.19) requires that the radiation, receive and transfer functions are computed and stored at discrete angles in the $\theta$ and $\phi$ dimensions on the unit sphere. We must choose the sampling rate in each dimension carefully, so that memory requirements are minimized while still meeting the Nyquist rate. As these functions are of finite (or *quasi-finite*) bandwidth [10], we need to determine the extent of the bandwidth, how it affects the sampling rate, and how the bandwidth changes when forming products of these functions.

Consider a function $f_1(x)$ having bandwidth $A$, sampled at a rate $N_1$, and a similar function $f_2(x)$ of bandwidth $B$ sampled at a rate $N_2$. To then calculate the integral

$$\int_{x_1}^{x_2} f_1(x) f_2(x)\, dx \tag{11.53}$$

we can use a numerical quadrature of the form

$$\int_{x_1}^{x_2} f_1(x) f_2(x)\, dx = \sum_{i=1}^{N_3} w(x_i) f_1(x_i) f_2(x_i)\,, \tag{11.54}$$

where we use a new sampling rate $N_3$. Because the bandwidth of the product $f_1(x) f_2(x)$ is the sum of the individual bandwidths, the sample rate $N_3$ must be sufficiently high to adequately capture the bandwidth of the product. This presents us with a dilemma, as the sampling rate needed to numerically evaluate the Green's function in (11.19) is greater than the rate needed to sample the radiation, receive, and transfer functions individually. We can save memory by storing these functions at a lower sampling rate and then upsampling them to the higher rate when needed. However, interpolation carries with it additional compute time for upsampling the functions, as well as interpolation error. To simplify the present discussion, we will assume that all functions are sampled at the rate required for integration. We will then return to the issue of interpolation when we discuss the MLFMA in Section 11.4.

### 11.3.4.1    Spherical Harmonic Representation

The optimal sampling rates and numerical quadrature rules for integrating expressions like (11.19) on the unit sphere are related to their spherical harmonic representation. Therefore, let us first consider the decomposition of a bandlimited function in terms of spherical harmonics. A function $f(\theta, \phi)$ is said to be *bandlimited* with bandwidth $L$ if it can be written as [11]

$$f(\theta, \phi) = \sum_{l=0}^{L} \sum_{m=-l}^{l} a_l^m Y_l^m(\theta, \phi)\,, \tag{11.55}$$

where $a_l^m$ are the spherical harmonic coefficients, $Y_l^m$ are the spherical harmonics given by

$$Y_l^m(\theta, \phi) = P_l^m(\cos\theta)e^{jm\phi}, \tag{11.56}$$

and $P_l^m(\cos\theta)$ is the normalized associated Legendre polynomial of degree $l$ and order $m$, given by

$$P_l^m(x) = \sqrt{\frac{2l+1}{4\pi}\frac{(l-m)!}{(l+m)!}}(1-x^2)^{m/2}\frac{d^m}{dx^m}P_l(x). \tag{11.57}$$

The coefficients $a_l^m$ are obtained via the integral

$$a_l^m = \int_0^{2\pi}\int_0^\pi f(\theta,\phi)Y_l^{m*}(\theta,\phi)\sin\theta\,d\theta d\phi, \tag{11.58}$$

and substituting (11.56) allows us to write (11.58) as

$$a_l^m = \int_0^{2\pi}e^{-jm\phi}\int_0^\pi f(\theta,\phi)P_l^m(\cos\theta)\sin\theta\,d\theta d\phi. \tag{11.59}$$

It was shown in [11] that for a function $f(\theta,\phi)$ of bandwidth $L$, the integral in (11.59) can be evaluated optimally using an $N_\theta = L$ point Gauss-Legendre quadrature rule in $\theta$, and an equally spaced $N_\phi = (2L+1)$ point quadrature (such as Simpson's rule) in $\phi$. However, we note that the bandwidth in the $\phi$ dimension scales as $\sin\theta$ with a maximum at $\theta = \frac{\pi}{2}$ [10]. Thus, for each sample in $\theta$, it is more efficient to choose $N_\phi(\theta) = \text{Int}(2L\sin\theta) + 1$ samples in $\phi$, where $\text{Int}(x)$ is the largest integer less than or equal to $x$.

### 11.3.4.2 Total Bandwidth

We can now determine the sample rate required to numerically integrate (11.19). Referring to Section 11.3.3, the transfer functions in region $R_l$ have a bandwidth of

$$L_T = k_l d + \beta(k_l d)^{1/3}, \tag{11.60}$$

and similarly, the radiation and receive functions have a bandwidth equal to

$$L_f = k_l d/2 + \beta(k_l d/2)^{1/3}. \tag{11.61}$$

Thus, the total bandwidth of (11.19) is $L_{tot} = L_T + 2L_f$. Following Section 11.3.4.1, all transfer functions, as well as temporary aggregated and transmitted fields are sampled and stored using $N_\theta = L_{tot}$ points in $\theta$ and $N_\phi(\theta) = \text{Int}(2L_{tot}\sin\theta) + 1$ samples in $\phi$, for a total of $N_k$ directions. Similarly, radiation and receive functions are sampled using $N_\theta = L_f$ points in $\theta$, and $N_\phi(\theta) = \text{Int}(2L_f\sin\theta) + 1$ samples in $\phi$, for a total of $N_f$ directions. However, the radiation and receive functions are potentially compressible, as we will see in Section 11.3.4.5.

### 11.3.4.3   Computation and Storage of Transfer Functions

The use of axis-aligned bounding cubes results in a set of vectors whose lengths are multiples of the box edge length $w$ in each cardinal direction. As a result, some vectors will be repeated between group pairs, and in those cases only one transfer function has to be stored. Thus, transfer functions for only those unique vectors $r_{ab}$ in each region are computed and stored in vectors $T_L(k_l, r_{ab})$, each having length $N_k$. We also pre-multiply the transfer functions by the quadrature weights $w_s(\hat{k})$ used for integration.

### 11.3.4.4   Computation of Radiation and Receive Functions

The radiation and receive functions in Section 11.2.3 are computed for each region separately. Inside each region, this is done on a *per-cube* basis, where we consider only basis functions having support in that region. As a result, cubes having no support in a region are considered empty and are assigned no radiation or receive functions. As a region may then be assigned only a subset of all basis functions, a global-to-local basis function map is constructed, which is used to access the correct elements in the input and output vectors in the matrix-vector product.

We continue using the rules outlined in Section 8.6 to enforce the required integral equations on each interface. In general, there would be separate radiation functions for the electric and magnetic basis functions, however when using RWG functions, only one radiation function is needed. Each testing function has two receive functions: one to receive fields radiated by electric basis functions, and one to receive fields radiated by magnetic basis functions.

For an individual cube $C$, the radiation and receive functions are stored in matrices $\mathbf{T}_{\theta,\phi}^{J,M}$ and $\mathbf{R}_{\theta,\phi}^{J,M}$. Each matrix has dimensions $N_f \times N_C$, where $N_f$ is the number of sample points and $N_C$ the number of basis functions in $C$, and comprises a full or reduced rank outer-product matrix (see Sections 9.1 and 9.6.1.2), depending on its compressibility. As mentioned previously, $\mathbf{T}_{\theta,\phi}^M = \eta_l \mathbf{T}_{\theta,\phi}^J$ in region $R_l$ when using RWG functions, and only $\mathbf{T}_{\theta,\phi}^J$ is actually stored.

### 11.3.4.5   Compression of Radiation and Receive Functions

We recall from Section 9.5 that when basis functions are clustered into spatially localized groups, and there are many closely-spaced RHS vectors, the RHS matrix blocks are compressible. By analogy, the radiation and receive functions comprise similar "RHS" blocks, where the RHS vectors now span the entire unit sphere. Therefore, it is reasonable to assume that these functions are potentially compressible via the SVD approach (Sections 9.1 and 11.3.2.1). We also note that since these functions represent *far* interactions, it should be possible to store them at reduced accuracy (via a smaller $\tau_{SVD}$) without significantly affecting the solution accuracy.

To illustrate, let us again consider the PEC Sphere and EMCC Cone-Sphere from Section 11.3.2.1. We again consider finest-level cube sizes of $0.25\lambda$, $0.375\lambda$, and $0.5\lambda$, and for each cube, we sample the radiation and receive functions, each comprising a matrix of size $N_f \times N_C$. We then perform an SVD compression to determine their compressibility. In Figures 11.7a–11.7f are plotted corresponding histograms of the compression, in percent, of the radiation and receive functions in each cube, for $\tau_{SVD} = 10^{-2}$. This $\tau_{SVD}$ was selected as it was found to deliver an excellent level of compression without any significant effects on solution accuracy. Numerical examples will demonstrate this later in Section 11.7.

As we saw when compressing the off-diagonal matrix blocks, the functions in some cubes are not compressible as they have a small number of basis functions. However, the compressibility is quite high across the board, which increases with cube size. In Table 11.2 is summarized the total size of the radiation and receive functions (in MB), in both compressed and uncompressed form. We note that for both objects, the compressed size is relatively flat versus cube size. In Table 11.3 is summarized the corresponding level of compression in each case (in percent), which increases with cube size.

In practice, computation and storage of radiation and receive functions is similar to the off-diagonal matrix blocks: they are filled completely, and then the SVD-base compression is applied. If it succeeds, we retain the compressed version, otherwise the full version is stored. As the dimensions of these matrices are typically small, this step is very fast and does not significantly increase compute time.

**TABLE 11.2:** Size of Radiation and Receive Functions (MB)

|  | PEC Sphere | | | EMCC Cone-Sphere | | |
|---|---|---|---|---|---|---|
|  | $0.25\,\lambda$ | $0.375\,\lambda$ | $0.5\,\lambda$ | $0.25\,\lambda$ | $0.375\,\lambda$ | $0.5\,\lambda$ |
| Uncompressed | 1290 | 1926 | 2362 | 313 | 404 | 517 |
| Compressed | 681 | 630 | 589 | 102 | 85 | 87 |

**TABLE 11.3:** Overall Compression of Radiation and Receive Functions (Percent)

| PEC Sphere | | | EMCC Cone-Sphere | | |
|---|---|---|---|---|---|
| $0.25\,\lambda$ | $0.375\,\lambda$ | $0.5\,\lambda$ | $0.25\,\lambda$ | $0.375\,\lambda$ | $0.5\,\lambda$ |
| 73 | 83 | 87 | 84 | 90 | 92 |

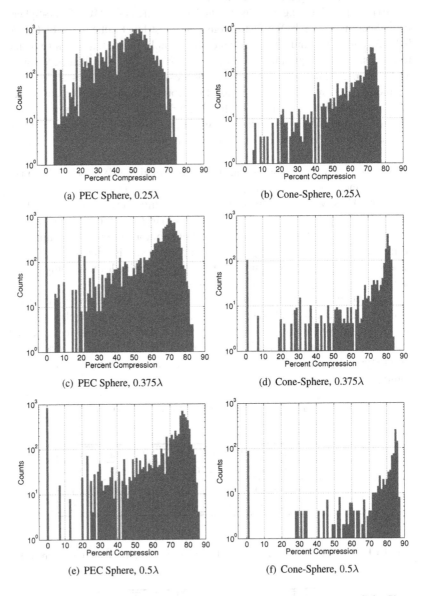

(a) PEC Sphere, 0.25λ

(b) Cone-Sphere, 0.25λ

(c) PEC Sphere, 0.375λ

(d) Cone-Sphere, 0.375λ

(e) PEC Sphere, 0.5λ

(f) Cone-Sphere, 0.5λ

**FIGURE 11.7:** Compression of Radiation and Receive Functions versus Cube Size

## 11.3.5 Matrix-Vector Product

We now discuss computation of the matrix-vector product in (11.1) using the one-level FMM, where the result vector $\mathbf{x}$ comprises the sum of near and and far products in (11.5). It was shown in [2] that the complexity of the one-level FMM is $O(N^{3/2})$.

### 11.3.5.1 Near Product

The near product comprises a straightforward matrix-vector product between the the near matrix $\mathbf{Z}^{near}$ and the right-hand side vector $\mathbf{y}$. As the near matrix is stored in block format, this operation is carried out most efficiently using Level 2 BLAS functions. Because fast access to individual matrix blocks is critical, their pointers should be stored in the CSR format (Section 11.6.1.2). As $\mathbf{Z}^{near}$ is very sparse, the time spent computing the near product is often less than the far product.

### 11.3.5.2 Far Product

To compute the far product, two complex vectors of length $N_k$ are allocated for each group. These vectors, denoted as $\mathbf{t}^{\theta}$ and $\mathbf{t}^{\phi}$, store the $\hat{\theta}$ and $\hat{\phi}$ components of the aggregated radiation functions. Two additional vectors $\mathbf{s}^{\theta}$ and $\mathbf{s}^{\phi}$ of length $N_k$ are allocated, comprising temporary storage for the $\hat{\theta}$ and $\hat{\phi}$ components of the far fields received by a group from all far groups. The one-level far-multiply, summarized in Algorithm 14, comprises an outer loop over all dielectric regions, where for each region $R_l$ there are two phases: an aggregation phase, and a transfer/disaggregation phase.

Step 4 comprises a mapping of the global right-hand side vector $\mathbf{y}$ of length $N$ to a *local* vector $\mathbf{y}_{c_j}$ of length $N_c$, where $P_{c_j}$ is the global to local basis function mapping for cube $c_j$. This is performed by simply populating $\mathbf{y}_{c_j}$ with the correct elements from $\mathbf{y}$. In steps 14 and 15, the products $T_L(k_l, \mathbf{r}_{ab}) \odot \mathbf{t}^{\theta}_{c_k}$ and $T_L(k_l, \mathbf{r}_{ab}) \odot \mathbf{t}^{\theta}_{c_k}$ comprise term-wise multiplication. Step 18 is a mapping of the local product vector $\mathbf{x}_{c_j}$ of length $N_c$, to the global product vector $\mathbf{x}$ of length $N$. As in the aggregation step, this is performed by simply adding to $\mathbf{x}$ the correct elements from $\mathbf{x}_{c_j}$.

All operations in the aggregation phase must be completed before moving to the disaggregation phase, and in general, two passes must be executed to yield the far product. In the first pass, the fields due to only the electric basis functions are aggregated and transferred. In the second pass, the fields due to the magnetic basis functions are aggregated and transferred. During the disaggregation step, the appropriate electric or magnetic receive function, respectively, is used for each testing function. In the purely conducting case, however, only one pass is required as there are no magnetic basis functions.

---

**Algorithm 14** One-Level FMM Far Product

---

1: **for all** Regions $R_l$ **do**
2:      Aggregation Phase:
3:      **for all** Non-Empty Cubes $c_j$ in $R_l$ **do**
4:         $\mathbf{y}_{c_j} = \text{map}(\mathbf{y}, P_{c_j})$
5:         $\mathbf{t}_{c_j}^\theta = \mathbf{T}_{\theta,c_j}^{J,M} \mathbf{y}_{c_j}$
6:         $\mathbf{t}_{c_j}^\phi = \mathbf{T}_{\phi,c_j}^{J,M} \mathbf{y}_{c_j}$
7:      **end for**
8:      Disaggregation Phase:
9:      **for all** Non-Empty Cubes $c_j$ in $R_l$ **do**
10:         $\mathbf{s}^\theta = 0$
11:         $\mathbf{s}^\phi = 0$
12:         **for all** Non-Empty Cubes $c_k$ far from $c_j$ **do**
13:            Find the $T_L(k_l, \mathbf{r}_{ab})$ that transfers from $c_k$ to $c_j$
14:            $\mathbf{s}^\theta = \mathbf{s}^\theta + T_L(k_l, \mathbf{r}_{ab}) \odot \mathbf{t}_{c_k}^\theta$
15:            $\mathbf{s}^\phi = \mathbf{s}^\phi + T_L(k_l, \mathbf{r}_{ab}) \odot \mathbf{t}_{c_k}^\theta$
16:         **end for**
17:         $\mathbf{x}_{c_j} = \mathbf{R}_\theta^{J,M} \mathbf{s}^\theta + \mathbf{R}_\phi^{J,M} \mathbf{s}^\phi$
18:         $\mathbf{x} = \mathbf{x} + \text{map}(\mathbf{x}_{c_j}, P_{c_j}^{-1})$
19:      **end for**
20: **end for**

---

## 11.4 Multi-Level Fast Multipole Algorithm (MLFMA)

In the one-level algorithm, the number of transfers grows exponentially with the size of the problem and the number of groups. Though it does offer a potential $O(N^{3/2})$ complexity and an improvement in memory utilization, a complexity of $O(N \log N)$ can be obtained from a multi-level algorithm [2, 12]. The Multi-Level Fast Multipole Algorithm (MLFMA) extends the one-level algorithm by dividing the geometry recursively into multiple levels with bounding cubes of progressively smaller size. As we will see, clustering basis functions this way greatly reduces the number of transfers, further accelerating the far product. However, there are some differences and subtleties that must be handled properly for the multi-level algorithm to be effective, which we will discuss in detail in this section.

### 11.4.1 MLFMA/SVD

So far in this this chapter, we have paid significant attention to the block near matrix and its compressibility via the SVD. We note that in the literature, though the block structure was recognized by several authors [12, 13], it does not appear that computation and storage of the near matrix was afforded special consideration. Most authors, such as those in [1], note that the near matrix is computed and stored in the "usual way", but provide no other details on how this is done. Our approach herein, where off-diagonal near matrix and preconditioner blocks are compressed via SVD, is novel in the literature and we denote it as MLFMA/SVD.

### 11.4.2 Spatial Subdivision and Clustering via Octree

Let us first consider what happens when we increase the size of the cubes in the one-level algorithm. This reduces the number of transfers at the expense of increasing the size of the near matrix, shifting more of the compute time to the near product. The sampling rates on the unit sphere also increase, increasing the storage requirements of the radiation, receive, and transfer functions. If we can devise a way to simultaneously use large and small groups, we can store the radiation and receive functions at a lower sample rate, while simultaneously reducing the number of transfers. The answer to this problem lies in a recursive subdivision of the geometry, based on the cube-based clustering strategy of the one-level FMM in Section 11.3.1. Using a tree-based scheme, larger groups can be divided into recursively smaller groups with a clear parent-child relationship. An excellent choice for this is the octree [14], a data structure commonly used in computer graphics. To generate an octree, the object is first

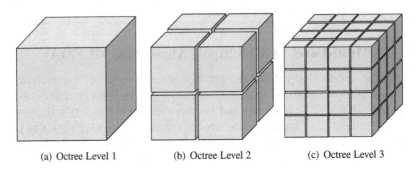

(a) Octree Level 1         (b) Octree Level 2         (c) Octree Level 3

**FIGURE 11.8:** Octree Levels 1–3

enclosed in a single large cube. This cube is then divided into eight smaller cubes, which are themselves divided into eight smaller cubes and so on, as illustrated in Figure 11.8. This approach is very similar to the one-level algorithm, where now the basis and testing functions are sorted and assigned to cubes on each level. The assignment process is very fast as only those basis functions assigned to a parent are assigned to its children. The cubes on the finest level of the tree are called *leaf* cubes.

Our approach herein is to fix the size $w$ of the smallest cubes at some fraction of a wavelength in $R_0$, such as $\lambda_0/2$ or $\lambda_0/4$. This length is then doubled until it exceeds the maximum dimension of the object's bounding box, thus defining $w$ for the top-level cube.

On levels 1 and 2, all groups are near groups. Starting then at the coarsest level (level 3), all near and far groups are identified as usual. On the higher levels, only the children of groups in the near zone of a parent group (see Section 11.3.1.1) are considered as candidate near and far groups for that parent's children. This greatly reduces the average number of far groups for each cube, and limits the number of unique transfer vectors to a maximum of 316 on each level. This is especially advantageous as the number of transfers can be very high in larger problems. As in the one-level algorithm, cubes that are non-empty in one region may be empty in another.

### 11.4.3   Near Matrix and Near Product

In the MLFMA, the nonzero part of the near matrix is defined by the near groups on the finest level of the octree. The near matrix is computed and stored the same as in the one-level FMM in Section 11.3.2. The same is true for the near product, which is carried out as described in Section 11.3.5.1.

## 11.4.4 Unit Sphere Sampling Rates

Transfer functions are computed for all unique transfer vectors (a maximum of 316) on octree levels $l \geq 3$, where cubes have a bounding diameter $d_l$, and the number of multipoles $L$ is computed via (11.51) and limited for closer groups as in Section 11.3.3.1. As the number of transfer functions per level is limited, their memory impact is relatively small. Thus, they are sampled and stored at the rate $N_{k,l}$ used for integration on each level, and do not need to be interpolated. Radiation and receive functions are computed and stored at the sample rate $N_f$ on the finest level of the octree following Section 11.3.4.4, where the bandwidth (11.61) depends on the diameter of the leaf cubes. A consequence of this choice is that these functions must be interpolated and phase shifted when moving between levels, as discussed in the following sections. The sample rates in $\theta$ and $\phi$ dimensions are computed following Section 11.3.4.2, with some oversampling in each dimension to minimize interpolation errors (see Section 11.4.6).

## 11.4.5 Far Product

The far product in the MLFMA operates on levels $l \geq 3$ in the octree. As in the one-level algorithm, the MLFMA comprises two phases: an aggregation phase and a transfer and disaggregation phase. However, the one-level algorithm must be extended, as the radiation and receive functions are stored on the finest level of the tree. In the aggregation phase, an upward pass is made through the octree, where radiation functions in each group are aggregated and passed up to their parents using interpolation and phase shifting. In the disaggregation phase, a downward pass is made through the octree, where transfers are made between all far groups and the received fields passed down to child groups through a combination of integration on the sphere, phase shifting, and "reverse interpolation" (anterpolation). The transfer scheme in the MLFMA is illustrated in Figures 11.9 - 11.11. Figure 11.9 depicts the transfers between groups on one of the coarser levels. On the next highest level in Figure 11.10, the far groups are children of groups that were considered near on the previous level. This is further emphasized on the finest level in Figure 11.11, where the number of transfers is now far less than it would have been in a one-level algorithm. We will now describe each phase in detail.

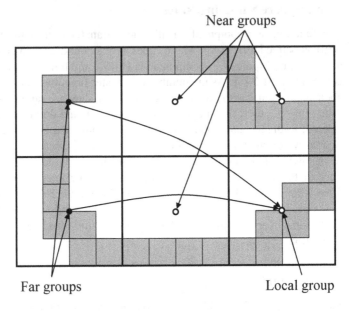

**FIGURE 11.9:** Transfers On MLFMA Level $M - 2$

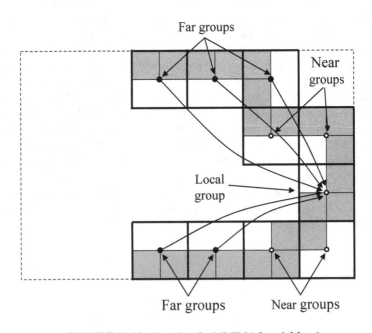

**FIGURE 11.10:** Transfers On MLFMA Level $M - 1$

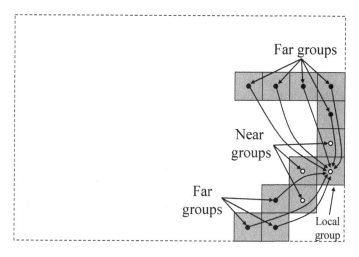

**FIGURE 11.11:** Transfers On MLFMA Level $M$

### 11.4.5.1 Upward Pass (Aggregation)

The aggregation phase comprises a recursive, upward pass through the tree. We begin on the finest level, where fields are aggregated in all leaf cubes following steps 3-7 in Algorithm 14. The aggregated field vector $\mathbf{t}_c^{\theta,\phi}$ for cube $c$ is computed via the matrix-vector product

$$\mathbf{t}_{c,l}^{\theta,\phi} = \mathbf{W}_{N_f}^{N_{k,l}} \mathbf{T}_{\theta,\phi,c}\, \mathbf{y}_c \, , \tag{11.62}$$

where $\mathbf{W}_{N_f}^{N_{k,l}}$ is an interpolation matrix of size $N_{k,l} \times N_f$ that upsamples from the rate $N_f$ of the radiation/receive functions to the rate $N_{k,l}$ used for integration on level $l$. Thus, $\mathbf{t}_{c,l}^{\theta,\phi}$, comprises a vector of length $N_{k,l}$. As the $\theta$ and $\phi$ component vectors are handled separately, we will defer that notation until the end of the disaggregation step and simply refer to vector $\mathbf{t}_{c,l}$.

It should be noted that the choice of upsampling to the rate $N_{k,l}$ is not arbitrary. It was found through numerical experimentation that the aggregation phase requires much less compute time than the transfer and disaggregation phase. Thus, immediately upsampling to this rate during the upward pass has little overhead, and avoids multiple interpolations at intermediate sample rates and their corresponding interpolation error.

On coarser levels, the aggregated field in each group comprises a sum of interpolated and phase-shifted aggregated fields from their children. To illustrate, consider a basis function located at $\mathbf{r}'$ and assigned to cube $c$, centered at $\mathbf{r}_c$. Following (11.23), its radiation function $T_l(\hat{\mathbf{k}})$ on level $l$ can be written as

$$T_l(\hat{\mathbf{k}}) = e^{jk\hat{\mathbf{k}}\cdot(\mathbf{r}'-\mathbf{r}_c)} \, . \tag{11.63}$$

This function can be passed up to the parent cube $p$ on level $l - 1$ via a phase shift, yielding a new radiation function $T_{l-1}(\hat{\mathbf{k}})$ given by

$$T_{l-1}(\hat{\mathbf{k}}) = T_l(\hat{\mathbf{k}})e^{jk\hat{\mathbf{k}}\cdot(\mathbf{r}_c - \mathbf{r}_p)} = T_l(\hat{\mathbf{k}})\,\psi(\mathbf{r}_p, \mathbf{r}_c, \hat{\mathbf{k}}) \qquad (11.64)$$

where the phase shift function $\psi(\mathbf{r}_p, \mathbf{r}_c, \hat{\mathbf{k}})$ is

$$\psi(\mathbf{r}_p, \mathbf{r}_c, \hat{\mathbf{k}}) = e^{jk\hat{\mathbf{k}}\cdot(\mathbf{r}_c - \mathbf{r}_p)}. \qquad (11.65)$$

Applying this phase shift, however, causes a bandwidth increase. Thus $\mathbf{t}_{c,l}$ must be first upsampled to the rate $N_{k,l-1}$ on level $l - 1$, then phase shifted before being added to the aggregated field vector $\mathbf{t}_{p,l-1}$ of parent cube $p$. Doing so, can now write $\mathbf{t}_{p,l-1}$ as a sum over all non-empty child cubes, yielding

$$\mathbf{t}_{p,l-1} = \sum_c \psi(\mathbf{r}_p, \mathbf{r}_c, \hat{\mathbf{k}}) \odot \left[\mathbf{W}_{N_{k,l}}^{N_{k,l-1}}\mathbf{t}_{c,l}\right], \qquad (11.66)$$

where the phase shift in (11.65) is sampled at $N_{k,l-1}$ directions. As we use the octree in the MLFMA, there are a total of 8 child-to-parent shift vectors per level, which are pre-computed and stored with little overhead.

### 11.4.5.2  Downward Pass (Disaggregation)

The disaggregation phase comprises a recursive, downward pass through the tree. To illustrate, let us first consider an MLFMA with only two levels: a coarse level (level $l - 1$) and a fine level (level $l$). We begin with a single parent group $p$ on the coarse level, and following steps 12-16 in Algorithm 14, we transfer the aggregated fields in all far groups to this one, yielding the local field vector $\mathbf{s}_{l-1}$. We now disaggregate this field with the receive functions $\mathbf{R}_{c,l}$ in a single child cube $c$ on level $l$. To do so, we must first upsample and phase shift the receive functions on the finer level. Following the same approach as in the upward pass, this operation can be written as

$$\mathbf{x}_{c,l-1} = \left[\mathbf{W}_{N_{k,l}}^{N_{k,l-1}}\mathbf{W}_{N_f}^{N_{k,l}}\mathbf{R}_{c,l}\right]^T \left[\psi^*(\mathbf{r}_p, \mathbf{r}_c, \hat{\mathbf{k}}) \odot \mathbf{s}_{l-1}\right], \qquad (11.67)$$

where after transposition the first term on the right-hand side is a matrix of size $N_{c,l} \times N_{k,l-1}$, and we have used the conjugate of (11.65) and re-ordered terms so the phase shift is applied to $\mathbf{s}_{l-1}$. Now consider child cube $c$, which has its own local field $\mathbf{s}_l$ transmitted from all far groups on level $l$. The disaggregation step on this level can be written as

$$\mathbf{x}_{c,l} = \mathbf{R}_{c,l}^T\mathbf{s}_l \qquad (11.68)$$

and the output vector $\mathbf{x}_c$ for cube $c$ is therefore

$$\mathbf{x}_c = \mathbf{x}_{c,l-1} + \mathbf{x}_{c,l}. \qquad (11.69)$$

It is clear that if there are many levels, repeated upsampling of the receive functions will yield a very inefficient algorithm. However, if we take advantage of the commutative property $(\mathbf{AB})^T = \mathbf{B}^T \mathbf{A}^T$, the operations in (11.67) can be reordered, yielding

$$\mathbf{x}_{c,l-1} = \mathbf{R}_{c,l}^T [\mathbf{W}_{N_f}^{N_{k,l}}]^T [\mathbf{W}_{N_{k,l}}^{N_{k,l-1}}]^T [\psi^*(\mathbf{r}_p, \mathbf{r}_c, \hat{\mathbf{k}}) \odot \mathbf{s}_{l-1}], \qquad (11.70)$$

where the matrix operations are carried out pairwise right to left. This operation still upsamples $\mathbf{R}_{c,l}^T$ to the rate $N_{k,l-1}$ on level $l-1$, but with a reversed order of operations. In the literature, this is often referred to as *adjoint interpolation* or *anterpolation* [2], though this author prefers the term *reverse interpolation* in this context.

Following (11.70), the disaggregation can now be done on level $l$, exclusively, where $\mathbf{x}_c$ can be written as

$$\mathbf{x}_c = \mathbf{R}_{c,l}^T \left[ \mathbf{s}_l + [\mathbf{W}_{N_f}^{N_{k,l}}]^T \mathbf{a} \right], \qquad (11.71)$$

which comprises disaggregation with the received fields $\mathbf{s}_l$ on level $l$ and the *anterpolated* fields $\mathbf{a}$ from the parent group on level $l-1$, given by

$$\mathbf{a} = [\mathbf{W}_{N_{k,l}}^{N_{k,l-1}}]^T [\psi^*(\mathbf{r}_p, \mathbf{r}_c, \hat{\mathbf{k}}) \odot \mathbf{s}_{l-1}]. \qquad (11.72)$$

Finally, we recall from Section 11.4.5.1 that there are $\theta$ and $\phi$-polarized field vectors, which are processed separately during both passes. Therefore, at the bottom of the downward pass, there will exist four field vectors $\mathbf{s}_l^{\theta,\phi}$ and $\mathbf{a}^{\theta,\phi}$, which must be multiplied with the appropriate receive functions. Thus, (11.71) becomes

$$\mathbf{x}_c = \mathbf{R}_{\theta,c,l}^T \left[ \mathbf{s}_l^\theta + [\mathbf{W}_{N_f}^{N_{k,l}}]^T \mathbf{a}^\theta \right] + \mathbf{R}_{\phi,c,l}^T \left[ \mathbf{s}_l^\phi + [\mathbf{W}_{N_f}^{N_{k,l}}]^T \mathbf{a}^\phi \right]. \qquad (11.73)$$

The disaggregation process can be generalized in a straightforward way to a tree with many levels, yielding a downward pass which is as follows: We begin on the coarsest level (3), and for an individual cube $p$, we perform transfers from all far groups, and as this is the coarsest level, there are no anterpolated fields. We then recurse into each child cube, where on the next finest level (4) we again perform transfers from all far groups. To the local field we now add the anterpolated fields from the parent group. This recursion continues until the finest level is reached, where (11.73) is evaluated. This yields a local output vector which is remapped and added to the global output vector. This process is then repeated for all other cubes on the coarsest level, completing the downward pass.

## 11.4.6 Interpolation Algorithms

The interpolation steps in the MLFMA comprise a significant portion of the far multiply time, particularly in the downward pass. Thus, we must make interpolation as fast as possible while minimizing the interpolation error. In this section we will discuss several different interpolation algorithms and their implementation.

### 11.4.6.1 Statement of the Problem

Given a function $f(\theta, \phi)$ on the unit sphere, let us compute $N_{k_1} = N_{\theta_1} \times N_{\phi_1}$ samples and store them in vector $\mathbf{s}_1$. The interpolation problem involves computing a new vector $\mathbf{s}_2$ at a different sample rate $N_{k_2} = N_{\theta_2} \times N_{\phi_2}$ using only the values from $\mathbf{s}_1$. This interpolation can be written as the matrix-vector product

$$\mathbf{s}_2 = \mathbf{W}\mathbf{s}_1, \tag{11.74}$$

where $\mathbf{W}$ is an interpolation matrix of size $N_{k_2} \times N_{k_1}$. If $N_{k_2} > N_{k_1}$, this operation *upsamples* $\mathbf{s}_1$, and if $N_{k_2} < N_{k_1}$ it *downsamples* $\mathbf{s}_1$.

Interpolation methods fall into one of two general categories: *global* or *local*. In a global interpolation, the matrix $\mathbf{W}$ is full and each sample in $\mathbf{s}_2$ is computed using all samples in $\mathbf{s}_1$. If $f(\theta, \phi)$ is bandlimited, this interpolation can be exact if it is sampled at or above the Nyquist rate. In a local interpolation, $\mathbf{W}$ is sparse and each sample in $\mathbf{s}_2$ is computed from a subset of $\mathbf{s}_1$. Local interpolations do not typically yield exact results, however they are often much faster and their error can be controlled. In our implementation of the MLFMA we use a local interpolation based on Lagrange polynomials, however we will also briefly consider global interpolation via spherical harmonics, which is useful in other areas of electromagnetics.

### 11.4.6.2 Global Interpolation by Spherical Harmonics

As we discussed in Section 11.3.4.1, if $f(\theta, \phi)$ is bandlimited, it can be decomposed into a sum of spherical harmonics (11.55) with the coefficients $a_l^m$ defined in (11.58). Once the coefficients are calculated, the value $f(\theta_p, \phi_p)$ can be computed as

$$f(\theta_p, \phi_p) = \sum_{l=0}^{L} \sum_{m=-l}^{l} a_l^m Y_l^m(\theta_p, \phi_p) \,. \tag{11.75}$$

If we now substitute (11.58) into (11.75), this yields

$$f(\theta_p, \phi_p) = \sum_{l=0}^{L} \sum_{m=-l}^{l} Y_l^m(\theta_p, \phi_p) \sum_{q=1}^{N_{k_1}} w_q Y_l^{m*}(\theta_q, \phi_q) f(\theta_q, \phi_q) \,, \tag{11.76}$$

where the original function $f(\theta, \phi)$ is sampled at $N_{k_1}$ points, and $w_q$ is the quadrature weight on the unit sphere. Exchanging the order of the summations allows us to write

$$f(\theta_p, \phi_p) = \sum_{q=1}^{N_{k_1}} w_q \sum_{l=0}^{L} \sum_{m=-l}^{l} Y_l^{m*}(\theta_p, \phi_p) Y_l^m(\theta_q, \phi_q) f(\theta_q, \phi_q) , \quad (11.77)$$

which can be re-written as

$$f(\theta_p, \phi_p) = \sum_{q=1}^{N_{k_1}} W_{pq} f(\theta_q, \phi_q) , \quad (11.78)$$

where the matrix elements $W_{pq}$ are

$$W_{pq} = w_q \sum_{l=0}^{L} \sum_{m=-l}^{l} Y_l^{m*}(\theta_p, \phi_p) Y_l^m(\theta_q, \phi_q) . \quad (11.79)$$

As the matrix **W** is full, it is not optimal for interpolation in the MLFMA. However, for many applications it is often efficient to store the coefficients and perform a reconstruction via (11.75). For example, this method is used in geopotential models of the Earth's gravitational field, allowing a fast and accurate reconstruction of the gravitational potential at any latitude or longitude. Two such models are the World Geodetic System 1984 (WGS84) [15] and the Earth Gravitational Model 1996 (EGM96) [16].

### 11.4.6.3 Local Interpolation by Lagrange Polynomials

Lagrange polynomials are an effective means of creating a very sparse interpolator with local support. Given $N$ samples of a function $f(x)$, this method generates a unique polynomial of order $N - 1$ that interpolates those samples. New values $f(x')$ can then be computed as

$$f(x') = \sum_{i=1}^{N} w_i(x') f(x_i) , \quad (11.80)$$

where the weights are

$$w_i(x') = \frac{(x' - x_1) \cdots (x' - x_{i-1})(x' - x_{i+1}) \cdots (x' - x_{N+1})}{(x_i - x_1) \cdots (x_i - x_{k-1})(x_i - x_{i+1}) \cdots (x_i - x_{N+1})} , \quad (11.81)$$

or

$$w_i(x') = \prod_{\substack{k=1 \\ k \neq i}}^{N} \frac{x' - x_k}{x_i - x_k} . \quad (11.82)$$

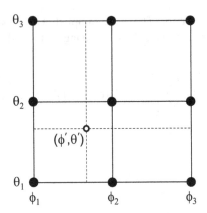

**FIGURE 11.12:** $3 \times 3$ Lagrange Interpolation Stencil

Interpolation on the unit sphere uses an $N \times N$ interpolation *stencil* as illustrated in Figure 11.12 for $N = 3$. In this case, the interpolation matrix $\mathbf{W}$ will have nine nonzero columns per row. This yields a very sparse interpolation, as there may be hundreds (or thousands) of sample points on the unit sphere. The use of a quadratic or cubic interpolator ($N = 3$ and $N = 4$, respectively) can provide reasonable results with small error provided that the function is oversampled. An error analysis of this method is discussed in [17]. In this author's experience, a $3 \times 3$ Lagrange stencil, with an oversampling factor of 2 in the $\theta$ and $\phi$ dimensions, yields sufficient accuracy. A similar investigation was also performed by other authors [18]. We also note that while the oversampling of functions increases the operation count in the upward and downward pass, it does not significantly affect the storage requirements of the radiation and receive functions, as they are stored in compressed form.

*Storage of Interpolation Matrix*

As the matrix $\mathbf{W}$ is very sparse, the CSR format (Section 11.6.1.2) is an ideal storage container. We note that the interpolation comprises a matrix-vector product of the form $\mathbf{b} = \mathbf{Wa}$, and as the elements in $\mathbf{W}$ are in row-major format, they can be accessed linearly. This is an ideal memory access pattern. However, the elements in $\mathbf{a}$ are not accessed linearly, as elements comprising the rows of the stencil are not completely contiguous in memory. Thus, there will be cache misses and the CPU will stall waiting on the data to be accessed from RAM. The same also occurs when performing anterpolations using $\mathbf{W}^T$. As the anterpolations in the downward pass comprise a significant portion of the compute time in the far product, special attention should be paid to optimizing the storage format and memory access patterns on the programmer's processing architecture [19, 20].

# 11.5 Preconditioners

The concept of preconditioning was introduced in Section 4.2.7. Preconditioners have several non-trivial sources of overhead, among which are their setup and compute time, storage space, and solve time per matrix-vector product, all of which should be minimized if possible. As only the near matrix is available explicitly in the FMM, we will consider preconditioners designed specifically for sparse matrices.

As discussed previously in Section 4.2.7, given a matrix $\mathbf{A}$ the preconditioner matrix $\mathbf{M}$ should approximate $\mathbf{A}$ in some way. Our approach herein is based on the idea that as $\mathbf{M}$ is only an approximation, it does not need to be as accurate as $\mathbf{A}$ to be effective. Thus, for an MoM matrix $\mathbf{A}$ with compressed off-diagonal blocks, we will also compress the off-diagonal blocks in $\mathbf{M}$ but at reduced accuracy. This will reduce the storage requirements for $\mathbf{M}$, and accelerate the preconditioner solve time as many blocks will be in compressed form. Numerical examples in Section 11.7 will show the effectiveness of this approach.

## 11.5.1 Information Content

The FMM near matrix is *very* sparse, representing only the interactions between the closest basis functions. Thus, only a limited amount of information is available for computing a preconditioner. Intuitively, we expect that as we decrease the size of groups on the finest MLFMA level, the information in the near matrix will also be reduced, making the preconditioner less effective. Therefore, the groups must be large enough so that the preconditioner remains effective, while minimizing the storage requirements of the near matrix, as well as the radiation and receive functions.

## 11.5.2 Diagonal Preconditioner

The diagonal or *Jacobi* preconditioner is the simplest form of preconditioner and is useful for illustrating the preconditioning concept. The elements of the diagonal preconditioner matrix $\mathbf{M}$ are

$$M_{ij} = A_{ij} \quad i = j \,, \tag{11.83}$$

and

$$M_{ij} = 0 \quad i \neq j \,. \tag{11.84}$$

We note that if we reduce the finest group size to a single basis function, the FMM near matrix now comprises only the diagonal. As noted in Section

11.5.1, the information content of the diagonal is quite small, and in practice this preconditioner is rarely effective.

### 11.5.3    Incomplete Block LU (ILU) Preconditioners

In this section we will consider several incomplete block LU (ILU) preconditioners. An ILU factorization of a sparse matrix $\mathbf{A}$ results in a sparse upper triangular matrix $\mathbf{U}$ and sparse lower unit triangular matrix $\mathbf{L}$, with a residual matrix $\mathbf{R} = \mathbf{LU} - \mathbf{A}$ that satisfies certain criteria [21]. In our approach, some off-diagonal blocks of the LU matrix are compressed using the SVD, following the block LU factorization specified in Algorithm 7. The primary differences between these preconditioners lies in the presumed zero patterns of $\mathbf{A}$ and the $\mathbf{LU}$ matrix, as well as the amount of fill-in.

#### 11.5.3.1    Block Diagonal

The block diagonal preconditioner, previously reported in [2, 12], is constructed using only the diagonal blocks of $\mathbf{A}$. Thus, for a $3 \times 3$ block matrix $\mathbf{A}$, the factorization $\mathrm{LU}(\mathbf{M})$ is simply

$$\mathrm{LU}(\mathbf{M}) = \begin{bmatrix} \mathrm{LU}(\mathbf{M}_{11}) & 0 & 0 \\ 0 & \mathrm{LU}(\mathbf{M}_{22}) & 0 \\ 0 & 0 & \mathrm{LU}(\mathbf{M}_{33}) \end{bmatrix}, \qquad (11.85)$$

where diagonal block is factored independently. Note that this preconditioner also uses very little information from $\mathbf{A}$, and as numerical examples will later show, it is also rarely effective in practice.

#### 11.5.3.2    Block ILU with Zero Fill-In (ILU(0))

We next consider an incomplete block LU factorization with no fill-in, denoted as ILU(0), where the $\mathbf{LU}$ matrix has the same zero pattern as $\mathbf{A}$ [21]. This preconditioner can be computed via Algorithm 7, where all fill-in blocks are simply ignored. Where for the off-diagonal near matrix blocks we would use a $\tau_{SVD} = 10^{-4}$, for the $\mathbf{L}$ and $\mathbf{U}$ blocks we instead use a $\tau_{SVD} = 10^{-3}$ or $10^{-2}$. This choice is based on the idea that an *approximation* of the $\mathbf{LU}$ matrix as produced by ILU(0) will perform almost as well as $\mathbf{LU}$, while requiring less storage space.

#### 11.5.3.3    Block ILU with Threshold (ILUT)

We can improve ILU(0) significantly by allowing fill-in blocks. The scalar approach by Saad in [21] computes the $\mathbf{LU}$ matrix by row, applying a dropping rule to the elements in each row based on their amplitude relative to the 2-norm of the row, as well as limiting the maximum number of elements per row. As

Algorithm 7 operates on blocks and not individual elements, this approach cannot be applied directly, but we will use a similar strategy. First, we update all trailing row and column blocks for diagonal $b$. Pre-existing (non fill-in) blocks are compressed using a similar $\tau_{SVD}$ as in the ILU(0), whereas fill-in blocks are compressed using a less accurate $\tau_{SVD} = 10^{-2}$ or $10^{-1}$ and are added to a temporary storage space. After the update is complete, we compute the matrix norm $\|\mathbf{Z}\|_1$ for all blocks on row $b$ and column $b$, including the fill-in blocks, and retain the maximums. We then drop all temporary fill-in blocks whose norm is less than some threshold $\tau_{ILUT}$ of the maximum row or column norm, where $\tau_{ILUT} = 10^{-4}$ or $10^{-3}$. The remaining fill-in blocks are added to secondary CSRMatrix structures (see Section 11.6.1.2), so that they can be accessed quickly during subsequent row and column updates. As these comprise a single trailing row or column of blocks, which progressively move to the right, they can be appended to an existing CSRMatrix with little performance penalty. Once the factorization is complete, these blocks are be combined with all other **LU** matrix blocks in a new, single CSRMatrix structure.

Our choices of $\tau_{SVD}$ and $\tau_{ILUT}$ should yield a compressed **LU** matrix that yields roughly the same improvement in iterative solver convergence as an uncompressed matrix, but with a smaller memory footprint. This will be examined and shown to be true in Section 11.7.8. It is also expected that with the added fill-in blocks, preconditioner compute time, memory requirements, and solve time will be greater than ILU(0). As we will see, our choice of $\tau_{SVD}$ for the fill-in blocks greatly minimizes the memory requirement.

### 11.5.4 Sparse Approximate Inverse (SAI)

Let us now consider a preconditioner matrix $\mathbf{M}$ constructed by minimizing the Frobenius norm of the residual matrix, given by

$$F(\mathbf{M}) = ||\mathbf{I} - \mathbf{AM}||_F^2 , \qquad (11.86)$$

for right-preconditioning. In this sense, $\mathbf{M}$ is called a sparse approximate inverse (SAI). The minimization problem can be decoupled into the sum of the squares of the 2-norms of the columns of the residual matrix, which is

$$||\mathbf{I} - \mathbf{AM}||_F^2 = \sum_{n=1}^{N} ||\mathbf{e}_n - \mathbf{Am}_n||_2^2 , \qquad (11.87)$$

where $\mathbf{e}_n$ and $\mathbf{m}_n$ are the columns of $\mathbf{I}$ and $\mathbf{M}$, respectively. The right-hand side of (11.87) comprises $N$ independent least-square sub-problems, which can be solved in parallel.

As the FMM near matrix is sparse, we must also determine a sparsity pattern for $\mathbf{M}$. Some treatments in the literature find this pattern through heuristic algorithms, or iterative refinement. In some cases, the elements in $\mathbf{A}$ are first down-selected (filtered) based on certain criteria (such as amplitude), yielding a further sparsified matrix $\tilde{\mathbf{A}}$ used to build $\mathbf{M}$. Approaches specifically tailored to the FMM problem such as those in [22, 23, 24] use the block structure of the near matrix to determine the pattern. Our approach herein assumes that $\mathbf{M}$ has the same zero pattern as $\mathbf{A}$.

In [21], Saad suggests solving for each column in (11.87) using a few iterations of a solver such as GMRES, which can be made fast by using sparse vectors and performing the matrix-vector product in a sparse-sparse mode. As the FMM near matrix is stored in block form, where some off-diagonal blocks are compressed, we will instead use a dense factorization.

### 11.5.4.1 Dense QR Factorization

We now consider the computation of an SAI preconditioner matrix $\mathbf{M}$ having the same the sparsity pattern as $\mathbf{A}$. For a block matrix with $M$ block rows and columns, this requires solving $M$ least-square problems of the form

$$\mathbf{AM}_{*k} = \mathbf{E}_k \qquad (11.88)$$

in the least-squares sense, where $\mathbf{M}_{*k}$ is block-column $k$ of $\mathbf{M}$, and $\mathbf{E}_k$ is a block-column comprising the identity matrix in block row $k$ and zero otherwise. To illustrate, consider a $9 \times 9$ sparse block matrix given by

$$\mathbf{A} = \begin{bmatrix} \mathbf{A}_{11} & \mathbf{A}_{12} & 0 & 0 & \mathbf{A}_{15} & 0 & 0 & 0 & 0 \\ \mathbf{A}_{21} & \mathbf{A}_{22} & 0 & \mathbf{A}_{24} & 0 & 0 & 0 & \mathbf{A}_{28} & 0 \\ 0 & 0 & \mathbf{A}_{33} & 0 & 0 & \mathbf{A}_{36} & 0 & 0 & \mathbf{A}_{39} \\ 0 & \mathbf{A}_{42} & 0 & \mathbf{A}_{44} & 0 & 0 & \mathbf{A}_{47} & 0 & 0 \\ \mathbf{A}_{51} & 0 & 0 & 0 & \mathbf{A}_{55} & 0 & 0 & 0 & 0 \\ 0 & 0 & \mathbf{A}_{63} & 0 & 0 & \mathbf{A}_{66} & 0 & \mathbf{A}_{68} & 0 \\ 0 & 0 & 0 & \mathbf{A}_{74} & 0 & 0 & \mathbf{A}_{77} & 0 & 0 \\ 0 & \mathbf{A}_{82} & 0 & 0 & 0 & \mathbf{A}_{86} & 0 & \mathbf{A}_{88} & \mathbf{A}_{89} \\ 0 & 0 & \mathbf{A}_{93} & 0 & 0 & 0 & 0 & \mathbf{A}_{98} & \mathbf{A}_{99} \end{bmatrix}, \qquad (11.89)$$

which is a simple but realistic representation of the FMM sparse near matrix. As each block column of $\mathbf{M}$ is computed independently of the others, let us compute block column $\mathbf{M}_{*4}$. In this case, the nonzero block rows of $\mathbf{M}_{*4}$ comprise rows 2, 4, and 7, which multiply block columns 2, 4, and 7 of $\mathbf{A}$ in the matrix-vector product in (11.88). Retaining only these block rows of $\mathbf{M}_{*4}$ and columns of **A** allows us to write the matrix-vector product as

$$
\begin{bmatrix}
\mathbf{A}_{12} & 0 & 0 \\
\mathbf{A}_{22} & \mathbf{A}_{24} & 0 \\
0 & 0 & 0 \\
\mathbf{A}_{42} & \mathbf{A}_{44} & \mathbf{A}_{47} \\
0 & 0 & 0 \\
0 & 0 & 0 \\
0 & \mathbf{A}_{74} & \mathbf{A}_{77} \\
\mathbf{A}_{82} & 0 & 0 \\
0 & 0 & 0
\end{bmatrix}
\begin{bmatrix}
\mathbf{M}_{24} \\
\mathbf{M}_{44} \\
\mathbf{M}_{74}
\end{bmatrix}
=
\begin{bmatrix}
0 \\
0 \\
0 \\
\mathbf{I} \\
0 \\
0 \\
0 \\
0 \\
0
\end{bmatrix} .
\tag{11.90}
$$

We note that many of the block rows in (11.90) are zero. Removing these rows from both sides, we are left with

$$
\tilde{\mathbf{A}}\mathbf{M} = \tilde{\mathbf{E}} =
\begin{bmatrix}
\mathbf{A}_{12} & 0 & 0 \\
\mathbf{A}_{22} & \mathbf{A}_{24} & 0 \\
\mathbf{A}_{42} & \mathbf{A}_{44} & \mathbf{A}_{47} \\
0 & \mathbf{A}_{74} & \mathbf{A}_{77} \\
\mathbf{A}_{82} & 0 & 0
\end{bmatrix}
\begin{bmatrix}
\mathbf{M}_{24} \\
\mathbf{M}_{44} \\
\mathbf{M}_{74}
\end{bmatrix}
=
\begin{bmatrix}
0 \\
0 \\
\mathbf{I} \\
0 \\
0
\end{bmatrix} ,
\tag{11.91}
$$

which comprises a dense, over-determined system. This can be solved in the least-squares sense using a QR factorization, such as the LAPACK function `cgels`. As each block column $\mathbf{M}_{*k}$ is independent, this is easily parallelized using threads.

When computing the block column $\mathbf{M}_{*k}$, it is straightforward to construct $\tilde{\mathbf{A}}$ by copying full matrix blocks or expanding compressed blocks from **A** into $\tilde{\mathbf{A}}$. After we have solved for $\tilde{\mathbf{M}}$, the corresponding blocks in **M** can be populated. Diagonal blocks are stored in full form, and an attempt is made to compress off diagonal blocks. As with the near matrix, if the compression succeeds we retain the compressed version, otherwise the full version is stored.

## 11.6 Software Implementation Notes

In this section we will discuss some elements of how the MLFMA/SVD was implemented in our *Serenity* MoM solver. As with the ACA/MLACA, what is presented here are only suggestions, based on the author's experience.

## 11.6.1 Software Class Support

In this section we will discuss some of the software classes implemented in support of the MLFMA. Fortunately, some of the classes written previously for the ACA can be reused as-is, or with only minor modifications.

### 11.6.1.1 Element Engine Class

The `ElementEngine` class from Section 9.6.1.1 needs no modifications for filling near-matrix blocks, as groups constructed using the octree (FMM) or $K$-means (ACA) are functionally identical if using the same data structures.

### 11.6.1.2 Sparse Block Matrix Class

The `BlockMatrix` class described in Section 9.6.1.2 is not optimal for storing the FMM near matrix. In the ACA, the groups are very large and may number several hundred to a thousand, whereas in the FMM the groups are much smaller and there may be tens (or hundreds) of thousands of groups. Thus, for $M$ groups, a row-major array of size $M \times M$ is not a viable approach for storing pointers to the near matrix blocks.

To solve this problem, we derive from the `BlockMatrix` class the `SparseBlockMatrix` class, which stores block pointers in two different ways. During matrix filling, pointers are stored in an `std::map` C++ STL container using the block row $sr$ and column $sc$ as the key. When filling is complete, a new structure is built that facilitates fast access to blocks via `getBlock(sr, sc)`, as `std::map` is less efficient in that role. This structure is the Compressed Sparse Row (CSR) format, described in the following section.

*Compressed Sparse Row Format*

The Compressed Sparse Row (CSR) format [21] is a popular and efficient way to store any sort of sparse matrix data. To illustrate, consider the sparse integer matrix $\mathbf{A}$ defined as

$$\mathbf{A} = \begin{bmatrix} 1 & 2 & 0 & 0 & 0 \\ 3 & 4 & 5 & 0 & 0 \\ 0 & 6 & 7 & 8 & 0 \\ 0 & 0 & 9 & 10 & 0 \\ 0 & 0 & 0 & 0 & 11 \end{bmatrix}, \qquad (11.92)$$

which is a $5 \times 5$ matrix with $N_z = 11$ nonzero entries. The CSR representation for $\mathbf{A}$ comprises three vectors $\mathbf{a}$, $\mathbf{j}$, and $\mathbf{i}$, which in this case are

$$\mathbf{a} = \begin{bmatrix} 1 & 2 & 3 & 4 & 5 & 6 & 7 & 8 & 9 & 10 & 11 \end{bmatrix}, \qquad (11.93)$$

$$\mathbf{j} = \begin{bmatrix} 1 & 2 & 1 & 2 & 3 & 2 & 3 & 4 & 3 & 4 & 5 \end{bmatrix}, \qquad (11.94)$$

and

$$\mathbf{i} = \begin{bmatrix} 1 \ 3 \ 6 \ 9 \ 11 \ 12 \end{bmatrix}. \qquad (11.95)$$

Vector $\mathbf{a}$ is of length $N_z$ and contains the nonzero elements of $\mathbf{A}$, stored row by row. Vector $\mathbf{j}$ is also of length $N_z$ and comprises the column indexes of the matrix elements in $\mathbf{a}$. Vector $\mathbf{i}$ is of length $N + 1$, and stores the offsets to each row in $\mathbf{a}$ and $\mathbf{j}$, allowing for fast access to the start and end of each row. The last element of $\mathbf{i}$ contains an offset to the fictional row $m = N + 1$, allowing a fast lookup of the number of nonzero columns for each row. Note that as the rows and columns are sorted, we can quickly find matrix element $\mathbf{A}_{mn}$ by performing a binary search for column $n$ within positions $\mathbf{i}(m)$ to $\mathbf{i}(m+1) - 1$ in vector $\mathbf{j}$.

### 11.6.1.3  FMM Region Class

The `FMMRegion` class is responsible for computing and storing all radiation, receive and transfer functions in a dielectric region. It also implements routines that perform the upward and downward passes in the MLFMA, and the mapping from global to local basis function vectors, and vice versa.

### 11.6.1.4  FMM Octree Class

The `FMMOctree` class groups basis functions recursively using the octree, and provides this data to the `ElementEngine` class for computing the near matrix blocks and to the `FMMRegion` class for computing radiation and receive functions for each group/cube. It also finds all near and far groups and unique transfer vectors on each octree level, information used by the `FMMRegion` class in performing the upward and downward passes.

## 11.6.2  Shared Memory Processing

Our implementation in *Serenity* is for shared memory systems, performing all operations in parallel on the CPU using POSIX Threads. For all level 2 BLAS functions and the SVD, the CPU-optimized Intel MKL BLAS and LAPACK libraries are used.

### 11.6.2.1  FMM CPU Thread Class

The `FMMCPUThread` class performs a variety of setup tasks as well as the matrix solution in parallel on the CPU. These operations are described in more detail in this section.

*Near Matrix Filling*

When filling the near matrix, each thread computes matrix blocks asynchronously from other threads, choosing from near matrix blocks that are not yet filled.

*Radiation, Receive and Transfer Functions*

Computation of transfer and radiation and receive functions in each region is parallelized by assignment of unique transfer vectors or cubes to threads.

*Preconditioner Matrix*

Construction of the ILU(0) or ILUT preconditioner follows Section 11.5.3, and re-purposes the code written for the ACA LU factorization described in Section 9.6.2.1. Key differences include the additional code needed to handle the fill-in blocks, which are not present in the ACA block matrix. Construction of the SAI preconditioner follows the dense QR factorization outlined in Section 11.5.4.1, where each thread computes block columns of $\mathbf{M}$ asynchronously from other threads, choosing from columns not yet processed.

*Parallelization of Matrix-Vector Product*

We have identified two strategies for parallelizing the matrix-vector product, as discussed in this section.

In the first method, each CPU thread is assigned a small, contiguous portion of all right-hand sides, which they solve asynchronously from other threads. In radar cross section problems with many incident angles, this approach is more efficient as the angles in each portion are often closely spaced, and the reuse of solution vectors from previous angles is more effective in reducing iteration count. However, this approach requires a separate MLFMA workspace for each CPU thread, increasing the total memory requirements. Additionally, iteration count is not always the same between angles, so some threads often finish their portion before the rest and go idle.

In the second method, all threads participate in the solution of each RHS, with each thread performing some portion of the calculation. Thus, thread synchronization is kept to a minimum. The near product is easily done in parallel by assignment of the block rows of $\mathbf{Z}^{near}$ to individual threads. We next consider the upward and downward passes. We start both passes on the coarsest octree level ($l = 3$), and note that the computations for each cube, and their children, can be done independently of other cubes if the upward pass is completed prior to starting the downward pass. Therefore, both passes can be performed in parallel asynchronously, with each thread starting on the coarsest level and selecting from cubes not already processed for that pass. A barrier is needed at the end of the upward pass, ensuring the aggregated fields across all levels have been computed. We note that on level ($l = 3$) there are a maximum of 64 cubes, and so this second method does not scale past 64 threads.

*MLFMA Workspace Management*

For the upward pass (aggregation), two complex vectors of length $N_{k,l}$ are required for each group on levels $l \geq 3$ to store the $\theta$- and $\phi$-polarized aggregated fields. For the downward pass (disaggregation), two vectors of length $N_{k,l}$ are required for each level $l \geq 3$ for storing the $\theta$- and $\phi$-polarized sum of anterpolated/received fields along the recursive path.

## 11.7  Numerical Examples

In this section we will compute the radar cross section of several test arti-cles using our *Serenity* MoM solver, which implements the MLFMA using the techniques outlined in this chapter. We again consider conducting, dielectric, and coated spheres, EMCC benchmark targets, and the monoconic RV with dielectric nose, again with larger electrical sizes than in prior chapters. We will also analyze the performance of the preconditioners presented in Section 11.5.

### 11.7.1  Compute Platform

Computations in this section were carried out on the Dell workstation pre-vious described in Section 8.8.2.

### 11.7.2  Run Parameters

All triangle meshes were constructed with at least $\lambda/10$ unknowns per wavelength in each dielectric region. The CFIE is applied on closed conductors with $\alpha = 0.5$, and PMCHWT applied to dielectric interfaces. For the examples in Sections 11.7.3–11.7.6, the finest-level cube size in the octree is $\lambda_0/2$. For SVD compression of radiation and receive functions, $\tau_{SVD} = 10^{-2}$, and for the near matrix, $\tau_{SVD} = 10^{-4}$. For each case an ILUT preconditioner is built, with $\tau_{SVD} = 10^{-3}$ for existing off-diagonal blocks, $\tau_{SVD} = 10^{-1}$ for fill-in blocks, and a block drop tolerance $\tau_{ILUT} = 10^{-4}$. For the iterative solution, GMRES iteration is used, with a residual tolerance $\epsilon = 10^{-3}$.

### 11.7.3  Spheres

In this section we will consider the bistatic RCS of conducting, fully dielec-tric, and coated spheres, having the same dimensions as the spheres in Section 8.8.3.

#### 11.7.3.1  Conducting Sphere

We first consider a conducting sphere with a radius of 0.5 meters at frequen-cies of 0.75, 1.5, 3.0 6.0, 12.0 and 24.0 GHz. For the MoM simulation, we use facet models comprising 5120, 20480, 81920, 327680, 1310720, and 5242880 triangles, resulting in 7680, 30720, 122880, 491520, 1966080, and 7864320 unknowns, respectively. The bistatic RCS from the MLFMA is compared to the Mie series in Figures 11.14a–11.14f and 11.15a–11.15d. The comparison is very good at all frequencies.

### 11.7.3.2 Dielectric Sphere

We next consider a lossless dielectric sphere ($\epsilon_r = 2.56$) with a radius of 0.5 meters at frequencies of 3.0, 6.0 and 12.0 GHz. For the MoM simulation, we re-use the facet models from Section 11.7.3.1, however there are now an equal number of magnetic basis functions, resulting in 245760, 983040 and 3932160 basis functions at 3.0, 6.0 and 12.0 GHz, respectively. The bistatic RCS from the MLFMA is compared to the Mie series in Figures 11.16a–11.16d, where the comparison is very good. At 12.0 GHz, there is a slight disagreement between 30 and 60 degrees in the VV-pol data, which is not seen in the HH-pol results. The cause of the discrepancy is not known.

### 11.7.3.3 Coated Sphere

We next consider the lossless coated sphere ($\epsilon_r = 2.56$) at 3.0, 6.0 and 12.0 GHz. At 3.0 GHz, the surface meshes for the conducting core and the coating comprise 20480 and 81920 triangles, resulting in 153600 electric and 122880 magnetic basis functions, respectively, for a total of 276480 unknowns. At 6.0, they comprise 81920 and 327680 triangles, yielding 614400 electric and 491520 magnetic basis functions, for a total of 1105920 unknowns. Finally, at 12.0 GHz, they comprise 327680 and 1310720 triangles, yielding 2457600 electric and 1966080 magnetic basis functions, for a total of 4423680 unknowns. The bistatic RCS from the FMM is compared to the Mie series in Figures 11.17a–11.17d, where the comparison is again very good at all frequencies.

## 11.7.4 Thin Square Plate

We next consider a thin, square conducting plate having length and width of 1 meter at 1.0, 2.0, 4.0, 8.0 and 16.0 GHz. For the MoM simulation, facet models with 3072, 12288, 49152, 196608, 786432, and 3145728 facets are used, resulting in 18560, 73984, 294400, 1178624, and 4716544 unknowns, respectively. The results from the MLFMA are compared to those from the single-level ACA for horizontal polarization in Figures 11.18a-11.18e, where the results are nearly indistinguishable.

## 11.7.5 Monoconic Reentry Vehicle

Next, we will examine the monoconic RV with lossless dielectric nose, previously considered in Section 10.4.4, and we will compare the results from the MLFMA to those from the MLACA (($L = 3$). We compute the monostatic RCS at 14.0, 16.0, 20.0, and 24.0 GHz, using facets models having 1065932 1167268, 2206972, and 3252068 triangles, respectively. These comprise the total number of facets used for the nose, base and septum interfaces, resulting in a total of 1645768, 1799000, 3403364, and 5022318 unknowns, respectively. The results are compared for vertical and horizontal polarizations in Figures 11.19a-11.19h, where the comparison is very good at all frequencies.

### 11.7.6   Business Jet

Finally, we consider a business jet, depicted in Figures 11.13a and 11.13b from front and rear viewing angles, respectively. The jet has a length of 17.6 meters (60 feet), a wingspan of 16.2 meters (53 feet), and a height of 3.8 meters (12 feet). The original model was obtained from an Internet database of freely available 3D models. To simplify the meshing process, the model was converted to a closed, conducting surface by removing features such as the windows and landing gear, and by closing off the engine nacelles. The model was then oriented in the $xy$ plane, with the nose aligned along the $+x$ axis.

We compute the monostatic RCS in the $xy$ plane for azimuthal angles $\phi$ from 0 to 180 degrees, at frequencies of 0.6, 1.2, 1.8 and 2.4 GHz. In support, facet models were constructed with 430392, 1606124, 3555630, and 6039472 triangles, resulting in 645588, 2409186, 5333445 and 9059208 unknowns, respectively. The results from the MLFMA are compared to those from the MLACA ($L = 3$) for vertical and horizontal polarizations in Figures 11.20a-11.20h, where the results are almost indistinguishable.

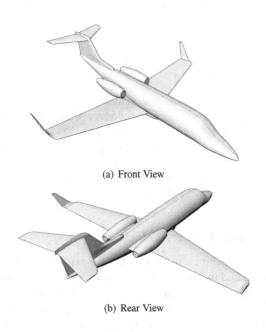

(a) Front View

(b) Rear View

**FIGURE 11.13:** Business Jet

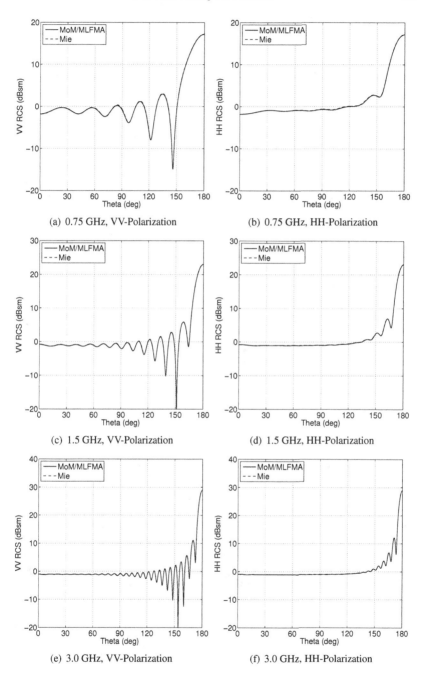

(a) 0.75 GHz, VV-Polarization

(b) 0.75 GHz, HH-Polarization

(c) 1.5 GHz, VV-Polarization

(d) 1.5 GHz, HH-Polarization

(e) 3.0 GHz, VV-Polarization

(f) 3.0 GHz, HH-Polarization

**FIGURE 11.14:** Conducting Sphere: Bistatic RCS (0.75 - 3.0 GHz)

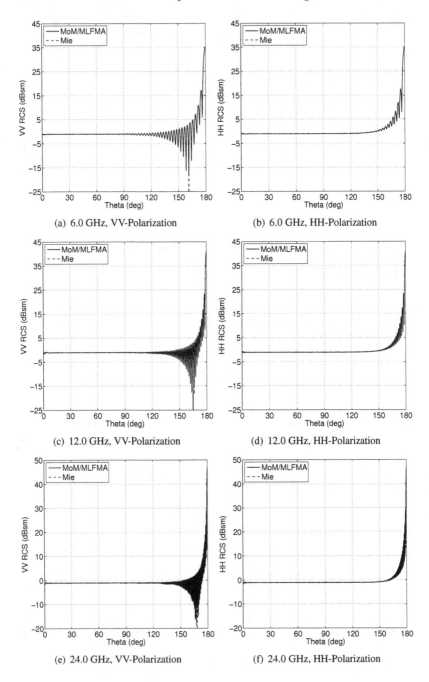

(a) 6.0 GHz, VV-Polarization

(b) 6.0 GHz, HH-Polarization

(c) 12.0 GHz, VV-Polarization

(d) 12.0 GHz, HH-Polarization

(e) 24.0 GHz, VV-Polarization

(f) 24.0 GHz, HH-Polarization

**FIGURE 11.15:** Conducting Sphere: Bistatic RCS (6.0 - 24.0 GHz)

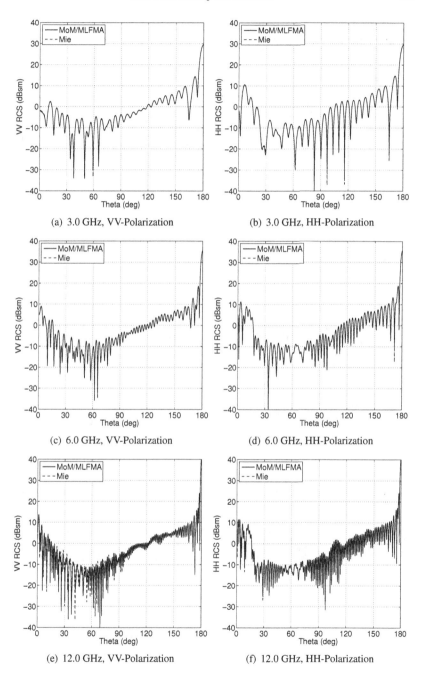

(a) 3.0 GHz, VV-Polarization

(b) 3.0 GHz, HH-Polarization

(c) 6.0 GHz, VV-Polarization

(d) 6.0 GHz, HH-Polarization

(e) 12.0 GHz, VV-Polarization

(f) 12.0 GHz, HH-Polarization

**FIGURE 11.16:** Dielectric Sphere: Bistatic RCS

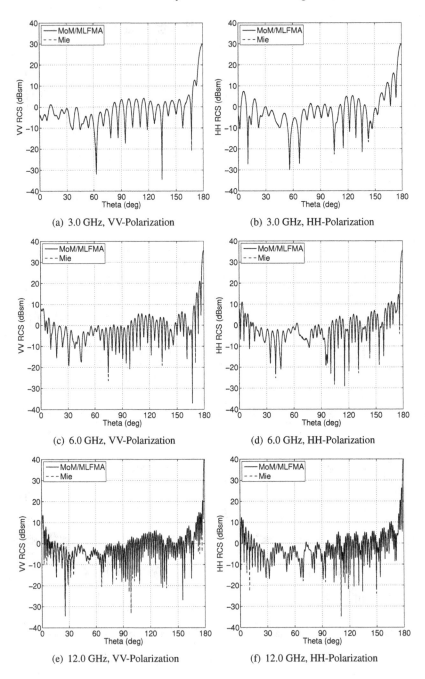

(a) 3.0 GHz, VV-Polarization

(b) 3.0 GHz, HH-Polarization

(c) 6.0 GHz, VV-Polarization

(d) 6.0 GHz, HH-Polarization

(e) 12.0 GHz, VV-Polarization

(f) 12.0 GHz, HH-Polarization

**FIGURE 11.17:** Coated Sphere: Bistatic RCS

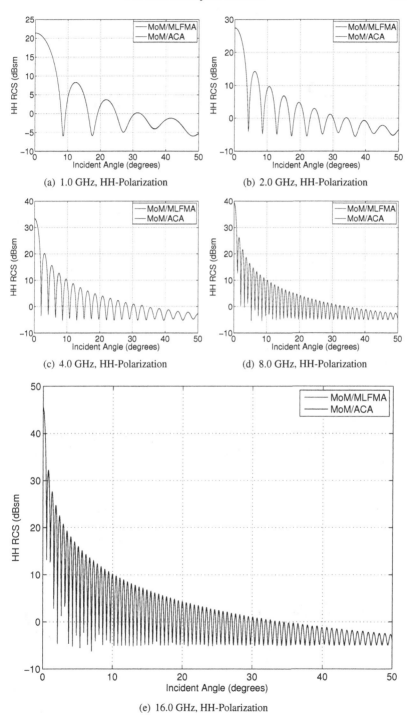

(a) 1.0 GHz, HH-Polarization

(b) 2.0 GHz, HH-Polarization

(c) 4.0 GHz, HH-Polarization

(d) 8.0 GHz, HH-Polarization

(e) 16.0 GHz, HH-Polarization

**FIGURE 11.18:** Thin Plate: HH RCS at 1.0, 2.0, 4.0, 8.0, and 16.0 GHz

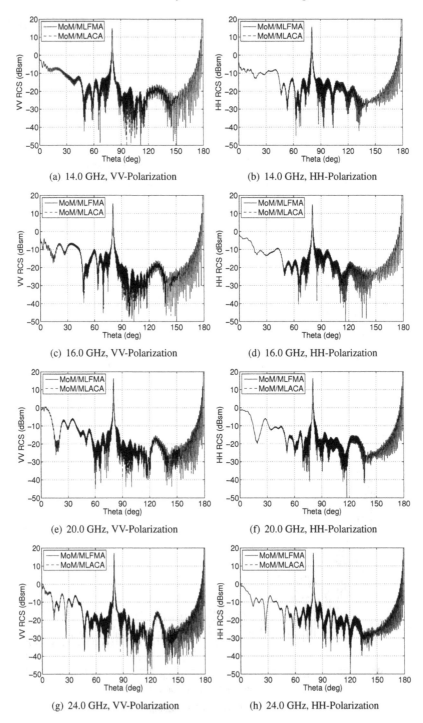

(a) 14.0 GHz, VV-Polarization

(b) 14.0 GHz, HH-Polarization

(c) 16.0 GHz, VV-Polarization

(d) 16.0 GHz, HH-Polarization

(e) 20.0 GHz, VV-Polarization

(f) 20.0 GHz, HH-Polarization

(g) 24.0 GHz, VV-Polarization

(h) 24.0 GHz, HH-Polarization

**FIGURE 11.19:** RV with Lossless Nose: RCS at 14.0, 16.0, 20.0, and 24.0 GHz

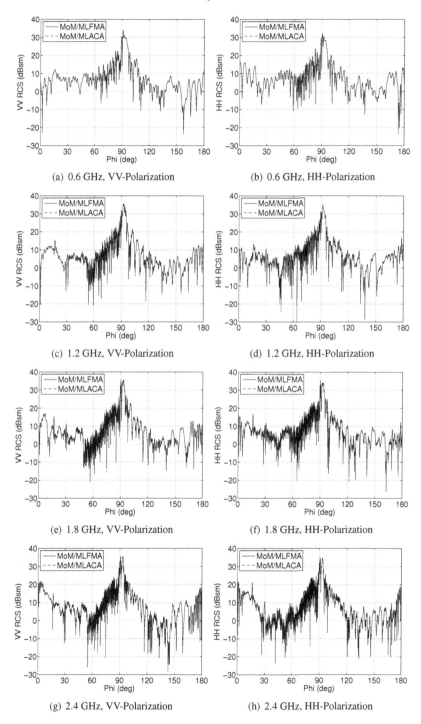

(a) 0.6 GHz, VV-Polarization

(b) 0.6 GHz, HH-Polarization

(c) 1.2 GHz, VV-Polarization

(d) 1.2 GHz, HH-Polarization

(e) 1.8 GHz, VV-Polarization

(f) 1.8 GHz, HH-Polarization

(g) 2.4 GHz, VV-Polarization

(h) 2.4 GHz, HH-Polarization

**FIGURE 11.20:** Business Jet: RCS at 0.6, 1.2, 1.8, and 2.4 GHz

## 11.7.7 Summary of Examples

Run metrics for the examples in this section are summarized in Table 11.4. Listed are the objects, operating frequency ($f_0$, in GHz), number of electric and magnetic basis functions ($N_J$ and $N_M$, respectively), storage (in GB) for the near matrix ($M_{near}$), preconditioner ($M_{pre}$), radiation, receive, and transfer functions ($M_{rad}$, $M_{rec}$, and $M_T$, respectively), and the number of octree levels ($N_O$).

In Figure 11.21 is plotted the total memory requirement ($M_{near} + M_{pre} + M_{rad} + M_{rec} + M_T$) versus the number of unknowns $N$ for all test cases in Table 11.4. We note that the storage requirement of our MLFMA/SVD algorithm scales approximately as $O(N \log N)$, which is consistent with the standard MLFMA [2].

**TABLE 11.4:** Run Metrics

| Object | $f_0$ | $N_J$ | $N_M$ | $M_{near}$ | $M_{pre}$ | $M_{rad}$ | $M_{rec}$ | $M_T$ | $N_O$ |
|---|---|---|---|---|---|---|---|---|---|
| PEC Sphere | 0.75 | 7680 | 0 | .039 | .037 | .005 | .004 | .005 | 4 |
| PEC Sphere | 1.5 | 30720 | 0 | .16 | .15 | .019 | .018 | .013 | 5 |
| PEC Sphere | 3.0 | 122880 | 0 | .63 | .60 | .075 | .072 | .035 | 6 |
| PEC Sphere | 6.0 | 491520 | 0 | 2.6 | 2.3 | .31 | .29 | .10 | 7 |
| PEC Sphere | 12.0 | 1966080 | 0 | 9.8 | 8.7 | 1.2 | 1.1 | .311 | 8 |
| PEC Sphere | 24.0 | 7864320 | 0 | 39 | 35 | 4.8 | 4.6 | 1.0 | 9 |
| Dielectric Sphere | 3.0 | 122880 | 122880 | 2.6 | 2.4 | .24 | .78 | .10 | 6 |
| Dielectric Sphere | 6.0 | 491520 | 491520 | 10 | 9.5 | .96 | 3.2 | .32 | 7 |
| Dielectric Sphere | 12.0 | 1966080 | 1966080 | 42 | 37 | 3.8 | 13 | 1.0 | 8 |
| Coated Sphere | 3.0 | 153600 | 122880 | 2.8 | 2.6 | .28 | .86 | .10 | 6 |
| Coated Sphere | 6.0 | 614400 | 491520 | 11 | 10 | 1.1 | 3.5 | .32 | 7 |
| Coated Sphere | 12.0 | 2457600 | 1966080 | 44 | 39 | 4.5 | 14 | 1.0 | 8 |
| Thin Plate | 1.0 | 18304 | 0 | .14 | .11 | .003 | .003 | .001 | 4 |
| Thin Plate | 2.0 | 73472 | 0 | .60 | .48 | .011 | .011 | .003 | 5 |
| Thin Plate | 4.0 | 294400 | 0 | 2.5 | 1.9 | .049 | .049 | .008 | 6 |
| Thin Plate | 8.0 | 1178624 | 0 | 10 | 7.8 | .19 | .19 | .023 | 7 |
| Thin Plate | 16.0 | 4716544 | 0 | 41 | 32 | .77 | .77 | .073 | 8 |
| Dielectric Nose RV | 14.0 | 1598780 | 46988 | 12 | 10 | .56 | 1.2 | .78 | 8 |
| Dielectric Nose RV | 16.0 | 1750780 | 48220 | 12 | 10 | .72 | 1.7 | .81 | 8 |
| Dielectric Nose RV | 20.0 | 3310288 | 93076 | 25 | 21 | 1.1 | 2.6 | 1.8 | 9 |
| Dielectric Nose RV | 24.0 | 4877898 | 144420 | 37 | 31 | 1.8 | 4.0 | 2.5 | 9 |
| Business Jet | 0.6 | 645588 | 0 | 6.2 | 5.7 | .22 | .21 | .12 | 8 |
| Business Jet | 1.2 | 2409186 | 0 | 20 | 19 | .83 | .80 | .40 | 9 |
| Business Jet | 1.8 | 5333445 | 0 | 41 | 37 | 1.8 | 1.7 | .59 | 9 |
| Business Jet | 2.4 | 9059208 | 0 | 66 | 59 | 3.2 | 3.0 | 1.4 | 10 |

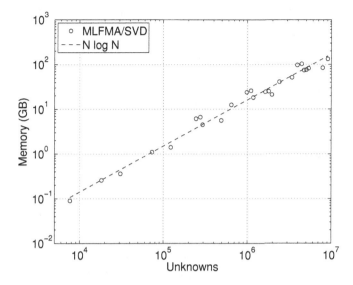

**FIGURE 11.21:** MLFMA/SVD Storage versus Number of Unknowns $N$

## 11.7.8 Preconditioner Performance

We will next compare the memory requirements and overall effectiveness of the preconditioners discussed in Section 11.5. We consider again the thin plate from Section 11.7.4 at 1.0, 2.0, 4.0 and 8.0 GHz, and as it is an open surface, only the EFIE is applied. As the EFIE is known to produce a poorly conditioned system matrix, this should be a good test article. Unless specified otherwise, all run settings are identical to those used in Sections 11.7.3-11.7.6.

### 11.7.8.1 Compressed versus Uncompressed Preconditioners

In practice we will compress the off-diagonal preconditioner blocks, so we must show that preconditioning effectiveness is not adversely impacted by the compression. To establish a baseline, we will first examine the performance of each preconditioner with no compression applied. For the ILUT preconditioner, fill-in blocks are dropped with a tolerance $\tau_{ILUT} = 10^{-4}$ but are not compressed, resulting in a much higher memory usage.

In Table 11.5 are summarized the iteration counts for the preconditioners at each frequency, for vertically and horizontally-polarized incident plane waves at an incident angle of 45 degrees off of broadside. GMRES iteration is used to a residual tolerance of $10^{-3}$, with 100 iterations allowed before restarting and no limit on the number of iterations. Compared to the non-preconditioned iterations, we note that the diagonal preconditioner reduces the iteration count a fair amount, up to 24 percent at 8.0 GHz. The block diagonal conditioner is

**TABLE 11.5:** Thin Plate Iteration Counts (Uncompressed Preconditioner)

| $f_0$ (GHz) | 0.5 | | 1.0 | | 2.0 | | 4.0 | | 8.0 | |
|---|---|---|---|---|---|---|---|---|---|---|
| Polarization | V | H | V | H | V | H | V | H | V | H |
| None | 384 | 346 | 524 | 502 | 718 | 544 | 884 | 709 | 1178 | 933 |
| Diagonal | 347 | 300 | 489 | 466 | 683 | 523 | 905 | 700 | 1198 | 997 |
| Block Diagonal | 121 | 93 | 355 | 255 | 1114 | 639 | 3223 | 1318 | fail | fail |
| ILU(0) | 9 | 9 | 17 | 18 | 29 | 30 | 50 | 54 | 66 | 68 |
| ILUT | 5 | 6 | 8 | 9 | 12 | 12 | 17 | 17 | 27 | 26 |
| SAI | 7 | 7 | 12 | 11 | 16 | 13 | 23 | 18 | 32 | 23 |

**TABLE 11.6:** Thin Plate Preconditioner Storage (Uncompressed, in GB)

| $f_0$ (GHz) | 0.5 | 1.0 | 2.0 | 4.0 | 8.0 |
|---|---|---|---|---|---|
| Block Diagonal | .0011 | .051 | .22 | .9 | 3.6 |
| ILU(0) | .075 | .40 | 1.8 | 7.7 | 32 |
| ILUT | .097 | .81 | 4.9 | 18.2 | 73 |
| SAI | .075 | .40 | 1.8 | 7.7 | 32 |

somewhat more effective at 0.5 and 1.0 GHz, however at 2.0 GHz and above the performance is harmed significantly, and at 8.0 GHz, the iteration did not converge after $10^4$ iterations and was halted manually. It is with the ILU and SAI preconditioners where convergence improves dramatically. Comparing ILU(0), ILUT and SAI, we note that ILUT is the clear winner among the three, with SAI only slightly behind ILUT, and ILU(0) falling far behind.

In Table 11.6 are shown the storage requirements (in GB) for the preconditioners in uncompressed form. The diagonal preconditioner is not listed as its storage requirements are negligible. Of particular note is the large requirement of the ILUT preconditioner compared to ILU(0) and SAI, due to added fill-in blocks. SAI and ILU(0) have the same sparsity pattern, thus their storage requirements are identical.

We next consider the same example, but with the ILU(0), ILUT, and SAI preconditioners in compressed form. For regular off diagonal blocks, $\tau_{SVD} = 10^{-3}$. For ILUT, $\tau_{SVD} = 10^{-1}$ for fill-in blocks, and the drop tolerance is $\tau_{ILUT} = 10^{-4}$. In Table 11.7 are summarized the iteration counts. We note that these are essentially unchanged from those in Table 11.5, confirming that the preconditioners continue to perform well after compression. In Table 11.8 are summarized the corresponding storage requirements ($M_{pre}$, in GB) for the compressed preconditioners, and the amount of compression ($C$, in percent). The compression level for each is significant, particularly for ILUT with its many fill-in blocks.

**TABLE 11.7:** Thin Plate Iteration Counts (Compressed Preconditioner)

| $f_0$ (GHz) | 0.5 | | 1.0 | | 2.0 | | 4.0 | | 8.0 | |
|---|---|---|---|---|---|---|---|---|---|---|
| Polarization | V | H | V | H | V | H | V | H | V | H |
| ILU(0) | 9 | 9 | 17 | 18 | 29 | 30 | 49 | 53 | 65 | 68 |
| ILUT | 5 | 6 | 8 | 9 | 12 | 12 | 17 | 17 | 29 | 28 |
| SAI | 7 | 7 | 12 | 11 | 16 | 13 | 23 | 18 | 31 | 23 |

**TABLE 11.8:** Thin Plate Preconditioner Storage (Compressed, in GB)

| $f_0$ (GHz) | 0.5 | | 1.0 | | 2.0 | | 4.0 | | 8.0 | |
|---|---|---|---|---|---|---|---|---|---|---|
| | $M_{pre}$ | C(%) | $M_{pre}$ | C(%) | $M_{pre}$ | C(%) | $M_{pre}$ | C(%) | $M_{pre}$ | C(%) |
| ILU(0) | .021 | 71 | .11 | 73 | .45 | 76 | 1.9 | 76 | 7.6 | 76 |
| ILUT | .022 | 78 | .11 | 86 | .48 | 91 | 1.9 | 89 | 7.8 | 89 |
| SAI | .024 | 68 | .12 | 69 | .53 | 71 | 2.2 | 72 | 8.9 | 72 |

### 11.7.8.2    Performance versus Incident Angle

We next compute the RCS of the thin plate for 361 incident angles from 0 to 90 degrees, where 0 degrees is at broadside. The finest level cube size is $0.5\lambda$, and the previous solution vector is used as the initial guess for subsequent iterations. In Figures 11.22a–11.22h are compared the iteration counts for the ILU(0), ILUT and SAI preconditioners at 1.0, 2.0, 4.0 and 8.0 GHz, for vertical and horizontal incidence. Again, ILU(0) performs the worst of the three. ILUT performs equal to or slightly better than SAI at 4.0 GHz or below, and is slightly behind at 8.0 GHz in horizontal polarization, and at higher angles in vertical polarization.

### 11.7.8.3    Performance versus Cube Size

Finally, we consider the performance of the ILU(0), ILUT, and SAI preconditioners at 8.0 GHz, for finest level cube sizes of $0.25\lambda$, $0.375\lambda$, $0.5\lambda$, and $0.75\lambda$. Summarized in Table 11.9 are the iteration counts for vertical and horizontal incidence. We observe that as the size of the cubes grows, the iteration count decreases. This result is expected, as the added information content in the near matrix results in a more effective preconditioner. Unexpected was the failure of ILU(0) at $0.25\lambda$ and $0.75\lambda$, where the iteration converged in a single iteration to a wrong result. This suggests that ILU(0) is not reliable and should not be used in practice. In Table 11.10 are summarized the corresponding storage requirements ($M_{pre}$, in GB) for the preconditioners, and the amount of compression ($C$, in percent). The level of compression increases with cube size, as noted previously in Section 11.3.2.1.

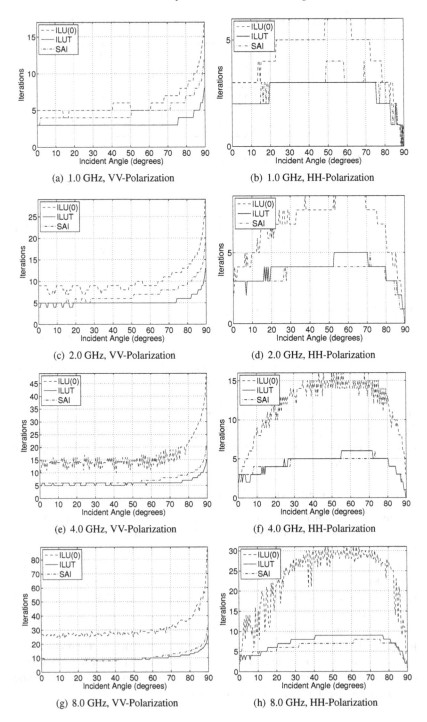

**FIGURE 11.22:** Thin Plate Iteration Counts versus Incident Angle

**TABLE 11.9:** Thin Plate Iteration Counts versus Cube Size

| $\lambda$ | 0.25 | | 0.375 | | 0.5 | | 0.75 | |
|---|---|---|---|---|---|---|---|---|
| Polarization | V | H | V | H | V | H | V | H |
| ILU(0) | 1* | 1* | 178 | 164 | 68 | 70 | 1* | 1* |
| ILUT | 47 | 38 | 65 | 59 | 29 | 28 | 20 | 22 |
| SAI | 48 | 36 | 35 | 27 | 31 | 23 | 20 | 17 |

\* Iteration failed.

**TABLE 11.10:** Thin Plate Preconditioner Storage (in GB) versus Cube Size

| $\lambda$ | 0.25 | | 0.375 | | 0.5 | | 0.75 | |
|---|---|---|---|---|---|---|---|---|
| | $M_{pre}$ | C(%) | $M_{pre}$ | C(%) | $M_{pre}$ | C(%) | $M_{pre}$ | C(%) |
| ILU(0) | 3.4 | 58 | 5.3 | 71 | 7.6 | 76 | 13.3 | 81 |
| ILUT | 3.7 | 83 | 5.6 | 88 | 7.8 | 89 | 13.7 | 92 |
| SAI | 4.3 | 47 | 6.4 | 64 | 8.9 | 72 | 15.0 | 79 |

## 11.7.9 Initial Guess in Iterative Solution

When using an iterative solver, there is often no *a priori* knowledge regarding the solution vector. Therefore, for an arbitrary right-hand side vector, the usual approach is to simply initialize the solution vector with zeros. In radar cross section problems, where incident angles are closely spaced, the right-hand side and solution vectors will be similar between angles. Therefore, the previous solution vector can be re-used, reducing the number of iterations. In [2], it was suggested that this can be improved further by adjusting the phase of the solution vector between incident angles. To illustrate, consider the plane-wave incident vector $\hat{\mathbf{r}}_1$, where the phase term at a point $\mathbf{r}$ in the free space region $R_0$ is

$$\psi_1(\mathbf{r}) = e^{jk_0\hat{\mathbf{r}}_1 \cdot \mathbf{r}} . \qquad (11.96)$$

Between angles, we can make following correction to the solution vector

$$\Delta\psi(\mathbf{r}) = e^{jk_0(\hat{\mathbf{r}}_2 - \hat{\mathbf{r}}_1) \cdot \mathbf{r}} . \qquad (11.97)$$

If using RWG basis functions, this can be implemented by adjusting the phase of the solution vector elements corresponding to the electric and magnetic basis functions assigned to each edge in $R_0$. For a given edge, the correction is

$$\Delta\psi(\mathbf{r}_c) = e^{jk_0(\hat{\mathbf{r}}_2 - \hat{\mathbf{r}}_1) \cdot \mathbf{r}_c} , \qquad (11.98)$$

where $\mathbf{r}_c$ is the center point of the edge.

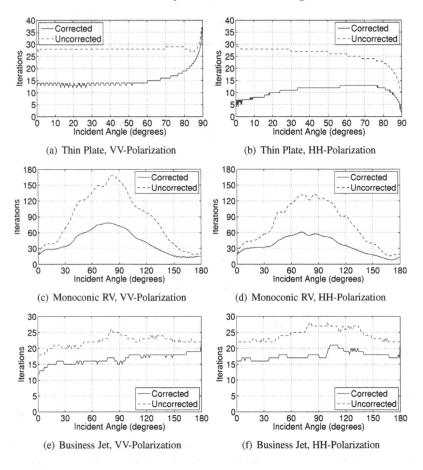

(a) Thin Plate, VV-Polarization

(b) Thin Plate, HH-Polarization

(c) Monoconic RV, VV-Polarization

(d) Monoconic RV, HH-Polarization

(e) Business Jet, VV-Polarization

(f) Business Jet, HH-Polarization

**FIGURE 11.23:** Phase Correction Impact on Iterative Solver Performance

We now consider several examples that demonstrate the effectiveness of this phase correction. We first consider the thin plate from Section 11.7.4 at 8.0 GHz, for 180 incident angles from 0 to 90 degrees (a .25 degree step size). The number of iterations per right-hand side are compared for vertical and horizontal incidence in Figures 11.23a and 11.23b, respectively. We see that the number of iterations has been decreased significantly by adding the correction. In Figures 11.23c and 11.23d is presented a similar comparison for the monoconic reentry vehicle from Section 11.7.5 at 16.0 GHz, for 361 incident angles from 0 to 180 degrees (a 0.5 degree step size). The improvement is again significant. Finally, in Figures 11.23e and 11.23f is shown a comparison for the business jet from Section 11.7.6 at 1.2 GHz, for 181 incident angles from 0 to 180 degrees (a 1 degree step size). The improvement is again very good.

# References

[1] R. Coifman, V. Rokhlin, and S. Wandzura, "The fast multipole method for the wave equation: A pedestrian prescription," *IEEE Antennas Propagat. Magazine*, vol. 35, pp. 7–12, June 1993.

[2] W. C. Chew, J. M. Jin, E. Michielssen, and J. Song, *Fast and Efficient Algorithms in Computational Electromagnetics*. Artech House, 2001.

[3] L. Greengard and V. Rokhlin, "A fast algorithm for particle simulations," *J. Comput. Phys.*, vol. 73, pp. 325–348, 1987.

[4] J. Stratton, *Electromagnetic Theory*. McGraw-Hill, 1941.

[5] D. E. Amos, "A subroutine package for Bessel functions of a complex argument and nonnegative order," Tech. Rep. SAND85-1018, Sandia National Laboratory, May 1985.

[6] P. Havé, "A parallel implementation of the fast multipole method for Maxwell's equations," *Int. J. Numer. Meth. Fluids*, vol. 43, no. 8, pp. 839–864, 2003.

[7] J. M. Song and W. C. Chew, "Error analysis for the truncation of multipole expansion of vector Green's functions," *IEEE Microwave Wireless Components Lett.*, vol. 11, no. 7, pp. 311–313, 2001.

[8] O. M. Bucci and G. Franceschetti, "On the spatial bandwidth of scattered fields," *IEEE Trans. Antennas Propagat.*, vol. 35, pp. 1445–1455, December 1987.

[9] N. Geng, A. Sullivan, and L. Carin, "Fast multipole method for scattering from an arbitrary PEC target above or buried in a lossy half space," *IEEE Trans. Antennas Propagat.*, vol. 49, pp. 740–748, May 2001.

[10] O. M. Bucci, C. Gennarelli, and C. Savarese, "Optimal interpolation of radiated fields over a sphere," *IEEE Trans. Antennas Propagat.*, vol. 39, pp. 1633–1643, November 1991.

[11] M. Böhme and D. Potts, "A fast algorithm for filtering and wavelet decomposition on the sphere," *Trans. Numer. Anal*, vol. 16, pp. 70–93, 2002.

[12] J. M. Song, C. C. Lu, and W. C. Chew, "Multilevel fast multipole algorithm for electromagnetic scattering by large complex objects," *IEEE Trans. Antennas Propagat.*, vol. 45, pp. 1488–1493, October 1997.

[13] S. Velamparambil and W. C. Chew, "Analysis and performance of a distributed memory multilevel fast multipole algorithm," *IEEE Trans. Antennas Propagat.*, vol. 53, pp. 2719–2727, August 2005.

[14] J. Foley, A. van Dam, S. Feiner, and J. Hughes, *Computer Graphics: Principles and Practice.* Addison-Wesley, 1996.

[15] B. Decker, "World geodetic system 1984," Tech. Rep. AD-A 167 570, Defense Mapping Agency, April 1996.

[16] F. Lemoine, S. Kenyon, J. Factor, R. Trimmer, N. Pavlis, D. Chinn, C. Cox, S. Klosko, S. Luthcke, M. Torrence, Y. Wang, R. Williamson, E. Pavlis, R. Rapp, and T. R. Olson, "The development of the joint NASA GSFC and the national imagery and mapping agency (NIMA) geopotential model EGM96," Tech. Rep. NASA/TP1998206861, National Aeronautics and Space Administration, July 1998.

[17] S. Koc, J. M. Song, and W. C. Chew, "Error analysis for the numerical evaluation of the diagonal forms of the spherical addition theorem," *SIAM J. Numer. Anal.*, vol. 36, no. 3, pp. 906–921, 1999.

[18] J. Song and W. Chew, "Interpolation of translation matrix in MLFMA," *Microw. Opt. Technol. Lett.*, vol. 30, pp. 109–113, July 2001.

[19] A. Elafrou, G. Goumas, and N. Koziris, "Performance analysis and optimization of sparse matrix-vector multiplication on modern multi- and many-core processors," in *2017 46th International Conference on Parallel Processing (ICPP)*, pp. 292–301, 2017.

[20] S. Williams, L. Oliker, R. Vuduc, J. Shalf, K. Yelick, and J. Demmel, "Optimization of sparse matrix-vector multiplication on emerging multicore platforms," *Supercomputing (SC07)*, 2007.

[21] Y. Saad, *Iterative Methods for Sparse Linear Systems.* PWS, 1st ed., 1996.

[22] T. Malas and L. Gurel, "Accelerating the multilevel fast multipole algorithm with the sparse-approximate-inverse (SAI) preconditioning," *SIAM J. Sci. Comput.*, vol. 31, no. 3, pp. 1968–1984, 2009.

[23] B. Carpentieri, I. Duff, L. Giraud, and G. Sylvand, "Combining fast multipole techniques and an approximate inverse preconditioner for large electromagnetism calculations," *SIAM J. Sci. Comput.*, vol. 27, no. 3, pp. 774–792, 2005.

[24] J. Lee, J. Zhang, and C. Lu, "Sparse inverse preconditioning of multilevel fast multipole algorithm for hybrid integral equations in electromagnetics," *IEEE Trans. Antennas Propagat.*, vol. 52, no. 9, pp. 2277–2287, 2004.

# Chapter 12

## Integration

In this book we have considered many integrals, where most were in one or two dimensions. Though some of these can be evaluated analytically, most have no analytic solution and numerical integration (quadrature) must be used instead. When using the Moment Method, the accuracy of the solution depends heavily on the accuracy of the matrix elements, thus a reliable and computationally efficient quadrature scheme is vital. In this chapter we will consider several one-dimensional quadrature schemes, as well as quadratures designed specifically for two-dimensional planar triangles.

## 12.1 One-Dimensional Integration

For the one-dimensional definite integral given by

$$I = \int_a^b f(x)\, dx \,, \tag{12.1}$$

a numerical quadrature generates an approximate solution using the sum

$$I = \int_a^b f(x)\, dx \approx \sum_{i=1}^{N} w(x_i) f(x_i) \,, \tag{12.2}$$

where the function $f(x)$ is evaluated at $N$ unique locations $x_i$ and multiplied by the quadrature weights $w(x_i)$. The weights and locations are dependent on the type and order of quadrature, often referred to as the quadrature *rule*. We will discuss several one-dimensional quadrature schemes in this section.

### 12.1.1 Centroidal Approximation

Some of the integrations in this book were evaluated using a centroidal approximation. This method simply evaluates the function once and assumes that its value is constant within the limits of integration. The evaluation is typically done at the midpoint of the domain, resulting in an approximation given by

$$I = \int_a^b f(x)\, dx \approx (b-a) f\big([b+a]/2\big) . \tag{12.3}$$

This "one-point" rule is useful in small domains where the integrand varies slowly, such as in computing the matrix elements between basis and testing function pairs that are far away from one another. In this case, the results from the centroidal rule often differ little from the results of a more accurate rule, but are much faster to compute. The error is often tolerable, since the magnitude of these elements iis usually small compared to those closer to the matrix diagonal.

### 12.1.2 Rectangular Rule

The rectangular rule uses a stair-step approximation to $f(x)$ throughout the range of integration. We subdivide the interval $[a, b]$ into $N$ small segments of equal length $h = (b-a)/N$ and compute the values $f(x_1)$, $f(x_2)$, ..., $f(x_N)$, as shown in Figure 12.1. The value of the integral within each segment is approximated as

$$\int_{x_i}^{x_{i+1}} f(x)\, dx \approx h f(x_i) . \tag{12.4}$$

Summing (12.4) across the entire interval, we derive the following expression

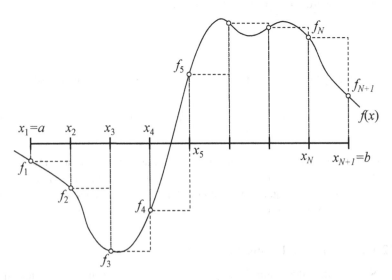

**FIGURE 12.1:** Rectangular (Stair-step) Approximation of $f(x)$

for the rectangular rule

$$I = \int_a^b f(x)\, dx \approx h \sum_{n=1}^{N} f(x_n) + E_R\,, \qquad (12.5)$$

where $E_R$ is the error. If $f''(x)$ is continuous within $[a, b]$, then the bound on the error is [1]

$$|E_R| \le \frac{(b-a)h^2}{24} f''(\xi) \qquad (12.6)$$

for some $\xi \in [a, b]$.

### 12.1.3 Trapezoidal Rule

The trapezoidal rule uses a piecewise linear approximation to $f(x)$ throughout the range of integration. We subdivide the interval $[a, b]$ into $N$ small segments of equal length $h = (b - a)/N$ and compute the values $f(x_1)$, $f(x_2), \ldots, f(x_N), f(x_{N+1})$, as shown in Figure 12.2. The value of the integral within each segment is approximated as

$$\int_{x_i}^{x_{i+1}} f(x)\, dx \approx \frac{h}{2}\left[f_i + f_{i+1}\right]. \qquad (12.7)$$

Summing (12.7) across the entire interval, we derive the following expression

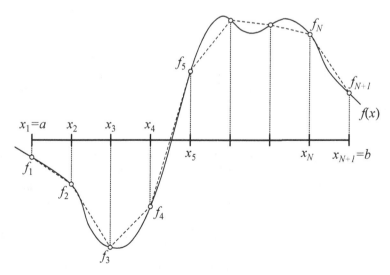

**FIGURE 12.2:** Piecewise Linear Approximation of $f(x)$

for the trapezoidal rule

$$I = \int_a^b f(x)\, dx \approx h\left[\frac{f_1}{2} + f_2 + \cdots + f_N + \frac{f_{N+1}}{2}\right] + E_T\,, \quad (12.8)$$

where $E_T$ is the error. If $f''(x)$ is continuous within $[a, b]$, then the bound on the error is [1]

$$|E_T| \leq \frac{(b-a)h^2}{12} f''(\xi) \quad (12.9)$$

for some $\xi \in [a, b]$. If the result is not accurate enough, the number of segments can be doubled and the previous samples reused, requiring only $N$ additional evaluations of $f(x)$ for the new rule. This process can be repeated as necessary until the desired accuracy is reached.

### 12.1.3.1   Romberg Integration

Repeated applications of the trapezoidal rule results in an error comprising even powers of $1/N$ [2], which can be written as

$$I = I_{n,1} + \frac{A}{N^2} + \frac{B}{N^4} + \frac{C}{N^6} + \cdots. \quad (12.10)$$

We note that the second-order term in the above decreases by a factor of four if we double the number of quadrature points. Using two applications of the rule, we can therefore write

$$I \approx I_{n,1} + \frac{K}{N^2} \quad (12.11)$$

and

$$I \approx I_{n+1,1} + \frac{K}{4N^2}\,. \quad (12.12)$$

If we solve the above two equations for $K$ we can eliminate the second order term and improve the previous solution to at least an $O(N^{-4})$ error. Doing so yields

$$I \approx I_{n+1,2} = \frac{4}{3}I_{n+1,1} - \frac{1}{3}I_{n,1}\,, \quad (12.13)$$

which is a Richardson's extrapolation. Combining successive applications of the rule with higher orders of extrapolation can greatly increase the accuracy of the solution and is known as *Romberg integration*. In this method, higher-order extrapolates have the form

$$I_{n,m} = I_{n+1,m-1} + \frac{I_{n+1,m-1} - I_{n,m-1}}{4^{m-1} - 1}\,. \quad (12.14)$$

An example hierarchy of extrapolates is illustrated in Figure 12.3 for $n \leq 4$. Applying Romberg integration using this hierarchy requires the results of the trapezoidal rule using 1, 2, 4, and 8 segments.

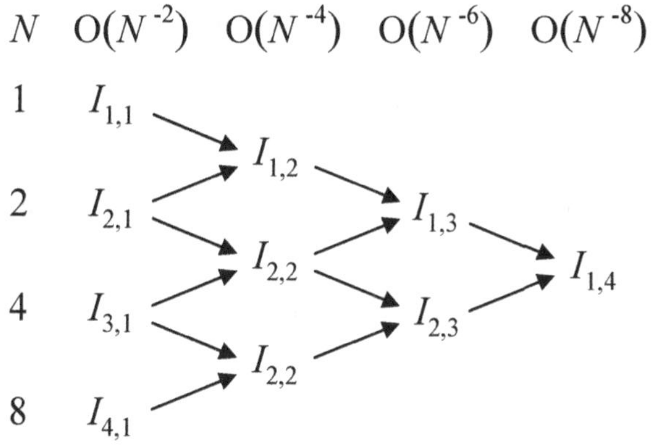

$$N \quad O(N^{-2}) \quad O(N^{-4}) \quad O(N^{-6}) \quad O(N^{-8})$$

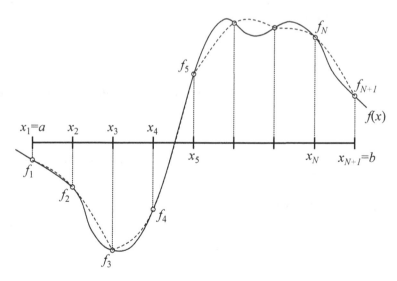

1   $I_{1,1}$

2   $I_{2,1}$   $I_{1,2}$   $I_{1,3}$

4   $I_{3,1}$   $I_{2,2}$   $I_{2,3}$   $I_{1,4}$

8   $I_{4,1}$   $I_{2,2}$

**FIGURE 12.3:** Hierarchy of Romberg Extrapolates

### 12.1.4 Simpson's Rule

Simpson's rule approximates $f(x)$ by a series of quadratic polynomials, as shown in Figure 12.4. The interval $[a, b]$ is divided into $N$ segments of equal length $h = (b - a)/N$, where $N \geq 2$ and even. Starting with the equation for a parabola

$$f(x) = Ax^2 + Bx + C \,, \tag{12.15}$$

**FIGURE 12.4:** Piecewise Parabolic Approximation of $f(x)$

we perform an integration over a two-segment interval $[-h, h]$, yielding

$$\int_{-h}^{h} \left(Ax^2 + Bx + C\right) dx = \left[A\frac{x^3}{3} + B\frac{x^2}{2} + Cx\right]\Bigg|_{-h}^{h} = \frac{h}{3}\left(2Ah^2 + 6C\right) .$$

(12.16)

Since the parabola passes through the points $(-h, y_1)$, $(0, y_2)$, and $(h, y_3)$, we can compute the values

$$f_1 = Ah^2 - Bh + C , \tag{12.17}$$

$$f_2 = C , \tag{12.18}$$

and

$$f_3 = Ah^2 + Bh + C , \tag{12.19}$$

from which we obtain

$$2Ah^2 = f_1 + f_3 - 2f_2 . \tag{12.20}$$

Inserting (12.18) and (12.20) into (12.16) yields

$$\int_{-h}^{h} \left(Ax^2 + Bx + C\right) dx = \frac{h}{3}\left[f_1 + 4f_2 + f_3\right]. \tag{12.21}$$

Summing (12.21) across the entire interval, we derive the following expression known as *Simpson's rule*

$$\int_{a}^{b} f(x)\, dx = \frac{h}{3}\left[f_1 + 4f_2 + 2f_3 + 4f_4 + \cdots + 2f_{N-1} + 4f_N + f_{N+1}\right] + E_S ,$$

(12.22)

where $E_S$ is the error. If $f^{(4)}(x)$ is continuous on $[a, b]$, then the bound on the error is [1]

$$|E_S| \leq \frac{b-a}{180} h^4 f^4(\xi) \tag{12.23}$$

for some $\xi \in [a, b]$. Note that the result of (12.21) is equal to that obtained by the extrapolation of the Trapezoidal rule in (12.13).

### 12.1.4.1 Adaptive Simpson's Rule

The Simpson's rule outlined thus far can be used in a recursive fashion similar to the Trapezoidal rule, where the number of segments is doubled and previous computations re-used. Writing the rule as

$$\int_{a}^{b} f(x)\, dx = S(a, b) + E_S , \tag{12.24}$$

and starting at the coarsest level (2 segments), we have

$$S(a,b) = \frac{b-a}{6}\left[f(a) + 4f\left(\frac{a+b}{2}\right) + f(b)\right] \qquad (12.25)$$

and

$$E_S^{(1)} \leq \frac{b-a}{180}h^4 M , \qquad (12.26)$$

resulting in

$$I^{(1)} = S(a,b) + E_S^{(1)} . \qquad (12.27)$$

If we now divide the interval $[a,b]$ in half at $c = (a+b)/2$, we get

$$I^{(2)} = S(a,c) + S(c,b) + E_S^{(2)} . \qquad (12.28)$$

The error is now

$$E_S^{(2)} \leq \frac{b-a}{180}\frac{h^4}{16}M = \frac{1}{16}E_S^{(1)} , \qquad (12.29)$$

hence

$$S^{(2)} - S^{(1)} = E_S^{(2)} - E_S^{(1)} = -15E^2 . \qquad (12.30)$$

Using the above, we can stop the recursion once we reach a user-specified tolerance $\epsilon$ when

$$\frac{1}{15}\left|S^{(2)} - S^{(1)}\right| \leq \epsilon . \qquad (12.31)$$

### 12.1.5 One-Dimensional Gaussian Quadrature

Using the trapezoidal or Simpson's rule, the interval $[a,b]$ is subdivided into equally spaced segments with fixed locations $x_i$. A more efficient integration rule can be developed if we allow the freedom to choose the optimal locations for integrating $f(x)$. Given the integral

$$\int_a^b W(x)f(x)\,dx , \qquad (12.32)$$

an $N$-point *Gaussian quadrature* integrates exactly a polynomial $f(x)$ of degree $2N - 1$ if the locations are chosen as the roots of the orthogonal polynomial for the same interval and *weighting function* $W(x)$. The most commonly used Gaussian quadrature is the *Gauss-Legendre* quadrature, of the form

$$\int_{-1}^{1} f(x)\,dx = \sum_{i=1}^{N} w(x_i)f(x_i) , \qquad (12.33)$$

where $W(x) = 1$. The $x_i$ for this rule are the roots of the Legendre polynomial of order $N$. The weights $w(x_i)$ are [3]

$$w(x_i) = \frac{2}{(1 - x_i^2)\left[P_N'(x_i)\right]^2} \, . \tag{12.34}$$

Routines that compute an $N$-point Gauss-Legendre rule determine the locations of the zeros of $P_m(x)$ by a Newton-Raphson iteration, and then the associated weights via (12.34). There are many short software routines for this available in books and via the Internet. Many of these retain the above range $[-1, 1]$, while some instead map to the range $[0, 1]$. To use the rule, a change of variable should be used to map the interval $[a, b]$ to $[-1, 1]$ or $[0, 1]$. Note that in this rule, the spacing between the quadrature locations is not constant. Therefore, previously computed values of $f(x)$ cannot be re-used if the number of quadrature points is changed. In some cases, Simpson's rule may be the better choice.

---

## 12.2   Integration over Triangles

Many of the integrals in Chapter 8 were performed over triangular subdomains, and so efficient analytic and numerical integration techniques for triangular elements are essential. In this section, we will consider these integrals in terms of *simplex coordinates*, also called *area coordinates* or *barycentric coordinates*. These coordinates comprise the transformation of a triangle of arbitrary shape to a canonical coordinate system. Analytic integrals are often easier to perform in simplex coordinates, and numerical quadrature rules are usually specified using them.

### 12.2.1   Simplex Coordinates

To develop the transformation to simplex coordinates, consider a triangle $T$ defined by the vertices $\mathbf{v}_1$, $\mathbf{v}_2$, and $\mathbf{v}_3$, with edges given by

$$\mathbf{e}_1 = \mathbf{v}_2 - \mathbf{v}_1 \, , \qquad \mathbf{e}_2 = \mathbf{v}_3 - \mathbf{v}_2 \, , \qquad \mathbf{e}_3 = \mathbf{v}_3 - \mathbf{v}_1 \, . \tag{12.35}$$

Any point $\mathbf{r}$ within the triangle can be written as a weighted sum of the three vertices as

$$\mathbf{r} = \gamma \mathbf{v}_1 + \alpha \mathbf{v}_2 + \beta \mathbf{v}_3 \, , \tag{12.36}$$

where $\alpha$, $\beta$, and $\gamma$ are the simplex coordinates

$$\alpha = \frac{A_1}{A}, \quad \beta = \frac{A_2}{A}, \quad \gamma = \frac{A_3}{A} \, , \tag{12.37}$$

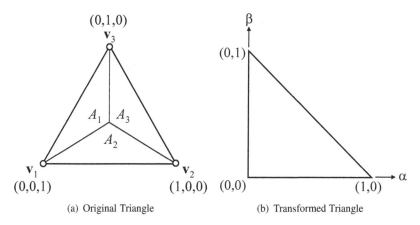

**FIGURE 12.5:** Simplex Coordinates for a Triangle

and where $A$ is the area of $T$, and $A_1, A_2$ and $A_3$ are the areas of the sub-triangles, as illustrated in Figure 12.5a. These coordinates are subject to the constraint

$$\alpha + \beta + \gamma = 1, \tag{12.38}$$

therefore,

$$\gamma = 1 - \alpha - \beta \tag{12.39}$$

and

$$\mathbf{r} = (1 - \alpha - \beta)\mathbf{v}_1 + \alpha\mathbf{v}_2 + \beta\mathbf{v}_3. \tag{12.40}$$

Note that (12.36) has the form of a linear interpolation. Indeed, simplex coordinates will also allow us to perform a linear interpolation of a function $f(x, y)$ at points inside the triangle if its values are known at the vertices. The resulting transformation of the integral to simplex coordinates is illustrated in Figure 12.5b, and is

$$\int_T f(\mathbf{r}) \, d\mathbf{r} = \int_{T'} f(\alpha, \beta) |J(\alpha, \beta)| \, d\alpha \, d\beta = 2A \int_0^1 \int_0^{1-\alpha} f(\alpha, \beta) \, d\beta \, d\alpha. \tag{12.41}$$

Given a point $\mathbf{r}$ inside a triangle, it is also desirable to obtain the simplex coordinates $(\alpha, \beta, \gamma)$. We can write the barycentric expansion of $\mathbf{r} = (x, y, z)$ in terms of the components of the triangle vertices as

$$x = \gamma x_1 + \alpha x_2 + \beta x_3, \tag{12.42}$$

$$y = \gamma y_1 + \alpha y_2 + \beta y_3, \tag{12.43}$$

$$z = \gamma z_1 + \alpha z_2 + \beta z_3. \tag{12.44}$$

Substituting $\gamma = 1 - \alpha - \beta$ into the above gives

$$x = (1 - \alpha - \beta)x_1 + \alpha x_2 + \beta x_3 , \tag{12.45}$$

$$y = (1 - \alpha - \beta)y_1 + \alpha y_2 + \beta y_3 , \tag{12.46}$$

$$z = (1 - \alpha - \beta)z_1 + \alpha z_2 + \beta z_3 . \tag{12.47}$$

Rearranging, this is

$$\alpha(x_2 - x_1) + \beta(x_3 - x_1) + x_1 - x = 0 , \tag{12.48}$$

$$\alpha(y_2 - y_1) + \beta(y_3 - y_1) + y_1 - y = 0 , \tag{12.49}$$

$$\alpha(z_2 - z_1) + \beta(z_3 - z_1) + z_1 - z = 0 , \tag{12.50}$$

and solving for $\alpha$ and $\beta$ gives us

$$\alpha = \frac{B(F + I) - C(E + H)}{A(E + H) - B(D + G)} , \tag{12.51}$$

and

$$\beta = \frac{A(F + I) - C(D + G)}{B(D + G) - A(E + H)} , \tag{12.52}$$

where

$$A = x_2 - x_1 , \tag{12.53}$$

$$B = x_3 - x_1 , \tag{12.54}$$

$$C = x_1 - x , \tag{12.55}$$

$$D = y_2 - y_1 , \tag{12.56}$$

$$E = y_3 - y_1 , \tag{12.57}$$

$$F = y_1 - y , \tag{12.58}$$

$$G = z_2 - z_1 , \tag{12.59}$$

$$H = z_3 - z_1 , \tag{12.60}$$

$$I = z_1 - z . \tag{12.61}$$

## 12.2.2 Radiation Integrals with a Constant Source

We will now consider radiation integrals having a constant-valued integrand, which can be written as

$$I = \int_S e^{j\mathbf{s}\cdot\mathbf{r}'} \, d\mathbf{r}' . \tag{12.62}$$

Let us evaluate the integral over a flat, planar triangle $T$ of arbitrary shape. Using (12.41), this integral becomes

$$I = \int_T e^{j\mathbf{s}\cdot\mathbf{r}'} \, d\mathbf{r}' = 2A \int_0^1 \int_0^{1-\alpha} e^{j\mathbf{s}(\alpha,\beta)\cdot\mathbf{r}'(\alpha,\beta)} d\beta \, d\alpha . \tag{12.63}$$

Expanding $\mathbf{r}'(\alpha, \beta)$ following (12.40), we can write (12.63) as

$$I = 2Ae^{js \cdot \mathbf{v}_1} \int_0^1 \int_0^{1-\alpha} e^{js \cdot \mathbf{e}_1 \alpha} e^{js \cdot \mathbf{e}_3 \beta} d\beta \, d\alpha \, , \qquad (12.64)$$

or

$$I = 2Ae^{js_1} \int_0^1 e^{js_2 \alpha} \int_0^{1-\alpha} e^{js_3 \beta} d\beta \, d\alpha \, , \qquad (12.65)$$

where

$$s_1 = \mathbf{s} \cdot \mathbf{v}_1 \, , \qquad s_2 = \mathbf{s} \cdot \mathbf{e}_1 \, , \qquad s_3 = \mathbf{s} \cdot \mathbf{e}_3 \, . \qquad (12.66)$$

Evaluating the integral versus $\beta$ yields

$$I = 2Ae^{js_1} \int_0^1 e^{js_2 \alpha} \left( \frac{e^{js_3(1-\alpha)}}{js_3} - \frac{1}{js_3} \right) d\alpha \, , \qquad (12.67)$$

which can be rewritten as

$$I = 2A \frac{e^{js_1}}{js_3} \int_0^1 \left( e^{js_3} e^{j\alpha(s_2 - s_3)} - e^{j\alpha s_2} \right) d\alpha \, . \qquad (12.68)$$

Integrating over $\alpha$ yields

$$I = 2A \frac{e^{js_1}}{js_3} \left( \frac{e^{js_2}}{j(s_2 - s_3)} - \frac{e^{js_2}}{js_2} - \frac{e^{js_3}}{j(s_2 - s_3)} + \frac{1}{js_2} \right) \, , \qquad (12.69)$$

and after some tedious but straightforward algebra, we obtain

$$I = -\frac{2Ae^{js_1}}{(s_2 - s_3)} \left( \frac{1 - e^{js_3}}{s_3} - \frac{1 - e^{js_2}}{s_2} \right) \, . \qquad (12.70)$$

If we now define the quantities

$$a_1 = \mathbf{s} \cdot \mathbf{v}_1 \, , \qquad a_2 = \mathbf{s} \cdot \mathbf{v}_2 \, , \qquad a_3 = \mathbf{s} \cdot \mathbf{v}_3 \, , \qquad (12.71)$$

we can then write

$$s_1 = a_1 \, , \qquad s_2 = a_2 - a_1 \, , \qquad s_3 = a_3 - a_1 \, . \qquad (12.72)$$

Substitution of the above into (12.70) then yields

$$I = \frac{2Ae^{ja_1}}{(a_2 - a_3)} \left( \frac{1 - e^{ja_2} e^{-ja_1}}{a_2 - a_1} - \frac{1 - e^{ja_3} e^{-ja_1}}{a_3 - a_1} \right) \, , \qquad (12.73)$$

which after some manipulation becomes

$$I = \frac{2A}{(a_3 - a_2)} \left( \frac{e^{ja_1} - e^{ja_2}}{a_1 - a_2} - \frac{e^{ja_1} - e^{ja_3}}{a_1 - a_3} \right) \, . \qquad (12.74)$$

### 12.2.2.1 Special Cases

When **s** is perpendicular to any of the three edges, (12.74) has a singularity and must be rewritten. When $\mathbf{s} \cdot \mathbf{e}_1 = a_2 - a_1 = 0$, we use the formula

$$I = \frac{2A}{(a_3 - a_2)} \left( j e^{ja_1} - \frac{e^{ja_1} - e^{ja_3}}{a_1 - a_3} \right). \tag{12.75}$$

Likewise, when $\mathbf{s} \cdot \mathbf{e}_3 = a_3 - a_1 = 0$,

$$I = \frac{2A}{(a_3 - a_2)} \left( \frac{e^{ja_1} - e^{ja_2}}{a_1 - a_2} - j e^{ja_1} \right), \tag{12.76}$$

and when $\mathbf{s} \cdot \mathbf{e}_2 = a_3 - a_2 = 0$,

$$I = \frac{2A}{(a_1 - a_2)} \left( j e^{ja_3} - \frac{e^{ja_1} - e^{ja_3}}{a_1 - a_3} \right). \tag{12.77}$$

When **s** is normal to the triangle, we are left with

$$I = A e^{ja_1}. \tag{12.78}$$

## 12.2.3 Radiation Integrals with a Linear Source

We next consider far-field radiation integrals with a linearly varying source on triangle $T$, such as the RWG function described in Section (8.2). These integrals can be written as

$$I = \int_T \rho(\mathbf{r}') e^{j\mathbf{s}\cdot\mathbf{r}'} \, d\mathbf{r}', \tag{12.79}$$

where the source function is defined as

$$\rho(\mathbf{r}') = \mathbf{r}' - \mathbf{v}_1. \tag{12.80}$$

This corresponds to the RWG function defined on the edge opposite vertex $\mathbf{v}_1$ on $T$. Using (12.41), we can write (12.79) as

$$I = 2A \int_0^1 \int_0^{1-\alpha} e^{j\mathbf{s}(\alpha,\beta)\cdot\mathbf{r}'(\alpha,\beta)} d\beta \, d\alpha, \tag{12.81}$$

and expanding $\mathbf{r}'(\alpha, \beta)$ following (12.40), we can write the above as

$$I = 2A e^{j\mathbf{s}\cdot\mathbf{v}_1} \int_0^1 \int_0^{1-\alpha} (\alpha\mathbf{e}_1 + \beta\mathbf{e}_3) e^{j\mathbf{s}\cdot\mathbf{e}_1\alpha} e^{j\mathbf{s}\cdot\mathbf{e}_3\beta} d\beta \, d\alpha, \tag{12.82}$$

or

$$I = 2A e^{js_1} \int_0^1 \int_0^{1-\alpha} (\alpha\mathbf{e}_1 + \beta\mathbf{e}_3) e^{js_2\alpha} e^{js_3\beta} d\beta \, d\alpha. \tag{12.83}$$

## 12.2.3.1 General Case

For the general case, (12.83) evaluates to

$$
I = -\frac{C}{s_3}\left[e^{js_2}\left(\frac{1-j(s_2-s_3)}{(s_2-s_3)^2}+\frac{js_2-1}{s_2^2}\right)-\frac{e^{js_3}}{(s_2-s_3)^2}+\frac{1}{s_2^2}\right]\mathbf{e}_1
$$
$$
-\frac{C}{s_2}\left[e^{js_3}\left(\frac{1-j(s_3-s_2)}{(s_3-s_2)^2}+\frac{js_3-1}{s_3^2}\right)-\frac{e^{js_2}}{(s_3-s_2)^2}+\frac{1}{s_3^2}\right]\mathbf{e}_3 ,
$$
$$(12.84)$$

where

$$
C = j2Ae^{js_1}. \tag{12.85}
$$

As in Section 12.2.2, this expression has singularities when **s** is perpendicular to any of the three edges. Therefore, we must also consider the special cases.

## 12.2.3.2 Special Cases

When **s** is perpendicular to $\mathbf{e}_1$ ($s_2 = 0$), we can write (12.83) as

$$
I = 2Ae^{js_1}\int_0^1\int_0^{1-\alpha}(\alpha\mathbf{e}_1+\beta\mathbf{e}_3)e^{js_3\beta}d\beta\,d\alpha , \tag{12.86}
$$

which evaluates to

$$
I = C\left[\frac{-1-js_3}{s_3^3}+\frac{1}{2s_3}+\frac{e^{js_3}}{s_3^3}\right]\mathbf{e}_1+C\left[e^{js_3}\left(\frac{js_3-2}{s_3^3}\right)+\frac{2}{s_3^3}+\frac{j}{s_3^2}\right]\mathbf{e}_3 . \tag{12.87}
$$

When **s** is perpendicular to $\mathbf{e}_3$ ($s_3 = 0$), we can write (12.83) as

$$
I = 2Ae^{js_1}\int_0^1\int_0^{1-\alpha}(\alpha\mathbf{e}_1+\beta\mathbf{e}_3)e^{js_2\alpha}d\beta\,d\alpha , \tag{12.88}
$$

which evaluates to

$$
I = C\left[e^{js_2}\left(\frac{js_2-2}{s_2^3}\right)+\frac{2}{s_2^3}+\frac{j}{s_2^2}\right]\mathbf{e}_1+C\left[\frac{-1-js_2}{s_2^3}+\frac{1}{2s_2}+\frac{e^{js_2}}{s_2^3}\right]\mathbf{e}_3 . \tag{12.89}
$$

When $s_2 = s_3 = s, s \neq 0$, we can write (12.83) as

$$
I = 2Ae^{js_1}\int_0^1\int_0^{1-\alpha}(\alpha\mathbf{e}_1+\beta\mathbf{e}_3)e^{js(\alpha+\beta)}d\beta\,d\alpha , \tag{12.90}
$$

which evaluates to

$$
I = C\left[e^{js}\left(\frac{1}{s^3}-\frac{1}{2s}-\frac{j}{s^2}\right)-\frac{1}{s^3}\right]\left[\mathbf{e}_1+\mathbf{e}_3\right]. \tag{12.91}
$$

When **s** is normal to the triangle, 12.83 becomes

$$
I = 2Ae^{js\cdot\mathbf{v}_1}\int_0^1\int_0^{1-\alpha}(\alpha\mathbf{e}_1+\beta\mathbf{e}_3)d\beta\,d\alpha , \tag{12.92}
$$

which evaluates to

$$
I = \frac{A}{3}e^{js_1}\left(\mathbf{e}_1+\mathbf{e}_3\right). \tag{12.93}
$$

## 12.2.4    Gaussian Quadrature on Triangles

The most commonly referenced Gauss-Legendre locations and weights for triangles are the symmetric quadrature rules presented in [4]. In this reference are tables varying from degrees 1 up to 20 (79 quadrature points). Tables for orders 2–5 are reproduced in Figures 12.1, 12.2, 12.3, and 12.4, respectively. The weights in these tables are normalized with respect to triangle area, i.e.

$$\int_S f(x,y) \, dx \, dy \approx A \sum_{i=1}^{N} w(\alpha_i, \beta_i) f(\alpha_i, \beta_i) \, . \qquad (12.94)$$

It is this author's experience that the three-point quadrature rules work well for the example problems presented in Chapters 8–11.

**TABLE 12.1:** Three-Point Quadrature Rule ($n = 2$)

| i | $\alpha$ | $\beta$ | $\gamma$ | $w$ |
|---|----------|---------|----------|-----|
| 1 | 0.66666667 | 0.16666667 | 0.16666667 | 0.33333333 |
| 2 | 0.16666667 | 0.66666667 | 0.16666667 | 0.33333333 |
| 3 | 0.16666667 | 0.16666667 | 0.66666667 | 0.33333333 |

**TABLE 12.2:** Four-Point Quadrature Rule ($n = 3$)

| i | $\alpha$ | $\beta$ | $\gamma$ | $w$ |
|---|----------|---------|----------|-----|
| 1 | 0.33333333 | 0.33333333 | 0.33333333 | -0.56250000 |
| 2 | 0.60000000 | 0.20000000 | 0.20000000 | 0.52083333 |
| 3 | 0.20000000 | 0.60000000 | 0.20000000 | 0.52083333 |
| 4 | 0.20000000 | 0.20000000 | 0.60000000 | 0.52083333 |

**TABLE 12.3:** Six-Point Quadrature Rule ($n = 4$)

| i | $\alpha$ | $\beta$ | $\gamma$ | $w$ |
|---|----------|---------|----------|-----|
| 1 | 0.10810301 | 0.44594849 | 0.44594849 | 0.22338158 |
| 2 | 0.44594849 | 0.10810301 | 0.44594849 | 0.22338158 |
| 3 | 0.44594849 | 0.44594849 | 0.10810301 | 0.22338158 |
| 4 | 0.81684757 | 0.09157621 | 0.09157621 | 0.10995174 |
| 5 | 0.09157621 | 0.81684757 | 0.09157621 | 0.10995174 |
| 6 | 0.09157621 | 0.09157621 | 0.81684757 | 0.10995174 |

**TABLE 12.4:** Seven-Point Quadrature Rule ($n = 5$)

| i | $\alpha$ | $\beta$ | $\gamma$ | $w$ |
|---|---|---|---|---|
| 1 | 0.33333333 | 0.33333333 | 0.33333333 | 0.22500000 |
| 2 | 0.05971587 | 0.47014206 | 0.47014206 | 0.13239415 |
| 3 | 0.47014206 | 0.05971587 | 0.47014206 | 0.13239415 |
| 4 | 0.47014206 | 0.47014206 | 0.05971587 | 0.13239415 |
| 5 | 0.79742698 | 0.10128650 | 0.10128650 | 0.12593918 |
| 6 | 0.10128650 | 0.79742698 | 0.10128650 | 0.12593918 |
| 7 | 0.10128650 | 0.10128650 | 0.79742698 | 0.12593918 |

### 12.2.4.1  Comparison with Analytic Solution

The formulas presented in Sections 12.2.3.1 and 12.2.3.2 are advantageous as they are analytic solutions which are valid at all frequencies. By comparison, an accurate numerical quadrature requires an increasing number of integration points as the frequency increases, which is undesirable. Let us compare the analytic solution to (12.83) to that obtained using a 7-point Gaussian quadrature rule (Table 12.4). The source triangle is equilateral and lies in the $xy$ plane with vertices $\mathbf{v}_1 = (0, 0)$, $\mathbf{v}_2 = (l, 0)$, and $\mathbf{v}_3 = (l/2, l\sqrt{3}/2)$. The direction of radiation is along the spherical angles $(\theta, \phi) = (\pi/2, \pi/4)$. In Figure 12.6a we compare the magnitude of the $\hat{\mathbf{x}}$ components of (12.83) versus the triangle edge length $l$ in wavelengths. In Figure 12.6b we plot the difference between the results. We see that the analytical and numerical results compare very well to just over $l = 1\lambda$, at which point the numerical results start to diverge.

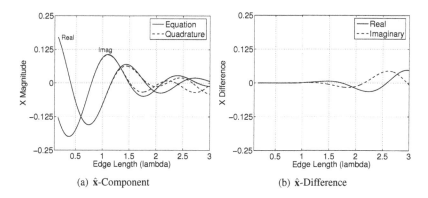

(a) $\hat{\mathbf{x}}$-Component     (b) $\hat{\mathbf{x}}$-Difference

**FIGURE 12.6:** Linear Source Radiation Integral: Equation vs. Quadrature

# References

[1] G. Thomas and R. Finney, *Calculus and Analytic Geometry*. Addison-Wesley, 1992.

[2] A. F. Peterson, S. L. Ray, and R. Mittra, *Computational Methods for Electromagnetics*. IEEE Press, 1998.

[3] W. H. Press, B. P. Flannery, S. A. Teukolsky, and W. T. Vetterling, *Numerical Recipes in C : The Art of Scientific Computing*. Cambridge University Press, 1992.

[4] D. Dunavant, "High degree efficient symmetrical Gaussian quadrature rules for the triangle," *Internat. J. Numer. Methods Engrg.*, vol. 21, pp. 1129–1148, 1985.

# Appendix A

## Scattering Using Physical Optics

We briefly discussed the Physical Optics (PO) approximation in Section 3.6.4, and we used the PO approximation for conducting interfaces (3.209) in several examples throughout this book. Because it is of vital importance in the treatment of electrically large objects, particularly in the area of radar cross section prediction, we discuss it in additional detail in this appendix. We will first consider the fields scattered from conducting surfaces, and then the fields scattered by dielectric interfaces.

### A.1   Field Scattered at a Conducting Interface

Recall again the surface equivalent problem of Figure 3.2c. Given a plane wave incident in Region $R_1$ with $\hat{\theta}^i$ and $\hat{\phi}^i$ components, the electric field in $R_1$ can be written as

$$\mathbf{E}_1^i(\mathbf{r}) = \left[ E_{1,\theta}^i \hat{\theta}^i + E_{1,\phi}^i \hat{\phi}^i \right] e^{jk_1 \hat{\mathbf{r}}^i \cdot \mathbf{r}}, \tag{A.1}$$

and the incident magnetic field as

$$\mathbf{H}_1^i(\mathbf{r}) = -\frac{1}{\eta_1} \hat{\mathbf{r}} \times \mathbf{E}_1^i(\mathbf{r}) = \frac{1}{\eta_1} \left[ E_{1,\phi}^i \hat{\theta}^i - E_{1,\theta}^i \hat{\phi}^i \right] e^{jk_1 \hat{\mathbf{r}}^i \cdot \mathbf{r}}. \tag{A.2}$$

For radar cross section problems, $R_1$ is the unbounded free space region $R_0$ where $k_1 = k_0 = w\sqrt{\mu_0 \epsilon_0}$ and $\eta_1 = \eta_0 = \sqrt{\mu_0 / \epsilon_0}$. Using the PO approximation, the interface $S_{12} = S$ is assumed to be perfectly flat and of infinite extent. If $R_2$ is perfectly conducting, $\mathbf{M}_1(\mathbf{r}) = 0$, and the electric surface current is given by (3.209). Following Section 3.5.3, the total scattered far electric field $\mathbf{E}_1^s(\mathbf{r})$ is then

$$\mathbf{E}_1^s(\mathbf{r}) = -j\omega \mathbf{A}(\mathbf{r}) = -j\frac{k_0}{4\pi} \frac{e^{-jk_0 r}}{r} \int_S \eta_0 \mathbf{J}_1(\mathbf{r}') e^{jk_0 \hat{\mathbf{r}}^s \cdot \mathbf{r}'} \, d\mathbf{r}', \tag{A.3}$$

where $\mathbf{E}_1^s(\mathbf{r})$ has only $\hat{\theta}$- and $\hat{\phi}$-polarized components. Thus, the scattered field

469

components are then related to the incident fields via the *scattering matrix*, written as

$$\begin{bmatrix} E^s_{1,\theta} \\ E^s_{1,\phi} \end{bmatrix} = \begin{bmatrix} S_{\theta\theta} & S_{\theta\phi} \\ S_{\phi\theta} & S_{\phi\phi} \end{bmatrix} \begin{bmatrix} E^i_{1,\theta} \\ E^i_{1,\phi} \end{bmatrix} S \ . \tag{A.4}$$

Using (A.2) and (3.209), the scattering matrix elements can be written as

$$S_{\theta\theta} = [\hat{\boldsymbol{\phi}}^i \times \hat{\boldsymbol{\theta}}^s] \cdot \hat{\mathbf{n}}_1 \ , \tag{A.5}$$

$$S_{\phi\theta} = [\hat{\boldsymbol{\phi}}^i \times \hat{\boldsymbol{\phi}}^s] \cdot \hat{\mathbf{n}}_1 \ , \tag{A.6}$$

$$S_{\theta\phi} = [\hat{\boldsymbol{\theta}}^s \times \hat{\boldsymbol{\theta}}^i] \cdot \hat{\mathbf{n}}_1 \ , \tag{A.7}$$

$$S_{\phi\phi} = [\hat{\boldsymbol{\phi}}^s \times \hat{\boldsymbol{\theta}}^i] \cdot \hat{\mathbf{n}}_1 \ , \tag{A.8}$$

where

$$S = j \frac{k_0}{2\pi} \frac{e^{-jk_0 r}}{r} \int_S e^{jk_0(\hat{\mathbf{r}}^i + \hat{\mathbf{r}}^s) \cdot \hat{\mathbf{r}}'} \ d\mathbf{r}' . \tag{A.9}$$

We note that for monostatic observations

$$S_{\theta\theta} = S_{\phi\phi} \ , \tag{A.10}$$

and

$$S_{\phi\theta} = S_{\theta\phi} = 0 \ . \tag{A.11}$$

The integral (A.9) is dependent only on the shape of the subdomain over which the integral is performed, and is often referred to as a *shape function*. For planar triangles, an analytic solution was presented in Section 12.2.2, and for general $N$-sided polygons, one can refer to [1].

---

## A.2 Plane Wave Decomposition at a Planar Interface

At a planar interface between two regions, the incident electric field $\mathbf{E}^i_1$ can be decomposed into parallel ($\parallel$) and perpendicular ($\perp$) components, also known as vertical ($\hat{\mathbf{v}}$) and horizontal ($\hat{\mathbf{h}}$) components, respectively, as shown in Figure A.1. The decomposed components can be written as

$$\mathbf{E}^i_1 = (\mathbf{E}^i_1 \cdot \hat{\mathbf{v}}^i)\hat{\mathbf{v}}^i + (\mathbf{E}^i_1 \cdot \hat{\mathbf{h}})\hat{\mathbf{h}} = E^i_{1,v}\hat{\mathbf{v}}^i + E^i_{1,h}\hat{\mathbf{h}} \ . \tag{A.12}$$

The reflected field components can be written in terms of the incident field components and the parallel and perpendicular reflection coefficients ($R_\parallel$ and $R_\perp$, respectively) as

$$\mathbf{E}^r_1 = R_\parallel E^i_{1,v}\hat{\mathbf{v}}^r + R_\perp E^i_{1,h}\hat{\mathbf{h}} \ . \tag{A.13}$$

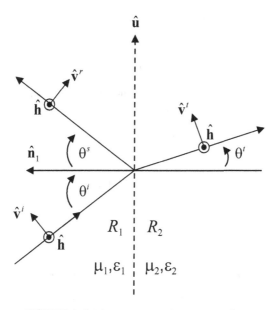

**FIGURE A.1:** Planar Interface between Regions

When $R_2$ is a dielectric, there may be fields transmitted into $R_2$. These can be written in terms of the incident field components and the parallel and perpendicular transmission coefficients ($T^\parallel$ and $T^\perp$, respectively) as

$$\mathbf{E}_2^t = T_\parallel E_{1,v}^i \hat{\mathbf{v}}^t + T_\perp E_{1,h}^i \hat{\mathbf{h}} . \tag{A.14}$$

The angle of incidence $\theta^i$ is equal to the angle of reflection $\theta^s$ (Snell's law), and the angle of transmission $\theta^t$ is related to $\theta^i$ via the relationship

$$k_1 \sin \theta^i = k_2 \sin \theta^t. \tag{A.15}$$

For perpendicular polarization, the reflection coefficient $R_\perp$ is

$$R_\perp = \frac{\eta_2 \cos \theta^i - \eta_1 \cos \theta^t}{\eta_2 \cos \theta^i + \eta_1 \cos \theta^t}, \tag{A.16}$$

and the transmission coefficient $T_\perp$ is

$$T_\perp = \frac{2\eta_2 \cos \theta^i}{\eta_2 \cos \theta^i + \eta_1 \cos \theta^t}. \tag{A.17}$$

For parallel polarization, the reflection coefficient $R_\parallel$ is

$$R_\parallel = \frac{\eta_2 \cos \theta^t - \eta_1 \cos \theta^i}{\eta_1 \cos \theta^i + \eta_2 \cos \theta^t}, \tag{A.18}$$

and the transmission coefficient $T_\parallel$ is

$$T_\parallel = \frac{\eta_2 \cos \theta^i}{\eta_1 \cos \theta^i + \eta_2 \cos \theta^t} . \qquad (A.19)$$

The expressions in (A.16)–(A.19) are known as the Fresnel coefficients for reflection and transmission [2]. Note that if region $R_2$ is a conductor, then

$$R_\perp = R_\parallel = -1 , \qquad (A.20)$$

and

$$T_\perp = T_\parallel = 0 . \qquad (A.21)$$

## A.3   Field Scattered at a Dielectric Interface

To compute the field scattered by a dielectric interface, we must find the equivalent currents $\mathbf{M}_1$ and $\mathbf{J}_1$ in (3.149) and (3.150). We note that only the components of the total electric and magnetic fields that lie in the plane of the interface contribute to the currents. Thus,

$$\mathbf{E} = E_{1,u}\hat{\mathbf{u}} + E_{1,h}\hat{\mathbf{h}} , \qquad (A.22)$$

and

$$\mathbf{H} = H_{1,u}\hat{\mathbf{u}} + H_{1,h}\hat{\mathbf{h}} , \qquad (A.23)$$

where $\hat{\mathbf{u}} = \hat{\mathbf{n}}_1 \times \hat{\mathbf{h}}$. The scalar components of the fields are

$$E_{1,h} = E_{1,h}^i(1 + R_\perp) , \qquad (A.24)$$

$$E_{1,u} = E_{1,u}^i(1 + R_\parallel) , \qquad (A.25)$$

$$H_{1,h} = H_{1,h}^i(1 - R_\parallel) , \qquad (A.26)$$

$$H_{1,u} = H_{1,u}^i(1 - R_\perp) . \qquad (A.27)$$

Substituting (A.22) and (A.23) into (3.149) and (3.150) yields

$$\mathbf{M}_1 = -\hat{\mathbf{n}}_1 \times \left[ E_{1,u}^i\hat{\mathbf{u}} + E_{1,h}^i\hat{\mathbf{h}} \right] , \qquad (A.28)$$

$$\mathbf{J}_1 = \hat{\mathbf{n}}_1 \times \left[ H_{1,u}^i\hat{\mathbf{u}} + H_{1,h}^i\hat{\mathbf{h}} \right] , \qquad (A.29)$$

which can be written as

$$\mathbf{M}_1 = -\hat{\mathbf{n}}_1 \times \mathbf{E}^i \cdot \mathbf{A} , \qquad (A.30)$$

$$\mathbf{J}_1 = \hat{\mathbf{n}}_1 \times \mathbf{H}^i \cdot \mathbf{B} \,, \tag{A.31}$$

where

$$\mathbf{A} = \hat{\mathbf{h}}(1 + R_\perp)\hat{\mathbf{h}} + \hat{\mathbf{u}}(1 + R_\parallel)\hat{\mathbf{u}} \,, \tag{A.32}$$

$$\mathbf{B} = \hat{\mathbf{h}}(1 - R_\parallel)\hat{\mathbf{h}} + \hat{\mathbf{u}}(1 - R_\perp)\hat{\mathbf{u}} \,. \tag{A.33}$$

Following Section 3.5.3, the total scattered far electric field $\mathbf{E}_1^s(\mathbf{r})$ is

$$\mathbf{E}_1^s(\mathbf{r}) = -j\omega \big[\mathbf{A}(\mathbf{r}) - \eta_0\, \hat{\mathbf{r}}^s \times \mathbf{F}(\mathbf{r})\big], \tag{A.34}$$

or

$$\mathbf{E}_1^s(\mathbf{r}) = -j\frac{k_0}{4\pi}\frac{e^{-jk_0 r}}{r} \int_S \big[\eta_0\, \mathbf{J}_1(\mathbf{r}') - \hat{\mathbf{r}} \times \mathbf{M}_1(\mathbf{r}')\big] e^{jk_0 \hat{\mathbf{r}}^s \cdot \mathbf{r}'}\, d\mathbf{r}' \,. \tag{A.35}$$

Using vector identities and some algebra, we can write the scattered fields in terms of the incident fields following (A.4), where the matrix elements are

$$S_{\theta\theta} = \frac{1}{2}\big[\hat{\boldsymbol{\theta}}^i \cdot \mathbf{A}' \times \hat{\boldsymbol{\phi}}^s + \hat{\boldsymbol{\phi}}^i \cdot \mathbf{B}' \times \hat{\boldsymbol{\theta}}^s\big], \tag{A.36}$$

$$S_{\phi\theta} = \frac{1}{2}\big[\hat{\boldsymbol{\phi}}^i \cdot \mathbf{A}' \times \hat{\boldsymbol{\phi}}^s - \hat{\boldsymbol{\theta}}^i \cdot \mathbf{B}' \times \hat{\boldsymbol{\theta}}^s\big], \tag{A.37}$$

$$S_{\theta\phi} = \frac{1}{2}\big[-\hat{\boldsymbol{\theta}}^i \cdot \mathbf{A}' \times \hat{\boldsymbol{\theta}}^s + \hat{\boldsymbol{\phi}}^i \cdot \mathbf{B}' \times \hat{\boldsymbol{\phi}}^s\big], \tag{A.38}$$

$$S_{\phi\phi} = \frac{1}{2}\big[-\hat{\boldsymbol{\phi}}^i \cdot \mathbf{A}' \times \hat{\boldsymbol{\theta}}^s - \hat{\boldsymbol{\theta}}^i \cdot \mathbf{B}' \times \hat{\boldsymbol{\phi}}^s\big], \tag{A.39}$$

and

$$\mathbf{A}' = -\hat{\mathbf{u}}(1 + R_\perp)\hat{\mathbf{h}} + \hat{\mathbf{h}}(1 + R_\parallel)\hat{\mathbf{u}} \,, \tag{A.40}$$

and

$$\mathbf{B}' = -\hat{\mathbf{u}}(1 - R_\parallel)\hat{\mathbf{h}} + \hat{\mathbf{h}}(1 - R_\perp)\hat{\mathbf{u}} \,. \tag{A.41}$$

---

## A.4  Layered Dielectrics over Conductor

While interfaces between dielectric half-spaces are not realistic in most general three-dimensional problems, conductors with thin dielectric coatings are encountered often in practice. Examples are reentry vehicles or aircraft which are often coated with a heat shield or a thin layer of radar-absorbing material (RAM). Such a geometry is illustrated in Figure A.2, where a conductor is coated with $N$ thin dielectric layers, each having its own thickness $d$ and dielectric parameters $(\mu, \epsilon)$. For incident and scattered fields in $R_1$, this

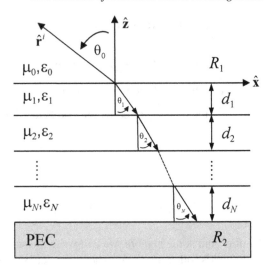

**FIGURE A.2:** Layered Dielectrics over Conductor

problem can be treated much the same way as in Section A.3, except that we compute *effective* reflection coefficients $(R_\perp, R_\parallel)$ at the outermost interface, taking into account the reflection and transmission inside of each layer. This can be done by starting at the innermost interface (PEC) and propagating the reflection coefficients at each interface to the surface via the expression [3]

$$\tilde{R}_{i,i+1} = \frac{R_{i,i+1} + \tilde{R}_{i+1,i+2}e^{-2jk_{i+1,z}d_{i+1}}}{1 + R_{i,i+1}\tilde{R}_{i+1,i+2}e^{-2jk_{i+1,z}d_{i+1}}} \; . \tag{A.42}$$

At each step, $\tilde{R}_{i,i+1}$ represents the generalized reflection coefficient at the interface between layers $i$ and $i + 1$, taking into account reflections and transmissions in all layers below it, and $R_{i,i+1}$ are the Fresnel coefficients (A.16) or (A.18) at the interface between layers $i$ and $i + 1$. The term $k_{i+1,z}$ comprises the $\hat{\mathbf{z}}$ component of the wavenumber in each layer, i.e.

$$k_{i,z} = k_i \cos \theta_i \; . \tag{A.43}$$

We can determine $k_{i,z}$ by noting that in the free space exterior region,

$$\mathbf{k}_0 = k_{x,0}\hat{\mathbf{x}} + k_{z,0}\hat{\mathbf{z}} = k_0 \sin \theta_0 \hat{\mathbf{x}} + k_0 \cos \theta_0 \hat{\mathbf{z}} \; , \tag{A.44}$$

and by Snell's law, $k_{x,0} = k_{x,1} = k_{x,2} \dots k_{x,N} = k_x$, therefore

$$k_{i,z} = \sqrt{k_i^2 - k_x^2} \; . \tag{A.45}$$

# References

[1] S. Lee and R. Mittra, "Fourier transform of a polygonal shape function and its application in electromagnetics," *IEEE Trans. Antennas Propagat.*, vol. 31, pp. 99–103, January 1983.

[2] C. A. Balanis, *Advanced Engineering Electromagnetics*. John Wiley and Sons, 1989.

[3] W. C. Chew, *Waves and Fields in Inhomogeneous Media*. IEEE Press, 1995.

# Index

Printed in the United States
by Baker & Taylor Publisher Services